Essentials of Probability Theory for Statisticians

CHAPMAN & HALL/CRC
Texts in Statistical Science Series

Statistical Theory: A Concise Introduction
F. Abramovich and Y. Ritov

Practical Multivariate Analysis, Fifth Edition
A. Afifi, S. May, and V.A. Clark

Practical Statistics for Medical Research
D.G. Altman

Interpreting Data: A First Course in Statistics
A.J.B. Anderson

Introduction to Probability with R
K. Baclawski

Linear Algebra and Matrix Analysis for Statistics
S. Banerjee and A. Roy

Mathematical Statistics: Basic Ideas and Selected Topics, Volume I, Second Edition
P. J. Bickel and K. A. Doksum

Mathematical Statistics: Basic Ideas and Selected Topics, Volume II
P. J. Bickel and K. A. Doksum

Analysis of Categorical Data with R
C. R. Bilder and T. M. Loughin

Statistical Methods for SPC and TQM
D. Bissell

Introduction to Probability
J. K. Blitzstein and J. Hwang

Bayesian Methods for Data Analysis, Third Edition
B.P. Carlin and T.A. Louis

Second Edition
R. Caulcutt

The Analysis of Time Series: An Introduction, Sixth Edition
C. Chatfield

Introduction to Multivariate Analysis
C. Chatfield and A.J. Collins

Problem Solving: A Statistician's Guide, Second Edition
C. Chatfield

Statistics for Technology: A Course in Applied Statistics, Third Edition
C. Chatfield

Analysis of Variance, Design, and Regression : Linear Modeling for Unbalanced Data, Second Edition
R. Christensen

Bayesian Ideas and Data Analysis: An Introduction for Scientists and Statisticians
R. Christensen, W. Johnson, A. Branscum, and T.E. Hanson

Modelling Binary Data, Second Edition
D. Collett

Modelling Survival Data in Medical Research, Third Edition
D. Collett

Introduction to Statistical Methods for Clinical Trials
T.D. Cook and D.L. DeMets

Applied Statistics: Principles and Examples
D.R. Cox and E.J. Snell

Multivariate Survival Analysis and Competing Risks
M. Crowder

Statistical Analysis of Reliability Data
M.J. Crowder, A.C. Kimber, T.J. Sweeting, and R.L. Smith

An Introduction to Generalized Linear Models, Third Edition
A.J. Dobson and A.G. Barnett

Nonlinear Time Series: Theory, Methods, and Applications with R Examples
R. Douc, E. Moulines, and D.S. Stoffer

Introduction to Optimization Methods and Their Applications in Statistics
B.S. Everitt

Extending the Linear Model with R: Generalized Linear, Mixed Effects and Nonparametric Regression Models, Second Edition
J.J. Faraway

Linear Models with R, Second Edition
J.J. Faraway

A Course in Large Sample Theory
T.S. Ferguson

Multivariate Statistics: A Practical Approach
B. Flury and H. Riedwyl

Readings in Decision Analysis
S. French

Discrete Data Analysis with R: Visualization and Modeling Techniques for Categorical and Count Data
M. Friendly and D. Meyer

Markov Chain Monte Carlo: Stochastic Simulation for Bayesian Inference, Second Edition
D. Gamerman and H.F. Lopes

Bayesian Data Analysis, Third Edition
A. Gelman, J.B. Carlin, H.S. Stern, D.B. Dunson, A. Vehtari, and D.B. Rubin

Multivariate Analysis of Variance and Repeated Measures: A Practical Approach for Behavioural Scientists
D.J. Hand and C.C. Taylor

Practical Longitudinal Data Analysis
D.J. Hand and M. Crowder

Logistic Regression Models
J.M. Hilbe

Richly Parameterized Linear Models: Additive, Time Series, and Spatial Models Using Random Effects
J.S. Hodges

Statistics for Epidemiology
N.P. Jewell

Stochastic Processes: An Introduction, Second Edition
P.W. Jones and P. Smith

The Theory of Linear Models
B. Jørgensen

Principles of Uncertainty
J.B. Kadane

Graphics for Statistics and Data Analysis with R
K.J. Keen

Mathematical Statistics
K. Knight

Introduction to Multivariate Analysis: Linear and Nonlinear Modeling
S. Konishi

Nonparametric Methods in Statistics with SAS Applications
O. Korosteleva

Modeling and Analysis of Stochastic Systems, Second Edition
V.G. Kulkarni

Exercises and Solutions in Biostatistical Theory
L.L. Kupper, B.H. Neelon, and S.M. O'Brien

Exercises and Solutions in Statistical Theory
L.L. Kupper, B.H. Neelon, and S.M. O'Brien

Design and Analysis of Experiments with R
J. Lawson

Design and Analysis of Experiments with SAS
J. Lawson

A Course in Categorical Data Analysis
T. Leonard

Statistics for Accountants
S. Letchford

Introduction to the Theory of Statistical Inference
H. Liero and S. Zwanzig

Statistical Theory, Fourth Edition
B.W. Lindgren

Stationary Stochastic Processes: Theory and Applications
G. Lindgren

Statistics for Finance
E. Lindström, H. Madsen, and J. N. Nielsen

The BUGS Book: A Practical Introduction to Bayesian Analysis
D. Lunn, C. Jackson, N. Best, A. Thomas, and D. Spiegelhalter

Introduction to General and Generalized Linear Models
H. Madsen and P. Thyregod

Time Series Analysis
H. Madsen

Pólya Urn Models
H. Mahmoud

Randomization, Bootstrap and Monte Carlo Methods in Biology, Third Edition
B.F.J. Manly

Introduction to Randomized Controlled Clinical Trials, Second Edition
J.N.S. Matthews

Statistical Rethinking: A Bayesian Course with Examples in R and Stan
R. McElreath

Statistical Methods in Agriculture and Experimental Biology, Second Edition
R. Mead, R.N. Curnow, and A.M. Hasted

Statistics in Engineering: A Practical Approach
A.V. Metcalfe

Statistical Inference: An Integrated Approach, Second Edition
H. S. Migon, D. Gamerman, and F. Louzada

Beyond ANOVA: Basics of Applied Statistics
R.G. Miller, Jr.

A Primer on Linear Models
J.F. Monahan

Applied Stochastic Modelling, Second Edition
B.J.T. Morgan

Elements of Simulation
B.J.T. Morgan

Probability: Methods and Measurement
A. O'Hagan

Introduction to Statistical Limit Theory
A.M. Polansky

Applied Bayesian Forecasting and Time Series Analysis
A. Pole, M. West, and J. Harrison

Statistics in Research and Development, Time Series: Modeling, Computation, and Inference
R. Prado and M. West

Essentials of Probability Theory for Statisticians
M.A. Proschan and P.A. Shaw

Introduction to Statistical Process Control
P. Qiu

Sampling Methodologies with Applications
P.S.R.S. Rao

A First Course in Linear Model Theory
N. Ravishanker and D.K. Dey

Essential Statistics, Fourth Edition
D.A.G. Rees

Stochastic Modeling and Mathematical Statistics: A Text for Statisticians and Quantitative Scientists
F.J. Samaniego

Statistical Methods for Spatial Data Analysis
O. Schabenberger and C.A. Gotway

Bayesian Networks: With Examples in R
M. Scutari and J.-B. Denis

Large Sample Methods in Statistics
P.K. Sen and J. da Motta Singer

Spatio-Temporal Methods in Environmental Epidemiology
G. Shaddick and J.V. Zidek

Decision Analysis: A Bayesian Approach
J.Q. Smith

Analysis of Failure and Survival Data
P. J. Smith

Applied Statistics: Handbook of GENSTAT Analyses
E.J. Snell and H. Simpson

Applied Nonparametric Statistical Methods, Fourth Edition
P. Sprent and N.C. Smeeton

Data Driven Statistical Methods
P. Sprent

Generalized Linear Mixed Models: Modern Concepts, Methods and Applications
W. W. Stroup

Survival Analysis Using S: Analysis of Time-to-Event Data
M. Tableman and J.S. Kim

Applied Categorical and Count Data Analysis
W. Tang, H. He, and X.M. Tu

Elementary Applications of Probability Theory, Second Edition
H.C. Tuckwell

Introduction to Statistical Inference and Its Applications with R
M.W. Trosset

Understanding Advanced Statistical Methods
P.H. Westfall and K.S.S. Henning

Statistical Process Control: Theory and Practice, Third Edition
G.B. Wetherill and D.W. Brown

Generalized Additive Models: An Introduction with R
S. Wood

Epidemiology: Study Design and Data Analysis, Third Edition
M. Woodward

Practical Data Analysis for Designed Experiments
B.S. Yandell

Texts in Statistical Science

Essentials of Probability Theory for Statisticians

Michael A. Proschan
National Institute of Allergy and Infectious Diseases, NIH

Pamela A. Shaw
University of Pennsylvania

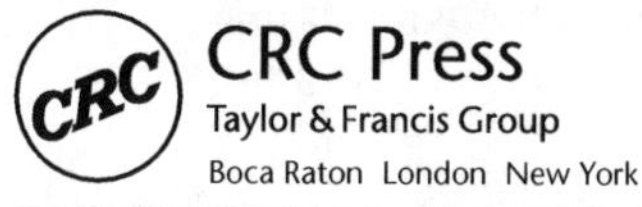

CRC Press
Taylor & Francis Group
Boca Raton London New York

CRC Press is an imprint of the
Taylor & Francis Group, an **informa** business
A CHAPMAN & HALL BOOK

CRC Press
Taylor & Francis Group
6000 Broken Sound Parkway NW, Suite 300
Boca Raton, FL 33487-2742

First issued in paperback 2019

ISBN-13: 978-1-4987-0419-9 (hbk)
ISBN-13: 978-0-367-87163-5 (pbk)

Library of Congress Cataloging-in-Publication Data

Names: Proschan, Michael A. | Shaw, Pamela, 1968-
Title: Essentials of probability theory for statisticians / Michael A. Proschan and Pamela A. Shaw.
Description: Boca Raton : Taylor & Francis, 2016. | Series: Chapman & hall/CRC texts in statistical science series | "A CRC title." | Includes
bibliographical references and index.
Identifiers: LCCN 2015042436 | ISBN 9781498704199 (alk. paper)
Subjects: LCSH: Probabilities--Textbooks. | Mathematical statistics--Textbooks.
Classification: LCC QA273 .P797 2016 | DDC 519.201--dc23
LC record available at http://lccn.loc.gov/2015042436

Visit the Taylor & Francis Web site at
http://www.taylorandfrancis.com

and the CRC Press Web site at
http://www.crcpress.com

Contents

Preface

Biostatistics and statistics departments are struggling with how much probability and measure theory to include in their curricula. The traditional statistics department model of a full year of probability using the texts of Billingsley or Chung, for example, is losing favor as students are in a rush to get through a graduate program and begin their careers. Some biostatistics departments have gone to the extreme of eliminating graduate-level probability altogether. Consequently, their students are left with a background that does not prepare them to make rigorous arguments. We wrote this book as a compromise between the two unpalatable extremes: overloading statistics students with extensive and mathematically challenging measure theory versus leaving them unprepared to prove their results. Rather than offering a comprehensive treatment of all of probability and measure theory, replete with proofs, we present the essential probability results that are used repeatedly in statistics applications. We also selectively present proofs that make repeated use of mathematical techniques that we continue to use in our statistical careers when rigor is needed. As biostatisticians, we have encountered numerous applications requiring careful use of probability. We share these in this book, whose emphasis is on being able to understand rigorously the meaning and application of probability results. While traditional graduate probability books are sometimes written for mathematics students with no knowledge of elementary statistics, our book is written with statisticians in mind. For example, we motivate characteristic functions by first discussing harmonic regression and its usefulness in understanding circadian rhythm of biological phenomena. Another example of our use of statistical applications to help understand and motivate probability is the study of permutation tests. Permutation tests provide fertile ground for understanding conditional distributions and asymptotic arguments. For example, it is both challenging and instructive to try to understand precisely what people mean when they assert the asymptotic equivalence of permutation and t-tests. In summary, we believe that this book is ideal for teaching students essential probability theory to make rigorous probability arguments.

The book is organized as follows. The first chapter is intended as a broad introduction to why more rigor is needed to take that next step to graduate-level probability. Chapter 1 makes use of intriguing paradoxes to motivate the need for rigor. The chapter assumes knowledge of basic probability and statistics. We return to some of these paradoxes later in the book. Chapter 2, on countability, contains essential background material with which some readers may already be familiar. Depending on the background of students, instructors may want to begin with this chapter or have students review it on their own and refer to it when needed. The remaining chapters are in the order that we believe is most logical. Chapters 3 and 4 contain the backbone of probability: sigma-fields, probability measures, and random variables and vectors. Chapter 5 introduces and contrasts Lebesgue-Stieltjes integration with the more familiar Riemann-Stieltjes integration, while Chapter 6 covers different modes of convergence. Chapters 7 and 8 concern laws of large numbers and central limit theorems. Chapter 9 contains additional results on convergence in distribution, including the delta method, while Chapter 10 covers the extremely important topic of conditional probability, expectation, and distribution. Chapter 11 contains many interesting applications from our actual experience. Other applications are interspersed throughout the book, but those in Chapter 11 are more detailed. Many examples in Chapter 11 rely on material covered in Chapters 9-10, and would therefore be difficult to present much earlier. The book concludes with two appendices. Appendix A is a brief review of prerequisite material, while Appendix B contains useful probability distributions and their properties. Each chapter contains a chapter review of key results, and exercises are intended to constantly reinforce important concepts.

We would like to express our extreme gratitude to Robert Taylor (Clemson University), Jie Yang (University of Illinois at Chicago), Wlodek Byrc (University of Cincinnati), and Radu Herbei (Ohio State University) for reviewing the book. They gave us very helpful suggestions and additional material, and caught typos and other errors. The hardest part of the book for us was constructing exercises, and the reviewers provided additional problems and suggestions for those as well.

Index of Statistical Applications and Notable Examples

In this text, we illustrate the importance of probability theory to statistics. We have included a number of illustrative examples that use key results from probability to gain insight on the behavior of some commonly used statistical tests, as well as examples that consider implications for design of clinical trials. Here, we highlight our more statistical applications and a few other notable examples.

Statistical Applications

1. Clinical Trials
   i. ECMO Trial: Example 10.45
   ii. Interim Monitoring: Example 8.51
   iii. Permuted block randomization: Example 1.1, 5.33, 7.3, 7.19, 10.20
   iv. Two-stage study design: Example 10.44
2. Test statistics
   i. Effect size: Section 11.10
   ii. Fisher's Exact test: Section 11.5.3
   iii. Goodness of fit test: Example 8.49
   iv. Likelihood ratio test: Section 11.3
   v. Logrank statistic: Section 11.12
   vi. Maximum test statistic: Example 9.1
   vii. Normal scores test: Example 4.58
   viii. Outlier test: Section 11.11
   ix. P-value: Example 9.1, 9.3
   x. Permutation test: Example 8.10, 8.20, 10.20, 10.45, 10.53; Sections 11.6, 11.7, 11.8
   xi. Pitman's test : Example 10.35
   xii. Sample mean: Example 4.44, 6.6, 10.34
   xiii. Sample median: Example 6.10, 7.4; Exercise: Section 7.1, #3

Other Notable Examples

Chapter 1

Introduction

This book is intended to provide a rigorous treatment of probability theory at the graduate level. The reader is assumed to have a working knowledge of probability and statistics at the undergraduate level. Certain things were over-simplified in more elementary courses because you were likely not ready for probability in its full generality. But now you are like a boxer who has built up enough experience and confidence to face the next higher level of competition. Do not be discouraged if it seems difficult at first. It will become easier as you learn certain techniques that will be used repeatedly. We will highlight the most important of these techniques by writing three stars (***) next to them and including them in summaries of key results found at the end of each chapter.

You will learn different methods of proofs that will be useful for establishing classic probability results, as well as more generally in your graduate career and beyond. Early chapters build a probability foundation, after which we intersperse examples aimed at making seemingly esoteric mathematical constructs more intuitive. Necessary elements in definitions and conditions in theorems will become clear through these examples. Counterexamples will be used to further clarify nuances in meaning and expose common fallacies in logic.

At this point you may be asking yourself two questions: (1) Why is what I have learned so far not considered rigorous? (2) Why is more rigor needed? The answers will become clearer over time, but we hope this chapter gives you some partial answers. Because this chapter presents an introductory survey of problems that will be dealt with in depth in later material, it is somewhat less formal than subsequent chapters.

1.1 Why More Rigor is Needed

You have undoubtedly been given the following simplified presentation. There are two kinds of random variables—discrete and continuous. Discrete variables have a probability mass function and continuous variables have a probability density function. In actuality, there are random variables that are not discrete, and yet do not have densities. Their distribution functions are said to be *singular*. One interesting example is the following.

Example 1.1. No univariate density Flip a biased coin with probability p of heads infinitely many times. Let $X_1, X_2, X_3, \ldots$ be the outcomes, with $X_i = 0$ denoting tails and

$X_i = 1$ denoting heads on the ith flip. Now form the random number

$$Y = 0.X_1X_2X_3\ldots, \tag{1.1}$$

written in base 2. That is, $Y = X_1 \cdot (1/2) + X_2 \cdot (1/2)^2 + X_3 \cdot (1/2)^3 + \ldots$ The first digit X_1 determines whether Y is in the first half $[0, 1/2)$ (corresponding to $X_1 = 0$) or second half $[1/2, 1]$ (corresponding to $X_1 = 1$). Whichever half Y is in, X_2 determines whether Y is in the first or second half of that half, etc. (see Figure 1.1).

```
          X1=0                 X1=1
[----------|----------|----------|----------]
   X2=0       X2=1       X2=0       X2=1
0          ¼          ½          ¾          1
```

Figure 1.1: Base 2 representation of a number $Y \in [0, 1]$. X_1 determines which half, $[0, 1/2)$ or $[1/2, 1]$, Y is in; X_2 determines which half of that half Y is in, etc.

What is the probability mass function or density of the random quantity Y? If $0.x_1x_2\ldots$ is the base 2 representation of y, then $P(Y = y) = P(X_1 = x_1)P(X_2 = x_2)\ldots = 0$ if $p \in (0, 1)$ because each of the infinitely many terms in the product is either p or $(1 - p)$. Because the probability of Y exactly equaling any given number y is 0, Y is not a discrete random variable. In the special case that $p = 1/2$, Y is uniformly distributed because Y is equally likely to be in the first or second half of $[0, 1]$, then equally likely to be in either half of that half, etc. But what distribution does Y have if $p \in (0, 1)$ and $p \neq 1/2$? It is by no means obvious, but we will show in Example 7.3 of Chapter 7 that for $p \neq 1/2$, the distribution of Y has no density!

Another way to think of the X_i in this example is that they represent treatment assignment ($X_i = 0$ means placebo, $X_i = 1$ means treatment) for individuals in a randomized clinical trial. Suppose that in a trial of size n, there is a planned imbalance in that roughly twice as many patients are assigned to treatment as to placebo. If we imagine an infinitely large clinical trial, the imbalance is so great that Y fails to have a density because of the preponderance of ones in its base 2 representation. We can also generate a random variable with no density by creating too much balance. Clinical trials often randomize using *permuted blocks*, whereby the number of patients assigned to treatment and placebo is forced to be balanced after every 2 patients, for example. Denote the assignments by $X_1, X_2, X_3, \ldots$, again with $X_i = 0$ and $X_i = 1$ denoting placebo or treatment, respectively, for patient i. With permuted blocks of size 2, exactly one of X_1, X_2 is 1, exactly one of X_3, X_4 is 1, etc. In this case there is so much balance in an infinitely large clinical trial that again the random number defined by Equation (1.1) has no density (Example 5.33 of Section 5.6). □

Example 1.2. No bivariate density Here we present an example of a singular bivariate distribution derived from independent normal random variables. Let X and Y be independent standard normal random variables (zero mean and unit variance), and consider the conditional distribution of (X, Y) given that $Y - X = 0$. If (X', Y') denote random variables from this conditional distribution, then all of the probability for (X', Y') is concentrated on the line $Y' - X' = 0$. Note that the distribution of (X', Y') cannot be discrete because $P(X' = x', Y' = y') = 0$ for every x' and y'. There also can be no joint density function for

(X', Y'). To see this, note that $P(-\infty < X' < \infty, Y' = X') = 1$. If there were a joint density function $f(x', y')$ for (X', Y'), then $P(-\infty < X' < \infty, Y' = X')$ would be the volume of the region $\{(x', y', z') : -\infty < x' < \infty, y' = x', 0 \le z' \le f(x', y')\}$. But this region is a portion of a plane, whose volume is 0 (See Figure 1.2). In other words, the probability that $-\infty < X' < \infty, Y' = X'$ would have to be 0 if there were a density $f(x', y')$ for (X', Y'). This contradiction shows that there can be no joint density $f(x', y')$ for (X', Y'). $\square$

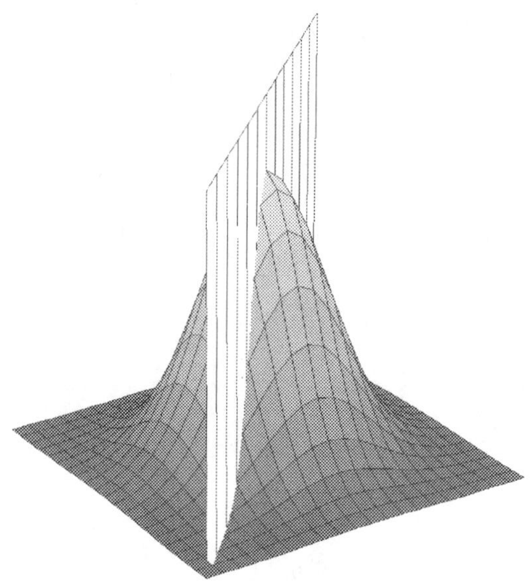

Figure 1.2: Conditioning on $Y - X = 0$ when (X, Y) are iid $N(0, 1)$.

Example 1.2 involved conditioning on an event of probability 0 ($Y - X = 0$), which always requires great care. Seemingly very reasonable arguments can lead to the wrong conclusion, as illustrated again by the next example.

Example 1.3. Borel's paradox: the equivalent event fallacy In biostatistics examples, patient characteristics may induce correlations in disease status or progression. For example, people with a specific gene might be more likely to get a certain type of cancer. Thus, two people with $X = 0$ (gene absent) or $X = 1$ (gene present) tend to have more similar outcomes (both cancer free or both with cancer) than two patients with opposite gene characteristics. This is an example with both X and Y discrete. Now suppose X and Y are both continuous. For example, X might be the expression level of a gene and Y might be the person's body surface area. Suppose we want to show that two people with the same value of X tend to have similar values of Y. One might postulate a model

$$Y = \beta_0 + \beta_1 X + \epsilon,$$

where ϵ is measurement error independent of X. Formulation 1 is to imagine that Nature generates a value X for each person, but for this pair, She generates a single X and applies it to both members of the pair. In that case, the covariance between Y measurements of two people with exactly the same X values is $\text{cov}(Y_1, Y_2) = \text{cov}(\beta_0 + \beta_1 X + \epsilon_1, \beta_0 + \beta_1 X + \epsilon_2) = \text{cov}(\beta_1 X, \beta_1 X) = \beta_1^2 \sigma_X^2 > 0$.

Formulation 2 is that Nature generates an X for each patient, but by serendipity, two people happen to have identical values of X. In other words, we observe $\beta_0 + \beta_1 X_1 + \epsilon_1$ and

$\beta_0 + \beta_1 X_2 + \epsilon_2$, and we condition on $X_1 = X_2 = X$. This seems equivalent to Formulation 1, but it is not. Conditioning on $X_1 = X_2 = X$ actually changes the distribution of X, but exactly how? Without loss of generality, take X_1 and X_2 to be independent standard normals, and consider the conditional distribution of X_2 given $X_1 = X_2$. One seemingly slick way to compute it is to formulate the event $\{X_1 = X_2\}$ as $\{X_2 - X_1 = 0\}$ and obtain the conditional distribution of X_2 given that $X_2 - X_1 = 0$. This is easy because the joint distribution of $(X_2, X_2 - X_1)$ is bivariate normal with mean vector $(0, 0)$, variances $(1, 2)$, and correlation coefficient $1/\sqrt{2}$. Using a standard formula for the conditional distribution of two jointly normal random variables, we find that the distribution of X_2 given that $X_2 - X_1 = 0$ is normal with mean 0 and variance 1/2; its density is

$$f(x_2 \,|\, X_2 - X_1 = 0) = \exp(-x_2^2)/\sqrt{\pi}. \tag{1.2}$$

Another way to think about the event $\{X_1 = X_2\}$ is $\{X_2/X_1 = 1\}$. We can obtain the joint density $g(u, v)$ of $(U, V) = (X_2, X_2/X_1)$ by computing the Jacobian of the transformation, yielding

$$g(u, v) = \frac{|u| \exp\left\{-\frac{u^2}{2}\left(\frac{v^2+1}{v^2}\right)\right\}}{2\pi v^2}.$$

Integrating over u from $-\infty$ to ∞ yields the marginal density of V:

$$h(v) = \frac{1}{\pi(v^2 + 1)}.$$

Therefore, the conditional density of U given $V = 1$ is $g(u, 1)/h(1) = |u| \exp(-u^2)/2$. That is, the conditional density of X_2 given $X_2/X_1 = 1$ is

$$\psi\,(x_2 \,|\, X_2/X_1 = 1) = |x_2| \exp(-x_2^2). \tag{1.3}$$

Expression (1.3) is similar, but not identical, to Equation (1.2). The two different conditional distributions of X_2 given $X_1 = X_2$ give different answers! Of course, there are many other ways to express the fact that $X_1 = X_2$. This example shows that, although we can define conditional distributions given the value of a random variable, there is no unique way to define conditional distributions given that two continuous random variables agree. Conditioning on events of probability 0 always requires great care, and should be avoided when possible. Formulation 1 is preferable because it sidesteps these difficulties. □

The following is an example illustrating the care needed in formulating the experiment.

Example 1.4. The two envelopes paradox: Improperly formulating the experiment Have you seen the commercials telling you how much money people who switch to their auto insurance company save? Each company claims that people who switch save money, and that is correct. The impression given is that you could save a fortune by switching from company A to company B, and then switching back to A, then back to B, etc. That is incorrect. The error is analogous to the reasoning in the following paradox.

Consider two envelopes, one containing twice as much money as the other. You hold one of the envelopes, and you are trying to decide whether to exchange it for the other one. You argue that if your envelope contains x dollars, then the other envelope is equally likely to contain either $x/2$ dollars or $2x$ dollars. Therefore, the expected amount of money you will have if you switch is $(1/2)(x/2) + (1/2)(2x) = (5/4)x > x$. Therefore, you decide you should switch envelopes. But the same argument can be used to conclude that you should switch again!

The problem is in your formulation of the experiment as someone handing you an envelope with x dollars in it, and then flipping a coin to decide whether to place $2x$ dollars or $(1/2)x$ dollars in the other envelope. If this had been the experiment, then you should switch, but then you should not switch again. The actual experiment is to first put x and $2x$ dollars in two envelopes and then flip a coin to decide which envelope to give you. Let U be the amount of money you have. Then U has the same distribution,

$$U = \begin{cases} x & \text{with probability } 1/2 \\ 2x & \text{with probability } 1/2, \end{cases}$$

whether or not you switch envelopes. Therefore, your expected value is $(x)(1/2)+(2x)(1/2) = (3/2)x$ whether or not you switch.

You might wonder what is wrong with letting X be the random amount of money in your envelope, and saying that the amount in the other envelope is

$$Y = \begin{cases} X/2 & \text{with probability } 1/2 \\ 2X & \text{with probability } 1/2. \end{cases} \tag{1.4}$$

Actually, this is true. Untrue is the conclusion that $\mathrm{E}(Y) = (1/2)\mathrm{E}(X/2) + (1/2)\mathrm{E}(2X) = (5/4)\mathrm{E}(X) > \mathrm{E}(X)$. This would be valid if the choice of either $X/2$ or $2X$ were independent of the value of X. In that case we could condition on $X = x$ and replace X by x in Equation (1.4). The problem is that the choice of either $X/2$ or $2X$ depends on the value x of X. Very small values of x make it less likely that your envelope contains the doubled amount, whereas very large values of x make it more likely that your envelope contains the doubled amount. To see why this invalidates the formula $\mathrm{E}(Y) = (1/2)\mathrm{E}(X/2)+(1/2)\mathrm{E}(2X)$, imagine generating a standard normal deviate Z_1 and setting

$$Z_2 = \begin{cases} -Z_1 & \text{if } Z_1 < 0 \\ +Z_1 & \text{if } Z_1 \geq 0. \end{cases} \tag{1.5}$$

Note that

$$Z_2 = \begin{cases} -Z_1 & \text{with probability } 1/2 \\ +Z_1 & \text{with probability } 1/2, \end{cases} \tag{1.6}$$

so you might think that conditioned on $Z_1 = z_1$, Equation (1.6) holds with Z_1 replaced by z_1. In that case $\mathrm{E}(Z_2|Z_1 = z_1) = (1/2)(-z_1) + (1/2)(z_1) = 0$ and $\mathrm{E}(Z_2) = 0$. But from Equation (1.5), $Z_2 = |Z_1| > 0$ with probability 1, so $\mathrm{E}(Z_2)$ must be strictly positive. The error was in thinking that once we condition on $Z_1 = z_1$, Equation (1.6) holds with Z_1 replaced by z_1. In reality, if $Z_1 = z_1 < 0$, then the probabilities in Equation (1.6) are 1 and 0, whereas if $Z_1 = z_1 \geq 0$, then the probabilities in Equation (1.6) are 0 and 1.

A similar error in reasoning applies in the auto insurance setting. People who switch from company A to company B do save hundreds of dollars, but that is because the people who switch are the ones most dissatisfied with their rates. If X is your current rate and you switch companies, it is probably because X is large. If you could save hundreds by switching, irrespective of X, then you would benefit by switching back and forth. The ads are truthful in the sense that **people who switch** do save money, but that does not necessarily mean that you will save by switching; that depends on whether your X is large or small. □

One thing we would like to do in advanced probability is define $x + y$ or $x - y$ when x or y is infinite. This is straightforward if only one of x and y is infinite. For instance, if $y = +\infty$ and x is finite, then $x + y = +\infty$. But is there a sensible way to define $x - y$ if $x = +\infty$ and $y = +\infty$? You may recall that $x - y$ is undefined in this case. The following puzzle illustrates very clearly why there is a problem with trying to define $x - y$ when x and y are both $+\infty$.

Example 1.5. Trying to define $\infty - \infty$ Suppose you have a collection of infinitely many balls and a box with an unlimited capacity. At 1 minute to midnight, you put 10 balls in the box and remove 1. At 1/2 minute to midnight, you put 10 more balls in the box and remove 1. At 1/4 minute to midnight, you put 10 more balls in the box and remove 1, etc. Continue this process of putting 10 in and removing 1 at $1/2^n$ minutes to midnight for each n. How many balls are in the box at midnight?

We must first dispel one enticing but incorrect answer. Some argue that we will never reach midnight because each time we halve the time remaining, there will always be half left. But this same line of reasoning can be used to argue that motion is impossible: to travel 1 meter, we must first travel 1/2 meter, leaving 1/2 meter left, then we must travel 1/4 meter, leaving 1/4 meter left, etc. This argument, known as Xeno's paradox, is belied by the fact that we seem to have no trouble moving! The paradox disappears when we recognize that there is a $1-1$ correspondence between distance and time; if it takes 1 second to travel 1 meter, then it takes only half a second to travel 1/2 meter, etc., so the total amount of time taken is $1 + 1/2 + 1/4 + \ldots = 2$ seconds. Assume in the puzzle that we take, at the current time, half as long to put in and remove balls as we took at the preceding time. Then we will indeed reach midnight.

Notice that the total number of balls put into the box is $10 + 10 + 10 \ldots = \infty$, and the total number taken out is $1 + 1 + 1 \ldots = \infty$. Thus, the total number of balls in the box can be thought of as $\infty - \infty$. But at each time, we put in 10 times as many balls as we take out. Therefore, it is natural to think that there will be infinitely many balls in the box at midnight. Surprisingly, this is not necessarily the case. In fact, there is actually no one right answer to the puzzle. To see this, imagine that the balls are all numbered $1, 2, \ldots$, and consider some alternative ways to conduct the experiment.

1. At 1 minute to midnight, put balls $1-10$ in the box and remove ball 1. At 1/2 minute to midnight, put balls $11-20$ in the box and remove ball 2. At 1/4 minute to midnight, put balls $21-30$ into the box and remove ball 3, etc. So how many balls are left at midnight? None. If there were a ball, what number would be on it? It is not number 1 because we removed that ball at 1 minute to midnight. It is not number 2 because we removed that ball at 1/2 minute to midnight. It cannot be ball number n because that ball was removed at $1/2^n$ minutes to midnight. Therefore, there are 0 balls in the box at midnight under this formulation.

2. At 1 minute to midnight, put balls $1-10$ in the box and remove ball 2. At 1/2 minute to midnight, put balls $11-20$ in the box and remove ball 3, etc. Now there is exactly one ball in the box at midnight because ball number 1 is the only one that was never removed.

3. At 1 minute to midnight, put balls $1-10$ in the box and remove ball 1. At 1/2 minute to midnight, put balls $11-20$ in the box and remove ball 11. At 1/4 minute to midnight, put balls $21-30$ in the box and remove ball 21, etc. Now there are infinitely many balls in the box because balls $2-10$, $12-20$, $22-30$, etc. were never removed.

It is mind boggling that the answer to the puzzle depends on which numbered ball is removed at each given time point. The puzzle demonstrates that there is no single way to define $\infty - \infty$. □

Examples 1.1–1.5 may seem pedantic, but there are real-world implications of insistence on probabilistic rigor. The following is a good illustration.

Example 1.6. Noticing trends after the fact A number of controversies result from noticing what seem in retrospect to be unlikely events. Examples include psychics seeming to know things about you that they should not know, amateur astronomers noticing what looks strikingly like a human face on another planet (Examples 3.37 and 4.51), or biblical scholars finding apparent coded messages when skipping letter sequences in the first 5 books of the Old Testament (Example 3.1). The problem is that even in the simple experiment of drawing a number randomly from the unit interval, every outcome is vanishingly rare in that it has probability 0. Therefore, it is always possible to make patterns noticed after the fact look as though they could not have happened by chance.

For the above reasons, clinical trialists insist on pre-specifying all analyses. For example, it is invalid to change from a t-test to a sign test after noticing that signs of differences between treatment and control are all positive. This temptation proved too great for your first author early in his career (see page 773 of Stewart et al., 1991). Neither is it valid to focus exclusively on the one subgroup that shows a treatment benefit. Adaptive methods allow changes after seeing data, but the rules for deciding whether and how to make such changes are pre-specified. An extremely important question is whether it is ever possible to allow changes that were not pre-specified. Can this be done using a permutation test in a way that maintains a rigorous probabilistic foundation? If so, then unanticipated and untoward events need not ruin a trial. We tackle this topic in portions of Chapter 11. □

We hope that the paradoxes in this chapter sparked interest and convinced you that failure to think carefully about probability can lead to nonsensical conclusions. This is especially true in the precarious world of conditional probability. Our goal for this book is to provide you with the rigorous foundation needed to avoid paradoxes and provide valid proofs. We do so by presenting classical probability theory enriched with illustrative examples in biostatistics. These involve such topics as outlier tests, monitoring clinical trials, and using adaptive methods to make design changes on the basis of accumulating data.

Chapter 2

Size Matters

2.1 Cardinality

Elementary probability theory taught us that there was a difference between discrete and continuous random variables, but exactly what does "discrete" really mean? Certainly X is discrete if it has only a finite number possible values, but it might be discrete even if it has infinitely many possible values. For instance, a Poisson random variable takes values $0, 1, 2, \ldots$ Still, $\Omega = \{0, 1, 2, \ldots\}$ feels very different from, say, $\Omega = [0, 1]$. This chapter will make precise how these different types of infinite sets differ. You will learn important counting techniques and arguments for determining whether infinite sets like $\{0, 1, 2, \ldots\}$ and $[0, 1]$ are equally numerous. In the process, you will see that $[0, 1]$ is actually more numerous.

For any set A we can talk about the number of elements in A, called the *cardinality* of A and denoted $\text{card}(A)$. If A has only finitely many elements, then cardinality is easy. But what is $\text{card}(A)$ if A has infinitely many elements? Answering "infinity" is correct, but only half the story. The fact is that some infinite sets have more members than other infinite sets. Consider the infinite set A consisting of the integers. We can imagine listing them in a systematic way, specifying which element is first, second, third, etc.:

$$
\begin{array}{cccccccc}
 & 1 & 2 & 3 & 4 & 5 & \ldots \\
 & \updownarrow & \updownarrow & \updownarrow & \updownarrow & \updownarrow & \\
A: & 0 & 1 & -1 & 2 & -2 & \ldots
\end{array}
$$

The top row shows the position on the list, with 1 meaning first, 2 meaning second, etc., and the bottom row shows the elements of A. Thus, the first element is 0, followed by 1, -1, etc. Such a list is a $1-1$ correspondence between the natural numbers (top row) and the elements of A (bottom row). We are essentially "counting" the elements of A. Even though A has infinitely many members, each one will be counted eventually. In fact, we can specify exactly where each integer will appear on the list: integer n is the $(2n)$th item if n is positive, and the $(2|n| + 1)$th item if n is negative or 0. This leads us to the following definition.

Definition 2.1. Countably infinite, countable, uncountably infinite, uncountable *A set A is said to be countably infinite if there is a $1-1$ correspondence between the elements of A and the natural numbers $1, 2, \ldots$, and countable if it is either finite or countably infinite. Otherwise A is called uncountable or uncountably infinite.*

$$\begin{array}{cccccc} & 1 & 2 & \ldots & n & \ldots \\ & \updownarrow & \updownarrow & & \updownarrow & \\ A: & a_1 & a_2 & \ldots & a_n & \ldots \end{array}$$

An example of a set whose elements cannot be listed systematically is the interval $[0, 1]$. To see that $[0, 1]$ is uncountable, suppose that you could construct an exhaustive list of its members:

$$\begin{array}{ccccccccccccc} x_1 & = & \boxed{.7} & 3 & 8 & 8 & 4 & 7 & 2 & 0 & 1 & \ldots \\ x_2 & = & .0 & \boxed{4} & 3 & 1 & 2 & 6 & 9 & 1 & 0 & \ldots \\ x_3 & = & .3 & 8 & \boxed{5} & 3 & 8 & 7 & 5 & 4 & 4 & \ldots \\ \vdots & & & & & & & & & & & \end{array}$$

To see that this list cannot be complete, let a_1 be any digit other than the first digit of x_1; the first digit of x_1 is 7 (inside box), so take $a_1 = 5$, for example. Now let a_2 be any digit other than the second digit of x_2; the second digit of x_2 is 4 (inside box), so take $a_2 = 0$, for example. Let a_3 be any digit other than the third digit of x_3, etc. It seems that the number $.a_1a_2a_3\ldots$ is in $[0, 1]$ but is not on the list. It differs from x_1 in the first digit; it differs from x_2 in the second digit; it differs from x_3 in the third digit, etc. It differs from all of the xs by at least one digit. This argument applies to any attempted list of $[0, 1]$, and therefore appears to prove that $[0, 1]$ is uncountable.

There is a slight flaw in the above reasoning. Some numbers in $[0, 1]$ have two different decimal representations. For example, 0.1 can be written as $0.1000\ldots$ or as $0.0999\ldots$ Therefore, two different representations of the same number can differ from each other in a given digit. To circumvent this problem, modify the above argument as follows. For any listing $x_1, x_2, \ldots$ of the numbers in $[0, 1]$, let a_1 be any digit other than 0, 9, or the first digit of x_1; let a_2 be any digit other than 0, 9, or the second digit of x_2, etc. Then $.a_1a_2\ldots$ is a number in $[0, 1]$ that is not equivalent to any of the $x_1, x_2, \ldots$ (because its digits cannot end in $.000\ldots$ or $.999\ldots$). This method of proof is the celebrated "diagonal technique" of G. Cantor, who made significant contributions to set theory in the late 1800s. We have established the following result.

Proposition 2.2. Uncountability of $[0, 1]$ *The interval $[0, 1]$ is uncountable.*

We now have a prototype for each type of infinite set, countable (the integers) and uncountable ($[0, 1]$). From these we can build new examples.

Proposition 2.3. The direct product of two countable sets is countable *The direct product $A \times B$ (the set of all ordered pairs (a_i, b_j)) of countable sets A and B is countable.*

Proof. We prove this when $A = \{a_i\}$ and $B = \{b_i\}$ are countably infinite. Form the matrix of ordered pairs (a_i, b_j), as shown in Figure 2.1.

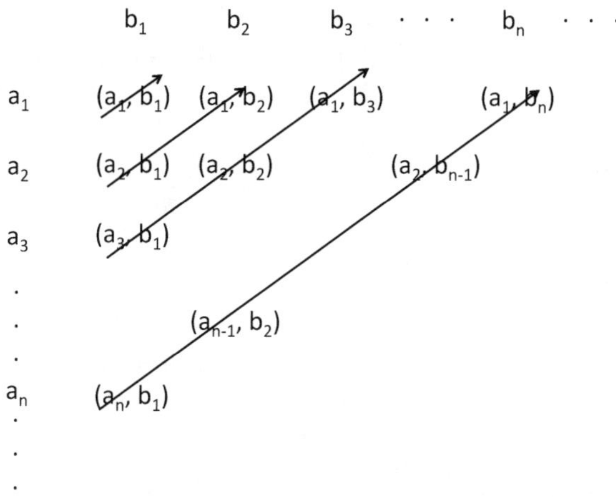

Figure 2.1: The direct product (set of ordered pairs) of two countably infinite sets.

List the elements of the matrix by traversing its diagonals, as shown in Figure 2.1 and Table 2.1.

Table 2.1:

1	2	3	4	5	6	...
$\updownarrow$	$\updownarrow$	$\updownarrow$	$\updownarrow$	$\updownarrow$	$\updownarrow$	
(a_1, b_1)	(a_2, b_1)	(a_1, b_2)	(a_3, b_1)	(a_2, b_2)	(a_1, b_3)	...

Each element of the direct product appears somewhere on the list. We can make this formal as follows. After traversing $k-1$ diagonals, the number of elements enumerated is $1 + 2 + \ldots + k - 1 = k(k-1)/2$. The jth element of the kth diagonal is (a_{k-j+1}, b_j), $j = 1, \ldots, k$. This is the $n = k(k-1)/2 + j$th element on the list. Substituting i for $k - j + 1$, we see that the pair (a_i, b_j) is the $n = (i+j-1)(i+j-2)/2 + j$th element on the list. We have established that the direct product of two countably infinite sets is in $1-1$ correspondence with the natural numbers. The extension of the proof to the case of A or B finite is straightforward and omitted. □

The following result is a helpful tool in building new countable and uncountable sets from old ones. Its proof is left as an exercise.

Proposition 2.4. A subset of a countable set is countable *If $A \subset B$, then if B is countable, so is A. Equivalently, if A is uncountable, so is B.*

We will make use of the following result repeatedly throughout this book.

Proposition 2.5. Rationals are countable *The set of rational numbers is countable.*

Proof. By Proposition 2.3, the set $D = \{(i, j), i = 1, 2, \ldots$ and j=1,2,...$\}$ is countable. By definition, the rational numbers are of the form i/j for (i, j) in a subset $C \subset D$, namely the set of (i, j) such that i and j have no common factors. By Proposition 2.4, C is countable.

Because there is a $1-1$ correspondence between the rationals and C, the rationals are also countable. □

Proposition 2.6. A countable union of countable sets is countable *A countable union* $\cup_{i=1}^{\infty} A_i$ *of countable sets* A_i *is countable.*

Proof. Assume first that each A_i is countably infinite. Let A be the matrix whose rows are A_i. If the a_{ij} are distinct, then $\cup_{i=1}^{\infty} A_i$ consists of the elements of A, which can be put in $1-1$ correspondence with the direct product (i,j), $i = 1,2,\ldots$, $j = 1,2,\ldots$ This direct product is countable by Proposition 2.3. If the a_{ij} are not all distinct, then $\cup_{i=1}^{\infty} A_i$ is a subset of the elements of A. By Proposition 2.4, $\cup_{i=1}^{\infty} A_i$ is countable.

Although we assumed that each A_i is countably infinite, we can extend the proof to the case when one or more of the A_i are finite. Simply augment each such A_i by countably infinitely many new members. The countable union of these extended sets is countable by the proof in the preceding paragraph. The countable union of the original sets is a subset of this countable set, so is countable by Proposition 2.4. □

We have seen that there are at least two different cardinalities for infinite sets—countable and uncountable. Are there infinite sets that are smaller than countable? Are some uncountable sets larger than others? To answer these questions we must first define what is meant by two infinite sets having the same cardinality.

Definition 2.7. Two sets having the same cardinality *Two sets* A *and* B *are said to have the same cardinality if there is a* $1-1$ *correspondence between* A *and* B.

Remark 2.8. *Two infinite sets can have the same cardinality even if one is a subset of the other, and even if the set difference between them has infinitely many elements. For example, the set of all integers and the set of negative integers have the same cardinality even though their set difference contains infinitely many elements, namely the nonnegative integers.*

The following result shows that no infinite set has smaller cardinality than countable infinity.

Proposition 2.9. Countable infinity is, in a sense, the "smallest" cardinality an infinite set can have *If* A *is an infinite set, then:*

1. *A has a countably infinite subset.*
2. *Augmenting A by a countably infinite set (or a finite set, for that matter) B does not change its cardinality.*

Proof. For the first result, select a countably infinite subset $A' = \{a_1', a_2', \ldots a_n', \ldots\}$ from A by first selecting a_1' from A, then selecting a_2' from $A \setminus \{a_1'\}$, then selecting a_3' from $A \setminus \{a_1', a_2'\}$, etc.

For the second result, we must prove that if A can be put in $1-1$ correspondence with a set C, then so can A augmented by $B = \{b_1, b_2, \ldots\}$. Let $a_s \leftrightarrow c_s$ denote the 1-1 correspondence between A and C. By part 1, we can select a countably infinite subset A' of A. Separating A into A' and $A'' = A \setminus A'$ and retaining the correspondence between A and C yields Table 2.2.

Table 2.2:

a'_1	a'_2	a'_3	$\ldots$	A''
$\updownarrow$	$\updownarrow$	$\updownarrow$	$\ldots$	$\updownarrow$
c'_1	c'_2	c'_3	$\ldots$	C''

Now retain the correspondence between A'' and C'', but modify the correspondence between A' and C' to obtain Table 2.3. This defines a $1-1$ correspondence between

Table 2.3:

a'_1	b_1	a'_2	b_2	a'_3	b_3	$\ldots$	A''
$\updownarrow$	$\updownarrow$	$\updownarrow$	$\updownarrow$	$\updownarrow$	$\updownarrow$	$\ldots$	$\updownarrow$
c'_1	c'_2	c'_3	c'_4	c'_5	c'_6	$\ldots$	C''

C and the augmentation of A with B, demonstrating that these two sets have the same cardinality. Therefore, augmenting an infinite set with a countably infinite set does not change its cardinality. We omit the straightforward extension for the case in which B is finite. □

We next address the question of whether some uncountable sets are larger than others by studying the countability of various strings of 0s and 1s. Doing so also allows us to see a connection between a countable set like the natural numbers and the uncountable set $[0,1]$.

Proposition 2.10. Connection between strings of 0s and 1s and $[0,1]$

1. *The set of strings $(x_1, x_2, \ldots)$ of 0s and 1s such that the number of 1s is finite is countable. Likewise, the set of strings $(x_1, x_2, \ldots)$ of 0s and 1s such that the number of 0s is finite is countable.*

2. *The set of all possible strings $(x_1, x_2, \ldots)$ of 0s and 1s has the same cardinality as $[0,1]$.*

Proof. Each string $(x_1, x_2, \ldots)$ of 0s and 1s with only finitely many 1s corresponds to the base 2 representation $0.x_1x_2\ldots$ of a unique rational number in $[0,1]$. Therefore, the set in part 1 is in $1-1$ correspondence with a subset of the rationals, and is therefore countable by Propositions 2.4 and 2.5. The second statement of part 1 follows similarly.

To prove the second part, let B be the set of all strings of 0s and 1s. It is tempting to argue that B is in $1-1$ correspondence with $[0,1]$ because $0.x_1x_2,\ldots x_n\ldots$ is the base 2 representation of a number in $[0,1]$. But some numbers in $[0,1]$ have two base 2 representations (e.g., $1/2 = .1000\ldots$ or $.0111\ldots$). If A is the set of strings of 0s and 1s that do not end in $111\ldots$, then A is in $1-1$ correspondence with $[0,1]$. Moreover, B is A augmented by the set C of strings of 0s and 1s ending in $111\ldots$ But C is countable because each $c \in C$ corresponds to a unique finite string of 0s and 1s, ending in 0, before the infinite string of 1s. By part 1, C is countable. Because augmenting an infinite set A by a countable set C does not change its cardinality (Proposition 2.9), B also has the cardinality of $[0,1]$. □

Proposition 2.10 can also be recast as a statement about subsets of a given set, as the following important remark shows.

Remark 2.11. Connection between subsets of a set and strings of 0s and 1s *If Ω is any set, the set of subsets of Ω can be put in $1-1$ correspondence with the set of strings of 0s and 1s, where the length of the strings is card(Ω). To see this, let $A \subset \Omega$. For each $x \in \Omega$, write a 1 below x if $x \in A$, and a 0 below x if $x \notin A$. For instance, if Ω is the set of natural numbers, then $A = \{1, 2, 4\}$ is represented by*

$$\begin{array}{ccccccc} 1 & 2 & 3 & 4 & 5 & 6 & \ldots \\ 1 & 1 & 0 & 1 & 0 & 0 & \ldots \end{array}$$

Remark 2.12. *Remark 2.11 and Proposition 2.10 imply that if Ω is any countably infinite set:*

1. *The set of* **finite** *subsets of Ω is countable.*
2. *The set of* **all** *subsets of Ω has the same cardinality as $[0, 1]$.*

We close this chapter with an axiom that at first blush seems totally obvious. For a finite number of nonempty sets $A_1, \ldots, A_n$, no one would question the fact that we can select one element, ω_i, from set A_i. But what if the number of sets is uncountably infinite? Can we still select one element from each set? The Axiom of Choice asserts that the answer is yes.

Axiom 2.13. Axiom of Choice *Let A_t, where t ranges over an arbitrary index set, be a collection of nonempty sets. From each set A_t, we can select one element.*

Most mathematicians accept the Axiom of Choice, though a minority do not. The following example illustrates why there might be some doubt, in the minds of some, about whether one can always pick one member from each of uncountably many sets.

Example 2.14. How not to choose For each $t \in (0, 1)$, let $A_t = \{s \in (0, 1) : s - t$ is rational$\}$. The Axiom of Choice asserts that we can create a set A consisting of one member of each A_t. Here is an interesting fallacy in trying to construct an explicit procedure for doing this. At step 0, make A empty. For $t \in (0, 1)$, step t is to determine whether t is in any of the sets A_s for $s < t$. If so, then we do not add t to A because we want A to include only one member of each A_t. If t is not in A_s for any $s < t$, then put t in A. Exercise 10 is to show that A constructed in this way is empty. No new members can ever be added at step t for any $t \in (0, 1)$ because potential new members have presumably already been added. This amusing example is reminiscent of Yogi Berra's saying, "Nobody goes to that restaurant anymore: it's too crowded." □

Exercises

1. Which has greater cardinality, the set of integers or the set of even integers? Justify your answer.
2. Prove that the set of irrational numbers in $[0, 1]$ is uncountable.
3. What is the cardinality of the following sets?
   (a) The set of functions $f : \{0, 1\} \longmapsto \{\text{integers}\}$.
   (b) The set of functions $f\{0, 1, \ldots, n\} \longmapsto \{\text{integers}\}$.

(c) The set of functions $f : \{\text{integers}\} \longmapsto \{0, 1\}$.

4. Which of the following arguments is (are) correct? Justify your answer.

   (a) Any collection $\mathcal{C}$ of nonempty, disjoint intervals (a, b) is countable because for each $C \in \mathcal{C}$, we can pick a rational number $x \in \mathcal{C}$. Therefore, we can associate the Cs with a subset of rational numbers.

   (b) $\mathcal{C}$ must be uncountable because for each $C \in \mathcal{C}$, we can pick an irrational number $x \in C$. Therefore, we can associate the Cs with irrational numbers.

   (c) *Any* collection $\mathcal{C}$ of nonempty intervals (not just disjoint) must be countable because for each $C \in \mathcal{C}$, we can pick a rational number. Therefore, we can associate $\mathcal{C}$ with a subset of rational numbers.

5. Prove that the set of numbers in $[0, 1]$ with a repeating decimal representation of the form $x = .a_1a_2 \ldots a_n a_1 a_2 \ldots a_n \ldots$ (e.g., $.111\ldots$ or $.976976976\ldots$, etc.) is countable. Hint: if $x = .a_1a_2 \ldots a_n a_1 a_2 \ldots a_n \ldots$, what is $10^n x - x$, and therefore what is x?

6. (a) Prove that any closed interval $[a, b]$, $-\infty < a < b < \infty$, has the same cardinality as $[0, 1]$.

   (b) Prove that any open interval (a, b), $-\infty < a < b < \infty$, has the same cardinality as $[0, 1]$.

   (c) Prove that $(-\infty, \infty)$ has the same cardinality as $[0, 1]$.

7. Prove Proposition 2.4.

8. Prove that if A is any set, the set of all subsets of A has larger cardinality than A. Hint: suppose there were a $1-1$ correspondence $a \leftrightarrow B_a$ between the elements of A and the subsets B_a of A. Construct a subset C of A as follows: if $a \in B_a$, then exclude a from C, whereas if $a \notin B_a$, include a in C.

9. The Cantor set is defined as the set of numbers in $[0, 1]$ whose base 3 representation $.a_1a_2 \ldots = a_1/3^1 + a_2/3^2 + \ldots +$ has $a_i \neq 1$ for each i. Show that the Cantor set has the same cardinality as $[0, 1]$.

10. Show that the set A constructed in Example 2.14 is empty.

11. Imagine a 1×1 square containing lights at each pair (r, s), where r and s are both rational numbers in the interval $[0, 1]$.

    (a) Prove that the set of lights is countable.

    (b) We can turn each light either on or off, and thereby create light artwork. Prove that the set of all possible pieces of light artwork is uncountable, and has the same cardinality as $[0, 1]$.

    (c) We create a line segment joining the point (r_1, s_1) to the point (r_2, s_2) by turning on the lights at all positions $(R, S) = (1 - \lambda)(r_1, s_1) + \lambda(r_2, s_2)$ as λ ranges over the rational numbers in $[0, 1]$. Show that the set of all such line segments is countable. Hint: think about the set of endpoints of the segments.

2.2 Summary

1. The cardinality of a set is, roughly speaking, its number of elements.

2. Infinite sets can have different cardinalities. Can the set be put in 1-1 correspondence with the natural numbers?

   (a) If yes, the set is countably infinite (the "smallest" infinite set). Prototypic example: rational numbers.

   (b) If no, the set is uncountably infinite. Prototypic example: $[0, 1]$.

3. A countable union of countable sets is countable.

4. Two sets have the same cardinality if there is a 1-1 correspondence between them.

5. *** There is an intimate connection between numbers in $[0, 1]$ and strings of countably many 0s and 1s:

   (a) The set S of strings $x_1, x_2, \ldots$, each $x_i = 0$ or 1, corresponds to base 2 representations of numbers in $[0, 1]$, so $\text{card}(S) = \text{card}([0, 1])$.

   (b) Strings of 0s and 1s correspond to subsets $A \subset \Omega$: record a 1 if $\omega \in A$ and 0 if $\omega \notin A$. Therefore, if $\Omega = \{a_1, a_2, \ldots\}$ is countably infinite:

      i. The set of all subsets of Ω has the same cardinality as $[0, 1]$.

      ii. The set of all subsets of Ω with a finite number of elements is countable.

Chapter 3

The Elements of Probability Theory

3.1 Introduction

One of the problems in the two-envelopes paradox in Example 1.4 is a misinterpretation about how the random experiment is carried out. Nothing is stated about how the amounts are selected other than that one envelope has twice as much money as the other. The person receiving the envelope incorrectly imagines that someone hands him x dollars and then flips a coin to decide whether to double or halve that amount for the other envelope. This may seem like a toy example, but there are a number of actual examples that have generated tremendous controversy at least in part because of the underlying problem illustrated by the two-envelope problem, namely difficulty in formulating the random experiment and accurately assessing probabilities. The following is a case in point.

Example 3.1. Bible code For hundreds of years, some people, including the great Sir Isaac Newton, have believed that there are prophecies encoded in the first 5 books of the Old Testament (the Torah) that can be revealed only by skipping letters in a regular pattern. For instance, one might read every 50th letter or every 100th letter, but the specified spacing must be maintained throughout the entire text. Most of the time this produces gibberish, but it occasionally produces tantalizing words and phrases like "Bin Laden," and "twin towers," that seem to foretell historic events. Books and DVDs about the so-called Bible code (Drosnin, 1998; The History Channel, 2003) maintain that the probability of the observed patterns is so tiny that chance has been ruled out as an explanation. But how does one even define a random experiment? After all, letters are not being drawn at random, but are chosen to form words that are in turn strung together to make sensible sentences. Where is the randomness?

A vocal skeptic (Brendan McKay) showed that similar messages could be found by skipping letters in any book of similar length such as *Moby-Dick*. In effect, he formulated the random experiment by treating the Torah as being drawn at random from the set of books of similar length. □

Examples such as the above inspire a very careful definition of the random experiment and probability theory presented in this chapter. We use the triple $(\Omega, \mathcal{F}, P)$ of probability theory, called the *probability space*, where Ω is the *sample space* of all possible outcomes of an experiment, $\mathcal{F}$ is the set of events (subsets of Ω) we are allowed to consider, and P is a probability measure. We also consider more general measures μ, not just probability measures, in which case $(\Omega, \mathcal{F}, \mu)$ is called a *measure space.* You probably have some familiarity with Ω and P, but not necessarily with $\mathcal{F}$ because you may not have been aware of the need to restrict the set of events you could consider. We will see that such restriction is sometimes necessary to ensure that probability satisfies key properties. These properties and the collection of events we may consider are intertwined, and we find ourselves in a Catch-22. We cannot understand the need for restricting the events we can consider without understanding the properties we want probability measures to have, and we cannot describe these properties without specifying the events for which we want them to hold. A more complete understanding of the elements of probability theory may come only after reading the entire chapter.

As you read this chapter, keep in mind three things.

Three key facts about probability spaces:

1. All randomness stems from the selection of $\boldsymbol{\omega} \in \boldsymbol{\Omega}$; once $\boldsymbol{\omega}$ is known, so too are the values of all random variables.
2. A single probability measure $\boldsymbol{P}$ determines the distribution of all random variables.
3. Without loss of generality, we can take the sample space $\boldsymbol{\Omega}$ to be the unit interval $[\mathbf{0}, \mathbf{1}]$ and $\boldsymbol{P}$ to be "Lebesgue measure," which in layman's terms corresponds to drawing $\boldsymbol{\omega}$ at random from $[\mathbf{0}, \mathbf{1}]$.

In fact, rigorous treatment of probability theory stemmed largely from one basic question: can we define a probability measure that satisfies certain properties and allows us to pick randomly and uniformly from $[0, 1]$? The surprising answer is "not without restricting the events we are allowed to consider." A great deal of thought went into how to restrict the allowable sets.

The chapter is organized as follows. Section 3.2 studies the collection $\mathcal{F}$ of events we are allowed to consider. Section 3.3 shows that this collection of allowable sets includes a very important one not usually covered in elementary probability courses, namely that infinitely many of a collection $A_1, A_2, \ldots$ of events occurred. Section 3.4 presents the key axioms of probability measures and the consequences of those axioms. Section 3.5 shows that to achieve a uniform probability measure, we must restrict the allowable sets. Section 3.6 discusses scenarios under which it is not possible to sample uniformly, while Section 3.7 concludes with the problems of trying to perform probability calculations after the fact.

3.2 Sigma-Fields

3.2.1 General Case

Imagine performing a random experiment like drawing a number randomly from the unit interval. The collection Ω of all possible outcomes is called the *sample space* of the experiment. These outcomes are sometimes called *simple events* or *elements* because they cannot be decomposed further. Other events (subsets of Ω) are *compound events* because they are composed of two or more simple events. Note the analogy in terminology with chemistry, where an element cannot be broken down, whereas a compound can be decomposed into two or more elements. An event E occurs if and only if the randomly selected ω belongs to E. Thus, when drawing randomly from $[0, 1]$, the sample space is $\Omega = [0, 1]$, each $\omega \in [0, 1]$ is a simple event, and the interval $E = (1/2, 1]$ is an example of a compound event.

Here is where we begin to diverge from what you have seen in more elementary courses. Rather than considering any event, we restrict the set of events we are allowed to consider to a certain pre-specified collection $\mathcal{F}$ of subsets of Ω. We want $\mathcal{F}$ to be large enough to contain all of the interesting sets we might want to consider, but not so large that it admits "pathological" sets that prevent probability from having certain desirable properties (see Section 3.5). The class of allowable events $\mathcal{F}$ must have certain properties. For instance, if E is allowable, then E^C should also be allowable because we naturally want to be able to consider the event that E did not occur. We would also like to be able to consider countable unions and intersections. It turns out that the tersest conditions yielding the sets we would like to be able to consider are given in the following definition.

Definition 3.2. Field and sigma-field *A nonempty collection of subsets of Ω is called a:*

1. *Field (also called an algebra) if $\mathcal{F}$ is closed under complements and unions of pairs of members. That is, whenever $E \in \mathcal{F}$, then $E^C \in \mathcal{F}$, and whenever $E_1, E_2 \in \mathcal{F}$, then $E_1 \cup E_2 \in \mathcal{F}$.*

2. *Sigma-field (also called a sigma-algebra) if $\mathcal{F}$ is closed under complements and countable unions. That is, whenever $E \in \mathcal{F}$, then $E^C \in \mathcal{F}$, and whenever $E_1, E_2, \ldots \in \mathcal{F}$, then $\cup_{i=1}^{\infty} E_i \in \mathcal{F}$.*

Remark 3.3.

1. *Any field is also closed under finite unions because we can apply the paired union result repeatedly. It is also closed under finite intersections by an argument similar to that given in item 2 below.*

2. *Any sigma-field $\mathcal{F}$ is closed under countable intersections because $\cap_{i=1}^{\infty} E_i = \left(\cup_{i=1}^{\infty} E_i^C\right)^C$ and each E_i^C belongs to $\mathcal{F}$.*

3. *Because any field or sigma-field $\mathcal{F}$ is nonempty, it contains a set E, and therefore E^C. Consequently, any field or sigma-field contains $\Omega = E \cup E^C$ and $\emptyset = \Omega^C$.*

A sigma-field is more restrictive than a field. Because we want to be able to consider countable unions, we require the allowable sets $\mathcal{F}$ to be a sigma-field. Therefore, it is understood in probability theory and throughout the remainder of this book that we are allowed to consider only the events belonging to the sigma-field $\mathcal{F}$. We will see that fields also play an important role in probability because a common technique is first to define a probability measure on a field and then extend it to a sigma-field.

Example 3.4. In 5-card poker, each simple event ω is a specific collection of 5 cards, such as $\omega = \{$5 of Diamonds, 8 of Hearts, 2 of Clubs, 3 of Spades, King of Spades$\}$. The sample space Ω is the set of all such collections of 5 cards. An example of a compound event is $\{$4 aces$\}$, which is $\{\{$Ace of Diamonds, Ace of Hearts, Ace of Clubs, Ace of Spades, 2 of Diamonds$\}$, $\{$Ace of Diamonds, Ace of Hearts, Ace of Clubs, Ace of Spades, 2 of Hearts$\}, \ldots, \{$Ace of Diamonds, Ace of Hearts, Ace of Clubs, Ace of Spades, King of Spades$\}\}$. The natural sigma-field to use here is $\mathcal{F}$ =the set of all possible unions of simple events; that way we can talk about any possible event. If instead we foolishly choose $\mathcal{F} = \{\{$full house$\}, \{$all hands other than a full house}, $\emptyset, \Omega\}$, then we would be allowed to consider only whether or not a full house occurred; we could not talk about events like a pair, a flush, etc. We could not even say whether a **specific** full house like {3 of Diamonds, 3 of Hearts, 5 of Hearts, 5 of Clubs, 5 of Spades} occurred, or whether any other **specific** set of 5 cards occurred. □

Example 3.5. Trivial and total sigma-fields For any sample space Ω, the smallest and largest sigma-fields are the trivial sigma-field $t(\Omega) = \{\emptyset, \Omega\}$ and the total sigma-field (also called the power sigma-field) $T(\Omega) = \{\text{all subsets of } \Omega\}$, respectively. □

Remember that we are allowed to determine only whether or not event E occurred for $E \in \mathcal{F}$, not for sets outside $\mathcal{F}$. The trivial sigma-field is not very useful because the only events that we are allowed to consider are the entire sample space and the empty set. In other words, we can determine only whether something or nothing happened. With the total sigma-field we can determine, for each subset $E \subset \Omega$, whether or not event E occurred. The trouble with $T(\Omega)$ is that it may be too large to ensure that the properties we desire for a probability measure hold for all $E \in \mathcal{F}$ (see Section 3.5). Therefore, we want $\mathcal{F}$ to be large, but not too large. We would certainly like it to contain each simple event $\{\omega\}$, $\omega \in \Omega$.

Example 3.6. For countable Ω, $\mathcal{F}$ should be the total sigma-field Suppose Ω is a countable set like $\{0, 1, 2, \ldots\}$. We certainly want $\mathcal{F}$ to contain each singleton. But any sigma-field that contains each singleton ω must contain all subsets of Ω. This follows from the fact that any subset E of Ω is a countable union $\cup_{i=1}^{\infty}\{\omega_i\}$ of singletons, and sigma-fields are closed under countable unions. Therefore, when the sample space is countable, we should always use the total sigma-field. □

3.2.2 Sigma-fields When $\Omega = R$

We next consider sigma-fields when Ω is the uncountable set R (or an interval). Again, any useful sigma-field must contain each singleton ω.

Example 3.7. Sigma-field of countable and co-countable sets: too small Suppose that Ω is an uncountable set like R or $[0, 1]$. If we want $\mathcal{F}$ to include each singleton ω, then $\mathcal{F}$ must contain each countable subset E. It must also contain the complement of each countable set. Each set whose complement is countable is called *co-countable.* Therefore, any sigma-field containing the singletons must contain all countable and co-countable sets. It is an exercise to prove that the collection of countable and co-countable sets is itself a sigma-field. Therefore, the smallest sigma-field containing the singletons is the collection of countable and co-countable sets. Unfortunately, this sigma-field is not large enough to be useful. For instance, if $\Omega = [0, 1]$, then we are not allowed to consider events like $[0, 1/2]$ because it is neither countable nor co-countable. □

We continue our quest for the smallest sigma-field that contains all of the "useful" subsets of the uncountable set R. We have seen that the collection of countable and co-countable sets is not a good choice because it excludes common sets like $[0, 1/2]$. We certainly want $\mathcal{F}$ to include all intervals. But remember that we do not want $\mathcal{F}$ to be too large, else it might contain "pathological sets" that cause probability to behave strangely. Therefore, we would like to find the **smallest** sigma-field $\mathcal{F}$ that contains all of the intervals. That is, we want $\mathcal{F}$ to contain all intervals, and if $\mathcal{G}$ is any other sigma-field containing the intervals, then $\mathcal{F} \subset \mathcal{G}$. The smallest sigma-field containing a collection $\mathcal{A}$, denoted $\sigma(\mathcal{A})$, is called the *sigma-field generated by $\mathcal{A}$.* We want to find σ(intervals). Any sigma-field containing the intervals must also contain all finite unions of disjoint sets of the form $(a, b]$,

$(-\infty, a]$, or (b, ∞). Therefore, if this latter collection is a sigma-field, it must be σ(intervals). Unfortunately, it is not a sigma-field:

Example 3.8. Very important field If $\Omega = R$, the collection $\mathcal{F}_0$ of finite unions of disjoint sets of the form $(a, b]$, $(-\infty, a]$, or (b, ∞) is a field, but not a sigma-field. It generates σ(intervals). □

The proof is an exercise.

Continuing our search for σ(intervals), consider the collection $\{\mathcal{F}_t\}_{t \in T}$ of all sigma-fields containing the intervals. There is at least one, namely the total sigma-field $T(R)$ of all subsets of the line. Let $\mathcal{F} = \cap_t \mathcal{F}_t$. We claim that $\mathcal{F} = \sigma$(intervals). We need only show that $\mathcal{F}$ is a sigma-field, because any other sigma-field containing the intervals clearly contains $\mathcal{F}$.

Proposition 3.9. Intersections of sigma-fields are sigma-fields *The intersection $\cap_t \mathcal{F}_t$ of any collection of sigma-fields of Ω is a sigma-field.*

Proof. First note that $\mathcal{F}$ is nonempty because $\emptyset \in \mathcal{F}_t$ and $\Omega \in \mathcal{F}_t$ for each t, and therefore $\emptyset$ and Ω belong to $\mathcal{F} = \cap_t \mathcal{F}_t$. Also, if $E \in \mathcal{F}$, then E belongs to each $\mathcal{F}_t$. Thus, E^C belongs to each $\mathcal{F}_t$, and therefore to $\mathcal{F} = \cap_t \mathcal{F}_t$. Similarly, if $E_1, E_2 \ldots$ belong to each $\mathcal{F}_t$, then $E = \cup_{i=1}^{\infty} E_i$ also belongs to each $\mathcal{F}_t$, so E belongs to $\mathcal{F} = \cap_t \mathcal{F}_t$. It follows that $\mathcal{F}$ is a sigma-field. □

Definition 3.10. 1-dimensional Borel sigma-field *The smallest sigma-field containing the intervals of R, namely the intersection of all sigma-fields containing the intervals, is called the Borel sigma-field and denoted $\mathcal{B}$ or $\mathcal{B}^1$. The sets in $\mathcal{B}$ are called Borel sets. The Borel sets of R that are subsets of $[0, 1]$ are denoted $\mathcal{B}_{[0,1]}$.*

The definition of the Borel sigma-field is somewhat dissatisfying because it is hard to picture the intersection of all sigma-fields containing the intervals. It would be more pleasing if we could begin with the intervals $\mathcal{I}_1$, throw in all complements and countable unions of sets in $\mathcal{I}_1$ to create $\mathcal{I}_2$, then throw in all complements and countable unions of sets in $\mathcal{I}_2$ to create $\mathcal{I}_3$, etc. It would still be difficult to picture all of the sets, but at least we could imagine how to construct them. Unfortunately, this process must be continued indefinitely. In fact, even $\cup_{i=1}^{\infty} \mathcal{I}_i$ does not produce all of the Borel sets (Billingsley, 2012).

Remark 3.11. The Borel sigma-field is very broad *The Borel sigma-field contains many sets other than intervals, including, for example:*

1. *Countable unions of intervals, and therefore all open sets by Proposition A.15.*
2. *All closed sets by part 1 and the fact that complements of sets in $\mathcal{B}$ are in $\mathcal{B}$.*
3. *The rational numbers by part 2 because each rational r is a closed set (it is easy to see that its complement is open), and the set of rational numbers is a countable union of these closed sets.*
4. *The irrational numbers, being the complement of the Borel set of rational numbers.*

It is actually difficult to construct a set that is **not** *a Borel set. Nonetheless, we will see later that an even larger sigma-field, called the Lebesgue sets, is also useful in probability and measure theory.*

It is sometimes helpful to consider the events $\{-\infty\}$ or $\{+\infty\}$. These sets are not in the Borel sigma-field because they are not subsets of $R = (-\infty, \infty)$. To consider them as events, we must extend the real line by augmenting it with $\{-\infty, +\infty\}$.

Definition 3.12. Extended line and extended intervals *The extended line $\bar{R}$ is $[-\infty, \infty]$, meaning $(-\infty, \infty) \cup \{-\infty, \infty\}$. An extended interval is of the form $[x, y]$, $[x, y)$, $(x, y]$, or (x, y), where x or y can be numbers or $\pm\infty$; $[a, \infty]$ means $[a, \infty) \cup \{\infty\}$, $[-\infty, b]$ means $(-\infty, b] \cup \{-\infty\}$, etc.*

When considering $\bar{R}$ instead of R, we must remember to take complements with respect to $\bar{R}$. For instance, the complement of $(0, 1]$ is $\bar{R} \setminus (0, 1] = [-\infty, 0] \cup (1, \infty]$, and the complement of $R = (-\infty, \infty)$ is $\bar{R} \setminus R = \{-\infty, +\infty\}$.

Definition 3.13. Extended Borel sigma-field *The extended Borel sigma-field $\bar{\mathcal{B}}$ is the smallest sigma-field of subsets of $\bar{R}$ that contains all of the extended intervals.*

Proposition 3.14. Equivalent formulations of extended Borel sigma-field *The following are equivalent.*

1. *$\bar{\mathcal{B}}$ is the extended Borel sigma-field.*

2. *$\bar{\mathcal{B}}$ is the smallest sigma-field that contains the two sets $\{-\infty\}$ and $\{+\infty\}$ and all Borel subsets of R.*

3. *$\bar{\mathcal{B}}$ consists of all Borel subsets B of R, together with all augmentations $\bar{B}$ of B by $-\infty$, $+\infty$, or both.*

Proof. Exercise.

3.2.3 Sigma-fields When $\Omega = R^k$

Many of the nuances introduced by generalizing from 1 to k dimensions can be illustrated when $k = 2$. Therefore, we begin by defining a two-dimensional analog of the one-dimensional Borel sets. Let R^2 be the product set $\{(x_1, x_2),\ x_1 \in R, x_2 \in R\}$. Consider the collection $\mathcal{A}$ of all rectangles of the form $I_1 \times I_2 = \{(x_1, x_2) : x_1 \in I_1, x_2 \in I_2\}$, where I_1 and I_2 are intervals (open, closed or half open, finite or infinite) of R. Then $\mathcal{A}$ is not a sigma-field or even a field because, for example, $[(0, 1) \times (0, 1)]^C \notin \mathcal{A}$. Nonetheless, there is a smallest sigma-field $\sigma(\mathcal{A})$ containing $\mathcal{A}$.

Definition 3.15. 2-dimensional Borel sigma-field $\mathcal{B}^2$ *The smallest sigma-field containing the set $\mathcal{A}$ of all possible rectangles is called the 2-dimensional Borel sigma-field, and denoted $\mathcal{B}^2$. Sets in $\mathcal{B}^2$ are called two-dimensional Borel sets. Figure 3.1 shows one such set.*

Proposition 3.16. $\mathcal{B}^2$ contains products of 1-dimensional Borel sets *$\mathcal{B}^2$ contains all sets of the form $B_1 \times B_2$, where B_1 and B_2 are one-dimensional Borel sets.*

Proof. We show this in two steps:

1. For any interval I, we show that $I \times B \in \mathcal{B}^2$ for every $B \in \mathcal{B}^1$.

2. Let $\mathcal{C} = \{C \in \mathcal{B}^1 : C \times B \in \mathcal{B}^2 \text{ for all } B \in \mathcal{B}^1\}$, and note that $\mathcal{C}$ contains all intervals by part 1. We show that $\mathcal{C}$ contains $\mathcal{B}^1$.

Step 1: For a given interval I, let $\mathcal{B}_I$ be the collection of one-dimensional Borel sets B such that $I \times B \in \mathcal{B}^2$. Then $\mathcal{B}_I$ contains all intervals because $\mathcal{B}^2$ contains products of intervals. Also, $\mathcal{B}_I$ is a sigma-field because $\mathcal{B}^2$ is. For instance, $\mathcal{B}_I$ is closed under countable unions because if $B_1, B_2, \ldots \in \mathcal{B}_I$, then $I \times \cup_i B_i = \cup_i (I \times B_i) \in \mathcal{B}^2$ because each $I \times B_i \in \mathcal{B}^2$. Also, the following argument shows that $\mathcal{B}_I$ is closed under complementation. If $B_i \in \mathcal{B}_i$, then

$$I \times B_i^C = (I \times B_i)^C \cap (I \times R). \tag{3.1}$$

By assumption, $I \times B_i \in \mathcal{B}^2$, so $(I \times B_i)^C \in \mathcal{B}^2$. Also, $I \times R \in \mathcal{B}^2$ because $\mathcal{B}^2$ contains all products of intervals, and $I \times R$ is a product of intervals. We have shown that both members of the intersection in (3.1) are in $\mathcal{B}^2$, so $I \times B_i^C$ is in $\mathcal{B}^2$ because $\mathcal{B}^2$ is a sigma-field. This means that $B_i^C \in \mathcal{B}_I$. We have shown that $\mathcal{B}_I$ is a sigma field containing all intervals, so $\mathcal{B}_I$ contains the smallest sigma-field containing the intervals, namely $\mathcal{B}^1$. Thus, $I \times B \in \mathcal{B}^2$ for each $B \in \mathcal{B}^1$, which completes the proof of step 1.

Step 2: Fix $B \in \mathcal{B}^1$ and define $\mathcal{C}$ to be the collection of one-dimensional Borel sets C such that $C \times B \in \mathcal{B}^2$. By step 1, $\mathcal{C}$ contains all intervals. An argument similar to that used in step 1 shows that $\mathcal{C}$ is a sigma-field. Because $\mathcal{C}$ is a sigma-field containing all intervals, $\mathcal{C}$ contains $\mathcal{B}^1$. This completes the proof of step 2. □

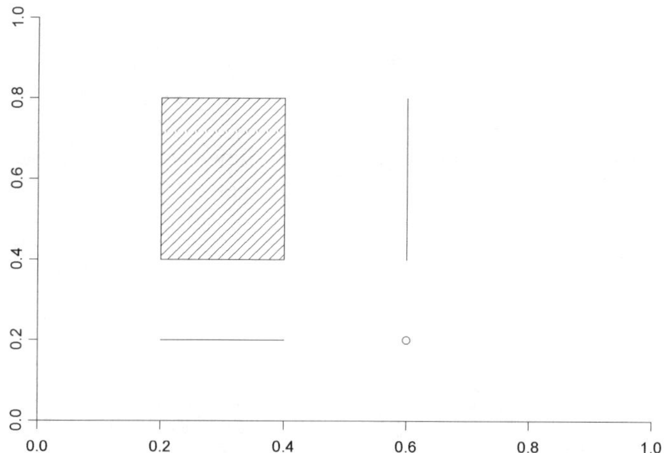

Figure 3.1: Two-dimensional Borel sets include direct products of one-dimensional Borel sets. Shown here is $B_1 \times B_2$, where $B_1 = \{(0.2, 0.4) \cup 0.6\}$ and $B_2 = \{0.2, \cup(0.4, 0.8)\}$.

The two-dimensional Borel sets form a very rich class that extends well beyond just direct products of one-dimensional Borel sets. For instance, $C = \{(x, y) \in (0,1) \times (0,1) : x + y < 1\}$ is a two-dimensional Borel set that is not a direct product. It is not a direct product because each $x \in (0,1)$ is the first element of at least one point in C, and likewise each $y \in (0,1)$ is the second element of at least one point in C. Therefore, if C were a direct product, it would have to contain the entire square $(0,1) \times (0,1)$, which it does not. To see that C is still a two-dimensional Borel set, note that C may be written as $\cup_r (0, r) \times (0, 1-r)$, where the union is over all rationals in $(0,1)$ (see Figure 3.2). Because C is a countable union of rectangles, it is in $\mathcal{B}^2$.

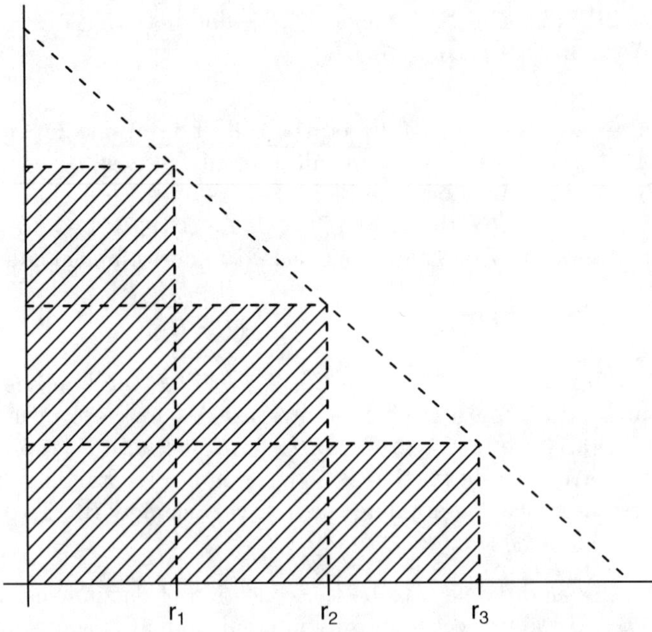

Figure 3.2: The union $\cup_r(0,r) \times (0,1-r)$ for $r = 1/4$, $1/2$, and $3/4$. Taking the union over **all** rationals $r \in (0,1)$ fills the entire triangle $\{(x,y) \in (0,1) \times (0,1) : \ y < 1-x\}$.

Any open set O in R^2 is in $\mathcal{B}^2$. To see this, note that each point $\mathbf{p} \in O$ is contained in a rectangle $(r,R) \times (s,S) \subset O$, where r, R, s and S are rational (Figure 3.3). Each such rectangle is a two-dimensional Borel set, so $O = \cup_{r,R,s,S}(r,R) \times (s,S)$, being a countable union of two-dimensional Borel sets, is also a two-dimensional Borel set. Therefore, $\mathcal{B}^2$ contains all open sets. It must also contain all complements of open sets, namely all closed sets. We have proven the following.

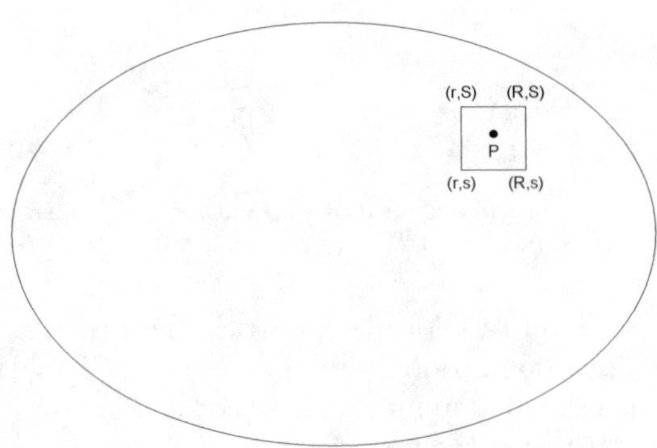

Figure 3.3: Each point P in an open set is encased in a rectangle $(r,R) \times (s,S)$ with r, R, s, and S rational.

Proposition 3.17. $\mathcal{B}^2$ contains all two-dimensional open and closed sets *$\mathcal{B}^2$ contains all open and closed sets in R^2.*

It is difficult for many people to construct a two-dimensional set that is not in $\mathcal{B}^2$.

The class of k-dimensional Borel sets $\mathcal{B}^k$ is defined as follows

Definition 3.18. k-dimensional Borel sigma-field $\mathcal{B}^k$ *The collection $\mathcal{B}^k$ of k-dimensional Borel sets is defined to be the smallest sigma-field containing all "hyper-rectangles" $\{(x_1, x_2, \ldots, x_k) \in R^k : x_i \in I_i, i = 1, \ldots, k\}$, where each I_i is an interval (of any type).*

The above argument for $k = 2$ is easily extended to show the following

Proposition 3.19. $\mathcal{B}^k$ contains all k-dimensional open and closed sets *$\mathcal{B}^k$ contains all open and closed sets in R^k.*

We saw that in one dimension, the collection of finite disjoint unions of intervals of the form $(a, b]$, $(-\infty, a]$, or (b, ∞) is a field generating $\mathcal{B}^1$ (see Example 3.8). This was very important because a common technique is to define a probability measure on a field and then extend it to a sigma-field. The same technique is useful in higher dimensions, so we need to find a field generating $\mathcal{B}^k$.

Proposition 3.20. Field generating $\mathcal{B}^k$ *The collection $\mathcal{F}_0$ of finite unions of disjoint sets of the form $B_1 \times B_2 \times \ldots \times B_k$, where each B_i is a one-dimensional Borel set, is a field generating $\mathcal{B}^k$.*

Proof. We established for $k = 2$ that products of Borel sets are in $\mathcal{B}^k$, and the proof is readily extended to higher k. Because each such product is in $\mathcal{B}^k$, finite unions of disjoint product sets are also in $\mathcal{B}^k$. The proof that $\mathcal{F}_0$ is a field is deferred to Section 4.5.2, where we prove a more general result about product spaces. □

Even though k-dimensional Borel sets include most sets of interest, we will see later that an even larger sigma-field, called k-dimensional Lebesgue sets, is useful in probability and measure theory.

Exercises

1. Which of the following are sigma fields of subsets of $\Omega = \{1, 2, 3, 4, 5, 6\}$?
   (a) $\{\{1, 2\}, \{1, 2, 3\}, \{4, 5, 6\}, \emptyset, \Omega\}$.
   (b) $\{\{1, 2\}, \{3, 4, 5, 6\}, \emptyset, \Omega\}$.
   (c) $\{\{1\}, \{3, 4, 5, 6\}, \{1, 3, 4, 5, 6\}, \Omega\}$.

2. Let $\Omega = \{1, 2, 3, \ldots\}$.
   (a) Is $\{1, 3, 5, \ldots\}, \{2, 4, 6, \ldots\}, \emptyset, \Omega\}$ a sigma-field of subsets of Ω?
   (b) Let R_i be the set of elements of Ω with remainder i when divided by 3. Is $\{R_0, R_1, R_2, R_0 \cup R_1, R_0 \cup R_2, R_1 \cup R_2, \emptyset, \Omega\}$ a sigma-field of subsets of Ω?
   (c) Let S_i be the set of elements of Ω with remainder i when divided by 4. What is $\sigma(S_0, S_1, S_2, S_3)$ (i.e., the smallest sigma-field containing each of these sets)?

3. Let $\Omega = [0, 1]$, $A_1 = [0, 1/2)$, and $A_2 = [1/4, 3/4)$. Enumerate the sets in $\sigma(A_1, A_2)$, the smallest sigma-field containing A_1 and A_2.

4. Let $\Omega = \{1, 2, \ldots\}$ and $A = \{2, 4, 6, \ldots\}$ Enumerate the sets in $\sigma(A)$, the smallest sigma-field containing A. What is $\sigma(A_1, A_2, \ldots)$, where $A_i = \{2i\}$, $i = 1, 2, \ldots$?

5. Let $B(0, r)$ be the two dimensional open ball centered at 0 with radius r; i.e., $B(0, r) = \{(\omega_1, \omega_2) : \omega_1^2 + \omega_2^2 < r^2\}$. If $\Omega = B(0, 1)$, what is $\sigma(B(0, 1/2), B(0, 3/4))$?

6. Show that if $\Omega = R^2$, the set of all open and closed sets of Ω is not a sigma-field.

7. Give an example to show that the union of two sigma-fields need not be a sigma-field.

8. Let Ω be a countably infinite set like $\{0, 1, 2, \ldots\}$, and let $\mathcal{F}$ be the set of finite and co-finite subsets (A is co-finite if A^C is finite). Show that $\mathcal{F}$ is a field, but not a sigma-field.

9. * Prove that if Ω is uncountable, then the set of countable and co-countable sets (recall that A is co-countable if A^C is countable) is a sigma-field.

10. * Show that the one-dimensional Borel sigma-field $\mathcal{B}$ is generated by sets of the form $(-\infty, x]$, $x \in R$. That is, the smallest sigma-field containing the sets $(-\infty, x]$, $x \in R$, is the Borel sigma-field. The same is true if $(-\infty, x]$ is replaced by $(-\infty, x)$.

11. * Prove that Example 3.8 is a field, but not a sigma-field.

12. Let $\mathcal{B}$ denote the Borel sets in R, and let $\mathcal{B}_{[0,1]}$ be the Borel subsets of $[0, 1]$, defined as $\{B \in \mathcal{B}, B \subset [0, 1]\}$. Prove that $\mathcal{B}_{[0,1]} = \{B \cap [0, 1],\ B \in \mathcal{B}\}$.

13. * The countable collection $\{A_i\}_{i=1}^\infty$ is said to be a *partition* of Ω if the A_i are disjoint and $\cup_{i=1}^\infty A_i = \Omega$. Prove that the sigma-field generated by $\{A_i\}_{i=1}^\infty$ consists of all unions of the A_i. What is its cardinality? Hint: think about Proposition 2.10.

14. ↑ Suppose that $\{A_i\}_{i=1}^\infty$ and $\{B_i\}_{i=1}^\infty$ are partitions of Ω and $\{B_i\}$ is *finer* than $\{A_i\}$, meaning that each A_i is a union of members of $\{B_i\}$. Then $\sigma(\{A_i\}_{i=1}^\infty) \subset \sigma(\{B_i\}_{i=1}^\infty)$.

15. Complete the following steps to show that a sigma-field $\mathcal{F}$ cannot be countably infinite. Suppose that $\mathcal{F}$ contains the nonempty sets $A_1, A_2 \ldots$.

    (a) Show that each set of the form $\cap_{i=1}^\infty B_i$, where each B_i is either A_i or A_i^C, is in $\mathcal{F}$.

    (b) What is wrong with the following argument: the set $\mathcal{C} = \{C = \cap_{i=1}^\infty B_i$, each B_i is either A_i or $A_i^C\}$ is in 1:1 correspondence with the set of all infinitely long strings of 0s and 1s because we can imagine A_i as a 1 and A_i^C as a 0. Therefore, by Proposition 2.10, $\mathcal{F}$ must have the same cardinality as $[0, 1]$. Hint: to see that this argument is wrong, suppose that the A_i are disjoint.

    (c) Show that any two non-identical sets in $\mathcal{C}$ are disjoint. What is the cardinality of the set of all countable unions of C sets? What can you conclude about the cardinality of $\mathcal{F}$?

16. Prove Proposition 3.14.

3.3 The Event That A_n Occurs Infinitely Often

If $\{A_n\} \in \mathcal{F}$, we can define other events involving the A_n. We have seen, for example, that $B = \cup_{n=1}^\infty A_n$ is the event that at least one A_n occurs, while $C = \cap_{n=1}^\infty A_n$ is the event that all of the A_n occur. Sometimes there is interest in seeing whether A_n occurs for infinitely many n. For instance, in a coin-flipping experiment, let A_n denote the event that the nth coin flip is heads. We might be interested in whether infinitely many flips are heads. We now show that for general $A_1, A_2, \ldots$, the event that A_n occurs for infinitely many n is a bona fide event in the sense that it is in $\mathcal{F}$. Note that $\omega \in A_n$ for infinitely many n if and only if no matter how large N is, $\omega \in A_n$ for at least one $n \geq N$. That is, $\omega \in \cup_{n \geq N} A_n$ for $N = 1$ and $N = 2$ and $N = 3$, etc. Because "and" means intersection, $\omega \in A_n$ for infinitely many n if and only if $\omega \in \cap_{N=1}^{\infty} \cup_{n \geq N} A_n = \cap_{N=1}^{\infty} B_N$, where $B_N = \cup_{n \geq N} A_n$. Because B_N is a countable union of $\mathcal{F}$-sets, $B_N \in \mathcal{F}$. This being true for each N, $\cap_{N=1}^{\infty} B_N \in \mathcal{F}$ because $\mathcal{F}$ is closed under countable intersections (see part 2 of Remark 3.3). Therefore, the event that A_n occurs for infinitely many n is in $\mathcal{F}$. We denote this event by $\{A_n \text{ i.o}\}$, which stands for A_n occurs infinitely often (in n).

In summary:

Proposition 3.21. The event $\{A_n$ i.o.$\}$ is in $\mathcal{F}$ *If $A_n \in \mathcal{F}$ for $n = 1, 2, \ldots$, the event $\{A_n$ i.o.$\}$, namely $\cap_{N=1}^{\infty} \cup_{n \geq N} A_n$, is in $\mathcal{F}$.*

The complement of $\{A_n \text{ i.o.}\}$, namely that A_n occurs only finitely often, is also in $\mathcal{F}$ because $\mathcal{F}$ is closed under complementation. This event is, by DeMorgan's laws (Proposition A.4), $\{\cap_{N=1}^{\infty} \cup_{n \geq N} A_n\}^C = \cup_{N=1}^{\infty} B_N^C = \cup_{N=1}^{\infty} \cap_{n \geq N} A_n^c$. This expression can also be deduced from general principles because if ω is in only finitely many of the A_n, then there must be an N for which ω is not in A_n for any $n \geq N$. That is, ω must be in A_n^C for all $n \geq N$. This must occur for at least one N, which translates into a union. Thus again we arrive at the expression $\cup_{N=1}^{\infty} \cap_{n \geq N} A_n^C$ for the event that A_n occurs only finitely often.

3.4 Measures/Probability Measures

A probability measure P is a function that assigns a probability to the allowable sets $\mathcal{F}$ and that satisfies certain conditions. Ideally, we would like P to satisfy $P(\Omega) = 1$ and $P(\cup_t E_t) = \sum_t P(E_t)$ for every collection of disjoint sets E_t, even an uncountable collection. Unfortunately, this *uncountable additivity* requirement is too much to ask because it precludes the existence of a "uniform" probability measure P on $\Omega = [0, 1]$. To see this, note that for such a measure, each $\omega \in \Omega$ has the same probability p, so if P were uncountably additive, then $1 = P(\Omega) = P([0, 1]) = \sum_{\Omega} p$. But this produces a contradiction because $\sum_{\Omega} p$ is 0 if $p = 0$ and ∞ if $p > 0$. Therefore, uncountable additivity is too strong a condition to require. It turns out the strongest property of this type that we can impose is *countable additivity*, as in the following definition.

Definition 3.22. Measure and probability measure *A measure μ is a nonnegative function with domain the allowable sets $\mathcal{F}$ such that if $E_1, E_2, \ldots \in \mathcal{F}$ are disjoint, then $\mu(\cup_{i=1}^{\infty} E_i) = \sum_{i=1}^{\infty} \mu(E_i)$. A probability measure P is a measure such that $P(\Omega) = 1$.*

Example 3.23. Counting measure Let $\Omega = \{x_1, x_2, \ldots\}$ be a countable set and $\mathcal{F} = T(\Omega)$ be the total sigma-field of all subsets of Ω. For $A \in \mathcal{F}$, define $\mu(A)$ =the number of elements of A. Then μ is clearly nonnegative and countably additive because the number of elements in the union of disjoint sets is the sum of the numbers of elements in the individual sets. Therefore, μ is a measure, called *counting measure*, on $\mathcal{F}$. Counting measure is not a probability measure if Ω has more than one element because $\mu(\Omega) > 1$. $\square$

Example 3.24. A probability mass function defines a measure Let Ω and $\mathcal{F}$ be as defined in Example 3.23. Suppose that $p_i \geq 0$ and $\sum_{i=1}^{\infty} p_i = 1$. For $A \in \mathcal{F}$, define $P(A) = \sum_{x_i \in A} p_i$. This is well defined because the summand is nonnegative, so we get the same value irrespective of the order in which we add (Proposition A.44). Then P is nonnegative and countably additive because if I_i and I index the elements of E_i and $E = \cup_j E_j$, respectively, where $E_1, E_2, \ldots$ are disjoint, then $P(\cup_j E_j) = \sum_{j \in I} p_j = \sum_i \sum_{j \in I_i} p_j$. Also, $P(\Omega) = \sum_{j=1}^{\infty} p_j = 1$. Therefore, P is a probability measure on $\mathcal{F}$. This example shows that any probability mass function–the binomial, Poisson, hypergeometric, etc.–specifies a probability measure. □

Examples 3.23 and 3.24 involved a countable sample space. The most important measure on the uncountable sample space R is the following

Example 3.25. Lebesgue measure There is a measure μ_L defined on a sigma-field $\mathcal{L}$ containing the Borel sets such that $\mu_L(C) = \text{length}(C)$ if C is an interval: μ_L is called Lebesgue measure and $\mathcal{L}$ are called Lebesgue sets. If we restrict attention to Lebesgue sets that are subsets of $[0, 1]$, μ_L is a probability measure. We can think of the experiment as picking a number at random from the unit interval. We will see in Section 4.6 that Lebesgue measure on $[0, 1]$ is the only probability measure we ever need. □

Courses in real analysis spend a good deal of effort proving the existence of Lebesgue measure. We present a very brief outline of the construction, which may be found in Royden (1968).

1. Define $\mu(I) = \text{length}(I)$ for any interval I.
2. Now extend μ to an arbitrary set A by defining an *outer measure* μ^* by $\mu^*(A) = \inf \sum_n \text{length}(I_n)$, where the infimum is over all countable unions $\cup_n I_n$ of intervals I_n such that $A \subset \cup_n I_n$. The term outer measure stems from the fact that we are approximating the measure of A from the outside, meaning that we consider the measure of a collection of sets that cover A. It is easy to see that μ^* agrees with μ if A is an interval. But μ^* is not necessarily countably additive on **all** sets, so the next step is:
3. Restrict the class of sets to $\mathcal{L} = \{L : \mu^*(A) = \mu^*(L \cap A) + \mu^*(L^C \cap A) \text{ for every set } A\}$. The collection $\mathcal{L}$, called Lebesgue sets, is then shown to be a sigma-field containing all Borel sets. The restriction of μ^* to $\mathcal{L}$ is Lebesgue measure μ_L.

Students are often puzzled that each possible outcome ω has probability 0, yet the probability of $[0, 1]$ is 1. Every event is "impossible" (i.e., has probability 0), yet something is sure to happen.

It turns out that the definition of probability measure leads to a myriad of consequences, the most basic of which are the following.

Proposition 3.26. Basic properties of probability measures *If P is a probability measure:*

1. $P(\emptyset) = 0$.
2. *If $E_1, E_2 \in \mathcal{F}$, $E_1 \subset E_2$, then $P(E_1) \leq P(E_2)$.*
3. $0 \leq P(E) \leq 1$ *for each $E \in \mathcal{F}$.*

4. *If* $E \in \mathcal{F}$, $P(E^C) = 1 - P(E)$.
5. *If* $E_1, E_2 \in \mathcal{F}$, $P(E_1 \cup E_2) = P(E_1) + P(E_2) - P(E_1 \cap E_2)$.

Proof. To see part 1, note that any event E may be written as the disjoint union $E \cup \emptyset$. By countable additivity, $P(E) = P(E) + P(\emptyset)$, from which it follows that $P(\emptyset) = 0$.

To see part 5, note that $E_1 \cup E_2 = E_1 \cup (E_2 \backslash E_1)$. Because E_1 and $E_2 \backslash E_1$ are disjoint,

$$P(E_1 \cup E_2) = P(E_1) + P(E_2 \backslash E_1). \tag{3.2}$$

Also, E_2 is the union of the disjoint sets $E_1 \cap E_2$ and $E_2 \backslash E_1$, so $P(E_1 \cap E_2) + P(E_2 \backslash E_1) = P(E_2)$. Thus, $P(E_2 \backslash E_1) = P(E_2) - P(E_1 \cap E_2)$. Substituting this expression into Equation (3.2) gives $P(E_1 \cup E_2) = P(E_1) + P(E_2) - P(E_1 \cap E_2)$. The proofs of the remaining three properties are left as exercises. □

The following result that any countable union can be written as a countable union of disjoint sets is extremely useful in probability theory.

Proposition 3.27. Any countable union can be written as a countable union of disjoint sets *Let* $E_1, E_2, \ldots \in \mathcal{F}$ *and define* $D_1 = E_1$, $D_2 = E_2 \backslash E_1$, $D_3 = E_3 \backslash (E_1 \cup E_2), \ldots$ *Then* $\{D_i\}$ *is a collection of disjoint sets and* $\cup_{i=1}^n E_i = \cup_{i=1}^n D_i$ *for* n *any positive integer or* $+\infty$.

Proof. It is clear that the D_i defined above are disjoint. We will prove that each ω in $\cup_{i=1}^n E_i$ is in $\cup_{i=1}^n D_i$ and vice versa. Note that each step in the following proof is valid whether n is a positive integer or $+\infty$. If $\omega \in \cup_{i=1}^n E_i$, then $\omega \in E_i$ for some finite $i \leq n$. Let I be the smallest $i \leq n$ for which $\omega \in E_i$. Then by definition, ω is in E_I but not in any of the previous E_i. That is, $\omega \in D_I$, so $\omega \in \cup_{i=1}^n D_i$. Thus, if $\omega \in \cup_{i=1}^n E_i$, then $\omega \in \cup_{i=1}^n D_i$. To prove the converse, note that if $\omega \in \cup_{i=1}^n D_i$, then $\omega \in D_i$ for some finite $i \leq n$. But then $\omega \in E_i$, and hence $\omega \in \cup_{i=1}^n E_i$. We have shown that each $\omega \in \cup_{i=1}^n E_i$ is in $\cup_{i=1}^n D_i$ and vice versa, and therefore $\cup_{i=1}^n E_i = \cup_{i=1}^n D_i$. □

One consequence of Proposition 3.27 is the following, which is called *countable subadditivity.* It is also called Boole's inequality, or, when there are only finitely many sets, Bonferroni's inequality.

Proposition 3.28. Countable subadditivity (also called Boole's or Bonferroni's inequality) *If* $E_1, E_2, \ldots$ *are any sets (not necessarily disjoint), then* $P(\cup_i E_i) \leq \sum_i P(E_i)$.

The proof is left as an exercise.

The importance of Proposition 3.28 in probability and statistics cannot be overstated. Researchers often perform many statistical tests in a single study, including comparisons of several treatments to the same control group, comparisons of two groups with respect to several outcomes, and testing of thousands of genes to determine their association with a given disease, to name a few. In such a setting there is a real danger of falsely rejecting at least one null hypothesis. In genetics testing, for example, it is quite common that associations found in one study cannot be replicated. To reduce the probability of this, researchers attempt to control the probability of falsely rejecting at least one null hypothesis, called the *familywise error rate (FWE)*. Bonferroni's inequality tells us that if we make the probability of a type I error on any given comparison α/k, the FWE will be $P(\cup_{i=1}^k\{\text{type I error on test } i\}) \leq \sum_{i=1}^k \alpha/k = \alpha$.

Another consequence of Proposition 3.27 is a certain continuity property of probability, namely that if a sequence of sets $E_1, E_2, \ldots$ "gets closer and closer" to a set E, then the probability of E_i also gets "closer and closer" to the probability of E. But what does it mean for a sequence of sets $(E_i)_{i=1}^{\infty}$ to get "closer and closer" to E?

Definition 3.29. Increasing and decreasing sets *We say that a sequence of sets* $(E_i)_{i=1}^{\infty}$ *increases to* E, *written* $E_i \uparrow E$, *if* $E_1 \subset E_2 \subset E_3 \subset \ldots$ *and* $\cup_{i=1}^{\infty} E_i = E$. *Similarly,* $(E_i)_{i=1}^{\infty}$ *decreases to* E, *written* $E_i \downarrow E$, *if* $E_1 \supset E_2 \supset E_3 \supset \ldots$ *and* $\cap_{i=1}^{\infty} E_i = E$.

For example, $E_i = (-\infty, i] \uparrow E = (-\infty, \infty)$ and $E_i = (-\infty, x + 1/i) \downarrow E = (-\infty, x]$ (see Figure 3.4).

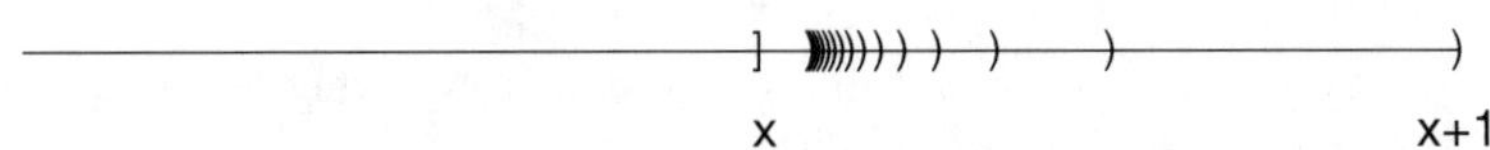

Figure 3.4: Sets $E_i = (-\infty, x + 1/i)$, $i = 1, 2, \ldots$ decreasing to $E = (-\infty, x]$.

Proposition 3.30. Continuity property of probability *If* $E_1, E_2, \ldots \in \mathcal{F}$ *and* $E_i \uparrow E$ *or* $E_i \downarrow E$, *then* $P(E_i) \rightarrow P(E)$. *The result about* $E_i \uparrow E$ *holds for arbitrary measures, not just probability measures (but see Problem 12 for the case* $E_i \downarrow E$*).*

Proof. Suppose first that $E_i \uparrow E$. By Proposition 3.27, $E = \cup_{i=1}^{\infty} E_i = \cup_{i=1}^{\infty} D_i$, where $D_i = E_i \backslash (\cup_{j=1}^{i-1} E_j)$ are disjoint. By countable additivity,

$$\begin{aligned} P(E) &= \sum_{i=1}^{\infty} P(D_i) = \lim_{n\to\infty} \sum_{i=1}^{n} P(D_i) \\ &= \lim_{n\to\infty} P(\cup_{i=1}^{n} D_i) = \lim_{n\to\infty} P(\cup_{i=1}^{n} E_i) \\ &= \lim_{n\to\infty} P(E_n), \end{aligned} \tag{3.3}$$

proving that if $E_i \uparrow E$, then $P(E_i) \rightarrow P(E)$. Note that each of the above steps is valid if P is replaced by an arbitrary measure μ.

Now suppose that $E_i \downarrow E$. Then $F_i = E_i^C \uparrow F = E^C$. By the result just proved, $P(F_i) \rightarrow P(F)$. But $P(F_i) = 1 - P(E_i)$ and $P(F) = 1 - P(E)$, so $P(E_i) \rightarrow P(E)$. □

We can use the continuity property of probability in conjunction with the following very useful technique to prove many interesting results. Suppose we know that a certain property, call it p, holds for all sets in a field, $\mathcal{F}_0$, and we want to prove that it holds for all sets in the sigma-field $\mathcal{F}$ generated by $\mathcal{F}_0$. Define $\mathcal{C}$ to be the collection of sets such that property p holds. Then $\mathcal{F}_0 \subset \mathcal{C}$ by assumption, and we want to prove that $\mathcal{F} \subset \mathcal{C}$. The technique is to prove that $\mathcal{C}$ is a *monotone class*, defined as follows.

Definition 3.31. Monotone class *A monotone class* $\mathcal{C}$ *is a nonempty collection of sets that is closed under countable increasing sets and countable decreasing sets; i.e., if* $E_1, E_2, \ldots \in \mathcal{C}$ *and* $E_i \uparrow E$ *or* $E_i \downarrow E$, *then* $E \in \mathcal{C}$.

The following result implies that $\mathcal{F} \subset \mathcal{C}$.

Theorem 3.32. Monotone class theorem *(Chung, 1974, page 18) Let $\mathcal{F}_0$ be a field, $\sigma(\mathcal{F}_0)$ be the smallest sigma-field containing $\mathcal{F}_0$, and $\mathcal{M}$ be a monotone class containing $\mathcal{F}_0$. Then $\sigma(\mathcal{F}_0) \subset \mathcal{M}$.*

In summary, the technique that will be used repeatedly to show that a property, say property p, holds for all sets in a sigma-field $\sigma(\mathcal{F}_0)$ generated by a field $\mathcal{F}_0$, is as follows.

Important technique for showing a property holds for all sets in a sigma-field

1. Let $\boldsymbol{\mathcal{C}}$ be the collection of sets in $\boldsymbol{\sigma(\mathcal{F}_0)}$ such that property $\boldsymbol{p}$ holds, and show that $\boldsymbol{\mathcal{C}}$ contains $\boldsymbol{\mathcal{F}_0}$.
2. Show that $\boldsymbol{\mathcal{C}}$ is a monotone class, namely nonempty and closed under countable increasing or decreasing sets.
3. Use Theorem 3.32 to deduce that property $\boldsymbol{p}$ holds for each set in $\boldsymbol{\sigma(\mathcal{F}_0)}$, thereby proving the result.

We illustrate the technique by proving the following result.

Proposition 3.33. Probability measures agreeing on a field agree on the generated sigma-field *Two probability measures P_1 and P_2 agreeing on a field $\mathcal{F}_0$ also agree on the sigma-field $\sigma(\mathcal{F}_0)$ generated by $\mathcal{F}_0$.*

Proof. Let $\mathcal{C}$ denote the collection of events E such that $P_1(E) = P_2(E)$, where P_1 and P_2 are the two probability measures. By assumption, $\mathcal{F}_0 \subset \mathcal{C}$. Now suppose that $E_1, E_2, \ldots \in \mathcal{C}$ and $E_i \uparrow E$ or $E_i \downarrow E$. By the continuity property of probability (Proposition 3.30), $P_1(E) = \lim_{i\to\infty} P_1(E_i) = \lim_{i\to\infty} P_2(E_i) = P_2(E)$. This shows that $P_1(E) = P_2(E)$, so $E \in \mathcal{C}$. That is, $\mathcal{C}$ is a monotone class containing $\mathcal{F}_0$. By Theorem 3.32, $\mathcal{C}$ contains $\sigma(\mathcal{F}_0)$, completing the proof. □

A very important result in probability theory (Caratheodory extension theorem) allows us first to define a probability measure on a field, and then extend it to the sigma-field generated by that field:

Theorem 3.34. Caratheodory extension theorem *A probability measure P on a field $\mathcal{F}_0$ has a unique extension to the sigma-field $\sigma(\mathcal{F}_0)$ generated by $\mathcal{F}_0$. That is, there exists a unique probability measure P' on $\sigma(\mathcal{F}_0)$ such that $P'(E) = P(E)$ for all F in $\mathcal{F}_0$.*

The uniqueness part follows from Proposition 3.33. The existence part is proven by defining an outer measure P^* analogous to that used to show the existence of Lebesgue measure (see discussion following Example 3.25). That is, $P^*(A)$ is defined to be $\inf \sum_i P(A_i)$, where the infimum is over all countable unions $\cup A_i$ such that $A \subset \cup_i A_i$ and $A_i \in \mathcal{F}_0$. Then P^* restricted to the sets A such that $P^*(A \cap B) + P^*(A^C \cap B) = P^*(B)$ for all sets B is the desired extension. We omit the details of the proof. See Billingsley (2012) for details.

An important application of the Caratheodory extension theorem concerns completion of a probability space. This is a technical detail of probability theory, but a very important one. Sometimes a certain property is known to hold outside a *null set* N, i.e., a set $N \in \mathcal{F}$ with $P(N) = 0$. We would like to declare that the set on which it fails to hold has probability 0, but there is a problem. All we really know is that it holds outside N; it might hold on part of N as well. The precise set on which it fails to hold is a subset $M \subset N$. If $M \in \mathcal{F}$, then $P(M) = 0$ by part 2 of Proposition 3.26. But if M is not in $\mathcal{F}$, we cannot define $P(M)$. This annoying problem does not occur if all subsets of sets of probability 0 are in $\mathcal{F}$. Fortunately, we can always *complete* a probability space by including all subsets of sets of probability 0.

Proposition 3.35. Completion of a measure space *Any measure space $(\Omega, \mathcal{F}, \mu)$ can be completed to $(\Omega, \bar{\mathcal{F}}, \bar{\mu})$, where $\mathcal{F} \subset \bar{\mathcal{F}}$ and $\bar{\mu}(F) = \mu(F)$ for all $F \in \mathcal{F}$.*

Definition 3.36. Lebesgue sigma-field *The completion of the k-dimensional Borel sets $\mathcal{B}^k$ is the Lebesgue sigma-field $\mathcal{L}^k$. The sets in $\mathcal{L}^k$ are called k-dimensional Lebesgue sets.*

The k-dimensional Lebesgue sets include all subsets of Borel sets of measure 0. It is an interesting fact that the cardinality of Lebesgue sets is strictly larger than that of Borel sets.

Exercises

1. * Prove that if $P(B) = 1$, then $P(A \cap B) = P(A)$.

2. Prove that if $\{A_n\}_{n=1}^{\infty}$ and $\{B_n\}_{n=1}^{\infty}$ are sequences of events such that $P(B_n) \to 1$ as $n \to \infty$, then $\lim_{n\to\infty} P(A_n \cap B_n)$ exists if and only if $\lim_{n\to\infty} P(A_n)$ exists, in which case these two limits are equal.

3. Prove properties 2-4 of Proposition 3.26.

4. Let $(\Omega, \mathcal{F}, P)$ be a probability space and $A \in \mathcal{F}$ be a set with $P(A) > 0$. Let $\mathcal{F}_A$ denote $\mathcal{F} \cap A = \{F \cap A,\ F \in \mathcal{F}\}$ and define $P_A(F) = P(F \cap A)/P(A)$ for each $F \in \mathcal{F}_A$; P_A is called the conditional probability measure given event A. Prove that $(A, \mathcal{F}_A, P_A)$ is indeed a probability space, i.e., that $\mathcal{F}_A$ is a sigma-field of subsets of A and P_A is a probability measure on $\mathcal{F}_A$.

5. Flip a fair coin countably infinitely many times. The outcome is an infinite string such as $0, 1, 1, 0, \ldots$, where 0 denotes tails and 1 denotes heads on a given flip. Let Γ denote the set of all possible infinite strings. It can be shown that each $\gamma \in \Gamma$ has probability $(1/2)(1/2)\ldots(1/2)\ldots = 0$ because the outcomes for different flips are independent (the reader is assumed to have some familiarity with independence from elementary probability) and each has probability 1/2. What is wrong with the following argument? Because Γ is the collection of all possible outcomes,
$$1 = P(\Gamma) = P(\cup_{\Gamma}\gamma) = \sum_{\gamma\in\Gamma} P(\gamma) = \sum_{\gamma\in\Gamma} 0 = 0.$$

6. $\uparrow$ In the setup of Problem 5, let A be a collection of infinite strings of 0s and 1s. Is
$$P(A) = P(\cup_{A}\gamma) = \sum_{\gamma\in A} P(\gamma) = \sum_{\gamma\in A} 0 = 0$$
if A is
   (a) The set of infinite strings with exactly one 1?
   (b) The set of infinite strings with finitely many 1s?
   (c) The set of infinite strings with infinitely many 1s?

7. If $(\Omega, \mathcal{F}, P)$ is a probability space, prove that the collection of sets A such that $P(A) = 0$ or 1 is a sigma-field.

8. Let $A_1, A_2, \ldots$ be a countable sequence of events such that $P(A_i \cap A_j) = 0$ for each $i \neq j$. Show that $P(\cup_{i=1}^{\infty} A_i) = \sum_{i=1}^{\infty} P(A_i)$.

9. Consider the following stochastic version of the balls in the box paradox in Example 1.5. Again there are infinitely many balls numbered $1, 2, \ldots$ Step 1 at 1 minute to midnight is to put balls numbered $1, 2$ into the box, and randomly remove one of the 2 balls. Step 2 at 1/2 minute to midnight is to add balls numbered $3, 4$, and randomly remove one of the 3 balls from the box. Step 3 at 1/4 minute to midnight is to add balls numbered $5, 6$ and randomly remove one of the 4 balls from the box. Step 4 at 1/8 minute to midnight is to add balls numbered $7, 8$ and randomly remove one of the 5 balls from the box, etc.

   (a) Show that the probability of the event A_n that ball number 1 is not removed at any of steps $1, 2, \ldots, n$ is $1/(n+1)$.

   (b) What is the probability of the event A_∞ that ball number 1 is never removed from the box? Justify your answer.

   (c) Show that with probability 1 the box is empty at midnight.

10. Use induction to prove the inclusion-exclusion formula:

$$P(\cup_{i=1}^{n} E_i)$$
$$= \sum_{i_1=1}^{n} P(E_{i_1}) - \sum_{1 \le i_1 < i_2 \le n} P(E_{i_1} \cap E_{i_2}) + \sum_{1 \le i_1 < i_2 < i_3 \le n} P(E_{i_1} \cap E_{i_2} \cap E_{i_3})$$
$$\ldots \quad + (-1)^{(n+1)} P(E_1 \cap E_2 \cap \ldots \cap E_n). \tag{3.4}$$

11. ↑ Suppose that n people all have distinct hats. Shuffle the hats and pass them back in random order. What is the probability that at least one person gets his or her hat back? Hint: apply the inclusion-exclusion formula with $E_i = \{$person i gets his or her own hat back$\}$. Show that the probability in question is a Taylor series approximation to $1 - \exp(-1)$.

12. Does an arbitrary measure have the continuity property of Proposition 3.30 for decreasing sets? Hint: consider counting measure (Example 3.23) and the sets $E_n = \{n, n+1, \ldots\}$.

13. * Let $r_1, r_2, \ldots$ be an enumeration of the rational numbers (such an enumeration exists because the rational numbers are countable). For given $\epsilon > 0$, let I_i be an interval of length $\epsilon/2^i$ containing r_i. What can you say about the Lebesgue measure of $\cup_i I_i$? What can you conclude from this about the Lebesgue measure of the rationals?

14. * ↑ What is the Lebesgue measure of any countable set?

15. Use Proposition 3.27 to prove Proposition 3.28.

3.5 Why Restriction of Sets is Needed

It seems natural to assume that we can draw a number randomly from the unit interval or unit circle. The circle arises naturally when testing hypotheses about biological phenomena, such as whether heart attacks are equally likely to occur at any time throughout the day/night (e.g., Muller et al., 1985). Taking time 0 to be midnight means that we should treat time 24 hours the same as time 0 hours. Bending the interval $[0, 24]$ into a circle makes these two time points line up. Without loss we can divide by 24 to make the circle C have radius 1.

What exactly does it mean to select randomly and uniformly from the unit circle C? The corresponding probability measure P should have the property that for any subset A of C, P should have the same value as for any rotation of A by an angle θ (Figure 3.5). We will show that there is no such probability measure if $\mathcal{F}$ consists of all subsets of C.

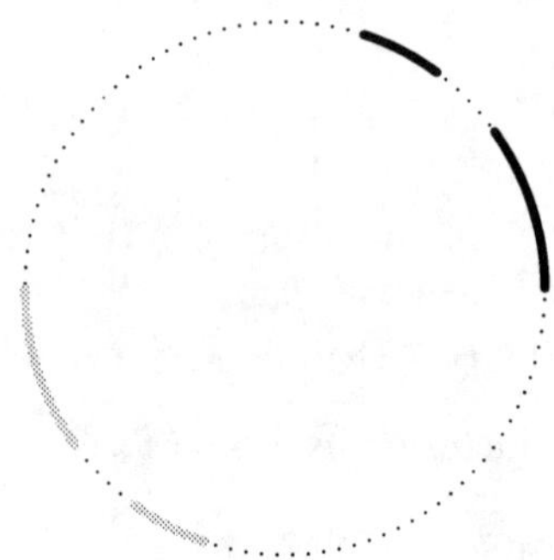

Figure 3.5: A rotation-invariant probability measure should assign the same probability to A (black region) as it does to any rotation of A (gray region). The gray region here is obtained by rotating the black region by an angle of π radians.

For each point x in C, consider the set of points A_x that can be obtained by rotating x clockwise by a rational angle. If $x \oplus r$ denotes clockwise rotation of x by the rational angle r, then $A_x = \{y \in C : \ y = x \oplus r, \ r \text{ rational}\}$. For each x, A_x has only countably many members. If x and y are two points in C, then A_x and A_y are either identical or disjoint, depending on whether y is or is not a rotation of x by a rational angle. Therefore, the entire unit circle C can be partioned into disjoint sets A_t, for t in some (uncountable) index set T, where $\cup_t A_t = C$. From each set A_t, select exactly one member, a_t, and let A be $\{a_t, t \in T\}$. This is possible by the Axiom of Choice (Axiom 2.13). What is the probability of A? Notice that $\cup_r (A \oplus r) = C$, where r ranges over the countable set of rational numbers. It follows that $1 = P(C) = \sum_r P(A \oplus r)$ by countable additivity. Moreover, each set $A \oplus r$ should have the same probability, p, in which case we have shown that $1 = \sum_r p$. But of course the right side is 0 if $p = 0$ and ∞ if $p > 0$.

What just happened here? We showed that if we try to define a rotation-invariant probability measure P defined for **every** subset of the unit circle, we get a contradiction. The same type of counterexample can be constructed when $\Omega = [0, 1]$, but it is slightly easier to understand on the unit circle. It turns out that if we restrict the sigma field $\mathcal{F}$, we **can** construct a rotation-invariant probability measure. For instance, when $\Omega = [0, 1]$, Lebesgue measure is rotation invariant. This example illustrates the need for making the sigma-field large, but not so large that it admits pathological sets that cause contradictions. As mentioned in Example 3.6, this issue does not arise when Ω is countable, in which case it always make sense to use the total sigma-field.

3.6 When We Cannot Sample Uniformly

There are some scenarios under which we cannot randomly and uniformly sample, and this can lead to some interesting problems outlined below.

Bayesian statisticians treat their uncertainty about the value of parameters by specifying prior distributions on them. For instance, to reflect complete uncertainty about a Bernoulli probability parameter p, a Bayesian might specify a uniform distribution on $[0, 1]$ (Lebesgue measure on the Borel subsets of $[0, 1]$). But how does one reflect complete uncertainty about the mean θ of a normal distribution? There is no uniform probability measure on the entire line; Lebesgue measure is not a probability measure because $\mu(R) = \infty$. Use of Lebesgue measure on R conveys complete certainty that θ is enormous; after all, the measure of any finite interval relative to the measure of the entire line is 0. It is very strange to think that θ has essentially no chance of being in any finite interval. Nonetheless, this so-called *improper prior* can be used to reproduce classical confidence intervals and tests (see Section 2.9 of Gelman, Carlin, Stern, and Rubin, 2004).

Another interesting scenario in which one cannot randomly and uniformly sample is from a countably infinite set. For instance, a Bayesian specifying a uniform prior for a Bernoulli parameter p has a problem if p is known to be rational. One cannot construct a uniform distribution on a countably infinite set because it would lead to a contradiction of countable additivity (see Exercise 1). Thus, there is no analog reflecting complete uncertainty about a countably infinite set like the rationals. Interestingly, if we had required probability measures to be only finitely additive instead of countably additive, we could define a uniform measure on a countably infinite set (Kadane and O'Hagan, 1995).

Another consequence of being unable to pick randomly from a countably infinite set concerns exchangeable random variables. The variables $X_1, \ldots, X_n$ are said to be *exchangeable* if $X_{\pi_1}, \ldots, X_{\pi_n}$ has the same distribution as $(X_1, \ldots, X_n)$ for any permutation $(\pi_1, \ldots, \pi_n)$ of $(1, 2, \ldots, n)$. An infinite set of random variables $X_1, X_2, \ldots$ is said to be exchangeable if every finite subset is exchangeable. Exchangeability is important for the validity of certain tests such as permutation tests.

The following argument suggests that we can always create exchangeable random variables by randomly permuting indices. For any finite set of random variables $(X_1, \ldots, X_n)$, exchangeable or not, randomly permute the observed values $x_1, \ldots, x_n$. The resulting random variables are exchangeable. For instance, if the observed values of X_1 and X_2 are $x_1 = 0$ and $x_2 = 1$, then the distribution of a random permutation (X_{π_1}, X_{π_2}) is

$$\begin{cases} (0, 1) & \text{with probability } 1/2 \\ (1, 0) & \text{with probability } 1/2, \end{cases} \tag{3.5}$$

which is exchangeable. Notice that the original distribution of (X_1, X_2) is arbitrary and irrelevant. Thus, the class of n-dimensional exchangeable binary random variables is very broad.

This technique of randomly permuting indices does not work for an infinite set of random variables. In fact, Exercise 2 asserts the impossibility of randomly permuting a countably infinite set such that each permutation has the same probability. The inability to apply the random permutation technique severely limits the set of possible distributions. In fact, it can be shown (DeFinetti's theorem—see Heath and Sudderth (1976) for an elementary proof) that every infinite set of exchangeable binary random variables has the property that each finite subset $X_1, \ldots, X_n$ are conditionally iid Bernoulli (P) given some random variable

P. That is, $X_1, \ldots, X_n$ have a *mixed Bernoulli distribution.* One consequence is that a pair (X_1, X_2) from an infinite string of exchangeable binary random variables cannot have the distribution (3.5).

Exercises

1. Prove the ironic fact that despite the existence of a uniform measure (Lebesgue measure) on the uncountable set $[0, 1]$, there is no uniform probability measure on a countably infinite set. That is, if Ω is countably infinite, there is no probability measure P assigning equal probability to each $\omega \in \Omega$.

2. $\uparrow$ Argue that there is no way to randomly permute the set of natural numbers such that each permutation has the same probability. Hint: if we could, then what would be the distribution of the first element?

3.7 The Meaninglessness of Post-Facto Probability Calculations

How often have you seen people notice what appears to be an unusual event and then ask you, the statistician, "What is the probability of that?" Their sense that they have witnessed something extremely rare fails to account for the fact that this was not a pre-specified event of interest. We know that in many experiments, each $\omega \in \Omega$ has extremely low probability. For instance, we have seen that **every** simple event ω when drawing a number randomly from the unit interval has probability 0. Likewise, if we pick randomly from the integers one through a million, each ω has probability one in a million. By not specifying events of interest in advance, one can always notice some pattern after the fact and make it appear that something incredible happened. We offer an example below.

Example 3.37. Face on Mars On July 26, 1976, NASA's Viking Orbiter spacecraft took a picture of a region of Mars that appears to show a human face (Figure 3.6). Some have concocted elaborate explanations involving ancient aliens who are trying to tell us they exist. After all, they argue, what is the probability of encountering an image that so closely resembles a human face? But if the image had looked like a bicycle, they would have inquired about the probability of so closely resembling a bicycle. Likewise, if the image had appeared on the moon, they would have asked the probability of seeing such a pattern on the moon. Therefore, even though each possible $\omega \in \Omega$ might have probability 0, the collection of ωs that would have triggered some recognized pattern may have non-negligible probability. The real question is what is the probability of finding, somewhere in the accessible part of the universe, some pattern that we would recognize? That probability may very well be high. We will examine this example more thoroughly in the next chapter. We will see, as with the Bible code controversy, the key is phrasing the question so that it represents a well-defined random experiment. □

3.8 Summary

1. A probability space is a triple $(\Omega, \mathcal{F}, P)$, where:

   (a) Ω is the sample space.

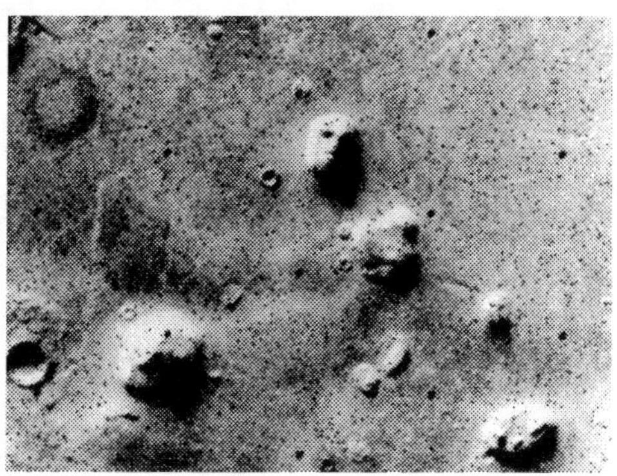

Figure 3.6: Mars face NASA image P1A01141.

(b) $\mathcal{F}$ is a sigma-field (nonempty, closed under complements and countable unions) of subsets of Ω: $\mathcal{F}$ are the events we are allowed to consider. We want $\mathcal{F}$ to be large enough to include useful sets but small enough to ensure that P is countably additive.

(c) P is a probability measure, a nonnegative, countably additive set function (meaning $P(\cup A_i) = \sum P(A_i)$ if the A_i are disjoint $\mathcal{F}$ sets) with $P(\Omega) = 1$.

2. *** A countable union of $\mathcal{F}$ events can be written as a countable union of disjoint $\mathcal{F}$ events.

3. *** Continuity property of probability: if $E_n \uparrow E$ or $E_n \downarrow E$ are all in $\mathcal{F}$, then $P(E_n) \to P(E)$.

4. Boole's inequality: $P(\cup A_i) \leq \sum_{i=1}^{\infty} P(A_i)$ for any $A_1, A_2, \ldots$ in $\mathcal{F}$. Also called subadditivity or, for a finite number of A_i, Bonferroni's inequality.

5. If $\mathcal{C}$ is a collection of subsets of Ω, $\sigma(\mathcal{C})$ is the smallest sigma field containing all sets in $\mathcal{C}$. It is the intersection of all sigma-fields containing $\mathcal{C}$.

6. (a) The Borel sets $\mathcal{B}^k$ of R^k are $\sigma\{(I_1 \times \ldots \times I_k), I_1, \ldots, I_k \text{ intervals}\}$. $\mathcal{B}^k$ contains all open and closed sets, and most subsets that could reasonably be encountered in practice.

   (b) The Lebesgue sets $\mathcal{L}^k$ of R^k are the completion of the k-dimensional Borel sets. $\mathcal{L}^k$ includes all subsets of Borel sets of measure 0.

7. *** A very useful technique for proving that a property p holds for all sets in $\sigma(\mathcal{F}_0)$, where $\mathcal{F}_0$ is a field, is as follows.

   (a) Define $\mathcal{C}$ to be the class of subsets of $\sigma(\mathcal{F})$ such that p holds, and show that $\mathcal{C}$ contains $\mathcal{F}_0$.

   (b) Prove that $\mathcal{C}$ is a monotone class (nonempty, closed under increasing unions and decreasing intersections).

   (c) Invoke Theorem 3.32.

Chapter 4

Random Variables and Vectors

4.1 Random Variables

4.1.1 Sigma-Fields Generated by Random Variables

We are now in a position to define a random variable. In a more elementary probability course, you may have learned that a random variable $X = X(\omega)$ is a real-valued function whose domain is Ω. For instance, in Example 3.4, where each ω is a specific set of 5 cards, let $X(\omega) = I\{\omega \text{ contains four aces}\}$ take the value 1 if ω contains 4 aces and 0 otherwise. Then $X(\omega)$ is a random variable. We calculate the probability of an event involving X by using the P measure of the set of ω that get mapped to the X event. In this example, the probability that $X = 1$ is just $P\{$Ace of Diamonds, Ace of Hearts, Ace of Clubs, Ace of Spades, 2 of Diamonds$\} + P\{$Ace of Diamonds, Ace of Hearts, Ace of Clubs, Ace of Spades, 2 of Hearts$\} + \ldots + P\{$Ace of Diamonds, Ace of Hearts, Ace of Clubs, Ace of Spades, King of Spades$\} = 48/\binom{52}{5}$ because each of the 48 disjoint simple events comprising $I\{\omega$ contains four aces$\} = 1$ has probability $1/\binom{52}{5}$.

One thing we took for granted in a calculation such as the one above is that any X event of interest corresponds to an allowable set, $E \in \mathcal{F}$; otherwise, we would not be allowed to calculate the probability of E. This motivates the following definition.

Definition 4.1. Measurable function/random variable *A function $f : \Omega \longmapsto R$ on a measure space $(\Omega, \mathcal{F}, \mu)$ is said to be $\mathcal{F}$-measurable (or measurable) if $f^{-1}(B) \in \mathcal{F}$ for each one-dimensional Borel set B. If the measure space $(\Omega, \mathcal{F}, \mu)$ is a probability space $(\Omega, \mathcal{F}, P)$, a measurable function $X(\omega)$ such that $P(|X| < \infty) = 1$ is called a random variable.*

Figure 4.1 illustrates the concept of an inverse image $X^{-1}(B)$ for the random variable $X(\omega) = \frac{1}{\omega(1-\omega)}$ on $(\Omega, \mathcal{F}, P) = ((0,1), \mathcal{B}_{(0,1)}, \mu_L)$. The Borel set B is $\{(6.25, \infty) \cup \{4\}\}$, and $X^{-1}(B) = \{(0, .20) \cup (.80, 1) \cup \{.5\}\}$ is in $\mathcal{B}_{(0,1)}$.

Example 4.2. Random variable examples Let $\Omega = [0, 1]$, and $\mathcal{F} = \mathcal{B}_{[0,1]}$ be the Borel subsets of Ω. Then the following are random variables:

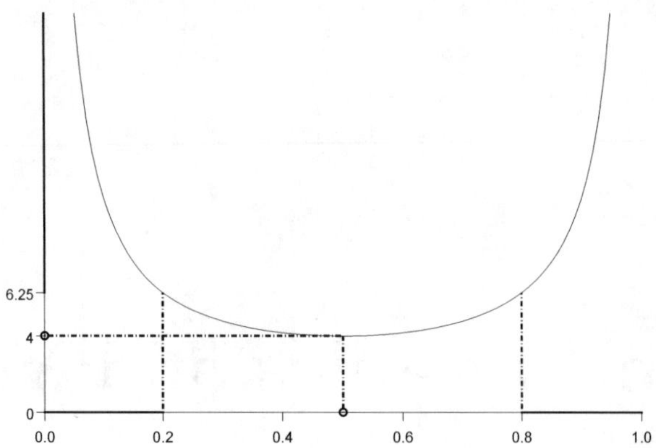

Figure 4.1: The random variable $X(\omega) = \frac{1}{\omega(1-\omega)}$ for $(\Omega, \mathcal{F}, P) = ((0,1), \mathcal{B}_{(0,1)}, \mu_L)$. If B is the Borel set $\{(6.25, \infty) \cup \{4\}\}$, then $X^{-1}(B) = \{(0, .20) \cup (.80, 1) \cup \{.5\}\}$ is in $\mathcal{B}_{(0,1)}$.

1. $X_1(\omega) \equiv c$.
2. $X_2(\omega) = I(\omega \geq 1/2)$.
3. $X_3(\omega) = \omega$.

To see that X_1 is a random variable, observe that if $B \in \mathcal{B}$, then $X_1^{-1}(B) = \Omega \in \mathcal{F}$ if B contains c and $X_1^{-1}(B) = \emptyset \in \mathcal{F}$ if B does not contain c. Therefore, X_1 is $\mathcal{F}$-measurable. Similarly,

$$X_2^{-1}(B) = \begin{cases} \emptyset \in \mathcal{F} & \text{if } B \text{ contains neither 0 nor 1} \\ \Omega \in \mathcal{F} & \text{if } B \text{ contains both 0 and 1} \\ [1/2, 1] \in \mathcal{F} & \text{if } B \text{ contains 1, but not 0} \\ [0, 1/2) \in \mathcal{F} & \text{if } B \text{ contains 0, but not 1.} \end{cases}$$

Thus, X_2 is $\mathcal{F}$-measurable. Also, $X_3(\omega)$ is a random variable because if $B \in \mathcal{B}$, then $X_3^{-1}(B) = B \cap [0,1] \in \mathcal{B}_{[0,1]} = \mathcal{F}$. □

To verify that each X in Example 4.2 was a random variable, we evaluated $X^{-1}(\mathcal{B}) = \{X^{-1}(B),\ B \in \mathcal{B}\}$. It is an exercise to show that $X^{-1}(\mathcal{B})$, which is a subset of $\mathcal{F}$, is itself a sigma-field. It is the smallest sigma-field with respect to which X is measurable.

Definition 4.3. Sigma-field generated by a random variable *If X is a random variable, the smallest sigma-field with respect to which X is measurable, namely $\{X^{-1}(B),\ B \in \mathcal{B}\}$, is called the sigma-field generated by X and denoted by $\sigma(X)$.*

After observing the random variable X, we can determine, for each $F \in \sigma(X)$, whether or not F occurred. In Example 4.2, $\sigma(X_1) = \{\emptyset, [0,1]\} \subset \{\emptyset, [0,1], [0,1/2), [1/2,1]\} = \sigma(X_2)$. This is an indication that X_2 gives more information than X_1 about which ω was actually drawn in the experiment: by observing X_1 we can tell only whether event F occurred for $F = \emptyset$ and $[0,1]$, whereas by observing X_2, we can tell whether F occurred for $F = \emptyset$, $[0,1]$,$[0,1/2)$, and $[1/2,1]$. In other words, we can narrow the possible values of ω somewhat. Also, $\sigma(X_2) \subset \sigma(X_3)$, so X_3 is even more informative than X_2. In fact, X_3 tells us exactly which ω was drawn in the experiment, whereas X_2 gives us only partial information about ω, namely whether $\omega < 1/2$ or $\omega \geq 1/2$.

When $\mathcal{F}$ comprises k-dimensional Borel or Lebesgue sets, a measurable function is called Borel measurable or Lebesgue measurable, respectively.

Definition 4.4. Borel measurable and Lebesgue measurable functions *A function $f : R^k \longmapsto R$ is said to be*

1. *Borel measurable, or simply a Borel function, if $f^{-1}(B)$ is a k-dimensional Borel set for each one-dimensional Borel set $B \in \mathcal{B}$.*
2. *Lebesgue measurable, or simply a Lebesgue function, if $f^{-1}(B)$ is a k-dimensional Lebesgue set for each one-dimensional Borel set $B \in \mathcal{B}$.*

The class of Borel functions is very broad. We will soon see that this class includes all continuous functions, but it also includes other types of functions typically encountered in practice.

The intuition from regarding sigma-fields generated by random variables as measures of their information about ω suggests that if Y is a function of X, the sigma-field generated by Y must be contained in the sigma-field generated by X. After all, if we know X, then we know Y, so Y must be less informative than X. Of course, if $Y = f(X)$ and $X = g(Y)$ for Borel functions f and g, then X and Y should be equally informative about ω. These sentiments are summarized by the following result.

Proposition 4.5. Functions of random variables generate smaller sigma-fields *Let $X(\omega)$ be a random variable and $Y(\omega) = f(X(\omega))$ for some Borel function $f : R \longmapsto R$. Then*

1. *Y is a random variable.*
2. *$\sigma(Y) \subset \sigma(X)$.*
3. *If, additionally, $X = g(Y)$ for a Borel function g, then $\sigma(X) = \sigma(Y)$.*

Proof. Exercise.

Throughout this book we will emphasize the importance of thinking about a random variable in terms of the sigma-field it generates.

It is somewhat cumbersome to verify $\mathcal{F}$-measurability of X because we must consider $X^{-1}(B)$ for every Borel set B. The following result makes it easier to verify $\mathcal{F}$-measurability.

Proposition 4.6. A simpler condition for $\mathcal{F}$-measurability *A necessary and sufficient condition for $\mathcal{F}$-measurability of $X(\omega)$ is that $X^{-1}(-\infty, x] \in \mathcal{F}$ for each $x \in R$. The statement remains true if we replace $(-\infty, x]$ by $(-\infty, x)$.*

Proof. The necessity part follows from the fact that $(-\infty, x]$ is a Borel set of R. To see the sufficiency part, assume that $X^{-1}(-\infty, x] \in \mathcal{F}$ for each x, and let $\mathcal{A}$ be the collection of Borel sets such that $X^{-1}(A) \in \mathcal{F}$. It is an exercise (Exercise 4) to show that $\mathcal{A}$ is a sigma-field. Therefore, $\mathcal{A}$ is a sigma-field containing all intervals $(-\infty, x]$, $x \in R$. This means $\mathcal{A}$ contains $\sigma\{(-\infty, x],\ x \in R\}$, the smallest σ-field containing $\{(-\infty, x],\ x \in R\}$. But by Exercise 10 of Section 3.2.3, $\sigma\{(-\infty, x],\ x \in R\}$ is the Borel sigma-field $\mathcal{B}$. Therefore,

$\mathcal{B} \subset \mathcal{A}$. That is, X is $\mathcal{F}$-measurable by Definition 4.1. We have shown that Definition 4.1 is equivalent to $X^{-1}(-\infty, x] \in \mathcal{F}$ for each $x \in R$.

The part about replacing $(-\infty, x]$ by $(-\infty, x)$ holds because if $X^{-1}(-\infty, x) \in \mathcal{F}$ for all $x \in R$, then $X^{-1}(-\infty, x] = \cap_n X^{-1}(-\infty, x + 1/n) \in \mathcal{F}$ for all $x \in R$ because $X^{-1}(-\infty, x]$ is a countable intersection of sets in the sigma-field $\mathcal{F}$. □

We are now in a position to prove that continuous functions are Borel measurable.

Proposition 4.7. Continuous implies Borel *A continuous function $f : R^k \longmapsto R$ is a Borel function.*

Proof. By Proposition 4.6, it suffices to show that $f^{-1}(-\infty, x) \in \mathcal{B}^k$ for each x. But $(-\infty, x)$ is an open set, so by Proposition A.67, $f^{-1}(-\infty, x)$ is an open set in R^k, which is a Borel set in R^k by Proposition 3.19. □

We next consider limits, infs, sups, liminfs and limsups of a sequence of random variables. This can lead to infinite values, and our definition of random variables requires them to be finite. We therefore extend the definition as follows.

Definition 4.8. Extended random variables *We say that X is an extended random variable if $X^{-1}(\bar{B}) \in \mathcal{F}$ for each $\bar{B}$ in the extended Borel sigma-field $\bar{\mathcal{B}}$ (see Definition 3.13).*

Proposition 4.9. Limiting operations on sequences of random variables produce extended random variables *If $X_n(\omega)$ is a random variable for each n, then the following are extended random variables.*

1. $\inf X_n(\omega)$.
2. $\sup X_n(\omega)$.
3. $\underline{\lim}\, X_n(\omega)$.
4. $\overline{\lim}\, X_n(\omega)$.
5. $\lim X_n$ *if it exists.*

Proof.

1. Event $\{\inf X_n(\omega) \geq x\}$ is equivalent to $X_n(\omega) \geq x$ for all n. That is, $\{\inf X_n(\omega) \geq x\} = \cap_n \{X_n(\omega) \geq x\}$. Also, $\{X_n(\omega) \geq x\} \in \mathcal{F}$ for each n because X_n is a random variable. Because $\mathcal{F}$ is closed under countable intersections, $\cap_n \{X_n(\omega) \geq x\} \in \mathcal{F}$. We have shown that $\{\inf X_n(\omega) \geq x\} \in \mathcal{F}$, so its complement, $\{\inf X_n(\omega) < x\}$, is also in $\mathcal{F}$. By Proposition 4.6, $\inf X_n(\omega)$ is a random variable.

2. Exercise.

3. Event $\{\underline{\lim}\, X_n(\omega) < x\}$ occurs if and only if, for at least one k, $X_n(\omega) \leq x - 1/k$ infinitely often in n. For fixed k, let A_k be the event that $X_n(\omega) \leq x - 1/k$ infinitely often in n. Then $\{\underline{\lim}\, X_n(\omega) < x\} = \cup_k A_k$. Proposition 3.21 shows that each $A_k \in \mathcal{F}$. Because $\mathcal{F}$ is closed under countable unions, $\cup_k A_k \in \mathcal{F}$, proving that $\{\underline{\lim}\, X_n < x\} \in \mathcal{F}$. By Proposition 4.6, $\underline{\lim}\, X_n$ is a random variable.

4. Exercise.

5. Exercise. □

Exercises

1. Let $\Omega = [0,1]$ and $\mathcal{F} = \{\emptyset, [0,1/2), [1/2,1), [0,1]\}$. Which of the following is (are) random variables: $X_1(\omega) = \omega^2$, $X_2(\omega) = I(\omega < 1/2)$, $X_3(\omega) = I(\omega \leq 1/2)$? Justify your answer.

2. In Example 3.4, suppose we foolishly defined $\mathcal{F}$ to be {{5-card hands with at least one pair}, {5-card hands not containing at least one pair}, $\emptyset, \Omega$}. Is $X(\omega) = I$(full house) a random variable?

3. * Prove that if X is any random variable, then $X^{-1}(\mathcal{B})$, defined as $\{X^{-1}(B),\ B \in \mathcal{B}\}$, is a sigma-field. More generally, if $f : \Omega \longmapsto \Gamma$ is a function and $\mathcal{G}$ is a sigma-field of subsets of Γ, then $\{f^{-1}(G),\ G \in \mathcal{G}\}$ is a sigma-field of subsets of Ω.

4. * Let $f : \Omega \longmapsto R$ be a finite-valued function and $\mathcal{A}$ be the collection $\{A \in \mathcal{B} : f^{-1}(A) \in \mathcal{F}\}$. Then $\mathcal{A}$ is a sigma-field of sets in $\mathcal{B}$.

5. Prove items 2,4, and 5 of Proposition 4.9.

6. Suppose that Ω is the set of integers, and $\mathcal{F} = T(\Omega)$ is the total sigma-field. Let $X(\omega) = \omega^2$ and $Y(\omega) = \omega$. What are $\sigma(X)$ and $\sigma(Y)$? What does this say about the information content of X and Y about which ω was drawn?

7. Suppose that X and Y are random variables on $(\Omega, \mathcal{F}, P)$, and suppose that $F \in \mathcal{F}$. Prove that
$$Z(\omega) = \begin{cases} X(\omega) & \text{if } \omega \in F \\ Y(\omega) & \text{if } \omega \in F^C \end{cases}$$
is a random variable on $(\Omega, \mathcal{F}, P)$.

8. Prove Proposition 4.5.

9. Suppose that X is a random variable with three possible values, say 0, 1, and 2. Let $A = \{\omega : X(\omega) = 0\}$ and $B = \{\omega : X(\omega) = 1\}$. Determine $\sigma(X)$ in terms of A, B, their complements, unions, and intersections.

10. Prove that if X is a discrete random variable (i.e., takes on only countably many possible values), then X generates a partition of Ω (see Exercise 13 of Section 3.2). What is $\sigma(X)$? What is the cardinality of $\sigma(X)$ if X takes a countably infinite number of values?

4.1.2 Probability Measures Induced by Random Variables

So far we have focused on the sigma-field generated by a random variable, but we now concentrate on the probability measure associated with a random variable. Probabilities of events involving X are calculated using the probability measure P on the original probability space $(\Omega, \mathcal{F}, P)$. We illustrated this with a discrete probability space at the beginning of Section 4.1. Here is an example with a continuous space.

Example 4.10. Let $(\Omega, \mathcal{F}, P) = ([0,1], \mathcal{B}_{[0,1],\mu_L})$, where μ_L is Lebesgue measure on $[0,1]$. If $X(\omega) = -\ln(\omega)$, then the probability that $X \le x$ is $\mu_L\{\omega : X(\omega) \le x\} = \mu_L\{\omega : \omega \ge \exp(-x)\} = 1 - \exp(-x)$ (Figure 4.2). Likewise, the probability that X is in the interval (a,b) is $\mu_L\{\omega : a < X(\omega) < b\} = \mu_L\{\omega : \exp(-b) < \omega < \exp(-a)\} = \exp(-a) - \exp(-b)$. In fact, we will see in Section 4.3 that once we know the probability that $X \le x$, we can calculate the probability that $X \in B$ for any Borel set B. In summary, the random variable on the original probability space induces a new probability measure P' on R, in this case defined by the exponential distribution with parameter 1. □

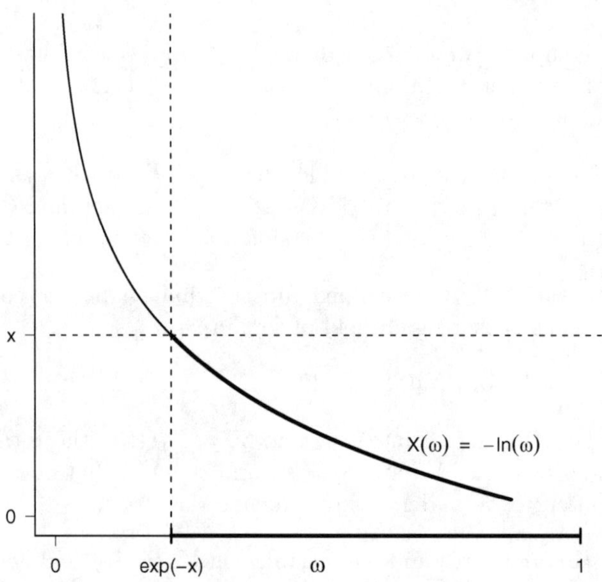

Figure 4.2: $X(\omega) = -\ln(\omega)$ and the probability that $X \le x$ is $\mu_L\{\omega : X(\omega) \le x\} = \mu_L\{\omega : \omega \ge \exp(-x)\} = 1 - \exp(-x)$.

Proposition 4.11. Probability measure induced by an r.v. *A random variable X on $(\Omega, \mathcal{F}, P)$ induces a probability measure P' on the probability space $(R, \mathcal{B}, P')$ through $P'(B) = P\{X^{-1}(B)\} = P\{\omega : X(\omega) \in B\}$.*

The proof that P' is a probability measure is left to Exercise 7.

To emphasize the correspondence between the original and induced probability spaces, we sometimes write $X : (\Omega, \mathcal{F}) \longmapsto (R, \mathcal{B})$. This does **not** mean that $X(F) \in \mathcal{B}$ for each $F \in \mathcal{F}$; this need not hold. Rather, it means that $X^{-1}(B) \in \mathcal{F}$ for each $B \in \mathcal{B}$ (i.e., X is $\mathcal{F}$-measurable).

Notice that the original probability measure is on a probability space of arbitrary objects. For example, Figure 4.3 illustrates the correspondence between the original and induced probability measures when Ω is the collection of trigonometric functions $\Omega = \{\cos(1x), \cos(2x), \cos(3x), \ldots\}$, $P = \{1/2, 1/2^2, 1/2^3, \ldots\}$, and $X(\omega)$ =the number of zeros of ω in $[0, 2\pi]$.

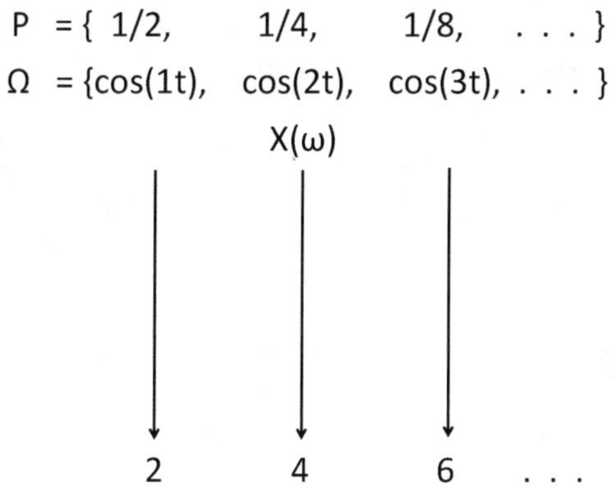

Figure 4.3: $\omega_i = \cos(it)$, $P(\omega_i) = 1/2^i$, and $X(\omega)$ is the number of zeros of ω in the interval $[0, 2\pi]$. The probability measure P on $(\Omega, \mathcal{F})$ induces a measure P' on $(R, \mathcal{B})$, namely the measure assigning probability $1/2^i$ to natural number $2i$, $i = 1, 2, \ldots$

Exercises

In the first 4 problems, let $(\Omega, \mathcal{F}, P) = ([0,1], \mathcal{B}_{[0,1]}, \mu_L)$, where μ_L denotes Lebesgue measure, and let P' be the probability measure on the line induced by the random variable X.

1. If $X(\omega) = 1 - \omega^2$, find $P'\{(1/2, 3/4)\}$.

2. If $X(\omega) = |\omega - 1/2| + 1$, find $P'\{(4/3, \infty)\}$.

3. If $X(\omega) = \omega/(1+\omega)$, find $P'\{(.25, .4)\}$.

4. If $X(\omega) = \omega/\{1 + I(\omega > 1/2)\}$, find $P'\{(.25, .75)\}$.

5. Let Ω be $\{1,2,3,4,5,6\}$, $\mathcal{F}$ be the total sigma-field of subsets of Ω, and $P\{\omega\} = 1/6$ for each ω. Let $X(\omega) = \{\sin(\omega\pi/2)\}^2$. Find $P'(\{0\})$, where P' is the probability measure on the line induced by X.

6. $\uparrow$ Suppose that in the preceding problem, $\mathcal{F} = \{\{1,3,5\}, \{2,4,6\}, \emptyset, \Omega\}$. Does your answer change? Is X still a random variable? What if X had been $\sin(\omega\pi/2)$ and $\mathcal{F} = \{\{1,3,5\}, \{2,4,6\}, \emptyset, \Omega\}$?

7. Prove that if X is a random variable on $(\Omega, \mathcal{F}, P)$, then P' defined by $P'(B) = P\{X^{-1}(B)\}$ is a probability measure on $(R, \mathcal{B})$.

8. Prove that the sum or difference of two discrete random variables (i.e., random variables taking on only countably many values) is discrete.

9. Prove that if Ω is countable, every random variable is discrete.

4.2 Random Vectors

4.2.1 Sigma-Fields Generated by Random Vectors

The k-dimensional analog of a random variable is a random vector $(X_1, \ldots, X_k)$, which is simply a collection of random variables. Just as we defined the sigma-field generated by a random variable, we can define the sigma-field generated by a random vector $X_1, \ldots, X_k$, even one consisting of (countably or uncountably) infinitely many random variables.

Definition 4.12. Sigma-field generated by an arbitrary collection of random variables *The sigma-field $\sigma\{X_t,\ t \in T\}$ generated by $\{X_t,\ t \in T\}$, where T is a countable or uncountable index set, is the smallest sigma-field for which each X_t is measurable.*

That is, $\sigma\{X_t,\ t \in T\}$ is the intersection of all sigma-fields containing all of the sets $X_t^{-1}(B),\ B \in \mathcal{B},\ t \in T$. Again it is difficult to picture this sigma-field, but as we add more random variables to a collection, the sigma-field generated by the collection enlarges. For example, with stochastic processes, we observe data X_t over time. The information up to and including time t is $\mathcal{F}_t = \sigma(X_s,\ s \le t)$. The longer we observe the process, the more information we get, reflected by the fact that $\mathcal{F}_s \subset \mathcal{F}_t$ for $s \le t$.

In one dimension we observed that $\sigma(X)$, the smallest sigma-field containing $X^{-1}(B)$ for all $B \in \mathcal{B}^1$, is just $\{X^{-1}(B),\ B \in \mathcal{B}^1\}$. This follows from the fact that this latter collection is a sigma-field (see Exercise 3 of Section 4.1.1). There is a k-dimensional analog of this result. Recall that we defined $\sigma(X_1, \ldots, X_k)$ to be the smallest sigma-field for which each X_i is measurable. We can also express $\sigma(X_1, \ldots, X_k)$ as follows.

Proposition 4.13. Alternative expression for $\sigma(X_1, \ldots, X_k)$

$$\begin{aligned}\sigma(X_1, \ldots, X_k) &= (X_1, \ldots, X_k)^{-1}(\mathcal{B}^k) \\ &= \{(X_1, \ldots, X_k)^{-1}(B),\ B \in \mathcal{B}^k\} \\ &= \{[\omega : (X_1(\omega), \ldots, X_k(\omega)) \in B],\ B \in \mathcal{B}^k\}. \end{aligned} \tag{4.1}$$

Proof. The first step is to show that $(X_1, X_2, \ldots, X_k)^{-1}(\mathcal{B}^k) \subset \sigma(X_1, \ldots, X_k)$. Let $\mathcal{C}$ denote the collection of k-dimensional Borel sets C such that $(X_1, X_2, \ldots,\ X_k)^{-1}(C) \in \sigma(X_1, \ldots, X_k)$. Because $X_1, \ldots X_k$ are random variables, $\mathcal{C}$ contains all sets of the form $B_1 \times B_2 \times \ldots \times B_k$, where each $B_i \in \mathcal{B}^1$. Furthermore, it is easy to see that $\mathcal{C}$ is a sigma-field, so $\mathcal{C}$ must contain $\mathcal{B}^k$, the sigma-field of sets in R^k generated by $\{B_1 \times B_2 \times \ldots \times B_k :\ B_i \in \mathcal{B}^1,\ i = 1, \ldots, k\}$ (see Proposition 3.20). Therefore $(X_1, \ldots, X_k)^{-1}(\mathcal{B}^k) \subset \sigma(X_1, \ldots, X_k)$. But $(X_1, \ldots, X_k)^{-1}(\mathcal{B}^k)$ is a sigma-field with respect to which each X_i is measurable, and $\sigma(X_1, \ldots, X_k)$ is the **smallest** such sigma-field. Therefore $(X_1, \ldots, X_k)^{-1}(\mathcal{B}^k) \supset \sigma(X_1, \ldots, X_k)$. Because $(X_1, \ldots, X_k)^{-1}(\mathcal{B}^k) \subset \sigma(X_1, \ldots, X_k)$ and $(X_1, X_2, \ldots, X_k)^{-1}(\mathcal{B}^k) \supset \sigma(X_1, \ldots, X_k)$, $(X_1, X_2, \ldots, X_k)^{-1}(\mathcal{B}^k) = \sigma(X_1, \ldots, X_k)$. □

Example 4.14. Roll a die and let $\mathcal{F} = T(\Omega)$ be the total sigma-field of all subsets of $\Omega = \{1, 2, 3, 4, 5, 6\}$. Let $X_1(\omega)$ be the indicator that the outcome ω of the random roll was an even number and $X_2(\omega)$ be the indicator that it was 3 or less. The sigma field generated by X_1 is $\{\emptyset, \{2, 4, 6\}, \{1, 3, 5\}, \{1, 2, 3, 4, 5, 6\}\}$, while the sigma-field generated by X_2 is $\{\emptyset, \{1, 2, 3\}, \{4, 5, 6\}, \{1, 2, 3, 4, 5, 6\}\}$. The sigma-field generated by (X_1, X_2) is the smallest sigma-field containing all of the sets $\emptyset, \{2, 4, 6\}, \{1, 3, 5\}, \{1, 2, 3\}, \{4, 5, 6\}, \{1, 2, 3, 4, 5, 6\}$. We can either directly find this sigma-field or use the fact that $\sigma(X_1, X_2) = (X_1, X_2)^{-1}(B)$, $B \in \mathcal{B}^2$. Consider the latter approach.

1. Suppose that B contains none of the four pairs $(0,0),(0,1),(1,0)$, or $(1,1)$. Then $(X_1, X_2)^{-1}(B) = \emptyset$.

2. Suppose B contains exactly one of the four pairs.

   (a) If B contains only $(0,0)$, then $(X_1, X_2)^{-1}(B) = \{5\}$.

   (b) If B contains only $(0,1)$, then $(X_1, X_2)^{-1}(B) = \{1,3\}$.

   (c) If B contains only $(1,0)$, then $(X_1, X_2)^{-1}(B) = \{4,6\}$.

   (d) If B contains only $(1,1)$, then $(X_1, X_2)^{-1}(B) = \{2\}$.

3. Suppose B contains exactly two of the pairs. Then $(X_1, X_2)^{-1}(B)$ is a union of the corresponding two sets in part 2. The possible unions are $\{2,5\}$, $\{1,2,3\}$, $\{1,3,5\}$, $\{2,4,6\}$, $\{4,5,6\}$, $\{1,3,4,6\}$.

4. Suppose B contains exactly three of the pairs. Then $(X_1, X_2)^{-1}(B)$ is a union of the corresponding three sets in part 2. The possible unions are $\{1,2,3,5\}$, $\{2,4,5,6\}$, $\{1,2,3,4,6\}$, $\{1,3,4,5,6\}$.

5. Suppose B contains all of the pairs. Then $(X_1, X_2)^{-1}(B) = \{1,2,3,4,5,6\}$.

When we include all of these sets, we get $\{\emptyset$, $\{2\}$, $\{5\}$, $\{1,3\}$, $\{4,6\}$, $\{2,5\}$, $\{1,2,3\}$, $\{1,3,5\}$, $\{2,4,6\}$, $\{4,5,6\}$, $\{1,3,4,6\}$, $\{1,2,3,5\}$, $\{2,4,5,6\}$, $\{1,2,3,4,6\}$, $\{1,3,4,5,6\}$, $\{1,2,3,4,5,6\}\}$. Notice that this sigma-field contains each of $\sigma(X_1)$ and $\sigma(X_2)$ because (X_1, X_2) is more informative about ω than either X_1 or X_2 alone. □

Exercises

1. Let Ω be the set of integers and $\mathcal{F} = T(\Omega)$ be the total sigma-field (all subsets of Ω). Let $X = 1(\omega \geq 0)$ and $Y = |\omega|$. Find $\sigma(X)$ and $\sigma(Y)$. Prove that $\sigma(X, Y) = \mathcal{F}$. Interpret these results in terms of the amount of information about ω contained in X, Y.

2. Let $(\Omega, \mathcal{F}, P)$ be the unit interval equipped with the Borel sets and Lebesgue measure. For $t \in [0,1]$, define $X_t(\omega)$ by $I(t = \omega)$. Is X_t a random variable? What is the sigma-field generated by X_t, $0 \leq t \leq 1$?

3. Flip two coins, so that $\Omega = \{(H,H),(H,T),(T,H),(T,T)\}$, where H and T mean heads and tails, respectively. Take $\mathcal{F} = T(\Omega)$, the total sigma-field. Let X_1 be the number of heads and X_2 be the number of tails. Show that $\sigma(X_1) = \sigma(X_2) = \sigma(X_1, X_2)$ and explain in terms of the information content of the random variables.

4. Let $\Omega = [0,1]$ and $\mathcal{F}$ be $\mathcal{B}_{[0,1]}$ be the Borel subsets of $[0,1]$.

   (a) Let $X_1(\omega) = I(\omega \in [1/2, 1))$ and $X_2 = I(\omega \in [1/4, 1/2) \cup [3/4, 1))$ be the first and second digits of the base 2 representation of ω (see Example 1.1 of Chapter 1). Determine $\sigma(X_1)$ and $\sigma(X_1, X_2)$ and comment on the information content of X_1 compared to that of X_1, X_2.

   (b) Let $X_1, X_2, \ldots$ be all of the digits of the base 2 representation of ω. Prove that $\sigma(X_1, X_2, \ldots)$ contains all intervals in $[0,1]$.

   (c) What can you conclude about how $\sigma(X_1,\ X_2, \ldots)$ compares to the Borel sets in $[0,1]$? Interpret this result in terms of the information content in $X_1, X_2, \ldots$ about ω.

4.2.2 Probability Measures Induced by Random Vectors

Just as a random variable X on $(\Omega, \mathcal{F})$ induces a probability measure P' on $(R, \mathcal{B})$, a random vector $(X_1, \ldots, X_k)$ induces a probability measure on $(R^k, \mathcal{B}^k)$.

Proposition 4.15. Probability measure induced by $(X_1, \ldots, X_k)$ *The random vector $(X_1, \ldots, X_k)$ induces a probability measure P' on $(R^k, \mathcal{B}^k)$ through*

$$\begin{aligned} P'(B) &= P\{(X_1, \ldots, X_k)^{-1}(B)\} \\ &= P\{\omega : (X_1(\omega), \ldots, X_k(\omega)) \in B\}, \ B \in \mathcal{B}^k. \end{aligned} \tag{4.2}$$

Example 4.16. Consider the probability space $([0,1], \mathcal{B}_{[0,1]}, \mu_L)$. Let $X_1(\omega) = \cos(2\pi\omega)$ and $X_2(\omega) = \omega^3$. If P' denotes the probability measure on $(R^2, \mathcal{B}^2)$ induced by (X_1, X_2), consider $P'\{[0,1] \times (1/125, 1/64)\}$. Note that $\cos(2\pi\omega) \in [0,1] \Leftrightarrow 2\pi\omega \in [0, \pi/2] \cup [3\pi/2, 2\pi]$. Equivalently, $\omega \in [0, 1/4] \cup [3/4, 1]$. Also, $\omega^3 \in (1/125, 1/64) \Leftrightarrow \omega \in (1/5, 1/4)$. Therefore,

$$\begin{aligned} P'\{[0,1] \times (1/125, 1/64)\} &= \mu_L\{([0,1/4] \cup [3/4,1]) \cap (1/5, 1/4)\} \\ &= \mu_L(1/5, 1/4) = 1/20. \end{aligned} \tag{4.3}$$

□

Example 4.17. Let $(\Omega, \mathcal{F}, P)$ again be $([0,1], \mathcal{B}_{[0,1]}, \mu_L)$, and let $X_1(\omega) = I\{\omega \in [0, 1/2)\}$, $X_2(\omega) = I\{\omega \in (1/4, 3/4)\}$, and $X_3(\omega) = I\{\omega \in (1/8, 1/2]\}$. If P' denotes the probability measure on $(R^3, \mathcal{B}^3)$ induced by (X_1, X_2, X_3), then

$$\begin{aligned} P'\{1,1,1\} &= \mu_L\{[0,1/2) \cap (1/4, 3/4) \cap (1/8, 1/2])\} \\ &= \mu_L(1/4, 1/2) = 1/4. \end{aligned} \tag{4.4}$$

□

Exercises

1. Let $(\Omega, \mathcal{F}, P)$ be $([0,1], \mathcal{B}_{[0,1]}, \mu_L)$, and $X_1(\omega) = \omega^2$ and $X_2(\omega) = I(\omega > 1/2)$. If P' denotes the probability measure on $(R^2, \mathcal{B}^2)$ induced by (X_1, X_2), find $P'((-\infty, .5] \times (.5, \infty))$.

2. Let $(\Omega, \mathcal{F}, P)$ be $([0,1], \mathcal{B}_{[0,1]}, \mu_L)$, and $X_1(\omega) = I(\omega \geq 1/2)$, $X_2(\omega) = I(\omega \in [1/4, 1/2) \cup [3/4, 1])$, $X_3(\omega) = I(\omega \in [1/8, 1/4) \cup [3/8, 1/2) \cup [5/8, 3/4) \cup [7/8, 1])$. If P' denotes the probability measure on $(R^3, \mathcal{B}^3)$ induced by (X_1, X_2, X_3), find $P'(\{1, 1, 0\})$.

3. Let P be a probability measure such that $P(X \in A, Y \in A) \geq P(X \in A)P(Y \in A)$ for all A in a field $\mathcal{A}$ generating the Borel sets $\mathcal{B}$. Use the monotone class theorem (Theorem 3.32) to prove that $P(X \in B, Y \in B) \geq P(X \in B)P(Y \in B)$ for all Borel sets B.

4.3 The Distribution Function of a Random Variable

Fundamental to probability theory is the distribution function of a random variable X, defined as follows.

Definition 4.18. Distribution function of a random variable *The distribution function (d.f.) of a random variable X is the function $F(x) = P\{\omega : X(\omega) \leq x\}$, $x \in R$.*

Proposition 4.19. Properties of distribution functions *A distribution function $F(x)$ satisfies the following conditions.*

1. *$F(x)$ is monotone increasing.*

2. *$F(x)$ is right-continuous at each x.*

3. *$\lim_{x\to-\infty} F(x) = 0$ and $\lim_{x\to\infty} F(x) = 1$.*

Proof. Monotonicity is apparent from the definition of a d.f. To prove right-continuity, it suffices to prove that if $x_n \downarrow x$, then $F(x_n) \to F(x)$. Note that $E_n = \{\omega : -\infty < X(\omega) \le x_n\} \downarrow E = \{\omega : -\infty < X(\omega) \le x\}$. By the continuity property of probability (Proposition 3.30), $P(E_n) \to P(E)$. But $P(E_n) = F(x_n)$ and $P(E) = F(x)$, so $F_n(x) \to F(x)$, proving that F is right-continuous at x.

To prove the limiting behavior at $\pm\infty$, we show that if $x_n \downarrow -\infty$, then $F(x_n) \to 0$, and if $x_n \uparrow \infty$, then $F(x_n) \to 1$. Observe that $\{\omega : X(\omega) \in (-\infty, x_n]\} \downarrow \emptyset$ as $x_n \to -\infty$ and $\uparrow \{\omega : -\infty < X(\omega) < \infty\}$ as $x_n \to \infty$. It follows from the continuity property of probability that $F(x_n) = P(-\infty < X(\omega) \le x_n) \to P(\emptyset) = 0$ as $x_n \to -\infty$ and $\to P(-\infty < X < \infty) = 1$ as $x_n \to \infty$. □

Remark 4.20. *The converse of Proposition 4.19 is also true. That is, any function satisfying the three conditions of Proposition 4.19 is the distribution function of a random variable X defined on some probability space $(\Omega, \mathcal{F}, P)$. We will see this in Section 4.6.*

The fact that $F(x)$ is monotone means that the left- and right-hand limits $\lim_{x\uparrow x_0} F(x)$ and $\lim_{x\downarrow x_0} F(x)$ both exist (why?). Thus, $F(x)$ cannot exhibit the kind of pathological behavior that, say, $\sin(1/x)$ does as $x \to 0$ (Figure 4.4). There are only two possibilities: the left and right limits either agree or disagree. If they agree, then $F(x)$ is continuous at x_0. If they disagree, then $F(x)$ has a *jump discontinuity* at x_0. A d.f. can have infinitely many jumps, but only countably many, as we will see in Proposition 4.21.

Proposition 4.21. Discontinuities of distribution functions *A distribution function F is continuous at x if and only if $P(X = x) = 0$. Furthermore, the collection of discontinuities of the distribution function $F(x)$ of a random variable X is countable.*

Proof. To prove the first statement, note that $P(X = x) = P(X \le x) - P(X < x) = F(x) - P(\cup_n\{X \le x_n\})$, where x_n is any sequence such that $x_n \uparrow x$. By the continuity property of probability (Proposition 3.30) and the fact that $E_n = \{\omega : X(\omega) \le x_n\} \uparrow E = \{\omega : X(\omega) < x\}$, $P(X = x) = 0$ is equivalent to $F(x) - \lim_n F(x_n) = 0$. This holds for any sequence $x_n \uparrow x$, so $P(X = x) = 0$ is equivalent to $F(x) - \lim_n F(x_n) = 0$ for **every** sequence $x_n \uparrow x$. That is, $P(X = x) = 0$ if and only if F is left-continuous at x. But F is right-continuous at x by part 2 of Proposition 4.19, so $P(X = x) = 0$ is equivalent to F being continuous at x.

We next prove that the set of discontinuities of F is countable. Because the smallest and largest possible values of $F(x)$ are 0 and 1, respectively, there cannot be more than n jumps of size $1/n$ or more. Thus, the set of jumps of size $1/n$ or more is countable. The set of all jumps is $\cup_n\{\text{jumps of size } 1/n \text{ or more}\}$, the countable union of countable sets. By Proposition 2.6, the set of discontinuities of F is countable. □

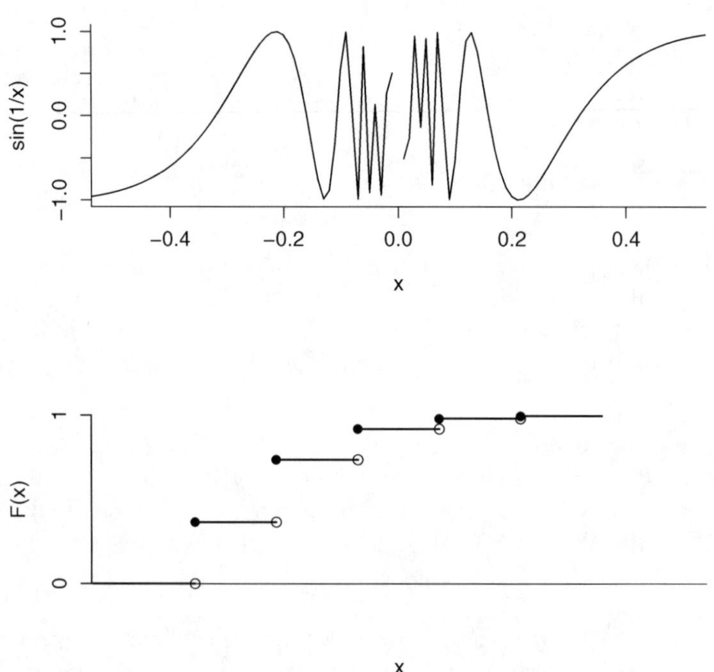

Figure 4.4: $f(x) = \sin(1/x)$ exhibits bizarre behavior with no limit as $x \to 0$ (top panel). A d.f. $F(x)$ cannot exhibit such behavior; the only type of discontinuity of $F(x)$ is a jump (bottom panel).

Remember from Section 4.1.2 that the probability measure P on $\mathcal{F}$ defines a probability measure P' on the Borel sets $\mathcal{B}$ of R via $P'(B) = P\{\omega : X(\omega) \in B\}$. We can calculate probabilities using either P on $\mathcal{F}$ or P' on $\mathcal{B}$; they give the same answer. For $B = (-\infty, x]$, $P'(B) = F(x)$. It turns out that once we know $F(x)$, we know $P'(B) = P\{\omega : X(\omega) \in B\}$ for each Borel set B, as the following result shows.

Proposition 4.22. Probability measure induced by X is determined by its d.f. *The probability measure $P'(B)$ on $(R, \mathcal{B})$ induced by a random variable X is uniquely determined by the distribution function $F(x)$ of X.*

Proof. The distribution function defines $P'(C)$ for all sets C of the form $(-\infty, a]$, $a \in R$. For C of the form $(a, b]$, $P'(C) = F(b) - F(a)$, so $P'(C)$ is also determined by F. For C of the form (b, ∞), $P'(C) = 1 - F(b)$ is determined by F. By countable additivity of probability measures, $P(C)$ is determined by F for all sets C in the field $\mathcal{F}_0$ of unions of a finite number of disjoint sets of the form $(a, b]$, $(-\infty, a]$, or (b, ∞) (see Example 3.8 of Section 3.2). But $\mathcal{F}_0$ generates the Borel sigma-field, so Proposition 3.33 shows that the probability $P'(B)$ of any Borel set B is uniquely determined from the probabilities of sets $C \in \mathcal{F}_0$, which we have shown are determined from F. This shows that the distribution function $F(x)$ of X completely and uniquely determines the probability measure P' induced by X. □

Proposition 4.23. Continuous distribution functions are uniformly continuous *If $F(x)$ is a distribution function that is continuous for all $x \in R$, then F is uniformly continuous.*

Proof. Let $\epsilon > 0$ be given. We must find δ such that $|y-x| < \delta \Rightarrow |F(y)-F(x)| < \epsilon$. Choose a positive number A such that $P(|X| \geq A) < \epsilon/2$. On the interval $[-A, A]$, F is uniformly continuous, so there exists a δ such that $|F(y) - F(x)| < \epsilon/2$ whenever $-A \leq x \leq A$, $-A \leq y \leq A$ and $|y - x| < \delta$. We will show that this same value of δ works for any $x \in R$ and $y \in R$, not just on the interval $[-A, A]$. Without loss of generality, assume that $\delta < 2A$, because we can always decrease δ.

Let $x \in R$ and $y \in R$ be any real numbers such that $|y - x| < \delta$, and without loss of generality, take $x < y$. Then $|F(y) - F(x)| < \epsilon/2 < \epsilon$ if x and y are both between $-A$ and A. On the other hand, if x and y are both less than $-A$, then

$$|F(y) - F(x)| \leq F(y) + F(x) < \epsilon/2 + \epsilon/2 = \epsilon.$$

A similar argument works if x and y both exceed A. If $x < -A$ and $-A \leq y \leq A$, then

$$\begin{aligned} |F(y) - F(x)| &= |F(y) - F(-A) + F(-A) - F(x)| \\ &\leq |F(y) - F(-A)| + |F(-A) - F(x)| \\ &< \epsilon/2 + \epsilon/2 = \epsilon. \end{aligned} \tag{4.5}$$

A similar argument works if $-A \leq x \leq A$ and $y > A$. We need not consider the case $x < -A$ and $y > A$ because we took $\delta < 2A$. This completes the proof that F is uniformly continuous. □

Exercises

1. Let $(\Omega, \mathcal{F}, P)$ be $((0,1), \mathcal{B}_{(0,1)}, \mu_L)$ and $X(\omega) = \{-\ln(1-\omega)\}^{1/\beta}$. Show that X has the Weibull distribution function $F(x) = 1 - \exp(-x^\beta)$.

2. Let $(\Omega, \mathcal{F}, P)$ be $([0,1], \mathcal{B}_{[0,1]}, \mu_L)$ and $X(\omega) = \omega/(1+\omega)$. Find the distribution function for X.

3. Prove that if $F_1(r) = F_2(r)$ for every rational number r, then $P_1'(B) = P_2'(B)$ for all Borel sets B.

4. Prove that any finite monotone function (not just a d.f.) has only countably many jumps.

5. Let $F(x)$ be a distribution function. For any constant $c \in (0,1)$, prove that the set of x such that $F(x) = c$ is either $\emptyset$, a single point, or an interval of the form $[a, b)$ or $[a, b]$, where $a < \infty$, $a < b \leq \infty$.

6. In the proof of Proposition 4.23, fill in some missing details:
   (a) How did we know that F is uniformly continuous on $[-A, A]$?
   (b) Show that $|F(y) - F(x)| < \epsilon$ for x and y both exceeding A.
   (c) Show that $|F(y) - F(x)| < \epsilon$ for $-A \leq x \leq A$ and $y > A$.

7. Suppose that $\theta \in \Theta$ is a parameter and X is a random variable whose distribution function $F_\theta(x)$ is the same for all $\theta \in \Theta$. Use the monotone class theorem (Theorem 3.32) to prove that, for each Borel set B, $P_\theta(X \in B)$ is the same for all $\theta \in \Theta$.

8. Let $r_1, r_2, \ldots$ be an enumeration of the rational numbers. Let $F(x) = \sum_{i=1}^{\infty}(1/2^i)I(x \geq r_i)$. Prove that $F(x)$ is a distribution function that has a jump discontinuity at every rational number. Can you concoct a distribution function F that has a jump discontinuity at every irrational number?

4.4 The Distribution Function of a Random Vector

The multivariate analog of a univariate distribution function is a multivariate distribution function, defined as follows.

Definition 4.24. Distribution function of a random vector *The multivariate distribution function (also called the joint distribution function)* $F(x_1, \ldots, x_k)$ *of* $(X_1, \ldots, X_k)$ *is defined to be* $P\{\omega : X_1(\omega) \le x_1, \ldots, X_k(\omega) \le x_k\}$.

The following example illustrates the calculation of a bivariate distribution function.

Example 4.25. Let $(\Omega, \mathcal{F}, P) = ([0,1], \mathcal{B}_{[0,1]}, \mu_L)$, and define $X_1(\omega) = I(\omega \in [1/2, 3/4])$ and $X_2(\omega) = \omega^2$. Consider the joint distribution function $F(x_1, x_2)$ for (X_1, X_2). If $x_1 < 0$, then $F(x_1, x_2) = 0$ because X_1 does not take negative values. If $0 \le x_1 < 1$, then $X_1 \le x_1$ is equivalent to $X_1 = 0$, which is equivalent to $\omega \in [0, 1/2) \cup (3/4, 1]$. Also, $X_2 \le x_2$ is impossible if $x_2 < 0$; if $x_2 \ge 0$, then $X_2 \le x_2$ is equivalent to $\omega \in [0, \min(1, x_2^{1/2})]$. Therefore, if $0 \le x_1 < 1$ and $x_2 \ge 0$,

$$\{X_1(\omega) \le x_1\} \cap \{X_2(\omega) \le x_2\} = \left\{\omega \in \Big([0,1/2) \cup (3/4,1]\Big) \cap [0, \sqrt{x_2}]\right\}. \tag{4.6}$$

The set on the right side of Equation (4.6) is

$$\begin{cases} [0, \sqrt{x_2}] & \text{if } 0 \le x_2 < 1/4. \\ [0, 1/2) & \text{if } 1/4 \le x_2 \le 9/16 \\ [0,1/2) \cup (3/4, \sqrt{x_2}] & \text{if } 9/16 < x_2 \le 1. \end{cases}$$

The Lebesgue measures of the three sets above are $x_2^{1/2}$, $1/2$, and $1/2 + x_2^{1/2} - 3/4 = x_2^{1/2} - 1/4$, respectively. If $x_1 \ge 1$, then the left side of Equation (4.6) is $X_2 \le x_2$, which has Lebesgue measure $\min(1, x_2^{1/2})$ if $0 \le x_2 < \infty$. Therefore, the distribution function for (X_1, X_2) is

$$F(x_1, x_2) = \begin{cases} 0 & \text{if } x_1 < 0 \text{ or } x_2 < 0 \\ \sqrt{x_2} & \text{if } 0 \le x_1 < 1,\ 0 \le x_2 < 1/4. \\ 1/2 & \text{if } 0 \le x_1 < 1,\ 1/4 \le x_2 \le 9/16 \\ \min(1, \sqrt{x_2}) - 1/4 & \text{if } 0 \le x_1 < 1,\ 9/16 < x_2 < \infty \\ \min(1, \sqrt{x_2}) & \text{if } 1 \le x_1 < \infty,\ 0 \le x_2 < \infty. \end{cases}$$

This example illustrates that careful bookkeeping is sometimes required to calculate a multivariate distribution function. □

The probability measure P' on R^k induced by $(X_1, \ldots, X_k)$ is completely determined from its multivariate distribution function:

Proposition 4.26. Probability measure induced by random vector X is determined by its d.f. *The probability measure* $P'(B)$ *on* $(R^k, \mathcal{B}^k)$ *induced by a random vector* $\mathbf{X}$ *is uniquely determined by the distribution function* $F(\mathbf{x})$ *of* $\mathbf{X}$.

The proof is left as an exercise.

In one dimension, a distribution function is right-continuous. The multivariate analog is that $F(x_1, \ldots, x_k)$ is *continuous from above*, meaning that if x_{in} approaches x_i from the right as $n \to \infty$, $i = 1, \ldots, k$, then $F(x_{1n}, \ldots, x_{kn}) \to F(x_1, \ldots, x_k)$.

Proposition 4.27. Multivariate analog of right-continuity *A multivariate distribution function is continuous from above.*

Proof. It is sufficient to show that $F(x_{1n},\ldots,x_{kn}) \rightarrow F(x_1,\ldots,x_k)$ for $x_{in} \downarrow x_i$, so assume that x_{in} is a decreasing sequence. Then $E_n = (-\infty, x_{1n}] \times (-\infty, x_{2n}] \times \ldots \times (-\infty, x_{kn}]$ decreases to $E = (-\infty, x_1] \times (-\infty, x_2] \times \ldots \times (-\infty, x_k]$ as $n \rightarrow \infty$, so the result follows from the continuity property of probability (Proposition 3.30). □

In one dimension, a d.f. $F(x)$ has only countably many discontinuities, so it is natural to explore whether this is true for multivariate distribution functions as well. The key question is then under what circumstances is a multivariate d.f. continuous? We already know that it must be continuous from above, so we could look for conditions under which it is continuous from below; i.e. $F(x_{1n},\ldots,x_{kn}) \rightarrow F(x_1,\ldots,x_k)$ for every sequence such that x_{in} approaches x_i from the left, or, equivalently, for every sequence such that $x_{in} \uparrow x_i$. Of course a general function f can be continuous from above and below, but not continuous, even if $k = 2$. For example:

Example 4.28. Continuous from above and below, but not continuous *Define a function f with domain $\{(x_1,x_2) : \ x_1 > 0, \ x_2 > 0\} \cup \{(0,0)\}$ as follows:*

$$f(x_1,x_2) = \begin{cases} +1 & \text{if } x_1x_2 > 0 \text{ or } (x_1,x_2) = (0,0) \\ -1 & \text{if } x_1x_2 < 0. \end{cases}$$

Then f is continuous at $(0,0)$ from above and below, but does not have the same limit if we approach $(0,0)$ from the "northwest" $(x_1 < 0, \ x_2 > 0)$ as it does if we approach $(0,0)$ from the "northeast" $(x_1 > 0, \ x_2 > 0)$, for example. Therefore, $f(x_1,x_2)$ is not continuous at the origin. □

The key difference between a function like f of Example 4.28 and a multivariate d.f. F is that F is increasing in each argument. This precludes counterexamples like Example 4.28.

Proposition 4.29. Continuous from below implies continuous for d.f.s *A multivariate d.f. is continuous at $(x_1,\ldots,x_k)$ if and only if it is continuous from below at $(x_1,\ldots,x_k)$.*

Proof. We know from Proposition 4.27 that F is continuous from above, so it suffices to prove that for a multivariate d.f., continuity from above and below is equivalent to continuity. Of course continuity implies continuity from above and below, so it suffices to prove the other direction. Suppose that F is continuous from above and below. If F were not continuous, then there would be an $\epsilon > 0$ and a sequence $\mathbf{x}_n = (x_{1n},\ldots,x_{kn})$ converging to $\mathbf{x}$ such that either $F(\mathbf{x}_n) \leq F(\mathbf{x}) - \epsilon$ for infinitely many n, or $F(\mathbf{x}_n) \geq F(\mathbf{x}) + \epsilon$ for infinitely many n. If $F(\mathbf{x}_n) \leq F(\mathbf{x}) - \epsilon$ for infinitely many n, then $F(\mathbf{x}'_n) \leq F(\mathbf{x}) - \epsilon$ for infinitely many n, where each component of $\mathbf{x}'_n$ is the minimum of the corresponding components of $\mathbf{x}_n$ and $\mathbf{x}$ (see Figure 4.5). Similarly, if $F(\mathbf{x}_n) \geq F(\mathbf{x}) + \epsilon$ for infinitely many n, then $F(\mathbf{x}''_n) \geq F(\mathbf{x}) + \epsilon$ for infinitely many n, where each component of $\mathbf{x}''_n$ is the maximum of the corresponding components of $\mathbf{x}_n$ and $\mathbf{x}$. Therefore, if there is any sequence violating continuity, then there is a sequence violating either continuity from below or continuity from above. This proves that for a multivariate d.f., continuity from above and below implies continuity. □

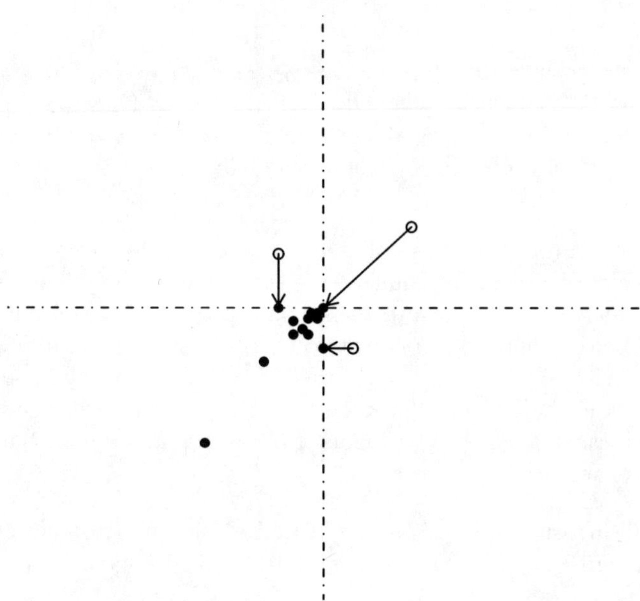

Figure 4.5: For any sequence (x_{1n}, x_{2n}) approaching (x_1, x_2) such that $F(x_{1n}, x_{2n}) \le F(x_1, x_2) - \epsilon$ for infinitely many n, we can find a sequence (x'_{1n}, x'_{2n}) approaching (x_1, x_2) from the "southwest" such that $F(x'_{1n}, x'_{2n}) \le F(x_1, x_2) - \epsilon$ for infinitely many n. Take $x'_{in} = \min(x_{in}, x_i)$, $i = 1, 2$, which projects points from other quadrants onto the southwest quadrant.

Proposition 4.30. Characterization of continuity of a multivariate d.f. at a point *Let F be a multivariate distribution function. For a given $(x_1, \ldots, x_k)$, let $p_i = P(X_1 \le x_1, \ldots, X_{i-1} \le x_{i-1}, X_i = x_i, X_{i+1} \le x_{i+1}, \ldots, X_k \le x_k)$. Then F is continuous at $(x_1, \ldots, x_k)$ if and only if $p_i = 0$ for all $i = 1, \ldots, k$.*

Proof. Let $p = P[(\cap_i \{X_i \le x_i\}) \cap (\cup_i \{X_i = x_i\})]$. Then $p = P(\cap_i \{X_i \le x_i\}) - P(\cap_i \{X_i < x_i\}) = F(x_1, \ldots, x_k) - \lim_{x_{1n} \to x_1^-, \ldots x_{kn} \to x_k^-} F(x_{1n}, \ldots, x_{kn})$. Therefore $p = 0$ if and only if F is continuous from below. By Proposition 4.29, F is continuous if and only if $p = 0$. Therefore, it suffices to establish that $p = 0$ is equivalent to $p_1 = 0, p_2 = 0, \ldots, p_k = 0$. Note that $p = P(\cup_i \{X_1 \le x_1, \ldots, X_{i-1} \le x_{i-1}, X_i = x_i, X_{i+1} \le x_{i+1}, \ldots, X_k \le x_k\})$, so countable subadditivity ensures that $p \le \sum_i p_i$. Therefore,

$$p_i \le p \le \sum_{i=1}^{k} p_i, \; i = 1, \ldots, k,$$

from which it follows that $p = 0$ if and only if $p_i = 0$ for all $i = 1, \ldots, k$. □

Corollary 4.31. Sufficient condition for continuity *A sufficient condition for continuity of the multivariate distribution function F at $(x_1, \ldots, x_k)$ is that $P(X_i = x_i) = 0$ for $i = 1, \ldots, k$.*

Proof. The stated condition clearly implies that $P(X_1 \le x_1, \ldots, X_{i-1} \le x_{i-1}, X_i = x_i, X_{i+1} \le x_{i+1}, \ldots, X_k \le x_k) = 0$. Now apply Proposition 4.30. □

Proposition 4.21 implies that there are only countably many x such that $P(X_i = x) > 0$ because each such x is a discontinuity point of the univariate d.f. of X_i. We call the line $x_i = x$ an axis line of discontinuity of a multivariate d.f. F if F is discontinuous at $(x_1, \ldots, x_i = x, \ldots, x_k)$ for one or more values of $x_1, \ldots, x_{i-1}, x_{i+1}, \ldots, x_k$. Let $\mathcal{L}_i$ denote the axis lines of discontinuity . Then Corollary 4.31 implies the following result (see Figure 4.6).

Proposition 4.32. Countably many axis lines of discontinuity *The collection* $\cup_i \mathcal{L}_i$ *of axis lines of discontinuity of a multivariate d.f. is countable.*

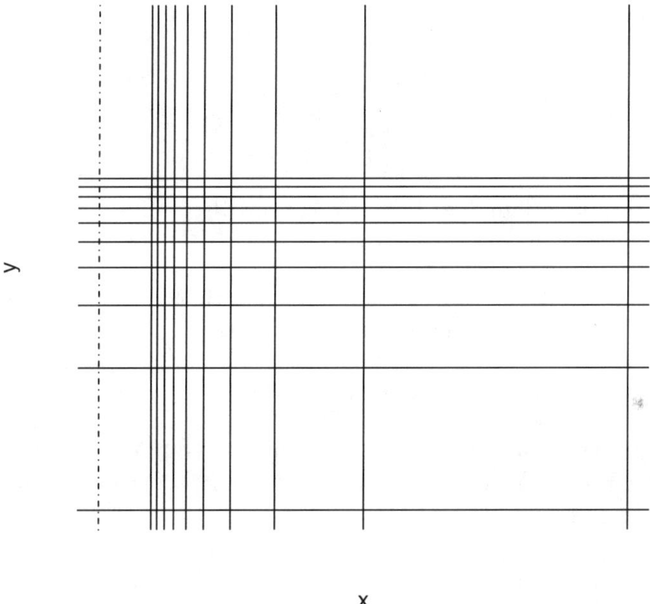

Figure 4.6: All of the discontinuities of a bivariate distribution function must lie on countably many horizontal/vertical lines.

Notice that Proposition 4.32 does not say that there are only countably many **points** of discontinuity of F. That is not true. For instance, let X_1 be Bernoulli (1/2) and independent of X_2 (again, the reader is assumed to have some familiarity with independence from elementary probability), which is standard normal; then $F(x_1, x_2) = \{(1/2)I(0 \le x_1 < 1) + I(x_1 \ge 1)\}\Phi(x_2)$, where Φ is the standard normal distribution function. Then F is discontinuous at the uncountably many points $(0, x_2)$ for all $x_2 \in R$ and $(1, x_2)$ for all $x_2 \in R$, though all points of discontinuity lie on only two lines, $x_1 = 0$ and $x_1 = 1$.

From a multivariate distribution function we can obtain the *marginal distribution function* for each X_i making up $\mathbf{X}$. For example, if $\mathbf{X} = (X_1, X_2)$ has joint distribution function $F(x_1, x_2)$, the marginal distribution function for X_1 is $P(X_1 \le x_1) = P(X_1 \le x_1, -\infty < X_2 < \infty) = \lim_{A \uparrow \infty} F(x_1, A)$. Similarly, if $\mathbf{X} = (X_1, \ldots, X_k)$, the joint distribution function for any subset of random variables can be obtained from the joint distribution function $F(x_1, \ldots, x_k)$ by setting the remaining x values to A and letting $A \uparrow \infty$. For instance, the joint distribution function for (X_1, X_2) is $\lim_{A\uparrow\infty} F(x_1, x_2, A, \ldots, A)$, and the marginal distribution for X_1 is $\lim_{A\uparrow\infty} F(x_1, A, \ldots, A)$.

Exercises

1. Let $(\Omega, \mathcal{F}, P)$ be $([0,1], \mathcal{B}_{[0,1]}, \mu_L)$, $X_1(\omega) = I(\omega \geq 1/2)$, and $X_2(\omega) = \omega$. What is the joint distribution function for (X_1, X_2)?

2. Let $(\Omega, \mathcal{F}, P)$ be $([0,1], \mathcal{B}_{[0,1]}, \mu_L)$, $X_1(\omega) = |\omega - 1/4|$, and $X_2(\omega) = \omega - 1/4$. What is the joint distribution function for (X_1, X_2)?

3. If $F(x, y)$ is the distribution function for (X, Y), find expressions for:

   (a) $P(X \leq x, Y > y)$.

   (b) $P(X \leq x, Y \geq y)$.

   (c) $P(X \leq x, Y = y)$.

   (d) $P(X > x, Y > y)$.

4. Use the monotone class theorem (Tbeorem 3.32) to prove Proposition 4.26.

5. Let $F(x, y)$ be a bivariate distribution function. Prove that $\lim_{A\uparrow\infty} F(x, A) = P(X \leq x)$ and $\lim_{A\uparrow\infty} F(A, y) = P(Y \leq y)$.

6. $\uparrow$ In the previous problem, can we take a sequence that tends to ∞, but is not monotone (such as $A_n = n^{1/2}$ if n is even and $A_n = \ln(n)$ if n is odd)? In other words, is $\lim_{n\to\infty} F(x, A_n) = P(X \leq x)$ for A_n not monotone, but still tending to ∞? If so, prove it. If not, find a counterexample.

7. To obtain the marginal distribution function $F(x_1)$ for X_1 from the joint distribution function for $(X_1, \ldots, X_k)$, can we use different A_is increasing to ∞? In other words, is $\lim_{A_2\uparrow\infty,\ldots,A_k\uparrow\infty} F(x_1, A_2, \ldots, A_k) = P(X_1 \leq x_1)$? Justify your answer.

4.5 Introduction to Independence

4.5.1 Definitions and Results

This section covers the very broad topic of independence, which includes independence of an arbitrary number of (even countably or uncountably infinitely many) events or random variables. Because there are so many different scenarios, we build from simple to more complicated situations. After doing that, we present an alternative definition using sigma-fields that unifies the theory. Keep in mind the main results we wish to prove, namely that if X_t are independent random variables, then functions $f_t(X_t)$ are also independent, as are vectors of non-overlapping components. For instance, if $X_1, X_2, \ldots$ are independent, then so are the collections $\{X_1, X_2, X_3\}, \{X_4, X_5\}$, and $\{X_8, X_{14}\}$, for example.

In elementary probability, we developed an intuitive understanding of independence starting with Bayes formula for conditional probability: if A_1 and A_2 are two events with $P(A_1) > 0$, the *conditional probability of* A_2 *given* A_1 is $P(A_2 \mid A_1) = P(A_1 \cap A_2)/P(A_1)$ (see Exercise 4 of Section 3.4). If this conditional probability is the same as the unconditional probability of A_2, then knowing whether A_1 occurred does not change the probability that

A_2 occurred, and A_1 and A_2 are said to be independent events. The condition $P(A_2 \mid A) = P(A_2)$ is equivalent to

$$P(A_1 \cap A_2) = P(A_1)P(A_2), \tag{4.7}$$

except that Equation (4.7) does not require $P(A_1) > 0$. Therefore, we adopt (4.7) as the preferred definition of independence of two events A_1 and A_2.

We would like to extend this definition to $n \geq 2$ events. It may seem like the natural extension of independence to n events $A_1, \ldots, A_n$ is to require $P(A_1 \cap \ldots \cap A_n) = P(A_1) \ldots P(A_n)$, but this is not sufficient. For example, roll a die and let A_1 be the event that the number is odd, A_2 be the event that the number is even, and A_3 be the empty set $\emptyset$. Then $P(A_1 \cap A_2 \cap A_3) = 0 = P(A_1)P(A_2)P(A_3)$, yet A_1 and A_2 are mutually exclusive, so knowledge of whether A_1 occurred gives us complete knowledge of whether A_2 occurred. Indeed, $0 = P(A_1 \cap A_2) \neq P(A_1)P(A_2)$. Therefore, we need to impose additional conditions in the definition of independence of multiple events.

Definition 4.33. Independence of events

1. *A finite collection of events $A_1, \ldots, A_n$ is defined to be independent if*

$$P(A_{i_1} \cap A_{i_2} \cap \ldots \cap A_{i_k}) = P(A_{i_1})P(A_{i2}) \ldots P(A_{i_k}) \tag{4.8}$$

   for every subcollection $\{i_1, \ldots, i_k\}$, $k = 1, \ldots, n$.

2. *A countably or uncountably infinite collection of events is defined to be independent if each finite subcollection is independent.*

Example 4.34. Pairwise independence is insufficient This example shows that independence of each pair (A_i, A_j) in the sense of Equation (4.7) is not sufficient to conclude independence of $A_1, \ldots, A_n$. Flip two coins, and let $A_i - \{\text{coin } i \text{ is heads}\}$, $i = 1, 2$, and $A_3 = \{\text{exactly one of the two coins is heads}\}$. It is clear that A_1 and A_2 are independent. Also, A_1 and A_3 are independent because $P(A_1 \cap A_3) = P(\text{coin 1 is heads, coin 2 is tails}) = 1/4 = P(\text{coin 1 is heads})P(\text{exactly one of the first two coins is heads})$. Likewise, A_2 and A_3 are independent. On the other hand, knowledge of whether A_1 occurred and whether A_2 occurred gives us complete information about whether A_3 occurred. Definition 4.33 correctly declares A_1, A_2, A_3 not to be independent because $P(A_1 \cap A_2 \cap A_3) = 0 \neq P(A_1)P(A_2)P(A_3)$. □

It is an important fact that the product rule for computing the probability of independent events holds for a countably infinite number of events as well.

Proposition 4.35. Product rule for countable number of independent sets *If $A_1, A_2, \ldots$ is a countably infinite set of independent events, then $P(\cap_{i=1}^{\infty} A_i) = \prod_{i=1}^{\infty} P(A_i)$.*

Proof. Let $B_n = \cap_{i=1}^{n} A_i$. Then $B_n \downarrow B = \cap_{i=1}^{\infty} A_i$, so the continuity property of probability (Proposition 3.30) implies that $P(B_n) \to P(B)$. But $P(B_n) = \prod_{i=1}^{n} P(A_i)$ because of the independence of the A_i. Therefore,

$$P(B) = \lim_{n\to\infty} P(B_n) = \lim_{n\to\infty} \prod_{i=1}^{n} P(A_i) = \prod_{i=1}^{\infty} P(A_i),$$

completing the proof. □

We now extend the definition of independence from events to random variables $X_1, \ldots, X_n$. Intuitively, independence of random variables $X_1, \ldots, X_n$ should mean independence of events associated with these random variables. That is, $P(X_{i_1} \in B_1 \cap \ldots \cap X_{i_k} \in B_k) = P(X_{i_1} \in B_1) \ldots P(X_{i_k} \in B_k)$ for every subcollection $X_{i_1}, \ldots, X_{i_k}$ and all one-dimensional Borel sets $B_1, \ldots, B_k$. It suffices that $P(X_1 \in B_1 \cap \ldots \cap X_n \in B_n) = P(X_1 \in B_1) \ldots P(X_n \in B_n)$ for all one-dimensional Borel sets because we can always write $P(X_{i_1} \in B_1 \cap \ldots \cap X_{i_k} \in B_k)$ as an intersection involving all $X_1, \ldots, X_n$ by augmenting the intersection with terms $\{X_i \in R\}$ for $i \notin \{i_1, \ldots, i_k\}$. For instance, if we are trying to determine whether $X_1, \ldots, X_5$ are independent, we can write the event

$$X_1 \in B_1 \cap X_2 \in B_2 \cap X_3 \in B_3 \tag{4.9}$$

involving the subcollection X_1, X_2, and X_3 as

$$X_1 \in B_1 \cap X_2 \in B_2 \cap X_3 \in B_3 \cap X_4 \in R \cap X_5 \in R, \tag{4.10}$$

and the probability of Expression (4.10) is $P(X_1 \in B_1 \cap X_2 \in B_2 \cap X_3 \in B_3)(1)(1)$. Therefore, the probability of Expression (4.9) is a product of its individual event probabilities if and only if the probability of Expression (4.10) is a product of its individual event probabilities. This motivates the following definition.

Definition 4.36. Independence of random variables

1. *A finite collection of random variables $X_1, \ldots, X_n$ is defined to be independent if $P(X_1 \in B_1, \ldots, X_n \in B_n) = \prod_{i=1}^n P(X_i \in B_i)$ for all one-dimensional Borel sets $B_1, \ldots, B_n$.*

2. *A countably or uncountably infinite collection of random variables is defined to be independent if each finite subcollection satisfies condition 1.*

It is helpful to have another way of checking for independence because Definition 4.36 requires consideration of all Borel sets. The following is a very useful result.

Proposition 4.37. Independence $\Leftrightarrow$ distribution function factors *$X_1, \ldots, X_n$ are independent by Definition 4.36 if and only if*

$$P(X_1 \le x_1, X_2 \le x_2, \ldots, X_n \le x_n) = F_1(x_1)F_2(x_2) \ldots F_n(x_n)$$

for all $x_1, x_2, \ldots, x_n$, where $F_1, F_2, \ldots, F_n$ are the d.f.s of $X_1, X_2, \ldots, X_n$.

Proof. It is clear that independence by Definition 4.36 implies $P(X_1 \le x_1, X_2 \le x_2, \ldots, X_n \le x_n) = F_1(x_1)F_2(x_2) \ldots F_n(x_n)$ for all $x_1, x_2, \ldots, x_n$ because $(-\infty, x_1], (-\infty, x_2], \ldots, (-\infty, x_n]$ are Borel sets. Therefore, it suffices to prove that if $P(X_1 \le x_1, X_2 \le x_2, \ldots, X_n \le x_n) = F_1(x_1)F_2(x_2) \ldots F_n(x_n)$ for all $x_1, x_2, \ldots, x_n$, then $X_1, \ldots, X_n$ are independent by Definition 4.36.

Step 1. Let $\mathcal{A}_1$ be the collection of Borel sets A_1 such that $P(X_1 \in A_1 \cap X_2 \le x_2 \cap \ldots \cap X_n \le x_n) = P(X_1 \in A_1)P(X_2 \le x_2) \ldots P(X_n \le x_n)$ for all $x_2, \ldots, x_n$. Then $\mathcal{A}_1$ contains all sets of the form $(-\infty, x_1]$ by assumption. It is an exercise to show that $\mathcal{A}_1$ contains the field $\mathcal{F}_0$ of Example 3.8, and that $\mathcal{A}_1$ is closed under increasing or decreasing sets. By the monotone class theorem (Theorem 3.32), $\mathcal{A}_1$ contains the smallest sigma-field containing $\mathcal{F}_0$, namely the Borel sets of R.

Step 2. Let A_1 be an arbitrary Borel set and define $\mathcal{A}_2$ to be the collection of Borel sets A_2 such that $P(X_1 \in A_1 \cap X_2 \in A_2 \cap X_3 \le x_3 \cap \ldots \cap X_n \le x_n) = P(X_1 \in A_1)P(X_2 \in$

$A_2)P(X_3 \le x_3)\ldots P(x_n \le x_n)$ for all $x_3, \ldots, x_n$. From what we proved in step 1, $\mathcal{A}_2$ contains all sets of the form $(-\infty, x_2]$. By the same technique that proved step 1, we can prove that $\mathcal{A}_2$ contains the field $\mathcal{F}_0$ of Example 3.8, and that it is a monotone class. By the monotone class theorem, it must contain the Borel sets.

Step 3. Let A_1 and A_2 be arbitrary Borel sets, and define $\mathcal{A}_3$ to be the collection of Borel sets A_3 such that $P(X_1 \in A_1 \cap X_2 \in A_2 \cap X_3 \in A_3 \cap X_4 \le x_4 \cap \ldots \cap X_n \le x_n) = P(X_1 \in A_1)P(X_2 \in A_2)P(X_3 \in A_3)P(X_4 \le x_4)\ldots P(X_n \le x_n)$ for all $x_4, \ldots, x_n$. The above technique shows that $\mathcal{A}_3$ contains all Borel sets.

Continuing in this fashion proves that $P(X_1 \in A_1 \cap \ldots \cap X_n \in A_n) = P(X_1 \in A_1)\ldots P(X_n \in A_n)$ for all Borel sets $A_1, \ldots, A_n$. □

Note that there is a more immediate proof if we were willing to accept the idea that, for given marginal distributions $F_1, \ldots, F_n$, there exist independent random variables $X_1, \ldots, X_n$ with those marginal distributions. In that case, we would know that there exist independent random variables with joint distribution function $F_1(x_1)F_2(x_2)\ldots F_n(x_n)$, and because the joint distribution function completely determines $P(X_1 \in B_1, \ldots, X_n \in B_n)$ for any Borel sets $B_1, \ldots, B_n$ (Proposition 4.26), this would prove the result. However, we do not establish the existence of independent random variables with given distribution functions until Section 4.6.

Example 4.38. Independent base 2 digits from randomly picking a number in $[0,1]$ Let $(\Omega, \mathcal{F}, P)$ be $([0,1], \mathcal{B}_{[0,1]}, \mu_L)$, and let $X_1, X_2, \ldots$ be the digits in the base 2 representation of ω. That is, $\omega = 0.X_1X_2\ldots = X_1/2^1 + X_2/2^2 + \ldots$, so $X_1(\omega)$ tells whether ω is in the left or right half of $[0,1]$, $X_2(\omega)$ tells whether ω is in the left or right half of that half, etc. The event

$$\{X_1 = x_1, \ldots, X_n = x_n\} \tag{4.11}$$

corresponds to ω being in a specific interval of width $1/2^n$, so the Lebesgue measure of event (4.11) is $1/2^n$. Also, each event $\{X_i = x_i\}$ corresponds to the union of $2^{(i-1)}$ disjoint intervals of width $1/2^i$, so its Lebesgue measure is $2^{(i-1)}/2^i = 1/2$ by countable additivity. Accordingly, the probability of (4.11) is the product of probabilities of the separate X_is. This actually shows that $X_{i_1}, \ldots, X_{i_n}$ are independent for any n and $i_1, \ldots, i_n$ because it shows that $X_1, \ldots, X_N$ are independent, where $N = \max(i_1, \ldots, i_n)$. Because the X_is corresponding to any finite set of indices are independent, $X_1, X_2, \ldots$ are independent. □

Proposition 4.39. Univariate functions of independent r.v.s are independent

1. *If $X_1, \ldots, X_n$ are independent and $f_1, \ldots, f_n$ are Borel functions, then $f_1(X_1), \ldots, f_n(X_n)$ are independent.*

2. *If X_t, $t \in T$, is a countably or uncountably infinite collection of independent random variables and f_t, $t \in T$ are Borel functions, then $f_t(X_t), t \in T$ are independent random variables.*

Proof. For item 1, let $B_1, \ldots, B_n$ be arbitrary Borel sets. Then $\cap\{f_i(X_i) \in B_i\} = \cap\{X_i \in f_i^{-1}(B_i)\} = \{X_i \in A_i\}$, where $A_i = f_i^{-1}(B_i)$ is a one-dimensional Borel set because f_i is a Borel function. It follows from this and independence of the X_i that $P[\cap\{f_i(X_i) \in B_i\}] = P[\cap\{X_i \in A_i\}] = \prod_{i=1}^n P\{X_i \in A_i\} = \prod P\{f_i(X_i) \in B_i\}$. Therefore, $f_1(X_1), \ldots, f(X_n)$ are independent random variables.

Item 2 follows from the fact that every finite subcollection is independent by the proof of item 1. □

Example 4.40. Independence of transformed r.v.s In statistics we sometimes apply transformations to data $X_1, \ldots, X_n$ to make their distribution more symmetric. For instance, if $X_1, \ldots, X_n$ appear skewed to the right, we might apply the log transformation $f(X_i) = \ln(X_i)$. Proposition 4.39 ensures that if the original observations $X_1, \ldots, X_n$ are independent, so are the transformed observations. □

We next proceed to the concept of independence of random vectors $\mathbf{X}_1, \ldots, \mathbf{X}_n$ with possibly different dimensions $k_1, \ldots, k_n$. The natural extension of Definition 4.36 is the following.

Definition 4.41. Independence of finite dimensional random vectors

1. *A finite collection of random vectors $\mathbf{X}_1, \ldots, \mathbf{X}_n$ of dimensions $k_1, \ldots, k_n$ is defined to be independent if $P(\{\mathbf{X}_1 \in B_1\} \cap \ldots \cap \{\mathbf{X}_n \in B_n\}) = \prod_{i=1}^n P(\mathbf{X}_i \in B_i)$ for arbitrary Borel sets $B_1, \ldots, B_n$ of respective dimensions $k_1, \ldots, k_n$.*

2. *A countably or uncountably infinite collection of random vectors of finite dimensions is defined to be independent if each finite subcollection of vectors satisfies condition 1.*

Note that Definition 4.41 does not imply independence within the components of $\mathbf{X}_i$, only between collections. For instance, suppose that a generic collection $\mathbf{X}_i$ consists of values of several covariates measured on person i; covariates on the same person are dependent, but the collections $\mathbf{X}_i$ are independent of each other because they are measured on different people.

Some results for independence of random variables have analogs for independence of random vectors. The following are two examples.

Proposition 4.42. Analog of Proposition 4.37 $\mathbf{X}_1, \ldots, \mathbf{X}_n$ *of respective dimensions $k_1, \ldots, k_n$ are independent by Definition 4.41 if and only if*

$$P(\mathbf{X}_1 \le \mathbf{x}_1, \mathbf{X}_2 \le \mathbf{x}_2, \ldots, \mathbf{X}_n \le \mathbf{x}_n) = F_1(\mathbf{x}_1)F_2(\mathbf{x}_2)\ldots F_n(\mathbf{x}_n)$$

for all $\mathbf{x}_1, \mathbf{x}_2, \ldots, \mathbf{x}_n$, where $F_1, F_2, \ldots, F_n$ are the multivariate distribution functions of $\mathbf{X}_1, \mathbf{X}_2, \ldots, \mathbf{X}_n$.

Proposition 4.43. Analog of Proposition 4.39 *If $\mathbf{X}_1, \ldots, \mathbf{X}_n$ are independent random vectors of respective dimensions $k_1, \ldots, k_n$ and $f_i : R^{k_i} \mapsto R$ are Borel functions (i.e., $\mathcal{B}^{k_i}$ measurable), $i = 1, \ldots, n$, then $f_1(\mathbf{X}_1), \ldots, f_{k_n}(\mathbf{X}_n)$ are independent random variables.*

Proof. The event $\cap\{f_i(\mathbf{X}_i) \in B_i\}$ is $\cap\{\mathbf{X}_i \in A_i\}$, where $A_i = f_i^{-1}(B_i)$ is a k_i-dimensional Borel set because f_i is $\mathcal{B}^{k_i}$ measurable. Because the $\mathbf{X}_i$ are independent, $P(\cap\{\mathbf{X}_i \in A_i\}) = \prod P(\mathbf{X}_i \in A_i) = \prod P(f_i(\mathbf{X}_i) \in B_i)$, showing that $f_1(\mathbf{X}_1), \ldots, f_n(\mathbf{X}_n)$ are independent. □

Example 4.44. Independence of sample mean and variance, and of fitted and residual Suppose that $(Y_1, \ldots, Y_n)$ are independent $\mathrm{N}(\mu, \sigma^2)$ random variables, and let $\mathbf{R} = (Y_1 - \bar{Y}, \ldots, Y_n - \bar{Y})$ be the "residual" vector. It is known that $(\bar{Y}, \mathbf{R})$ are independent (more generally, the residual vector in regression is independent of the fitted vector). It follows from Proposition 4.43 that $f(\mathbf{R})$ is independent of $g(\bar{Y}) = \bar{Y}$ for any function $f : R^n \mapsto R$ that is $\mathcal{B}^n$-measurable. Taking $f(\mathbf{R}) = (n-1)^{-1}\sum R_i^2$ shows that the sample variance $(n-1)^{-1}\sum_{i=1}^n (Y_i - \bar{Y})^2$ of a normal sample is independent of the sample mean. Interestingly, the normal distribution is the only one with this property. □

Let $X_1,\ldots,X_n$ be any random variables, and let $I_1,\ldots,I_k$ be sets of indices, i.e., subsets of $\{1,2,\ldots,n\}$. Let $\mathbf{X}_i$ consist of the components whose indices correspond to I_i. For instance, with $n=4$, $k=2$, $I_1=\{1,2,3\}$ and $I_2=\{3,4\}$, then $\mathbf{X}_1=(X_1,X_2,X_3)$ and $\mathbf{X}_2=(X_3,X_4)$. In this case $\mathbf{X}_1$ and $\mathbf{X}_2$ have overlapping components because X_3 is in both $\mathbf{X}_1$ and $\mathbf{X}_2$. If $I_1,\ldots,I_k$ had been disjoint, we would say that $\mathbf{X}_1,\ldots,\mathbf{X}_k$ are non-overlapping collections of $(X_1,\ldots,X_n)$.

Proposition 4.45. Non-overlapping collections of independent variables are independent *Non-overlapping collections of the independent random variables $X_1,\ldots,X_n$ are independent.*

Proof. Let $\mathbf{X}_1,\ldots,\mathbf{X}_k$ be the vectors of non-overlapping components. Then

$$
\begin{aligned}
&P(\mathbf{X}_1\le\mathbf{x}_1,\ldots,\mathbf{X}_k\le\mathbf{x}_k)\\
&\quad= P(X_{11}\le x_{11},\ldots,X_{1n_1}\le x_{1n_1},\ldots,X_{k1}\le x_{k1},\ldots,X_{kn_k}\le x_{kn_k})\\
&\quad= \prod_{i=1}^{k}\prod_{j=1}^{n_i}P(X_{ij}\le x_{ij})\\
&\quad= \prod_{i=1}^{k}P(\mathbf{X}_i\le\mathbf{x}_i). \qquad (4.12)
\end{aligned}
$$

The second and third steps follow from the fact that subcollections of independent components are also independent. □

We have not yet defined independence of infinite-dimensional random vectors. For instance, what does it mean for $(X_1,\ldots,X_n,\ldots)$ to be independent of $(Y_1,\ldots,Y_n,\ldots)$?

Definition 4.46. Independence of infinite dimensional random vectors

1. *A finite collection of random vectors $\mathbf{X}_1,\ldots,\mathbf{X}_n$ whose dimensions could be infinite is defined to be independent if $\mathbf{X}'_1,\ldots,\mathbf{X}'_n$ are independent by Definition 4.41 for each finite subcollection $\mathbf{X}'_i$ from $\mathbf{X}_i$, $i=1,\ldots,n$.*

2. *A countably or uncountably infinite collection of random vectors whose dimensions could be infinite is defined to be independent if each finite subcollection of vectors satisfies condition 1.*

Proposition 4.47. *Proposition 4.45 remains true if $(X_1,\ldots,X_n,\ldots)$ is infinite.*

Instead of giving separate definitions for independence of random variables/vectors in different scenarios, as we did in Definitions 4.36, 4.41, and 4.46, we could seek a unified definition of independence that applies to both random variables and random vectors. We will see that this has advantages and disadvantages. Remember that this book emphasizes the importance of thinking about random variables in terms of the sigma-fields they generate. Sigma-fields allow for a very natural, unified definition of independence. Recall that the sigma-field $\sigma(X_i)$ generated by X_i, namely the smallest sigma-field with respect to which X_i is measurable, is just $\{X_i^{-1}(B),\ B\in\mathcal{B}\}$ (Exercise 3 of Section 4.1.1). Similarly, the sigma-field generated by any collection of random variables is the smallest sigma-field with respect to which all of the random variables are measurable.

Definition 4.48. Independence of fields and sigma-fields

1. *A finite collection of fields or sigma-fields* $\mathcal{F}_1, \mathcal{F}_2, \ldots, \mathcal{F}_n$ *is called independent if* $F_1, F_2, \ldots, F_n$ *are independent events for each* $F_1 \in \mathcal{F}_1, F_2 \in \mathcal{F}_2, \ldots F_n \in \mathcal{F}_n$.

2. *An infinite collection of fields or sigma-fields is called independent if each finite sub-collection is independent.*

Proposition 4.49. Independent fields implies independent sigma-fields *If* $\mathcal{F}_1, \ldots, \mathcal{F}_n$ *are independent fields, then* $\sigma(\mathcal{F}_1), \ldots, \sigma(\mathcal{F}_n)$ *are independent sigma-fields.*

Proof when $n = 2$. Let $F \in \mathcal{F}_1$ and define $\mathcal{G}$ to be the collection of $G \in \sigma(\mathcal{F}_2)$ such that F and G are independent. Then $\mathcal{G}$ contains $\mathcal{F}_2$ by assumption. Now suppose that $G_1, G_2, \ldots$ are $\mathcal{G}$-sets that increase or decrease to G. Then $F \cap G_1, F \cap G_2, \ldots$ increase or decrease to $F \cap G$. The continuity property of probability (Proposition 3.30) implies that $P(F \cap G) = \lim_{n\to\infty} P(F \cap G_n) = \lim_{n\to\infty} P(F)P(G_n) = P(F)P(G)$. Therefore, F and G are independent, so $G \in \mathcal{G}$. This shows that $\mathcal{G}$ is a monotone class. By the monotone class theorem, $\mathcal{G}$ contains $\sigma(\mathcal{F}_2)$. Therefore, every set in $\sigma(\mathcal{F}_2)$ is independent of every set in $\mathcal{F}_1$. Now let $G \in \sigma(\mathcal{F}_2)$ and define $\mathcal{F}$ to be the collection of $F \in \sigma(\mathcal{F}_1)$ such that F and G are independent. By what we just proved, $\mathcal{F}$ contains $\mathcal{F}_1$. The continuity property of probability (Proposition 3.30) implies that $\mathcal{F}$ is a monotone class. By the monotone class theorem, $\mathcal{F}$ contains $\sigma(\mathcal{F}_1)$. Therefore, each set in $\sigma(\mathcal{F}_1)$ is independent of each set in $\sigma(\mathcal{F}_2)$, completing the proof when $n = 2$. □

Remark 4.50. *We could have defined independence of arbitrary collections of random vectors in terms of whether the sigma-fields generated by those collections are independent. It can be shown that this definition agrees with Definitions 4.36, 4.41, and 4.46. The disadvantage of defining things this way is that it obscures the fact that independence depends only on the measures on* R^k *induced by the random vectors, not on the original sigma-field.*

Example 4.51. Face on Mars revisited Recall Example 3.6 purporting to show a human face on Mars, spurring some to conclude that ancient astronauts must be responsible. We argued that the probability of seeing some recognizable pattern somewhere in the universe could be quite high. The actual process by which patterns on planets are formed is very complicated, but imagine the over-simplified scenario of constructing a pattern completely at random using the lights of Exercise 11 of Section 2.1. Recall that lights are placed at positions (r, s) for all rational r and s, and each light may be turned on or off.

Proponents of the ancient astronaut theory might argue that even a very simple pattern such as a specific line segment has probability 0 because it requires a **specific** infinite set of lights to be turned on. If $X_1, X_2, \ldots$ are the iid Bernoulli (1/2) indicators of those specific lights being on, then $P(X_1 = 1, X_2 = 1, \ldots) = (1/2)(1/2)\ldots = 0$ by Proposition 4.35. This same argument can be applied to any pattern requiring a specific infinite set of lights to be turned on. For instance, a circle C_R of radius R must have probability 0. Even the set $\{C_R\}$ of circles of all rational radii is countable. Moreover, the number N of distinct **recognizable** patterns (i.e., the set of objects on earth), though very large, is finite. Even if we allow different sizes through all rational scale factors, the set of recognizable patterns is countable. Therefore, the probability of seeing some recognizable pattern is 0 by countable additivity. Ancient astronaut theorists might conclude from this that the probability of seeing such a recognizable pattern is 0.

One problem with the above reasoning is that the human eye will perceive a line segment even if not all of the lights are turned on. For instance, if enough equally spaced lights along a line segment are turned on, the human eye will perceive the entire line segment (see Figure 4.7). Therefore, because the probability of any finite set of lights being turned on is nonzero, the probability of perceiving a given line segment is nonzero. Similarly, the probability of stringing together multiple apparent line segments is also nonzero, though it may be tiny. A sufficiently long string of apparent line segments can mimic any pattern. For instance, Figure 4.8 shows 8, 16, and 64 line segments strung together to give the appearance of a circle. Therefore, in any specific square of a given size, the probability of seeing what appears to be a human face is $\epsilon > 0$, though it may be tiny. Now consider a countably infinite set of non-overlapping squares. Under the random lighting scenario, the patterns observed in these different squares are independent. The probability of seeing what appears to be a human face on at least one of these squares is $1-(1-\epsilon)(1-\epsilon)\ldots = 1-0 = 1$. Non-believers in the ancient astronaut theory can argue that not only is seeing such a pattern not unusual, it is guaranteed!

Figure 4.7: From left to right: 11, 101, and 5001 lights turned on along the line segment $y = 2x$, $x \in [0,1]$.

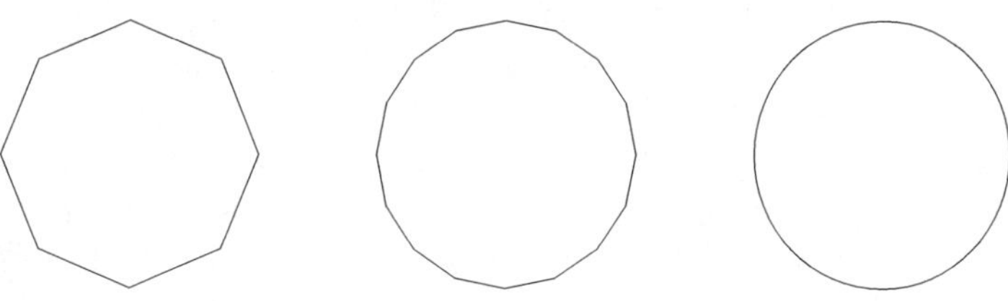

Figure 4.8: From left to right: 8, 16, or 64 consecutive line segments joined together to give the appearance of a circle.

□

4.5.2 Product Measure

How do we know that independent random variables with given marginal distributions exist? That is, how do we know there is some probability space $(\Omega, \mathcal{F}, P)$ such that $X_1(\omega), X_2(\omega), \ldots, X_n(\omega)$ are independent and have respective marginal distribution functions $F_1, F_2, \ldots$? One way to show this uses *product measure*. We illustrate product measure when $n = 2$.

Let $(\Omega_i, \mathcal{F}_i, P_i)$, $i = 1, 2$ be probability spaces, and define $\Omega = \Omega_1 \times \Omega_2$ to be the product set $\{\omega = (\omega_1, \omega_2) : \omega_i \in \Omega_i, i = 1, 2\}$. Similarly, if $F_i \in \mathcal{F}_i$, $i = 1, 2$, $F = F_1 \times F_2$ is called a product set. Let E be the union of a finite number of disjoint product sets,

$E = \cup_{i=1}^k (A_{i1} \times A_{i2})$, where $A_{i1} \in \mathcal{F}_1$, $A_{i2} \in \mathcal{F}_2$ and $(A_{i1} \times A_{i2}) \cap (A_{j1} \times A_{j2}) = \emptyset$ for $i \neq j$, which is equivalent to $A_{i1} \cap A_{j1} = \emptyset$ or $A_{i2} \cap A_{j2} = \emptyset$. We will show that the collection $\mathcal{F}_0$ of these unions of a finite number of disjoint product sets is a field.

We show first that $\mathcal{F}_0$ is closed under complements:

$$\begin{aligned} E^C &= \left\{\cup_{i=1}^k (A_{i1} \times A_{i2})\right\}^C = \cap_{i=1}^k [(A_{i1}^C \times \Omega_2) \cup (\Omega_1 \times A_{i2}^C)] \\ &= \cap_{i=1}^k [\{A_{i1}^C \times (A_{i2} \cup A_{i2}^C)\} \cup \{(A_{i1} \cup A_{i1}^C) \times A_{i2}^C\}] \\ &= \cap_{i=1}^k \left\{(A_{i1}^C \times A_{i2}) \cup (A_{i1}^C \times A_{i2}^C) \cup (A_{i1} \times A_{i2}^C)\right\} \\ &= \cap_{i=1}^k (D_{i1} \cup D_{i2} \cup D_{i3}), \end{aligned} \tag{4.13}$$

where D_{i1}, D_{i2}, D_{i3} correspond to $A_{i1}^C \times A_{i2}$, $A_{i1}^C \times A_{i2}^C$, and $A_{i1} \times A_{i2}^C$, respectively.

Expression (4.13) consists of the set of ω in at least one of D_{i1}, D_{i2}, and D_{i3} for each i. Consider the matrix whose ith row consists of D_{i1}, D_{i2}, D_{i3}, and imagine each possible way to pick one of the three Ds from each of the k rows. Let j_i be the index (1, 2, or 3) of the D selected in row i. For instance, the boldfaced Ds below correspond to the choice $j_1 = 2, j_2 = 1, \ldots, j_k = 3$.

$$\begin{pmatrix} D_{11} & \mathbf{D_{12}} & D_{13} \\ \mathbf{D_{21}} & D_{22} & D_{23} \\ \vdots & \vdots & \vdots \\ D_{k1} & D_{k2} & \mathbf{D_{k3}} \end{pmatrix}$$

There are 3^k different ways to pick $j_1, \ldots, j_k$. For each, form the intersection $D_{1j_1} \cap D_{2j_2} \cap \ldots \cap D_{kj_k}$. Expression (4.13) consists of all possible unions of these intersections:

$$\cap_{i=1}^k (D_{i1} \cup D_{i2} \cup D_{i3}) = \cup_{j_1, \ldots j_k} \cap_{i=1}^k D_{ij_i}, \tag{4.14}$$

where each j_i ranges over the set $\{1, 2, 3\}$. Because the three sets D_{i1}, D_{i2}, D_{i3} are disjoint, $\cap_{i=1}^k D_{ij_i}$ and $\cap_{i=1}^k D_{ij_i'}$ are disjoint unless each $j_i' = j_i$. Moreover, each $\cap_{i=1}^k D_{ij_i}$ is clearly a product set because $\cap_{i=1}^k (C_{i1} \times C_{i2}) = (\cap_{i=1}^k C_{i1}) \times (\cap_{i=1}^k C_{i2})$ for any C_{i1}, C_{i2}, $i = 1, \ldots, k$. Therefore, E^C is a union of a finite number of disjoint product sets. That is, $\mathcal{F}_0$ is closed under complements.

We show next that $\mathcal{F}_0$ is closed under intersections of pairs, which will establish that $\mathcal{F}_0$ is a field because closure under complements and paired intersections also implies closure under paired unions. Note that

$$\begin{aligned} &\left\{\cup_{i=1}^k (A_{i1} \times A_{i2})\right\} \cap \left\{\cup_{j=1}^m (B_{j1} \times B_{j2})\right\} \\ &\qquad = \cup_{i,j} (A_{i1} \times A_{i2}) \cap (B_{j1} \times B_{j2}) \\ &\qquad = \cup_{i,j} \left\{(A_{i1} \cap B_{j1}) \times (A_{i2} \cap B_{j2})\right\} \end{aligned} \tag{4.15}$$

is the union of a finite number of disjoint product sets. Thus, $\mathcal{F}_0$ is closed under intersection of pairs, completing the proof that $\mathcal{F}_0$ is a field. □

Definition 4.52. Product sigma-field $\mathcal{F}_1 \times \mathcal{F}_2$ *The sigma-field $\sigma(\mathcal{F}_0)$ generated by $\mathcal{F}_0$ is called the product sigma-field of $\mathcal{F}_1$ and $\mathcal{F}_2$ and is denoted $\mathcal{F}_1 \times \mathcal{F}_2$.*

Define P on $\mathcal{F}_0$ by $P\{\cup_{i=1}^k (A_{1i} \times A_{2i})\} = \sum_{i=1}^k P_1(A_{1i}) P_2(A_{2i})$. It can be established that P is a probability measure on $\mathcal{F}_0$. Therefore, we can extend P to $\mathcal{F}_1 \times \mathcal{F}_2$ by Proposition

3.34. This probability measure on $\mathcal{F}_1 \times \mathcal{F}_2$ is called product measure and denoted $P_1 \times P_2$. If $F_1 \in \mathcal{F}_1$ and $F_2 \in \mathcal{F}_2$, then $P(F_1 \times F_2) = P_1(F_1)P_2(F_2)$.

We are now in a position to define independent random variables on $\mathcal{F}_1 \times \mathcal{F}_2$. If $X_1(\omega_1)$ and $X_2(\omega_2)$ are random variables on $(\Omega_1, \mathcal{F}_1, P_1)$ and $(\Omega_2, \mathcal{F}_2, P_2)$, define $\tilde{X}_1(\omega_1, \omega_2) = X_1(\omega_1)$ and $\tilde{X}_2(\omega_1, \omega_2) = X_2(\omega_2)$. Then $(\tilde{X}_1, \tilde{X}_2)$ are random variables on $(\Omega = \Omega_1 \times \Omega_2, \mathcal{F} = \mathcal{F}_1 \times \mathcal{F}_2, P = P_1 \times P_2)$ with

$$\begin{aligned} P\{(\omega) : \tilde{X}_1(\omega) \in B_1 \cap \tilde{X}_2(\omega) \in B_2\} &= P\{(X_1^{-1}(B_1) \times X_2^{-1}(B_2)\} \\ &= P_1\{X_1^{-1}(B_1)\}P_2\{X_2^{-1}(B_2)\} \\ &= P(\tilde{X}_1 \in B_1)P(\tilde{X}_2 \in B_2). \end{aligned} \tag{4.16}$$

That is, $\tilde{X}_1$ and $\tilde{X}_2$ are independent random variables on $(\Omega_1 \times \Omega_2, \mathcal{F}_1 \times \mathcal{F}_2, P_1 \times P_2)$ with the same marginal distributions as X_1 and X_2.

We can use a similar construction to create n independent random variables. In the next section, we show how to embed a product structure on $\Omega = [0, 1]$ with P being Lebesgue measure. This will allow us to create infinitely many independent random variables with given marginal distributions.

Exercises

1. Suppose that X and Y are independent random variables. Can X and Y^2 be dependent? Explain.

2. Let X have a standard normal distribution and

$$Y = \begin{cases} -1 & \text{if } |X| \le 1 \\ +1 & \text{if } |X| > 1. \end{cases}$$

Show that X and Y^2 are independent, but X and Y are not independent. What is wrong with the following argument: since X and Y^2 are independent, X and $Y = \sqrt{Y^2}$ are independent by Proposition 4.39.

3. Prove that if $A_1, \ldots, A_n$ are independent events and each B_i is either A_i or A_i^C, then $B_1, \ldots, B_n$ are also independent.

4. Prove that if X_1 and X_2 are random variables each taking values 0 or 1, then X_1 and X_2 are independent if and only if $P(X_1 = 1, X_2 = 1) = P(X_1 = 1)P(X_2 = 1)$. That is, two binary random variables are independent if and only if they are uncorrelated.

5. If $A_1, A_2, \ldots$ is a countably infinite sequence of independent events, then $P(\cup_{i=1}^{\infty} A_i) = 1 - \prod_{i=1}^{\infty}\{1 - \Pr(A_i)\}$.

6. Let $\Omega = \{\omega_1, \omega_2, \omega_3, \omega_4\}$, where $\omega_1 = (-1, -1)$, $\omega_2 = (-1, +1)$, $\omega_3 = (+1, -1)$, $\omega_4 = (+1, +1)$. Let $X(\omega)$ be the indicator that the first component of ω is $+1$, and $Y(\omega)$ be the indicator that the second component of ω is $+1$. find a set of probabilities p_1, p_2, p_3, p_4 for $\omega_1, \omega_2, \omega_3, \omega_4$ such that X and Y are independent. Find another set of probabilities such that X and Y are not independent.

7. Let X be a Bernoulli random variable with parameter p, $0 < p < 1$. What are necessary and sufficient conditions for a Borel function $f(X)$ to be independent of X?

8. Suppose that X is a random variable taking only 10 possible values, all distinct. The $\mathcal{F}$ sets on which X takes those values are $F_1, \ldots, F_{10}$. You must determine whether X is independent of another random variable, Y. Does the determination of whether they are independent depend on the set of possible values $\{x_1, \ldots, x_{10}\}$ of X? Explain.

9. Flip a fair coin 3 times, and let X_i be the indicator that flip i is heads, $i = 1, 2, 3$, and X_4 be the indicator that the number of heads is even. Prove that each pair of random variables is independent, as is each trio, but X_1, X_2, X_3, X_4 are not independent.

10. Prove that a random variable X is independent of itself if and only if $P(X = c) = 1$ for some constant c.

11. Prove that a sigma-field is independent of itself if and only if each of its sets has probability 0 or 1.

12. Prove that if X is a random variable and f is a Borel function such that X and $f(X)$ are independent, then there is some constant c such that $P(g(X) = c) = 1$.

13. Let $(\Omega, \mathcal{F}, P) = ([0,1], \mathcal{B}_{[0,1]}, \mu_L)$, and for $t \in [0,1]$, let $X_t = I(\omega = t)$. Are $\{X_t,\ t \in [0,1]\}$ independent?

14. Prove that the collection $\mathcal{A}_1$ in step 1 of the proof of Proposition 4.37 contains the field $\mathcal{F}_0$ of Example 3.8.

15. It can be shown that if $(Y_1, \ldots, Y_n)$ have a multivariate normal distribution with $\mathrm{E}(Y_i^2) < \infty$, $i = 1, \ldots, n$, then any two subcollections of the random variables are independent if and only if each correlation of a member of the first subcollection and a member of the second subcollection is 0. Use this fact to prove that if $Y_1, \ldots, Y_n$ are iid normals, then $(Y_1 - \bar{Y}, \ldots, Y_n - \bar{Y})$ is independent of $\bar{Y}$. What can you conclude from this about the sample mean and sample variance of iid normals?

16. Show that if Y_i are iid from any non-degenerate distribution F (i.e., Y_i is not a constant), then the residuals $R_1 = Y_1 - \bar{Y}, \ldots, R_n = Y_n - \bar{Y}$ cannot be independent.

17. Prove that the random variables $X_1, X_2, \ldots$ are independent by Definition 4.36 if and only if the sigma fields $\mathcal{F}_1 = \sigma(X_1), \mathcal{F}_2 = \sigma(X_2), \ldots$ are independent by Definition 4.48.

4.6 Take $(\Omega, \mathcal{F}, P) = ((0,1), \mathcal{B}_{(0,1)}, \mu_L)$, Please!

In elementary probability, we did not think much about the domain of a random variable X; we focused instead on its range and the probabilities of the different values. Now we are being more rigorous, defining the probability triple $(\Omega, \mathcal{F}, P)$. It turns out that without loss of generality, we can assume that $\Omega = (0,1)$, $\mathcal{F}$ =the Borel subsets of $(0,1)$, and P is Lebesgue measure μ_L. With this triple, we can generate a random variable with **any** distribution function F. You may have seen the technique before when $F(x)$ is continuous and strictly increasing. If U is uniformly distributed on $(0,1)$ (i.e., $P(U \le u) = u$ for $0 < u < 1$), then $F^{-1}(U)$ has distribution function F because

$$\begin{aligned} P(F^{-1}(U) \le x) &= P[F\{F^{-1}(U)\} \le F(x)] \\ &= P(U \le F(x)) = F(x). \end{aligned} \tag{4.17}$$

This argument breaks down if F is not continuous or strictly increasing. For instance, let $F(x) = 0$ for $x < 1/2$ and 1 for $x \ge 1/2$. Then $F^{-1}(u)$ is not even defined for $0 < u < 1$.

Fortunately, a modification of the above technique works even when $F(x)$ is discontinuous or not strictly increasing. The trick is to define $F^{-1}(u)$ slightly differently:

Definition 4.53. Inverse probability transformation *Let $F(x)$ be any distribution function and $0 < u < 1$. Then the inverse probability transformation $F^{-1}(u)$ is defined to be* $\inf\{x : F(x) \geq u\}$ *(Figure 4.9).*

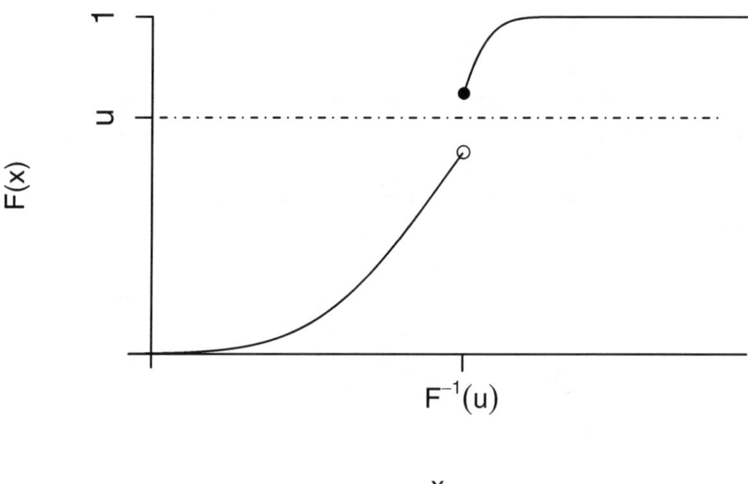

Figure 4.9: $F^{-1}(u) = \inf\{x : F(x) \geq u\}$. Note that $F(t) < u$ for $t < F^{-1}(u)$, while $F(t) \geq u$ for $t \geq F^{-1}(u)$. Therefore, $t \geq F^{-1}(u)$ is equivalent to $F(t) \geq u$.

Proposition 4.54. $F(t) \geq u \Leftrightarrow t \geq F^{-1}(u)$.

Proof. We will prove the result in 2 steps, namely showing:

$$F(t) < u \quad \text{for } t < F^{-1}(u), \text{ and} \tag{4.18}$$
$$F(t) \geq u \quad \text{for } t \geq F^{-1}(u). \tag{4.19}$$

Inequality (4.18) follows from the fact that $F^{-1}(u)$ is a lower bound on the set

$$A_u = \{x : F(x) \geq u\}.$$

In other words, all x such that $F(x) \geq u$ are at least as large as $F^{-1}(u)$, so $t < F^{-1}(u) \Rightarrow F(t) < u$. This concludes the proof of inequality (4.18).

To see inequality (4.19), assume first that t is strictly larger than $F^{-1}(u)$. Then t is not a lower bound on A_u (because $F^{-1}(u)$ is the greatest lower bound on A_u), so there exists a point x_n in A_u with $x_n < t$. Therefore, $F(x_n) \leq F(t)$. Because $x_n \in A_u$, $F(x_n) \geq u$. We have shown that

$$u \leq F(x_n) \leq F(t),$$

proving that $F(t) \geq u$ for t strictly larger than $F^{-1}(u)$. To see that the inequality also holds if $t = F^{-1}(u)$, note the following.

1. By what we have just proven, $F(t+1/n) \geq u$ because $t+1/n$ is strictly larger than $F^{-1}(u)$. Take the limit of both sides of this inequality to see that

2. $u \leq \lim_{n\to\infty}\{F(t+1/n)\} = F(t)$ by right-continuity of F. Thus, $u \leq F(t)$.

This completes the proof of inequality (4.19), and therefore, of Proposition 4.54. □

Proposition 4.55. $F^{-1}(\omega)$ has distribution function $F(x)$ *Let $\Omega = (0,1)$, $\mathcal{F} = B_{(0,1)}$, and P be Lebesgue measure μ_L. Then $F^{-1}(\omega)$ is a random variable with distribution function $F(x)$.*

Proof. That $F^{-1}(\omega)$ is a random variable follows from the monotonicity of F^{-1}, the fact that monotone functions are Borel measurable, and Proposition 4.5. By Proposition 4.54, the event $F^{-1}(\omega) \leq t$ is equivalent to $F(t) \geq \omega$, so $P\{\omega : F^{-1}(\omega) \leq t\} = P\{\omega \in (0, F(t)]\} = F(t)$ because the Lebesgue measure of an interval is its length. □

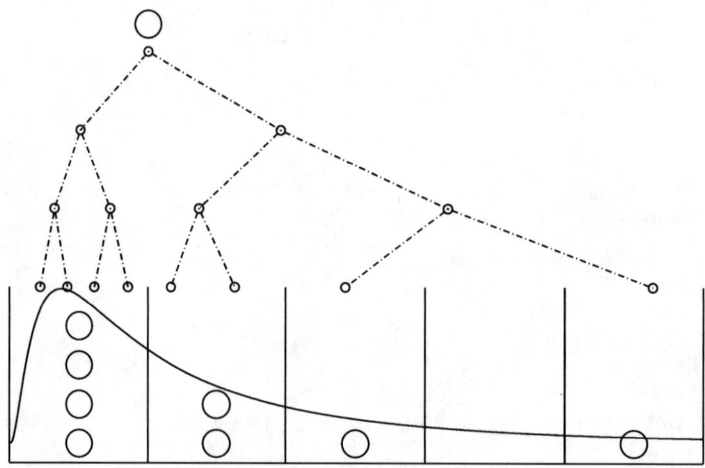

Figure 4.10: Quincunx depiction of $Y = F^{-1}(U)$, where $U = 0.\tau_1\tau_2\ldots$ and the τ_i are iid Bernoulli (1/2). The deflection in row i is left or right when τ_i is 0 or 1, respectively.

Here is a visual way to understand the fact that $Y = F^{-1}(\omega) \sim F(y)$. Suppose that F is a distribution function with density function f with respect to Lebesgue measure. Imagine a device called a *quincunx* with balls rolling down a board, hitting nails, and bouncing either to the left or right with equal probability. We use quincunxes in Chapter 8 to help understand the central limit theorem. Suppose that the first row has a single nail at horizontal position corresponding to the median of F, namely $F^{-1}(1/2)$. This is shown in Figure 4.10, where F is the lognormal distribution function. The ball bounces to either the first quartile ($F^{-1}(1/4)$) or third quartile ($F^{-1}(3/4)$) in the second row, with equal probability. If it hits the nail at $F^{-1}(1/4)$ in the second row, it is equally likely to bounce to either $F^{-1}(1/8)$ or $F^{-1}(3/8)$ in the third row. Likewise, if it hits the nail at $F^{-1}(3/4)$ in the second row, it is equally likely to bounce to either $F^{-1}(5/8)$ or $F^{-1}(7/8)$ in the third row. The dotted lines in Figure 4.10 show the possible paths of a ball rolling down a quincunx with 4 rows. Approximately 4 out of every 8 balls fall into the leftmost of the equal-width bins at the bottom because 4 of the 8 equally likely paths lead to that bin. The shape of the empirical histogram formed by the balls mimics the shape of the density function f. Indeed, as we increase the number of rows and equal-width bins at the bottom, the shape of the empirical density function becomes indistinguishable from f.

The above quincunx is simply a visual corroboration of the fact that $F^{-1}(\omega)$ has distribution function F. After all, if ω has base 2 representation $0.\tau_1\tau_2\ldots = \tau_1/2 + \tau_2/2^2 + \ldots$, then $F^{-1}(\omega) < 1/2$ corresponds to $\tau_1 = 0$ and $F^{-1}(\omega) > 1/2$ corresponds to $\tau_1 = 1$. Therefore, left or right deflections in the first row correspond to $\tau_1 = 0$ or 1, respectively. Likewise, left or right deflections in the second row correspond to $\tau_2 = 0$ or 1, respectively, etc. The randomly selected ω dictates, through its base 2 representation, the full set of left or right deflections in the rows of the quincunx. Note that we ignored the possibility that the base 2 representation of ω terminates because this event has probability 0.

So far we have shown that we can define a random variable on $((0,1), \mathcal{B}_{(0,1)}, \mu_L)$ with distribution F. The next step is to show that we can define independent random variables X_1 and X_2 on $((0,1), \mathcal{B}_{(0,1)}, \mu_L)$ with respective distributions F and G. Let $\omega_1, \omega_2, \ldots$ represent digits in the base 2 representation of the number $\omega \in (0,1)$ drawn at random using Lebesgue measure. Then $\omega_1, \omega_2, \ldots$ are independent and identically distributed (iid) Bernoulli random variables with parameter 1/2 (see Example 4.38). By Proposition 4.47, the collection $(\omega_1, \omega_3, \omega_5, \ldots)$ is independent of $(\omega_2, \omega_4, \omega_6, \ldots)$. Moreover, $U_1 = \omega_1 \times (1/2) + \omega_3 \times (1/2)^2 + \omega_5 \times (1/2)^3 + \ldots$ and $U_2 = \omega_2 \times (1/2) + \omega_4 \times (1/2)^2 + \omega_6 \times (1/2)^3 + \ldots$ each have the same distribution as the original random variable $\omega = \omega_1 \times (1/2) + \omega_2 \times (1/2)^2 + \ldots$, namely a uniform on $(0,1)$. It follows that $F^{-1}(U_1)$ and $G^{-1}(U_2)$ are independent with respective distribution functions F and G.

Can we extend this argument to obtain a countably infinite collection $(X_1, X_2, \ldots)$ of independent random variables defined on $((0,1), \mathcal{B}_{(0,1)}, \mu_L)$ with respective distribution functions $F_1, F_2, \ldots$? To create two independent random variables, we used two disjoint subsets of indices, namely the even and odd numbers. To generate an infinite number of independent random variables, we must define infinitely many disjoint subcollections $I_1, I_2, \ldots$, each with a countably infinite number of indices. One way to do this is to let $I_1 = \{2^1, 2^2, 2^3, \ldots\}$, $I_2 = \{3^1, 3^2, 3^3, \ldots\}$, $I_3 = \{5^1, 5^2, 5^3, \ldots\}$, etc., so that $I_n = \{p_n^1, p_n^2, p_n^3, \ldots\}$, where p_n is the nth prime. We claim that $I_1, I_2, I_3, \ldots$ are disjoint collections. If not, then there would be an integer k such that $k = p_i^r = p_j^q$, where $i \neq j$. But this would violate the fact that every positive integer can be written **uniquely** as a product of primes. Define $U_j = Y_{p_j} \times (1/2) + Y_{p_j^2} \times (1/2)^2 + \ldots$ Then $U_1, U_2, \ldots$ are independent by Proposition 4.47, and each is uniformly distributed, so $F_1^{-1}(U_1), F_2^{-1}(U_2), \ldots$ are independent with respective distributions $F_1, F_2, \ldots$

The preceding discussion shows that whenever we consider a sequence of independent random variables, there is no loss of generality by assuming that the underlying probability space is $((0,1), B_{(0,1)}, \mu_L)$. We have established the following result.

Proposition 4.56. Can always assume $((0,1), \mathcal{B}_{(0,1)}, \mu_L)$ *There exist independent random variables $X_1, \ldots, X_n, \ldots$ defined on $((0,1), \mathcal{B}_{(0,1)}, \mu_L)$ with respective distribution functions $F_1, \ldots, F_n, \ldots$*

Remark 4.57. *Note that to create the independent random variables in the development above, we needed to ensure that the indices corresponding to the base 2 representations of the different variables were disjoint. We could still create random variables X_1 and X_2 with distributions F_1 and F_2 if the indices were overlapping, but X_1 and X_2 would not necessarily be independent. For instance, ω and $1-\omega$ are both uniformly distributed, so $X_1 = F_1^{-1}(\omega)$ and $X_2 = F_2^{-1}(1-\omega)$ have distributions F_1 and F_2, but are not independent.*

Example 4.58. Normal scores test

One application of $F^{-1}(U) \sim F$ is the normal scores rank test comparing the location parameters of two groups. Before describing this particular test, we note that rank tests in general can be an attractive alternative to parametric tests when there may be outliers or the distributions have heavy tails. Unlike the original data, ranks cannot be too extreme, so a rank test can confer a substantial power advantage over the t-test if the true distribution has heavy tails. The first step of a rank test combines data from the treatment and control groups and ranks them from 1 (smallest) to n (largest). A commonly used statistic, the Wilcoxon rank sum statistic (Hollander and Wolfe, 1973), sums the ranks of treatment observations and compares it to its null distribution. If the distribution of the data is skewed or has fat tails, the Wilcoxon test can have substantially higher power than the t-test. On the other hand, it can be shown that if the data really are normally distributed, the power of the Wilcoxon rank sum test is approximately 5% lower than that of the t-test, asymptotically.

An ingenious alternative to the Wilcoxon test is the following procedure. Assume the data come from a continuous distribution function. Generate standard normal observations randomly, and replace the ith rank statistic of the original data with the ith order statistic from the standard normal data. The replacement data are from a normal distribution, so a t-test is automatically valid. This almost magical method can be viewed as a two-step procedure. The first step is conceptual; if we knew the distribution function F for the X_i, we could imagine replacing the original data $X_1, \ldots, X_n$ by $U_1 = F(X_1), \ldots, U_n = F(X_n)$. It can be shown (Exercise 7) that each U_i has a uniform distribution on $(0,1)$. The second step replaces the U_i by the standard normal deviates $Z_i = \Phi^{-1}(U_i)$. The downside of the method is that inference depends on the normal data randomly generated. Two different people applying the same test will get different answers. Even the same researcher repeating the test will get a different p-value. An alternative is to repeat many times the procedure of randomly generating standard normal order statistics and use the sample mean of each order statistic. That would reduce variability and make the procedure more repeatable. As the number of repetitions tends to infinity, the sample mean of the ith order statistic tends to $\mathrm{E}\{Z_{(i)}\}$, the expected value of the ith order statistic. Therefore, we can dispense with generating random deviates and just replace the ith order statistic from the original data with the expected value of the ith order statistic from a standard normal distribution. This adaptation of the procedure is known as the normal scores test. Unlike the Wilcoxon test, the normal scores test loses no power compared to a t-test if the data are normally distributed and the sample size tends to ∞. □

Example 4.59. Copulas Another application of the inverse probability transform is in generating data from a multivariate distribution with arbitrary marginal distributions using a copula model (Nelson, 1999). For example, suppose we are modeling the joint distribution of two variables measured on the same person. We want to account for the fact that the variables are correlated. We could use the bivariate normal distribution, but suppose one or both variables have heavy tails or skewed distributions, or are discrete. Is there a flexible way to model such data?

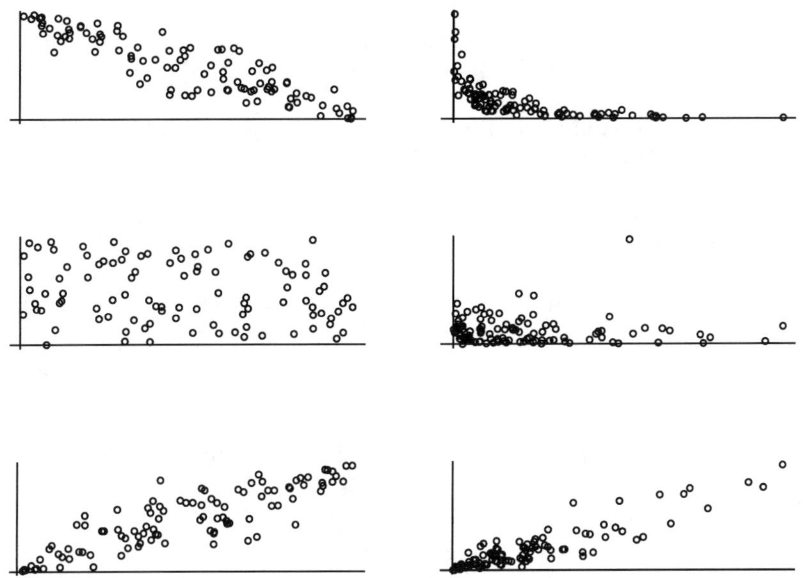

Figure 4.11: Copula model. The left side shows correlated uniforms (U_1, U_2) generated as $U_i = \Phi(V_i)$, $i = 1, 2$, where (V_1, V_2) are bivariate normal with correlation $\rho = -0.9$ (top), $\rho = 0$ (middle), and $\rho = 0.9$ (bottom). The right side shows correlated exponentials generated by $X = F^{-1}(U_1)$ and $Y = F^{-1}(U_2)$, where $F(t) = 1 - \exp(-t)$.

Let $X \sim F$ and $Y \sim G$. We would like to generate a pair (X, Y) with arbitrary marginal distributions F and G, such that X and Y are correlated. One way begins by generating a pair (V_1, V_2) from a bivariate normal distribution with standard normal marginals and correlation ρ. Expression (8.35) of Section 8.6 shows how to transform iid standard normals to achieve bivariate normals with the desired correlation matrix. Next, compute $U_1 = \Phi(V_1)$, $U_2 = \Phi(V_2)$. The marginal distribution of each U_i is uniform $[0, 1]$, although the Us are correlated because (V_1, V_2) are correlated. Now take $X = F^{-1}(U_1)$ and $Y = G^{-1}(U_2)$. Then X has marginal distribution F and Y has marginal distribution G. Also, X and Y are correlated because (V_1, V_2) are correlated and Φ, F, and G are monotone functions. Larger values of ρ generate more correlated values of (X, Y), as seen in Figure 4.11. Pairs (U_1, U_2) and (X, Y) are shown on the left and right, respectively, for $\rho = -0.9$ (top), 0.0 (middle) and 0.9 (bottom) and X and Y marginally exponential with parameter 1. The middle panel corresponds to independent data, while the top and bottom show highly negatively and positively correlated pairs, respectively.

In this example, the function $F^{-1}(U_1), G^{-1}(U_2))$ converting correlated uniforms (U_1, U_2) to correlated observations with marginal distribution functions F and G is called a *copula*. Sklar's theorem (Sklar, 1959) asserts that it is always possible to generate an arbitrary $(X_1, \ldots, X_n)$ from correlated uniforms using a copula. This is easy to prove when the marginal distribution functions $F_1, \ldots, F_k$ are continuous: $(U_1, \ldots, U_k) = (F_1(X_1), \ldots, F_k(X_k))$ are correlated uniforms and $(F^{-1}(U_1), \ldots, F^{-1}(U_k)) = (X_1, \ldots, X_k)$ with probability 1 (exercise). It follows that $(F^{-1}(U_1), \ldots, F^{-1}(U_k))$ has the same joint distribution as $(X_1, \ldots, X_k)$. $\square$

Exercises

1. Use the inverse probability transformation to construct a random variable that has a uniform distribution on $[0, a]$: $F(x) = x/a$ for $0 \le x \le a$.

2. Use the inverse probability transformation to construct a random variable that has an exponential distribution with parameter λ : $F(x) = 1 - \exp(-\lambda x)$.

3. Use the inverse probability transformation to construct a random variable that has a Weibull distribution: $F(x) = 1 - \exp\{-(x/\eta)^\beta\}$, $x \ge 0$.

4. Use the inverse probability transformation to construct a random variable X on $((0,1), \mathcal{B}_{(0,1)}, \mu_L)$ with the following distribution function:

$$F(x) = \begin{cases} 0 & \text{if } x < 0 \\ x & \text{if } 0 \le x < .5 \\ .75 & \text{if } .5 \le x < 10 \\ 1 & \text{if } x \ge 10. \end{cases}$$

5. Give an explicit formula for using the copula method to construct two negatively correlated exponential random variables with parameter 1.

6. Give an explicit formula for using the copula method to construct two strongly positively correlated random variables with the following marginal probability mass function:

$$p(x) = \begin{cases} 1 & \text{w.p. } 1/4 \\ 2 & \text{w.p. } 1/2 \\ 3 & \text{w.p. } 1/4. \end{cases}$$

7. Suppose that the distribution function $F(x)$ for a random variable is continuous. Prove that $F(X)$ is a random variable and has a uniform distribution. That is, $P(F(X) \le u) = u$ for each $u \in (0,1)$. Show this first when F is strictly increasing, and then extend the proof to arbitrary continuous F.

8. We showed how to find countably infinitely many disjoint sets of indices $\{p_1, p_1^2, \ldots\}$, $\{p_2, p_2^2 \ldots\}, \ldots$, where the ps are primes, and this allowed us to generate countably infinitely many independent random variables from a single random draw of ω from $(0,1)$. Can we find uncountably many disjoint sets of indices, and thereby generate uncountably many independent random variables from a single random draw of ω from $(0,1)$? Explain.

4.7 Summary

1. (a) A random variable is a function $X : \Omega \longmapsto R$: $X^{-1}(B) \in \mathcal{F}$ for each $B \in \mathcal{B}$.

   (b) A random vector is a function $\mathbf{X} : \Omega \longmapsto R^k$: $\mathbf{X}^{-1}(B) \in \mathcal{F}$ for each $B \in \mathcal{B}^k$.

2. A random vector induces a probability measure P' on R^k: $P'(B) = P\{\omega : X_1(\omega), \ldots, X_k(\omega) \in B\}$ for $B \in \mathcal{B}^k$.

3. The distribution function $F(x_1, \ldots, x_k) = P(X_1 \le x_1, \ldots, X_k \le x_k)$ completely determines $P'(B)$ for every $B \in \mathcal{B}^k$.

(a) A univariate distribution function is right-continuous; it is continuous at x if and only if it is left-continuous at x, and there are only countably many points of discontinuity.

(b) A multivariate distribution function is continuous from above; it is continuous at $\mathbf{x}$ if and only if it is continuous from below at $\mathbf{x}$, and there are only countably many axis lines of discontinuity.

4. The sigma-field $\sigma\{X_t,\ t \in T\} \subset \mathcal{F}$ generated by $\{X_t\ t \in T\}$ is the smallest sigma-field containing all sets $X_t^{-1}(B)$, $B \in \mathcal{B}^1$, $t \in T$.

   (a) Sigma-fields generated by random variables give an indication of the information content about ω contained in the random variables.

   (b) If $\sigma(X) \subset \sigma(Y)$, then Y is more informative than X.

   (c) If $\mathbf{X}$ is a k-dimensional random vector, then an equivalent expression for $\sigma(\mathbf{X})$ is $\{\mathbf{X}^{-1}(B),\ B \in \mathcal{B}^k\}$.

5. If $A_1, A_2, \ldots$ are independent events, then $P(\cap_{i=1}^{\infty} A_i) = \prod_{i=1}^{\infty} P(A_i)$, and the same is true for all sub-collections of the A_i.

6. Arbitrary collections of random variables are independent if and only if the sigma-fields generated by those collections are independent.

7. If $\{X_t\}$ are independent, then:

   (a) $\{f_t(X_t)\}$ are independent for Borel functions f_t.

   (b) non-overlapping collections of the X_t are independent.

8. *** Inverse probability transformation: For arbitrary F, $X = F^{-1}(\omega) \sim F$, where $(\Omega, \mathcal{F}, P) = ((0,1), \mathcal{B}_{(0,1)}, \mu_L)$ and $F^{-1}(\omega) = \inf\{x :\ F(x) \geq \omega\}$.

9. *** $((0,1), \mathcal{B}_{(0,1)}, \mu_L)$ is the only probability space we ever need.

   (a) Without loss of generality, we can assume that the experiment consists of drawing a single number ω at random from the unit interval.

   (b) Drawing $\omega = 0.X_1X_2\ldots$ randomly is equivalent to flipping countably infinitely many coins: $X_1 = 1$ (heads on coin 1) means ω is in the right half of $(0,1)$, $X_2 = 1$ (heads on coin 2) means ω is in the right half of that half, etc.

   (c) We can use the inverse probability transformation to define independent random variables $X_1, X_2, \ldots$ on $((0,1), \mathcal{B}_{(0,1)}, \mu_L)$ with arbitrary distributions $F_1, F_2, \ldots$

Chapter 5

Integration and Expectation

This chapter is about the connection between probability measures and integration, also called expectation. The reader may already be familiar with this connection for Riemann integration defined in calculus, but we will see that there is another type of integral called the Lebesgue integral. The Lebesgue integral can exist even when the integrand is very irregular, and is thus the preferred method of integration that we use for the rest of the book.

5.1 Heuristics of Two Different Types of Integrals

We adopt a somewhat informal style here to motivate Lebesgue integration, which will be defined more rigorously in the next section.

In calculus you learned that the *Riemann* integral of a function $f(x)$ over an interval $[a, b]$ is defined as a limit of partial sums. We partition $[a, b]$ into $a = x_0 < x_1 < \ldots < x_n = b$ and form the sum

$$\sum_{i=1}^{n} f(\xi_i)\Delta x_i, \tag{5.1}$$

where ξ_i is a point in the ith interval, $[x_{i-1}, x_i)$ and $\Delta x_i = x_i - x_{i-1}$. Note that ξ_i and Δx_i depend on n as well, but we have suppressed the notation for simplicity. Then f is said to be Riemann integrable if we get the same limit as $n \to \infty$ and $\max_{\{1 \le i \le n\}} \Delta x_i \to 0$ regardless of the intermediate point ξ_i selected (Figure 5.1).

Riemann integrability requires f to be fairly regular. A function such as

$$f(x) = \begin{cases} 0 & \text{if } x \text{ is rational} \\ 1 & \text{if } x \text{ is irrational} \end{cases} \tag{5.2}$$

is not Riemann integrable on $[0, 1]$ because the value of the limit (5.1) for $\int_0^1 f(x)dx$ depends on which intermediate point is selected; if ξ_i is rational for each i, then the limit is 0, whereas if ξ_i is irrational for each i, the limit is 1. If the ξ_i alternate between rational and irrational, the limit does not exist.

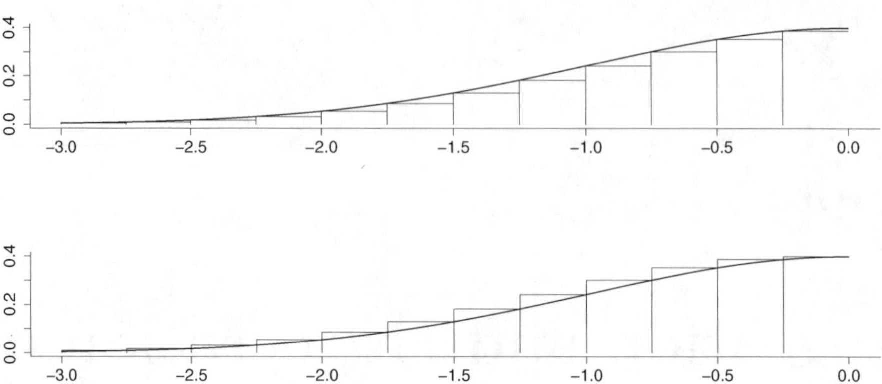

Figure 5.1: Partial sums of the form 5.1, where the intervals are the same size and the intermediate point ξ_i is the leftmost (upper panel) or rightmost (bottom panel) point in each interval.

An alternative way to define an integral divides the y-axis, rather than the x-axis, into equal-sized intervals. Form the partial sum

$$\sum_{i=1}^{\infty} y_i \mu\{f^{-1}(A_i)\}, \tag{5.3}$$

where y_i is a point in the ith y-interval and μ is Lebesgue measure. That is, for each y-interval we select an intermediate value y_i and multiply by the Lebesgue measure of the set of xs that get mapped into the y-interval (see Figure 5.2). This assumes that f is a (Lebesgue) measurable function, so that the Lebesgue measure of $f^{-1}(A_i)$ is defined. If we get the same limiting value I as the common width of the intervals forming the y-partition tends to 0, irrespective of the intermediate value y_i selected, then I is an alternative way to define the integral of $f(x)$. This alternative method is called the *Lebesgue* integral.

The advantage of partitioning the y-axis instead of the x-axis is that the y values within a given interval are automatically close to each other. There is no longer any need for $f(x)$ to be a well-behaved function of x. We pay a small price in that a simple width, Δx_i, in the Riemann sum (5.1) is replaced by a more intimidating expression, $\mu\{f^{-1}(A_i)\}$, in the Lebesgue sum (5.3). Nonetheless, we are able to compute the Lebesgue measure of even bizarre sets $f^{-1}(A_i)$.

Now return to the function (5.2) and consider its Lebesgue integral $\int_0^1 f(x)dx$. The function f takes only two possible values, 0 or 1. If the common width of the y-intervals is small enough, then 0 and 1 will be in separate intervals (Figure 5.3). Therefore, if A_i and A_j are the intervals containing $y = 0$ and $y = 1$, respectively, then Expression (5.3) is $y_i\mu\{f^{-1}(A_i)\} + y_j\mu\{f^{-1}(A_j)\}$, where y_i and y_j are intermediate values in A_i and A_j. But $f^{-1}(A_i)$ and $f^{-1}(A_j)$ consist of the rationals and irrationals, respectively, in $[0, 1]$, whose respective Lebesgue measures are 0 and 1. Therefore, Expression (5.3) is y_j. As the common width of the intervals tends to 0, $y_j = y_{j,n}$ tends to 1. Therefore, the Lebesgue integral $\int_0^1 f(x)dx$ exists and equals 1.

We can define the integral of a function f with respect to an arbitrary measure μ (not just Lebesgue measure) on an arbitrary space Ω (not just the line). If μ is a measure on a

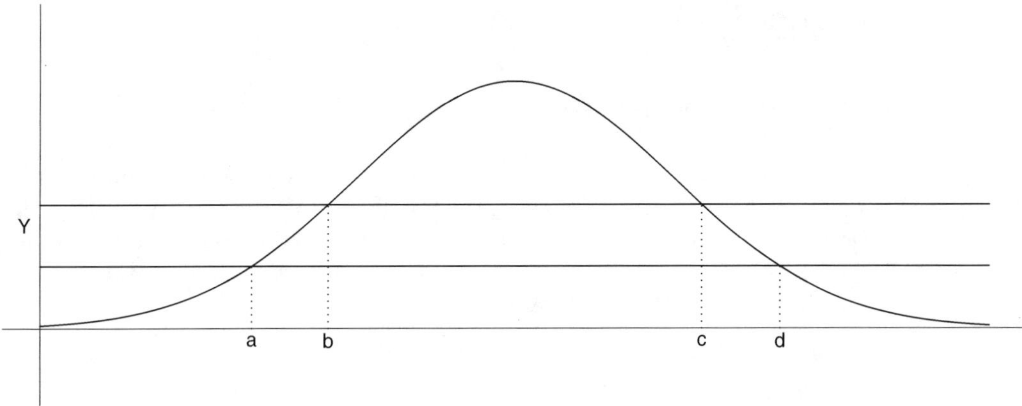

Figure 5.2: The horizontal lines show one interval A_i of a partition of the y-axis. The set $f^{-1}(A_i)$ of xs that get mapped into A_i is $[a, b] \cup [c, d]$. Each term of Expression (5.3) is the product of an intermediate y value in A_i and $\mu\{f^{-1}(A_i)\} = b - a + d - c$.

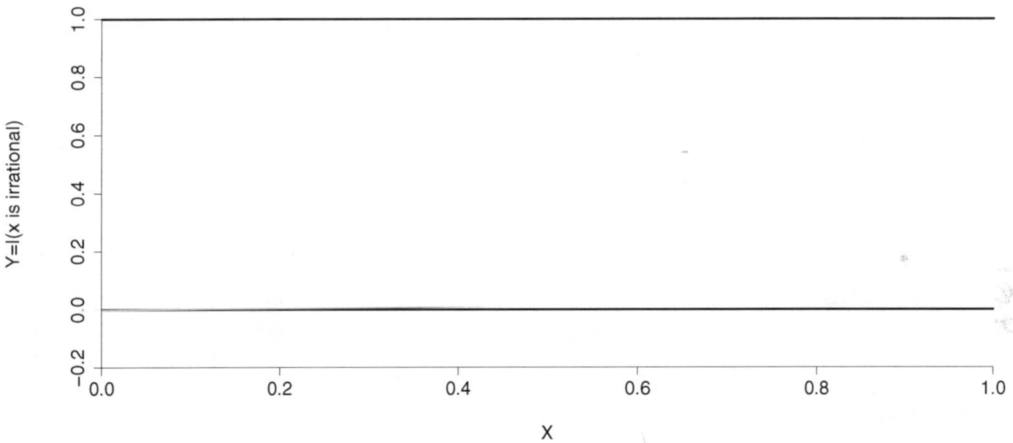

Figure 5.3: Partitioning the y-axis to form a Lebesgue partial sum for the function (5.2). When the common interval width is small enough, the intervals containing 0 and 1 are distinct.

probability space Ω, we can define the integral

$$\int f(\omega)d\mu(\omega) \tag{5.4}$$

by partitioning the y-axis into intervals of equal width, forming the sum of Expression (5.3), and then taking the limit as the common width tends to 0. There are two potential problems: 1) the limit may not exist and 2) the limit may depend on the intermediate point y_i selected. We avoid both of these problems if f is nonnegative and we choose the intervals in a clever way, as we see in the next section.

5.2 Lebesgue-Stieltjes Integration

5.2.1 Nonnegative Integrands

Let $(\Omega, \mathcal{F}, \mu)$ be a measure space, and let $f(\omega) : \Omega \mapsto R$ be a nonnegative, finite-valued $\mathcal{F}$-measurable function; i.e., $f^{-1}(B) \in \mathcal{F}$ for every Borel set B of the line. We define the Lebesgue-Stieltjes integral of f using a stepwise process. Step 0 is to divide the y-axis into intervals of length 1, $A_{0i} = [i-1, i)$, $i = 1, 2, \ldots$ and form partial sums of the form of Expression (5.3), where y_i is the smallest value, $i-1$, in A_{0i}:

$$S_0 = \sum_{i=1}^{\infty}(i-1)\mu_{0i}, \quad \text{where } \mu_{0i} = \mu\{\omega : f(\omega) \in A_{0i}\}. \tag{5.5}$$

Step 1 divides each A_{0i} in half, yielding $A_{1i} = [(i-1)/2, i/2)$, $i = 1, 2, \ldots$. Again form partial sums of the form of Expression (5.3), where y_i is the smallest value of interval A_{1i}. This yields

$$S_1 = \sum_{i=1}^{\infty}\{(i-1)/2\}\mu_{1i}, \quad \text{where } \mu_{1i} = \mu\{\omega : f(\omega) \in A_{1i}\}. \tag{5.6}$$

Notice that $S_0 \le S_1$.

Continue dividing intervals in half and forming partial sums of the form of Expression (5.3), where y_i is the smallest value in the interval. At step m, we get $A_{mi} = [(i-1)/2^m, i/2^m)$ and

$$S_m = \sum_{i=1}^{\infty}\{(i-1)/2^m\}\mu_{mi} \quad \text{where } \mu_{mi} = \mu\{\omega : f(\omega) \in A_{mi}\}. \tag{5.7}$$

Also, $S_0 \le S_1 \le \ldots S_m \le \ldots$ It follows from Proposition A.33 that $\lim_{m\to\infty} S_m$ exists, although it might be infinite.

Definition 5.1. Lebesgue integral of a nonnegative function *The Lebesgue-Stieltjes integral $\int f(\omega)d\mu(\omega)$ of the nonnegative, finite-valued, $\mathcal{F}$-measurable function $f(\omega)$ with respect to the measure μ is defined to be $\lim_{m\to\infty} S_m$, where S_m is defined by Expression (5.7). If μ is a probability measure P and the nonnegative function $f(\omega)$ is a random variable $X(\omega)$, then $\int X(\omega)dP(\omega)$ is called the expectation or expected value of X, denoted* $\mathrm{E}(X)$.

Remark 5.2. When f can be infinite *We have assumed that $f(\omega) < \infty$ because if $f(\omega) = \infty$, then $f(\omega)$ is not contained in any of the intervals A_{mi}. We now extend the definition of integration when $f(\omega)$ is a nonnegative function such that $f^{-1}(B) \in \mathcal{F}$ for each B in the extended Borel sigma-field $\bar{\mathcal{B}}$ (see Remark 3.13). Suppose the μ measure of the set of ω such that $f(\omega) = \infty$ is 0. We adopt the convention that $\infty \cdot 0 = 0$, so the value of the integral does not change if f is infinite only on a set of measure 0. On the other hand, if the nonnegative function $f(\omega)$ satisfies $f(\omega) = \infty$ on $\omega \in A$, where $\mu(A) > 0$, then the Lebesgue-Stieltjes integral $\int f(\omega)d\mu(\omega)$ is defined to be ∞.*

The reader may wonder why y_i of Expression (5.3) is chosen to be the smallest number in the given y-interval instead of requiring the sum to converge to the same limit irrespective of the intermediate value y_i selected in the given interval. It turns out that this happens automatically if μ is a finite measure such as a probability measure. To see this, suppose that we use instead the largest value, $y_i = i/2^m$, in each interval and take the limit of

$$\sum_{i=1}^{\infty}(i/2^m)\mu_{mi}. \tag{5.8}$$

If Expression (5.7) tends to ∞, then so does Expression (5.8). On the other hand, suppose that Expression (5.7) tends to a finite limit L. Then

$$\begin{aligned}
\sum_{i=1}^{\infty}(i/2^m)\mu_{mi} &= \sum_{i=1}^{\infty}\{(i-1+1)/2^m\}\mu_{mi} \\
&= \sum_{i=1}^{\infty}\{(i-1)/2^m\}\mu_{mi} + (1/2)^m\sum_{i=1}^{\infty}\mu_{mi} \\
&= \sum_{i=1}^{\infty}\{(i-1)/2^m\}\mu_{mi} + (1/2)^m\mu\{\omega : f(\omega) < \infty\} \\
&\to L + 0 = L.
\end{aligned} \tag{5.9}$$

The last step follows from the fact that $\mu\{\omega : f(\omega) < \infty\}$ is no greater than $\mu(\Omega) < \infty$. Because the limit of Expression (5.3) for an arbitrary choice of intermediate value y_i is bounded below by the limit when the smallest y_i is chosen and bounded above by the limit when the largest y_i is chosen, this shows that for finite measures, we get the same limit for Expression (5.3) regardless of which intermediate value y_i is selected.

For non-finite measures such as Lebesgue measure on the line, we do not necessarily get the same answer when we use the largest value in each interval instead of the smallest. For instance, consider the function $f(\omega) \equiv 0$ and Lebesgue measure μ. By any reasonable definition, $\int f(\omega)d\mu(\omega)$ should be 0, yet that is not what we get if we use the largest value in each interval. Expression (5.8) is ∞ for each m because the measure of the first interval, μ_{m1}, is $\mu(-\infty,\infty) = \infty$. That is why this formulation of the Lebesgue-Stieltjes integral approximates f from **below** by using the smallest number in each interval. It should be noted, however, that there are several equivalent alternative definitions of the Lebesgue integral.

Proposition 5.3. If f is nonnegative, $\int f d\mu = 0$ if and only if $f = 0$ a.e. *If $f(\omega)$ is a nonnegative, $\mathcal{F}$-measurable function, then $\int f(\omega)d\mu(\omega) = 0$ if and only if $\mu\{\omega : f(\omega) > 0\} = 0$.*

Proof. Exercise.

Exercises

1. Suppose that $\Omega = [0,1]$ and $\mathcal{F}$ consists of Borel sets of $[0,1]$. Let

$$f(\omega) = \begin{cases} 0 & \text{if } \omega \text{ is rational} \\ 5 & \text{if } \omega \text{ is irrational.} \end{cases}$$

What is $\int f(\omega)d\mu(\omega)$? What if $\Omega = R$ and $\mathcal{F} = \mathcal{B}$?

2. Suppose that $\Omega = R$, $\mathcal{F} = \mathcal{B}$, and μ is Lebesgue measure. Let $E \in \mathcal{F}$ be any countable set, and f be a nonnegative $\mathcal{F}$-measurable function. What is $\int f(\omega)d\mu(\omega)$?

3. Show using the definition of the integral that $\int I(A)d\mu(\omega) = \mu(A)$. Prove that the integral of a nonnegative function taking only finitely many values $a_1, \ldots, a_k$ on respective $\mathcal{F}$ sets $A_1, \ldots, A_k$ is $\sum_{i=1}^{k} a_i\mu(A_i)$.

4. Using the fact that expression Expression (5.7) is equivalent to $(1/2^m)\sum_{i=2}^{\infty}(i-1)\mu_{mi}$, and

$$\begin{array}{rcccccccc} \sum_{i=2}^{\infty}(i-1)\mu_{mi} & = & \mu_{m2} & + & \mu_{m3} & + & \mu_{m4} & + & \cdots \\ & & & + & \mu_{m3} & + & \mu_{m4} & + & \cdots \\ & & & + & & + & \mu_{m4} & + & \cdots, \end{array} \tag{5.10}$$

show that the integral of the nonnegative measurable function f is

$$\int f(\omega)d\mu(\omega) = \lim_{m\to\infty}(1/2^m)\sum_{i=1}^{\infty}\mu\{\omega : f(\omega) \geq i/2^m\}. \tag{5.11}$$

5. $\uparrow$ Prove that if f and g are nonnegative measurable functions with $f(\omega) \leq g(\omega)$ for all ω, then $\int f(\omega)d\mu(\omega) \leq \int g(\omega)d\mu(\omega)$.

6. Prove Proposition 5.3.

5.2.2 General Integrands

Now suppose that $f(\omega)$ is any $\mathcal{F}$-measurable function. Let $f^+(\omega) = f(\omega)I\{f(\omega) \geq 0\}$ and $f^-(\omega) = -f(\omega)I\{f(\omega) < 0\}$. Despite the somewhat misleading notation, f^+ and f^- are both nonnegative functions. Also, they are $\mathcal{F}$-measurable and $+f(\omega) = f^+(\omega) - f^-(\omega)$ and $|f(\omega)| = f^+(\omega) + f^-(\omega)$. We have already defined the integral for nonnegative functions, so $\int f^+(\omega)d\mu(\omega)$ and $\int f^-(\omega)d\mu(\omega)$ are well-defined. This leads naturally to the following definition.

Definition 5.4. Lebesgue integral for general integrands *If f is any $\mathcal{F}$-measurable function, $\int f(\omega)d\mu(\omega)$ is defined as $\int f^+(\omega)d\mu(\omega) - \int f^-(\omega)d\mu(\omega)$, provided this expression is not of the form $\infty - \infty$. If it is of the form $\infty - \infty$, then we say that $\int f(\omega)d\mu(\omega)$ does not exist. When μ is a probability measure P and $f(\omega)$ is a random variable $X(\omega)$, the integral $\int X(\omega)dP(\omega)$ is called the expectation or expected value of X, denoted $E(X)$.*

Notice that Definition 5.4 allows the integral to be infinite. For instance, the integral is $+\infty$ if $\int f^+(\omega)d\mu(\omega) = \infty$ and $\int f^-(\omega)d\mu(\omega) < \infty$. The integral is $-\infty$ if $\int f^+(\omega)d\mu(\omega) < \infty$ and $\int f^-(\omega) = \infty$.

One common expectation encountered in statistics is of the squared deviation from the mean, $\mathrm{E}\{X - \mathrm{E}(X)\}^2$, called the variance of X. The standard deviation of X is the square root of the variance. Statisticians often use the following notation:

$$\begin{array}{rcl} \textbf{mean} : \mu & = & \mathrm{E}(X) \\ \textbf{variance} : \sigma^2 & = & \mathrm{E}(X-\mu)^2 \\ \textbf{standard deviation} : \sigma & = & \sqrt{\sigma^2}. \end{array} \tag{5.12}$$

Finally, if X and Y are random variables with means μ_X, μ_Y and finite variances σ_X^2 and σ_Y^2, the covariance σ_{XY} and correlation ρ_{XY} between X and Y are

$$\begin{array}{rcl} \textbf{covariance} : \sigma_{XY} & = & \mathrm{E}\{(X-\mu_X)(Y-\mu_Y)\} = \int \{X(\omega)-\mu_X\}\{Y(\omega)-\mu_Y\}dP(\omega) \\ & = & \mathrm{E}(XY) - \mu_X\mu_Y \\ \textbf{correlation} : \rho_{XY} & = & \sigma_{XY}/(\sigma_X\sigma_Y). \end{array}$$

We have defined integration over the entire sample space Ω, though we have suppressed the notation $\int_\Omega f(\omega)d\mu(\omega)$ to simply $\int f(\omega)d\mu(\omega)$. We define integration over any set $A \in \mathcal{F}$ as follows.

Definition 5.5. Integration over a specific set *If $A \in \mathcal{F}$, $\int_A f(\omega)d\mu(\omega)$ is defined to be $\int_\Omega f(\omega)I(\omega \in A)d\mu(\omega)$.*

This definition ensures that we can always consider the region of integration to be the entire space Ω.

The way we have defined the expectation of a random variable X involves the measure P on the original probability space $(\Omega, \mathcal{F}, P)$, but more elementary courses define expectation without ever specifying $(\Omega, \mathcal{F}, P)$. If we look more carefully, we see that we do not really need to know the original space. Recall that the random variable X induces a probability measure P' on the line through $P'(B) = P\{\omega : X(\omega) \in B\}$ for each one-dimensional Borel set B. Therefore, the approximating sum $S_m = \sum_{i=1}^{\infty}\{(i-1)/2^m\}P[(i-1)/2^m \le X < i/2^m]$ of Expression (5.7) for a nonnegative random variable $X(\omega)$ can be written in terms of the induced measure P' as $S_m = \sum_{i=1}^{\infty}\{(i-1)/2^m\}P'[(i-1)/2^m, i/2^m)$. Similarly, $\lim_{m\to\infty} S_m$ depends only on P'. Consequently:

Proposition 5.6. Law of the unconscious statistician *The expectation $E(X) = \int_\Omega X(\omega)dP(\omega)$ of a random variable X is equivalent to $\int_R x dP'(x)$, where P' is the probability measure on $(R, \mathcal{B})$ induced by X.*

In other words, for the computation of E(X), we can take the probability space to be $(R, \mathcal{B}, P')$ and the random variable to be $f(x) = x$. This shows that the expectation of X depends only on the measure P' on R induced by X.

Remark 5.7. Another way to denote E(X) *Because the induced probability measure P' is completely determined by the distribution function $F(x)$ of X (see Proposition 4.22), it is common to denote the expectation of X by $\int x dF(x)$.*

5.3 Properties of Integration

We began this chapter by reviewing Riemann integration and its limitations. We then introduced the Lebesgue integral and saw that it exists even when the integrand is very irregular. It can be shown that when the Riemann and Lebesgue integrals both exist, they are equal. Therefore, it is not surprising that the same properties hold for Riemann and Lebesgue integrals. We review some of these important properties below. Because the reader has undoubtedly encountered these in the context of Riemann integration, we omit some of the proofs. We focus instead on how to use these elementary properties to deduce other results. We give each result in two forms. Part (a) is the general result for arbitrary measures, while part (b) expresses the result in random variable notation for probability measures.

Elementary properties of integration

Assume that the functions below are all $\mathcal{F}$-measurable.

1. **Linearity** If c_1 and c_2 are constants,

(a) $\int\{c_1f_1(\omega)+c_2f_2(\omega)\}d\mu(\omega) = c_1\int f_1(\omega)d\mu(\omega)+c_2\int f_2(\omega)d\mu(\omega)$ whenever the right side is not of the form $\infty-\infty$ or $-\infty+\infty$.

(b) $\mathrm{E}(c_1X_1+c_2X_2)=c_1\mathrm{E}(X_1)+c_2\mathrm{E}(X_2)$ whenever the right side is not of the form $\infty-\infty$ or $-\infty+\infty$.

2. **Integrable⇔absolutely integrable**

   (a) $\int f(\omega)d\mu(\omega)$ exists and is finite if and only if $\int|f(\omega)|d\mu(\omega)<\infty$.

   (b) $\mathrm{E}(X)$ exists and is finite if and only if $\mathrm{E}(|X|)<\infty$.

3. **Modulus inequality***

   (a) $|\int f(\omega)d\mu(\omega)|\le\int|f(\omega)|d\mu(\omega)$.

   (b) $|\mathrm{E}(X)|\le\mathrm{E}(|X|)$.

4. **Preservation of ordering***

   (a) If $f(\omega)\le g(\omega)$ for $\omega\notin N$, where $\mu(N)=0$, then $\int f(\omega)d\mu(\omega)\le\int g(\omega)d\mu(\omega)$.

   (b) If $P(X\le Y)=1$, then $\mathrm{E}(X)\le\mathrm{E}(Y)$.

5. **Integration term by term**

   (a) If either f_n is nonnegative for all n or $\sum_{n=1}^{\infty}\int|f_n(\omega)|d\mu(\omega)<\infty$, then $\int\sum_{n=1}^{\infty}f_n(\omega)d\mu(\omega)=\sum_{n=1}^{\infty}\int f_n(\omega)d\mu(\omega)$.

   (b) If either X_n is nonnegative for all n or $\sum_{n=1}^{\infty}\mathrm{E}(|X_n|)<\infty$, then $\mathrm{E}\left(\sum_{n=1}^{\infty}X_n\right)=\sum_{n=1}^{\infty}\mathrm{E}(X_n)$.

* Holds even if one or both sides are infinite.

Proof of select items.

2. By definition, $\int f(\omega)d\mu(\omega)$ is finite if and only if $\int f^-(\omega)d\mu(\omega)<\infty$ and $\int f^+(\omega)d\mu(\omega)<\infty$. On the other hand,

$$\begin{aligned}\int|f(\omega)|d\mu(\omega) &= \int\{f^-(\omega)+f^+(\omega)\}d\mu(\omega)\\ &= \int f^-(\omega)d\mu(\omega)+\int f^+(\omega)d\mu(\omega)\end{aligned}$$

   by property 1, so $\int|f(\omega)|d\mu(\omega)<\infty$ if and only if $\int f^-(\omega)d\mu(\omega)<\infty$ and $\int f^+(\omega)d\mu(\omega)<\infty$. Therefore, $\int f(\omega)d\mu(\omega)$ is finite if and only if $\int|f(\omega)|d\mu(\omega)<\infty$.

3. The result is trivially true if the right side is ∞, so it suffices to consider f such that $\int|f(\omega)|d\mu(\omega)<\infty$, and therefore both $\int f^-(\omega)d\mu(\omega)<\infty$ and $f^+(\omega)d\mu(\omega)<\infty$. In that case,

$$\begin{aligned}\left|\int f(\omega)d\mu(\omega)\right| &= \left|\int\{f^+(\omega)-f^-(\omega)\}d\mu(\omega)\right|\\ &= \left|\int f^+(\omega)d\mu(\omega)-\int f^-(\omega)d\mu(\omega)\right| \text{ (property 1)}\\ &\le \left|\int f^+(\omega)d\mu(\omega)\right|+\left|\int f^-(\omega)d\mu(\omega)\right| \text{ (triangle inequality)}\\ &= \int f^+(\omega)d\mu(\omega)+\int f^-(\omega)d\mu(\omega)\end{aligned}$$

$$= \int \{f^+(\omega)d\mu(\omega) + f^-(\omega)d\mu(\omega)\} \text{ (property 1)}$$
$$= \int \left|f(\omega)\right| d\mu(\omega). \qquad \square$$

Note that the modulus inequality is analogous to the triangle inequality; if we replace the integral in the modulus inequality with a finite sum, we get the triangle inequality.

A common technique is to use the above properties in conjunction with indicator functions to prove certain results, as in the following example.

Example 5.8. Use of indicator functions We would like to use the above properties to prove that if $\mathrm{E}(|X|^p) < \infty$ for some real number $p \geq 1$, then $\mathrm{E}(|X|) < \infty$. It is tempting to argue that $|X| \leq |X|^p$, so $\mathrm{E}(|X|) \leq \mathrm{E}(|X|^p) < \infty$ by the preservation of ordering property, but this is not quite right because if $X(\omega) < 1$, then $|X| > |X|^p$. We can modify the argument by using indicator functions.

$$|X| = |X|I(|X| \leq 1) + |X|I(|X| > 1).$$

If we can show that $\mathrm{E}\{|X|I(|X| \leq 1)\} < \infty$ and $\mathrm{E}\{|X|I(|X| > 1)\} < \infty$, then the result will follow from property 1. To show that $\mathrm{E}\{|X|I(|X| \leq 1)\} < \infty$, note that $\{|X|I(|X| \leq 1)\} \leq 1$, so by the preservation of ordering property, $\mathrm{E}\{|X|I(|X| \leq 1)\} \leq \mathrm{E}(1) = 1 < \infty$. To show that $\mathrm{E}\{|X|I(|X| > 1)\} < \infty$, note that $|X|I(|X| > 1) \leq |X|^p I(|X| > 1) \leq |X|^p$. Therefore, $\mathrm{E}\{|X|I(|X| > 1)\} \leq \mathrm{E}(|X|^p) < \infty$. Property 1 now shows that

$$\mathrm{E}(|X|) = \mathrm{E}\{|X|I(|X| \leq 1)\} + \mathrm{E}\{|X|I(|X| > 1)\} < \infty,$$

completing the proof. Thus, for example, if the second moment $\mathrm{E}(X^2)$ is finite, then $\mathrm{E}(|X|)$ is finite, so $\mathrm{E}(X)$ exists and is finite by property 2. Hence, for random variables with a finite second moment, the mean exists and is finite. $\square$

A very important result whose proof also uses indicator functions is the following.

Proposition 5.9. Necessary and sufficient condition for $\mathrm{E}(|X|^k) < \infty$ *If X is a random variable and k is a positive integer, then $\mathrm{E}(|X|^k) < \infty$ if and only if $\sum_{i=1}^{\infty} i^{k-1} P(|X| \geq i) < \infty$.*

Proof. We prove the result for $k = 1$. By the preservation of ordering property,

$$\begin{aligned} \mathrm{E}(|X|) &= \sum_{i=1}^{\infty} \mathrm{E}\{|X|I(i-1 \leq |X| < i)\} \\ &\leq \sum_{i=1}^{\infty} iP(i-1 \leq |X| < i) = \sum_{i=1}^{\infty} iP_i, \end{aligned}$$

where $P_i = P(i-1 \leq |X| < i)$. Similarly, $\mathrm{E}(|X|) \geq \sum_{i=1}^{\infty}(i-1)P_i = (\sum_{i=1}^{\infty} iP_i) - 1$. Therefore, $\mathrm{E}(|X|^k) < \infty$ if and only if $\sum_{i=1}^{\infty} iP_i < \infty$. Also,

$$\begin{array}{rcccccccl} \sum_{i=1}^{\infty} iP_i &=& P_1 &+& P_2 &+& P_3 &+& \ldots \\ &&&+& P_2 &+& P_3 &+& \ldots \\ &&&&&+& P_3 &+& \ldots, \end{array} \tag{5.13}$$

etc. The ith row sums to $P(|X| \geq i-1)$. Summing these totals over rows shows that $\sum_{i=1}^{\infty} iP_i = \sum_{i=1}^{\infty} P(|X| \geq i-1) = 1 + \sum_{i=1}^{\infty} P(|X| \geq i)$. Thus, $\mathrm{E}(|X|) < \infty$ if and only if $\sum_{i=1}^{\infty} P(|X| \geq i) < \infty$. $\square$

A more mundane way to view the trick of rearranging $\sum_{i=1}^{\infty} iP_i$ into rows and columns using Expression (5.13) is as follows.

$$\begin{aligned}\sum_{i=1}^{\infty} iP_i &= \sum_{i=1}^{\infty} P_i \left[\sum_{h=1}^{i}\{h-(h-1)\}\right] = \sum_{i=1}^{\infty}\sum_{h=1}^{i} P_i \\ &= \sum_{h=1}^{\infty}\left\{\sum_{i=h}^{\infty} P_i\right\} = \sum_{h=1}^{\infty} P(|X| \geq h-1). \qquad (5.14)\end{aligned}$$

Likewise, to prove the result for $k > 1$, write $\sum_{i=1}^{\infty} i^k P_i$ as $\sum_{i=1}^{\infty} P_i\{\sum_{h=1}^{i} h^k - (h-1)^k\}$ (the outer sum telescopes, meaning that almost all terms cancel each other out) and change the order of summation (exercise).

One question that comes up repeatedly in probability theory and real analysis is under what conditions the pointwise convergence of $f_n(\omega)$ to $f(\omega)$ implies that

$$\lim_{n\to\infty} \int f_n(\omega)d\mu(\omega) = \int \lim_{n\to\infty} f_n d\mu(\omega) = \int f(\omega)d\mu(\omega). \qquad (5.15)$$

That is, when can we interchange the limit and the integral? To see that some conditions are needed, let μ be Lebesgue measure on $\Omega = (0,1)$ and consider the measurable functions

$$f_n(\omega) = \begin{cases} n & \text{if } 0 < \omega \leq 1/n \\ 0 & \text{if } 1/n < \omega < 1. \end{cases}$$

Then $f_n(\omega)$ converges to $f(\omega) = 0$ for every $\omega \in (0,1)$, yet

$$\begin{aligned}\int f_n(\omega)d\mu(\omega) &= n \cdot \mu\{\omega : 0 < \omega \leq 1/n\} + 0 \cdot \mu\{\omega : 1/n < \omega < 1\} \\ &= n(1/n) + 0 = 1 \qquad (5.16)\end{aligned}$$

for each n. Therefore, $\lim_{n\to\infty} \int f_n(\omega)d\mu(\omega) = 1$ but $\int \lim_{n\to\infty} f_n(\omega)d\mu(\omega) = \int 0\, d\mu(\omega) = 0$.

Notice that in the above example, $\int \underline{\lim}\, f_n(\omega) \leq \underline{\lim} \int f_n(\omega)d\mu(\omega)$. In fact, this always holds:

Lemma 5.10. Fatou's lemma

1. *If $f_n(\omega) \geq 0$ for all $\omega \in \Omega \setminus N$, where $\mu(N) = 0$, then $\int \underline{\lim}\, f_n(\omega)d\mu(\omega) \leq \underline{\lim} \int f_n(\omega)d\mu(\omega)$.*
2. *In random variable terminology, if $P(X_n \geq 0) = 1$, then $\mathrm{E}\{\underline{\lim}\,(X_n)\} \leq \underline{\lim}\, \mathrm{E}(X_n)$.*

The importance of Fatou's lemma is that when we ponder whether Equation (5.15) holds, we no longer have to wonder whether the integral on the right is finite; if the limit on the left is finite, then the integral on the right is finite.

In Section A.6.2 we discussed, in the context of Riemann integration over a closed interval, a sufficient condition for interchanging limits and integration (see Proposition A.66). We see from Fatou's lemma that we did not need to impose the condition that $\int_A^B f(x)dx$ exists. Actually, the conditions of Proposition A.66 are quite strong. We would like to be able to interchange limits and integrals under weaker conditions. Two more useful results along these lines are the monotone convergence theorem (MCT) and dominated convergence theorem (DCT). We give one version of these results now, strengthening them later.

Theorem 5.11. monotone convergence theorem (MCT)

1. *Suppose that, for ω outside a null set N (i.e., $\mu(N) = 0$), $f_n(\omega) \geq 0$ for all n and $f_n(\omega) \uparrow f(\omega)$. Then $\int f_n(\omega)d\mu(\omega) \uparrow \int f(\omega)d\mu(\omega)$.*

2. *In random variable terminology, if $P\{\omega : X_n(\omega) \geq 0$ for all n and $X_n(\omega) \uparrow X(\omega)\} = 1$, then* $\mathrm{E}(X_n) \uparrow \mathrm{E}(X)$.

Theorem 5.12. Dominated convergence theorem (DCT)

1. *Suppose that for all ω outside a null set, $f_n(\omega) \to f(\omega)$ and $|f_n(\omega)| \leq g(\omega)$, where $\int g(\omega)d\mu(\omega) < \infty$. Then $\int f_n(\omega)d\mu(\omega) \to \int f(\omega)d\mu(\omega)$.*

2. *In random variable terminology, if $P\{\omega : X_n(\omega) \to X(\omega)$ and $|X_n(\omega)| \leq Y(\omega)$* for all n$\}$ $= 1$, *where* $\mathrm{E}(Y) < \infty$, *then* $\mathrm{E}(X_n) \to \mathrm{E}(X)$.

An immediate corollary to the DCT is the following.

Theorem 5.13. Bounded convergence theorem (BCT) *Let X_n be a random variable such that $P\{X_n(\omega) \to X(\omega)\} = 1$ and $P\{|X_n| \leq c\} = 1$ for each n, where c is a constant. Then* $\mathrm{E}(X_n) \to \mathrm{E}(X)$.

Remark 5.14. MCT and DCT also apply to sums *It is important to realize that the MCT and DCT apply to* **all** *integrals, and because sums are just integrals with respect to counting measure, they apply to sums as well.*

The following example illustrates Remark 5.14.

Example 5.15. Suppose we want to evaluate $\lim_{n\to\infty} \sum_{k-1}^{\infty} (-1)^n/(nk^2)$. Write the sum in more suggestive integral notation as follows. Let μ be counting measure (see Example 3.23) on $\Omega = \{1, 2, \ldots\}$ and set $f_n(\omega) = (-1)^n/(n\omega^2)$ for $\omega = 1, 2, \ldots$ Then $f_n(\omega) \to f(\omega) = 0$ as $n \to \infty$ for each ω. Furthermore, $|f_n(\omega)| \leq 1/\omega^2$ for each n, where $\int_\Omega (1/\omega^2)d\mu(\omega) = \sum_{\omega=1}^{\infty} 1/\omega^2 < \infty$. By the DCT, $\lim_{n\to\infty} \sum_{k=1}^{\infty} (-1)^n/(nk^2) = \lim_{n\to\infty} \int_\Omega f_n(\omega)d\mu(\omega) = \int_\Omega 0\, d\mu(\omega) = 0$. □

Example 5.16. Here is an example of the use of the dominated convergence theorem to show that if $E(|X|) < \infty$ and $a_n \to \infty$, then $\int_{a_n}^{\infty} X(\omega)dP(\omega) \to 0$. By definition, $\int_{a_n}^{\infty} X(\omega)dP(\omega)$ means $\int_\Omega X(\omega)I\{X(\omega) \geq a_n\}dP(\omega) = \mathrm{E}\{XI(X \geq a_n)\}$. Then $Y_n(\omega) = X(\omega)I(X(\omega) \geq a_n)$ converges to 0 for each ω for which $X(\omega) < \infty$. Moreover, $|Y_n| \leq |X|$, and $E(|X|) < \infty$ by assumption. By the DCT, $\mathrm{E}(Y_n) \to 0$. □

Exercises

1. Use elementary properties of integration to prove that $\mathrm{E}\{(X - \mu_X)(Y - \mu_Y)\} = \mathrm{E}(XY) - \mu_X\mu_Y$, assuming the expectations are finite.

2. Explain why the notation $\int_a^b f(\omega)d\mu(\omega)$ is ambiguous unless $\mu(a) = \mu(b) = 0$. How should we write the integral if we mean to include the left, but not the right, endpoint of the interval?

3. Let $f(\omega)$ be an $\mathcal{F}$-measurable function such that $\int |f(\omega)|d\mu(\omega) < \infty$. Prove that if $A_1, A_2, \ldots$ are disjoint sets in $\mathcal{F}$, then $\int_{\cup_i A_i} f(\omega)d\mu(\omega) = \sum_{i=1}^{\infty} \int_{A_i} f(\omega)d\mu(\omega)$.

4. Show that the dominated convergence theorem (DCT) implies the bounded convergence theorem (Theorem 5.13).

5. Use the monotone convergence theorem (MCT) to prove part of Elementary Property 5, namely that if X_n are nonnegative random variables, then $\mathrm{E}(\sum_n X_n) = \sum \mathrm{E}(X_n)$.

6. Prove that if $\mathrm{E}(X) = \mu$, where μ is finite, then $\mathrm{E}\{XI(|X| \leq n)\} \to \mu$ as $n \to \infty$.

7. Prove that $\lim_{n\to\infty} \int_{[0,1]} \{\cos(nx)/n\}dx = 0$.

8. Find $\lim_{n\to\infty} \sum_{k=1}^{\infty} 1/\{k(1+k/n)\}$ and justify your answer.

9. Find $\lim_{n\to\infty} \sum_{k=1}^{\infty} (1-1/n)^{kn}/\{k!\ln(1+k)\}$ and justify your answer.

10. The dominated convergence theorem (DCT) has the condition that "$P\{\omega : X_n(\omega) \to X(\omega)$ and $|X_n(\omega)| \leq Y(\omega)$ for all n$\} = 1$." Show that this is equivalent to "$P\{\omega : X_n(\omega) \to X(\omega)\} = 1$ and $P\{\omega :\ |X_n(\omega)| \leq Y(\omega)\} = 1$ for each n.

11. * The DCT and MCT apply to limits involving $t \to t_0$ as well (see Definition A.54 of the Appendix). For example, show the following. Let f_t, f, and g be measurable functions and $A = \{\omega : |f_t(\omega)| \leq g(\omega)$ for all t and $f_t(\omega) \to f(\omega)$ as $t \to t_0\}$, where $\int g(\omega)d\mu < \infty$. If $A \in \mathcal{F}$ and $\mu(A^C) = 0$, then $\lim_{t\to t_0} \int f_t(\omega)d\mu(\omega) = \int f(\omega)d\mu(\omega)$.

12. $\uparrow$ Is the result of the preceding problem correct if it is stated as follows? Let $A_t = \{\omega : |f_t(\omega)| \leq g(\omega)\}$ and $B = \{\omega :\ f_t(\omega) \to f(\omega)$ as $t \to t_0\}$, where $\int g(\omega)d\mu < \infty$. If $\mu(A_t^C) = 0$ and $\mu(B^C) = 0$, then $\lim_{t\to t_0} \int f_t(\omega)d\mu(\omega) = \int f(\omega)d\mu(\omega)$.

13. Suppose that X is a random variable with density function $f(x)$, and consider $\mathrm{E}(X) = \int_{-\infty}^{\infty} xf(x)dx$. Assume further that f is symmetric about 0 (i.e., $f(-x) = f(x)$ for all $x \in R$). Is the following argument correct?

$$\begin{aligned}\int_{-\infty}^{\infty} xf(x)dx &= \lim_{A\to\infty} \int_{-A}^{A} xf(x)dx \\ &= \lim_{A\to\infty} 0 = 0\end{aligned}$$

because $g(x) = xf(x)$ satisfies $g(-x) = -g(x)$. Hint: consider $f(x) = \{\pi(1+x^2)\}^{-1}$; are $\mathrm{E}(X^-)$ and $\mathrm{E}(X^+)$ finite?

14. Show that if f_n are nonnegative measurable functions such that $f_n(\omega) \downarrow f(\omega)$, then it is not necessarily the case that $\lim_{n\to\infty} \int f_n(\omega)d\mu(\omega) = \int \lim_{n\to\infty} f_n(\omega)d\mu(\omega)$ for an arbitrary measure μ. Hint: let μ be counting measure and $f_n(\omega)$ be the indicator that $\omega \geq n$.

15. Use the preservation of ordering property to give another proof of the fact that if f is a nonnegative, measurable function, then $\int f(\omega)d\mu(\omega) = 0$ if and only if $\mu(A) = 0$, where $A = \{\omega : f(\omega) > 0\}$. Hint: if $\mu(A) > 0$, then there must be a positive integer n such that $\mu\{\omega : f(\omega) > 1/n\} > 0$.

16. Use elementary integration properties and Fatou's lemma to prove the monotone convergence theorem.

5.4 Important Inequalities

One of the most basic inequalities concerning expectation is Markov's inequality.

Proposition 5.17. Markov's inequality *If X is any random variable and $c > 0$, then $P(|X| \geq c) \leq \mathrm{E}(|X|)/c$.*

Proof.

$$\begin{aligned} |X| &= |X|I(|X| < c) + |X|I(|X| \geq c) \\ &\geq 0 + |X|I(|X| \geq c) \geq cI(|X| \geq c) \end{aligned}$$

$$\begin{aligned} \mathrm{E}(|X|) &\geq \mathrm{E}\{cI(|X| \geq c)\} \\ &= c\,P(|X| \geq c), \text{ so} \end{aligned}$$

$$P(|X| \geq c) \leq \mathrm{E}(|X|)/c. \qquad \square \tag{5.17}$$

Applying Markov's inequality to the random variable $(X - \mu)^2$ yields Chebychev's inequality:

Corollary 5.18. Chebychev's inequality) *If X is a random variable with mean μ and variance σ^2, and $c > 0$, then $P(|X - \mu| \geq c) \leq \sigma^2/c^2$.*

The result follows immediately from Markov's inequality and the fact that $|X - \mu| \geq c$ is equivalent to $(X - \mu)^2 \geq c^2$.

Chebychev's inequality is important in statistics because it shows that a random variable X is unlikely to be "too far" from its mean. Restated, Chebychev's inequality says that the probability of X being at least c standard deviations away from its mean is no greater than $1/c^2$.

Recall that a function $f(x)$ is said to be convex if $f(\lambda x_1 + (1 - \lambda)x_2) \leq \lambda f(x_1) + (1 - \lambda)f(x_2)$ for all $x_1 \leq x_2$ and $0 \leq \lambda \leq 1$. Geometrically, this condition says that the secant line joining $(x_1, f(x_1))$ to $(x_2, f(x_2))$ lies on or above the curve. An equivalent condition is that for each x, there exists a line passing through $(x, f(x))$ that lies entirely on or below the curve (see top panel of Figure 5.4). Similarly, a function is concave if its secant line lies on or below the curve. Equivalently, for each point x there exists a line passing through $(x, f(x))$ that lies entirely on or above the curve (see bottom panel of Figure 5.4).

Proposition 5.19. Jensen's inequality *Suppose that $f(x)$ is a convex function and that X is a random variable with finite mean μ. If $\mathrm{E}(|f(X)|) < \infty$, then $E\{f(X)\} \geq f\{\mathrm{E}(X)\}$.*

Proof. Because $f(x)$ is convex, there exists a line $y = f(\mu) + b(x - \mu)$ passing through $(\mu, f(\mu))$ that lies entirely below or on the curve. That is, $f(x) \geq f(\mu) + b(x - \mu)$. Now replace x by the random variable X and take the expected value of both sides to get, by the preservation of ordering property, $\mathrm{E}\{f(X)\} \geq f(\mu) + b\{\mathrm{E}(X) - \mu\} = f(\mu)$. $\square$

Corollary 5.20. *If X is any random variable and $p \geq 1$, then $E(|X|) \leq \{E(|X|^p)\}^{1/p}$.*

Proof. Exercise.

Note that Corollary 5.20 is another way to see that if $E(|X|^p)$ is finite for some $p \geq 1$, then $\mathrm{E}(|X|) < \infty$ (see also Example 5.8).

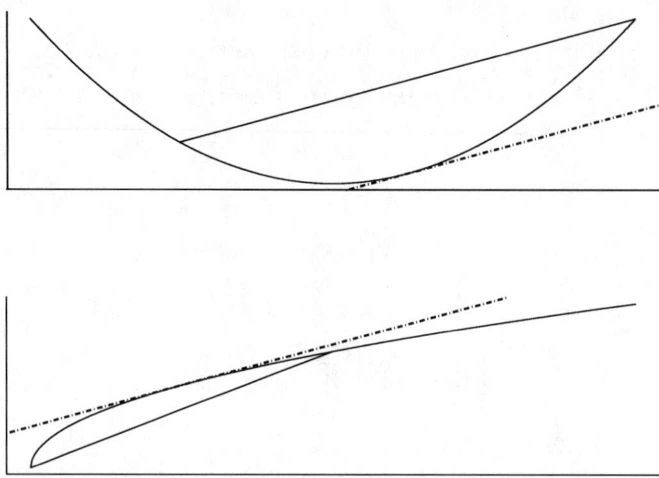

Figure 5.4: Top: convex function. The secant line (solid line) joining any two points on the curve lies on or above the curve. Equivalently, for each x there is a line through $(x, f(x))$ that lies entirely on or below the curve (dotted line). Bottom: concave function. The secant line (solid line) joining any two points on the curve lies on or below the curve. Equivalently, for each x there is a line through $(x, f(x))$ that lies entirely on or above the curve (dotted line).

In some cases we want to bound the expectation of a product of two random variables X and Y. One basic inequality comes from the fact that $(x-y)^2 = x^2 + y^2 - 2xy$, from which $2xy = x^2 + y^2 - (x-y)^2 \le x^2 + y^2$. That is,

$$xy \le x^2/2 + y^2/2. \tag{5.18}$$

If we replace x and y with random variables $|X|$ and $|Y|$, we deduce that

$$\mathrm{E}(|XY|) \le (1/2)\mathrm{E}(X^2) + (1/2)\mathrm{E}(Y^2).$$

Thus, if X and Y have finite second moments, then $\mathrm{E}(XY)$ is finite as well.

Another way to obtain inequality (5.18) is as a special case of another important inequality. Because $\ln(x)$ is a concave function on $x > 0$,

$$\lambda \ln(x_1) + (1-\lambda)\ln(x_2) \le \ln\{\lambda x_1 + (1-\lambda)x_2\}$$

for $0 < x_1 \le x_2$. Now exponentiate both sides to get

$$x_1^{\lambda} x_2^{1-\lambda} \le \lambda x_1 + (1-\lambda)x_2. \tag{5.19}$$

Let $p = 1/\lambda$ and $q = 1/(1-\lambda)$. Substituting $x = x_1^{1/p}$ and $y = x_2^{1/q}$ into Equation (5.19), we get

$$xy \le x^p/p + y^q/q \tag{5.20}$$

for positive x and y and $1/p + 1/q = 1$. Inequality (5.18) is the special case when $p = q = 2$, although (5.18) holds for positive or negative x, y, whereas (5.20) requires positive x and y.

Proposition 5.21. Hölder's inequality *If $p > 0$, $q > 0$, and $1/p + 1/q = 1$, then*

$$\mathrm{E}(|XY|) \le \{\mathrm{E}(|X|^p)\}^{1/p}\{\mathrm{E}(|Y|^q)\}^{1/q}. \tag{5.21}$$

Proof. Replace x and y in inequality (5.20) with $|X|$ and $|Y|$, random variables with $\mathrm{E}(|X|^p) = \mathrm{E}(|Y|^q) = 1$. Then $\mathrm{E}(|XY|) \le \{\mathrm{E}(|X|^p)\}/p + \{\mathrm{E}(|Y|^q)\}/q = 1/p + 1/q = 1$. This proves that (5.21) holds when $\mathrm{E}(|X|^p) = \mathrm{E}(|Y|^q) = 1$. More generally, if $\mathrm{E}(|X|^p) = \Gamma_{Xp} > 0$ and $\mathrm{E}(|Y|^q) = \Gamma_{Yq} > 0$, let $X' = X/\Gamma_{Xp}^{1/p}$ and $Y' = Y/\Gamma_{Yq}^{1/q}$. Then the pth moment of $|X'|$ and qth moment of $|Y'|$ are both 1 and

$$\begin{aligned} \mathrm{E}(XY) &= \mathrm{E}\{(\Gamma_{Xp}^{1/p}|X'|)(\Gamma_{Yq}^{1/q}|Y'|)\} \\ &= \Gamma_{Xp}^{1/p}\Gamma_{Yq}^{1/q}\mathrm{E}(|X'Y'|) \\ &\le \Gamma_{Xp}^{1/p}\Gamma_{Yq}^{1/q}, \end{aligned} \tag{5.22}$$

which proves the result whenever $\mathrm{E}(|X|^p) > 0$ and $\mathrm{E}(|Y|^q) > 0$. The result is immediate if either $\mathrm{E}(|X|^p) = 0$ (which implies that $X = 0$ with probability 1) or $\mathrm{E}(|Y|^q) = 0$ (which implies that $Y = 0$ with probability 1). □

Corollary 5.22. Schwarz's inequality *If X and Y are random variables,*

$$\mathrm{E}(|XY|) \le \sqrt{\mathrm{E}(X^2)\mathrm{E}(Y^2)}.$$

Another important inequality allows us to conclude that whenever $|X|^p$ and $|Y|^p$ are integrable, then so is $(X+Y)^p$ or $(X-Y)^p$.

Proposition 5.23. *$(|X| + |Y|)^p \le 2^p(|X|^p + |Y|^p)$ and $E(|X| + |Y|)^p \le 2^p\{\mathrm{E}(|X|^p) + \mathrm{E}(|Y|^p)\}$.*

Proof.

$$\begin{aligned} (|X| + |Y|)^p &\le (2\max(|X|, |Y|)^p \\ &= 2^p \max(|X|^p, |Y|^p) \\ &\le 2^p(|X|^p + |Y|^p). \end{aligned} \tag{5.23}$$

Taking the expectation of both sides yields the second result. □

Proposition 5.24. Minkowski's inequality *If $p \ge 1$, then*

$$\{\mathrm{E}(|X+Y|^p)\}^{1/p} \le \{\mathrm{E}(|X|^p)\}^{1/p} + \{\mathrm{E}(|Y|^p)\}^{1/p}. \tag{5.24}$$

Proof. Note that

$$|X+Y|^p \le (|X| + |Y|)|X+Y|^{p-1} = |X||X+Y|^{p-1} + |Y||X+Y|^{p-1}. \tag{5.25}$$

Now apply Hölder's inequality to $|X||X+Y|^{p-1}$ to conclude that

$$\mathrm{E}(|X||X+Y|^{p-1}) \le \{\mathrm{E}(|X|^p)\}^{1/p}\{\mathrm{E}(|X+Y|^{p-1})^q\}^{1/q}$$

whenever $1/p + 1/q = 1$. But $1/p + 1/q = 1$ means that $(p-1)q = p$. Therefore,

$$\mathrm{E}(|X||X+Y|^{p-1}) \le \{\mathrm{E}(|X|^p)\}^{1/p}\{\mathrm{E}(|X+Y|^p)\}^{1/q}. \tag{5.26}$$

Similarly,

$$\mathrm{E}(|Y||X+Y|^{p-1}) \le \{\mathrm{E}(|Y|^p)\}^{1/p}\{\mathrm{E}(|X+Y|^p)\}^{1/q}. \tag{5.27}$$

Combining Equation (5.25) with Equations (5.26) and (5.27) yields

$$\{\mathrm{E}(|X+Y|^p)\} \leq \left[\{\mathrm{E}(|X|^p)\}^{1/p} + \{\mathrm{E}(|Y|^p)\}^{1/p}\right] \{\mathrm{E}(|X+Y|^p)\}^{1/q} \tag{5.28}$$

Dividing both sides of Equation (5.28) by $\{\mathrm{E}(|X+Y|^p)\}^{1/q}$ and noting that $1 - 1/q = 1/p$ yields the stated result. □

Exercises

1. State and prove a result analogous to Jensen's inequality, but for concave functions.
2. Prove that if X has mean 0, variance σ^2, and finite fourth moment $\mu_4 = \mathrm{E}(X^4)$, then $\mu_4 \geq \sigma^4$.
3. Prove the Schwarz inequality.
4. Prove Corollary 5.20.
5. Prove that Markov's inequality is strict if $P(|X| > c) > 0$ or $\mathrm{E}\{|X|I(|X| < c\} > 0$. Does this imply that Markov's inequality is strict unless $|X| = c$ with probability 1? (Hint: consider X taking values c and 0 with probabilities p and $1 - p$).
6. Prove that the inequality in Jensen's inequality is strict unless $f(X) = f(\mu) + b(X - \mu)$ with probability 1 for some constant b.
7. Prove that if $0 < \sigma_X < \infty$ and $0 < \sigma_Y < \infty$, the correlation coefficient $\rho = \mathrm{E}(XY)/\sigma_X\sigma_Y$ between X and Y is between -1 and $+1$.
8. Suppose that x_i are positive numbers. The sample geometric mean is defined by $G = (\prod_{i=1}^n x_i)^{1/n}$. Note that $\ln(G) = (1/n)\sum_{i=1}^n \ln(x_i)$. Using this representation, prove that the arithmetic mean is always at least as large as the geometric mean.
9. The sample harmonic mean of numbers $x_1, \ldots, x_n$ is defined by $\{(1/n)\sum_{i=1}^n (1/x_i)\}^{-1}$. Show that the following ordering holds for positive numbers: harmonic mean $\leq$ geometric mean $\leq$ arithmetic mean. Does this inequality hold without the restriction that $x_i > 0$, $i = 1, \ldots, n$?
10. Let $f_0(x)$ and $f_1(x)$ be density functions with $\int |\ln\{f_1(x)/f_0(x)\}| f_0(x)dx < \infty$. Then $\int \ln\{f_1(x)/f_0(x)\} f_0(x)dx \leq 0$.
11. Suppose that X and Y are independent nonnegative, nonconstant random variables with mean 1 and both $U = X/Y$ and $V = Y/X$ have finite mean. Prove that U and V cannot both have mean 1.

5.5 Iterated Integrals and More on Independence

We pointed out the connection between integration over the original sample space Ω with its probability measure P and integration over R using the induced probability measure μ defined by its distribution function $F(x)$. The same connection exists for integrals of a function of more than one variable. For example, suppose we wish to compute the expectation of the random variable $g(X, Y)$, where X and Y are arbitrary random variables on

$(\Omega, \mathcal{F}, P)$ and $g : R^2 \mapsto R$ is a Borel function of (x, y). We get the same answer whether we integrate over the original sample space Ω using P or over the product space $R \times R$ using the induced measure μ:

$$\int_{\Omega} g(X(\omega), Y(\omega))dP(\omega) = \int\int_{R\times R} g(x, y)d\mu(x, y).$$

The integral on the right is defined the same way as was the integral on the original probability space, with Ω replaced by $R \times R$ and P replaced by μ.

If X and Y are independent with respective marginal probability measures μ_X and μ_Y, then the probability measure μ on $R \times R$ is product measure $\mu_X \times \mu_Y$ as defined in Section 4.5.2. A natural question when evaluating $\mathrm{E}\{g(X, Y)\}$ is whether we can integrate first over, say x, and then integrate the resulting expression over y. That is, we want to know whether

$$\int\int_{R\times R} g(x, y)d(\mu_X \times \mu_Y) = \int \left[\int g(x, y)d\mu_X(x)\right] d\mu_Y(y). \tag{5.29}$$

In particular, for this to make sense, $\int g(x, y)d\mu_X(x)$ must be a measurable function of y.

Remark 5.25. *The following technical presentation of conditions ensuring the validity of the interchange of order of integration is somewhat painful in that it requires the completion of probability spaces discussed in Proposition 3.35. We need this to ensure that certain subsets of sets of measure* 0 *are in the relevant sigma-fields. Loosely speaking, the interchange of order of integration is permitted if* $g(x, y)$ *is measurable with respect to product measure and either* g *is nonnegative (Tonelli's theorem) or* $\int\int_{R\times R} |g(x, y)|d(\mu_X \times \mu_Y) < \infty$ *(Fubini's theorem).*

For precise statement of the theorems in their general form, consider the following measure spaces.

1. Let $(A_1, \mathcal{A}, \mu_1)$ and $(A_2, \mathcal{A}_2, \mu_2)$ be arbitrary measure spaces.
2. Let $(A_1 \times A_2, \mathcal{A}_1 \times \mathcal{A}_2, \mu_1 \times \mu_2)$ be the product measure space associated with the spaces in part 1.
3. Let $(A_i, \overline{\mathcal{A}_i}, \overline{\mu_i})$, $i = 1, 2$, and $(A_1 \times A_2, \overline{\mathcal{A}_1 \times \mathcal{A}_2}, \overline{\mu_1 \times \mu_2})$ be the completions of these spaces, as in Proposition 3.35.

We need one other definition to state the result in full generality.

Definition 5.26. Sigma-finite *A measure* μ *on a measure space* $(A, \mathcal{A}, \mu)$ *is said to be sigma-finite if* $A = \cup_{i=1}^{\infty} A_i$, *where* $\mu(A_i) < \infty$.

Of course any probability measure or other finite measure is sigma-finite.

One important scenario under which we can interchange the order of integration is when g is nonnegative.

Theorem 5.27. Tonelli's theorem *If each* μ_i *is sigma finite and* $g(a_1, a_2)$ *is nonnegative and measurable with respect to* $\overline{\mathcal{A}_1 \times \mathcal{A}_2}$, *then* $\int g(a_1, a_2)d\overline{\mu_1}(a_1)$ *is measurable with respect to* $\overline{\mathcal{A}_2}$, *and*

$$\int\int_{A_1\times A_2} g(a_1, a_2)d(\overline{\mu_1 \times \mu_2}) = \int \left[\int g(a_1, a_2)d\overline{\mu_1}(a_1)\right] d\overline{\mu_2}(a_2) \tag{5.30}$$

holds regardless of whether the integrals are finite.

Another useful result is Fubini's theorem:

Theorem 5.28. Fubini's theorem *If* $\int\int_{A_1\times A_2} |g(a_1,a_2)|d(\overline{\mu_1\times\mu_2}) < \infty$ *and* μ_1 *and* μ_2 *are sigma-finite measures, then* $\int g(a_1,a_2)d\overline{\mu_1}(a_1)$ *is measurable with respect to* $\overline{\mathcal{A}_2}$ *and Equation (5.30) holds.*

These powerful results have implications for independence. In particular, they can be used to show that independent random variables are uncorrelated.

Proposition 5.29. Independent r.v.s are uncorrelated *If* X *and* Y *are independent random variables such that* $\mathrm{E}(|X|)<\infty$ *and* $\mathrm{E}(|Y|)<\infty$*, then* $\mathrm{E}(XY)=\mathrm{E}(X)\mathrm{E}(Y)$*; i.e.,* X *and* Y *are uncorrelated.*

Proof. Let μ_X and μ_Y be the marginal distribution functions for X and Y, respectively, and $\mu_X\times\mu_Y$ be product probability measure on $R\times R$. By Tonelli's theorem,

$$\begin{aligned}
\mathrm{E}(|XY|) &= \int\int_{R\times R}(|xy|)d(\mu_X\times\mu_Y) \\
&= \int_y\left\{\int_x(|x|\,|y|)d\mu_X(x)\right\}d\mu_Y(y) = \int_y |y|\left\{\int_x |x|d\mu_X(x)\right\}d\mu_Y(y) \\
&= \int_y |y|\mathrm{E}(|X|)d\mu_Y(y) = \mathrm{E}(|X|)\int_y |y|d\mu_Y(y) \\
&= \mathrm{E}(|X|)\mathrm{E}(|Y|). \qquad (5.31)
\end{aligned}$$

Therefore, XY is integrable over the product measure. By Fubini's theorem, the above steps can be repeated without the absolute values, yielding $\mathrm{E}(XY)=\mathrm{E}(X)\mathrm{E}(Y)$. Because $\mathrm{cov}(X,Y)=\mathrm{E}(XY)-\mathrm{E}(X)\mathrm{E}(Y)$, $\mathrm{E}(XY)=\mathrm{E}(X)\mathrm{E}(Y)$ is equivalent to $\mathrm{cov}(X,Y)=0$. □

The reverse direction does not hold. That is, X and Y can be uncorrelated without being independent. For instance, if $Z\sim \mathrm{N}(0,1)$, then Z and Z^2 are uncorrelated because $\mathrm{E}(ZZ^2)=\mathrm{E}(Z^3)=0=\mathrm{E}(Z)\mathrm{E}(Z^2)$. On the other hand, Z and Z^2 are clearly not independent. If they were, then with $p=P(1\le Z\le 1)$,

$$\begin{aligned}
p &= P(1\le Z\le 1\cap Z^2\le 1) \\
&= P(1\le Z\le 1)P(Z^2\le 1)=p^2, \qquad (5.32)
\end{aligned}$$

which implies that $p=0$ or 1. But $p\approx .68$ is not 0 or 1.

Suppose that not only are X and Y uncorrelated, but **every** Borel function of X is uncorrelated with **every** Borel function of Y (whenever the correlation exists). Then X and Y are independent.

Proposition 5.30. Independence is equivalent to 0 correlation between all functions of X and Y such that the covariance exists X *and* Y *are independent if and only if* $f(X)$ *and* $g(Y)$ *are uncorrelated (i.e.,* $\mathrm{E}\{f(X)g(Y)\}=\mathrm{E}\{f(X)\}\mathrm{E}\{g(Y)\}$*) for all Borel functions* f *and* g *such that* $\mathrm{cov}\{f(X),g(Y)\}$ *exists.*

Proof. If X and Y are independent, then so are $f(X)$ and $g(Y)$ by Proposition 4.39. By Proposition 5.29, $\mathrm{cov}\{f(X),g(Y)\}=0$, proving the $\rightarrow$ direction. Now suppose that $f(X)$ and $g(Y)$ are uncorrelated for all Borel functions f and g such that the correlation exists. Take $f(X)=I(X\le x)$ and $g(Y)=I(Y\le y)$. Then $\mathrm{cov}\{f(X),g(Y)\}=P(X\le x,Y\le y)-P(X\le x)P(Y\le y)$. This holds for arbitrary x and y, so the joint distribution

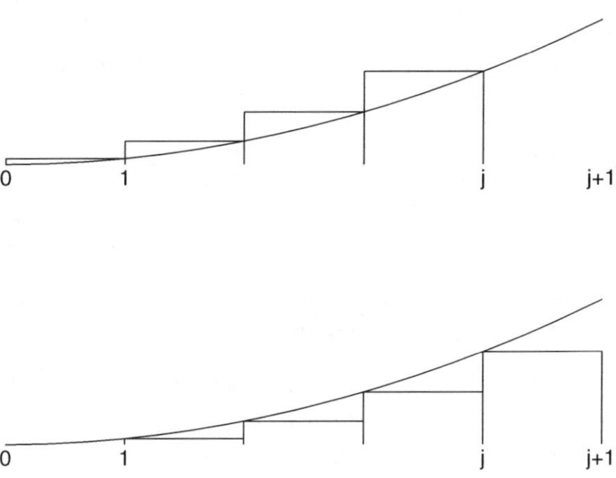

Figure 5.5: The sum $s = \sum_{i=1}^{j} i^k$ is the area of the bars embedded in the graph of $f(t) = t^k$, $j = \lfloor |x| \rfloor$; s is at least as large as the area under the curve between 0 and j (top panel), and no greater than the area under the curve between 1 and $j + 1$ (bottom panel).

function $F(x, y)$ for X and Y factors into $F_X(x)F_Y(y)$. By Proposition 4.37, X and Y are independent. This completes the proof of the $\leftarrow$ direction. □

Reversing the order of integration can be helpful in other settings as well. For instance, Proposition 5.9 states that if X is a random variable and k is a positive integer, then $\mathrm{E}(|X|^k) < \infty$ if and only if $\sum_{i=1}^{\infty} i^{k-1} P(|X| \geq i) < \infty$. One way to prove this is as follows.

$$\begin{aligned} \sum_{i=1}^{\infty} i^{k-1} P(|X| \geq i) &= \sum_{i=1}^{\infty} i^{k-1} \int_{\Omega} I(|X| \geq i) dP \\ &= \int_{\Omega} \sum_{i=1}^{\infty} i^{k-1} I(|X| \geq i) dP = \sum_{i=1}^{\lfloor |X| \rfloor} i^{k-1}, \end{aligned} \tag{5.33}$$

where $\lfloor |X| \rfloor$ denotes the greatest integer less than or equal to $|X|$. The reversal of sum and integral follows from Tonelli's theorem (the sum is an integral with respect to counting measure) Also, we see from Figure 5.5 that

$$\frac{\lfloor |X| \rfloor^k}{k} = \int_0^{\lfloor |X| \rfloor} t^{k-1} dt \leq \sum_{i=1}^{\lfloor |X| \rfloor} i^{k-1} \leq \int_1^{\lfloor |X| \rfloor + 1} t^{k-1} dt = \frac{(\lfloor |X| \rfloor + 1)^k}{k} \leq \frac{2^k (\lfloor |X| \rfloor^k + 1)}{k}$$

(the rightmost inequality is by Proposition 5.23). Therefore, $\int_{\Omega} \sum_{i=1}^{\lfloor |X| \rfloor} i^{k-1}$ is finite if and only if $\int_{\Omega} \lfloor |X| \rfloor^k dP$ is finite, which in turn is finite if and only if $\int_{\Omega} |X|^k dP$ is finite.

Exercises

1. Suppose that $\mathrm{cov}\{Y, f(X)\} = 0$ for every Borel function f such that $\mathrm{cov}\{Y, f(X)\}$ exists. Show that this does not necessarily imply that X and Y are independent. Hint: let Z be $\mathrm{N}(0, 1)$, and set $Y = Z$ and $X = Z^2$.

2. Let X be a nonnegative random variable with distribution function $F(x)$. Prove that $\int_0^\infty \{1 - F(x)\}dx = \int_0^\infty x dF(x) = \mathrm{E}(X)$.

3. Prove that $\int_{-\infty}^\infty \{F(x+a) - F(x)\}dx = a$, for any distribution function F and constant $a \geq 0$.

5.6 Densities

Elementary probability courses usually assume that a continuous random variable X has a probability density function $f(x)$, defined as follows.

Definition 5.31. Density *If there is a nonnegative function $f(x)$ such that the distribution function $F(x)$ of the random variable X satisfies $F(x) = \int_{-\infty}^x f(t)dt$ for all $x \in R$, then f is said to be a probability density function (density for short).*

Some familiar examples include the following.

1. **Normal** The normal density with mean $\mu \in (-\infty, \infty)$ and variance $\sigma^2 \in (0, \infty)$ is

$$f(x; \mu, \sigma^2)) = \frac{\exp\left\{\frac{-(x-\mu)^2}{2\sigma^2}\right\}}{\sqrt{2\pi\sigma^2}}, \quad -\infty < x < \infty,$$

2. **Beta** The beta density with parameters $\alpha \in (0, \infty)$ and $\beta \in (0, \infty)$ is

$$f(p; \alpha, \beta) = \frac{\Gamma(\alpha+\beta)}{\Gamma(\alpha)\Gamma(\beta)} p^{\alpha-1}(1-p)^{\beta-1}; \quad 0 < p < 1, \text{ and}$$

3. **Exponential** The exponential density with mean $1/\lambda$ is

$$f(x; \lambda) = \lambda \exp(-\lambda x), \quad 0 \leq x < \infty.$$

Proposition 5.32. Density defines probability of any Borel set *If $f(x)$ is a probability density function for the random variable X, then $P(X \in B) = \int_B f(x)dx$ for every one-dimensional Borel set B.*

Proof. Let $\mathcal{A}$ be the collection of Borel sets A such that $P(X \in A) = \int_A f(x)dx$. Then $\mathcal{A}$ contains all sets of the form $(-\infty, x]$ by Definition 5.31. It is easy to show that $\mathcal{A}$ must contain all intervals, and that it contains the field in Example 3.8. Moreover, $\mathcal{A}$ is a monotone class because if $A_n \in \mathcal{F}$ and $A_n \uparrow A$ then $\int_{A_n} f(x)dx = \int_R f(x)I(x \in A_n)dx \uparrow \int f(x)I(x \in A)dx$ (DCT) $= \int_A f(x)dx$. Therefore, $A \in \mathcal{A}$. The same argument works if $A_n \downarrow A$. Therefore, $\mathcal{A}$ is a monotone class containing the field in Example 3.8. The result now follows from the monotone class theorem (Theorem 3.32). □

Not all continuous random variables have a density, as the following example shows.

Example 5.33. No density Put a blue and red ball in an urn, and then draw them without replacement. If X_i is the indicator that the ith draw is blue, $i = 1, 2$, then each of the outcomes $(X_1 = 1, X_2 = 0)$ and $(X_1 = 0, X_2 = 1)$ has probability $1/2$. Now repeat the experiment and let (X_3, X_4) be the result of the next two draws, and continue in this way ad infinitum. Now form the random number $Y = 0.X_1X_2X_3\ldots = X_1/2 + X_2/2^2 + X_3/2^3 + \ldots$

That is, the Xs are the base 2 representation of the random number Y in the unit interval. Notice that Y is not a discrete random variable because the probability of each outcome $0.x_1x_2x_3\ldots$ is 0. Therefore, you might think that Y must have a density $f(y)$. In that case,

$$P(Y \in A) = \int I(A)f(y)dy \tag{5.34}$$

for every Borel set A. Let A be the set of numbers in $[0,1]$ whose base 2 representation has precisely one 1 and one 0 in the first two digits, the next two digits, the next two digits, etc. Because of the way Y was constructed, $P(Y \in A) = 1$. But under Lebesgue measure, the probability of exactly one 1 and one 0 in any given pair is $1/2$, so the Lebesgue measure of A is $(1/2)(1/2)(1/2)\ldots = 0$ by Proposition 4.35. Moreover, if the Lebesgue measure of A is 0, then the integral in Equation (5.34) is 0 by Proposition 5.3. Therefore,

$$1 = P(Y \in A) = \int I(A)f(y)dy = 0,$$

a contradiction. Therefore, Equation (5.34) cannot hold for every Borel set A. In other words, Y is not discrete, yet has no density. Note that Y is equivalent to the random number formed using permuted block randomization, with blocks of size 2, discussed at the end of Example 1.1. □

In Example 5.33, the key to showing that there was no density function was to construct a Borel set A of Lebesgue measure 0, yet $P(Y \in A) > 0$. If every Borel set A with Lebesgue measure 0 had $P(Y \in A) = 0$, we would not have been able to construct a counterexample of the type in Example 5.33. In fact, we will soon see that we cannot construct a counterexample of any kind if this condition holds. This motivates the following definition.

Definition 5.34. Absolute continuity *A probability measure μ_X for a random variable X is said to be absolutely continuous with respect to Lebesgue measure if $\mu_X(A) = 0$ for every Borel set A with Lebesgue measure* 0.

The following is a famous theorem of probability and measure theory stated for the special case of Lebesgue measure. Chapter 10 presents the more general Radon-Nikodym theorem for arbitrary measures.

Theorem 5.35. Special case of Radon-Nikodym theorem *The probability measure μ_X for a random variable X is absolutely continuous with respect to Lebesgue measure if and only if there is a density function f such that $\mu_X(B) = \int_B f(x)dx$ for all Borel sets B.*

There is another way to view a density function. The fundamental theorem of calculus for Riemann integrals asserts that if f is any function that is continuous at b, and F is a function such that $F(b) - F(a) = \int_a^b f(s)ds$, then F is differentiable at b and $F'(b) = f(b)$. This holds for Lebesgue integrals as well. In fact, a stronger result holds:

Theorem 5.36. *Suppose that f is a nonnegative function such that $F(b) = F(a) + \int_a^b f(s)ds$. Then:*

1. *$F'(b) = f(b)$ except on a set of Lebesgue measure* 0.
2. **Fundamental theorem of calculus** *$F'(b) = f(b)$ at every b such that f is continuous.*

Proof. We prove only the fundamental theorem of calculus. Suppose that f is continuous at b, and let $\Delta > 0$.

$$\begin{aligned}\frac{F(b+\Delta)-F(b)-\Delta f(b)}{\Delta} &= \frac{\left(\int_b^{b+\Delta} f(s)ds\right)-\Delta f(b)}{\Delta}\\ &= \frac{1}{\Delta}\int_b^{b+\Delta}\{f(s)-f(b)\}ds\\ \left|\frac{F(b+\Delta)-F(b)}{\Delta}-f(b)\right| &\le \sup_{b\le s\le b+\Delta}|f(s)-f(b)|. \qquad (5.35)\end{aligned}$$

If f is continuous at b, then $\sup_{b\le s\le b+\Delta}|f(s)-f(b)|\to 0$ as $\Delta\to 0$. The same thing can be shown if $\Delta < 0$. □

We can view Theorem 5.36 as saying that a density function $f(x)$ is the derivative (called the *Radon-Nikodym derivative*) of the distribution function $F(x)$ with respect Lebesgue measure. We can generalize this result in different ways. For now we consider two-dimensional density functions.

Definition 5.37. Bivariate density *The pair (X, Y) is said to have density function $f(x, y)$ with respect to two-dimensional Lebesgue measure if $f(x, y)$ is nonnegative and $P(X \le x, Y \le y) = \int_{R^2} f(s, t)I(s \le x, t \le y)d(s\times t)$ for all $x \in R$, $y \in R$, where $s \times t$ denotes two-dimensional Lebesgue measure.*

A very common bivariate density function is the bivariate normal:

$$f(x,y)=\frac{1}{2\pi\sigma_X\sigma_Y\sqrt{1-\rho^2}}\exp\left[-\frac{1}{2(1-\rho^2)}\left\{\frac{(x-\mu_X)^2}{\sigma_X^2}-\frac{2\rho(x-\mu_X)(y-\mu_Y)}{\sigma_X\sigma_Y}+\frac{(y-\mu_Y)^2}{\sigma_Y^2}\right\}\right]$$

Marginally, $X \sim \mathrm{N}(\mu_X, \sigma_X^2)$ and $Y \sim \mathrm{N}(\mu_Y, \sigma_Y^2)$. The correlation coefficient ρ dictates the degree to which X and Y tend to track together, as seen from the level curves in Figure 5.6. The three curves correspond to $\rho = -0.9$, $\rho = 0$, and $\rho = 0.9$, respectively.

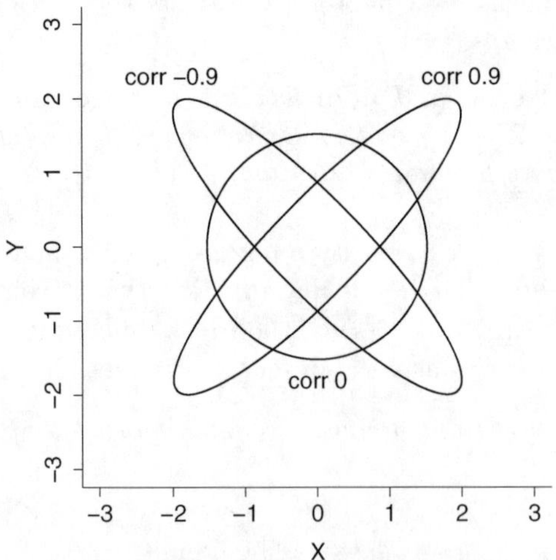

Figure 5.6: Level curve $f(x, y) = 0.05$ of the bivariate normal density function $f(x, y)$ for $\rho = 0$ (circle), $\rho = 0.9$ (upward slanting ellipse) and $\rho = -0.9$ (downward slanting ellipse).

Because a probability density function is nonnegative, Tonelli's theorem implies that (1) we can write $\int_{R^2} f(s,t)I(s \le x, t \le y)d(s \times t)$ as an iterated integral, and (2) it makes no difference whether we integrate first over s or t.

The proof of the following result is similar to that of Proposition 5.32, and is omitted.

Proposition 5.38. Bivariate density defines probability of any 2-dimensional Borel set *If $f(x,y)$ is a density function for the random vector (X,Y) with respect to two-dimensional Lebesgue measure $x \times y$, then $P\{(X,Y) \in B\} = \int_B f(x,y)d(x \times y)$ for every two-dimensional Borel set B.*

Proposition 5.39. *Let $f(x,y)$ and $F(x,y)$ be the density and distribution functions for the pair (X,Y). Then $\partial^2 F(x,y)/\partial x \partial y = \partial^2 F(x,y)/\partial y \partial x = f(x,y)$ except on a set of (x,y) values of two-dimensional Lebesgue measure 0.*

Proof. The function $g(x,y) = \int_{-\infty}^{y} f(x,s)ds$ satisfies $\int_a^b g(x,y)dx = F(b,y) - F(a,y)$ for all $a \in R$, $b \in R$. It follows from Theorem 5.36 that except on a set $N_y(x)$ of x points with one-dimensional Lebesgue measure 0, $\partial F(x,y)/\partial x = g(x,y)$. The two-dimensional Lebesgue measure of the exceptional set $\cup_y N_y(x)$ is $\int_{-\infty}^{\infty} \int_{N_y(x)} dxdy = 0$ because $\int_{N_y(x)} dx = 0$.

Also, $g(x,b) - g(x,a) = \int_a^b f(x,y)dy$ for all $a \in R, b \in R$, so except on a set $N_x(y)$ of y points with one-dimensional Lebesgue measure 0, $\partial g/\partial y = f(x,y)$. The two-dimensional Lebesgue measure of the exceptional set $\cup_x N_x(y)$ is $\int_{-\infty}^{\infty} \int_{N_x(y)} dydx = 0$ because $\int_{N_x(y)} dy = 0$. We have shown that, except on the set $\{\cup_y N_y(x)\} \cup \{\cup_x N_x(y)\}$ of two-dimensional Lebesgue measure at most $0+0=0$, $\partial\{\partial F(x,y)/\partial x\}/\partial y = f(x,y)$. A similar proof shows that the same result holds if we reverse the order of differentiation. □

5.7 Keep It Simple

Our definition of Lebesgue-Stieltjes integration is conceptually, but not operationally, convenient. It illustrates the key difference between Riemann and Lebesgue integration, namely that Lebesgue integration partitions the y-axis rather than the x-axis. However, proving results about integration is easier using another equivalent definition.

To motivate the second definition, we first recast Definition 5.1. If f is a nonnegative *elementary function,* meaning that f takes on only countably many values a_i, $i = 1, 2, \ldots$ then Definition 5.1 implies $\int f(\omega)d\mu(\omega) = \sum_{i=1}^{\infty} a_i \mu\{\omega : f(\omega) = a_i\}$. For a more general nonnegative f, approximate f by the elementary functions

$$f_m = \sum_{i=1}^{\infty} (i-1)/2^m I\{(i-1)/2^m \le f(\omega) < i/2^m\} \tag{5.36}$$

that approach f from below. Definition 5.1 says that $\int f(\omega)d\mu(\omega)$ is $\lim_{m\to\infty} \int f_m(\omega)d\mu(\omega)$, the limit of integrals of the elementary functions (5.36) approximating f.

It is somewhat more convenient to approximate nonnegative f from below by functions f_m taking on only finitely many values. We can do this by truncating Expression (5.36):

$$f_m = \sum_{i=1}^{m2^m} (i-1)/2^m I\{(i-1)/2^m \le f(\omega) < i/2^m\} + mI\{f(\omega) \ge m\}. \tag{5.37}$$

Then $f_m(\omega) \uparrow f(\omega)$ even if $f(\omega) = \infty$. These f_m are called *simple* functions, defined as follows.

Definition 5.40. Simple random variable/function *A function $f(\omega) : (\Omega, \mathcal{F})$ is said to be simple if it is measurable and takes only finitely many values. Also, a random variable $X(\omega)$ that is a simple function of ω is called a simple random variable.*

Definition 5.1 implies that the integral of any nonnegative simple function taking values a_i on set A_i, $i = 1, \ldots, k$ is $\sum_{i=1}^{k} a_i \mu(A_i)$ (Exercise 3 of Section 5.2.1). By the MCT, the integral of any nonnegative measurable function f is the limit of integrals of the f_m of (5.37). In fact, the MCT ensures that we would get the same answer using *any* nonnegative simple functions, not just the specific functions in Expression (5.37). This leads to the following equivalent definition of integration.

Definition 5.41. Equivalent definition of Lebesgue-Stieltjes integration

1. *If $f(\omega)$ is a nonnegative simple function taking values $a_1, \ldots, a_n$, define $\int f(\omega) d\mu(\omega)$ to be $\sum_{i=1}^{n} a_i \mu\{\omega : f(\omega) = a_i\}$.*
2. *If $f(\omega)$ is any nonnegative measurable function, define $\int f(\omega) d\mu(\omega)$ by $\lim \int f_m(\omega) d\mu(\omega)$, where the f_m are any simple functions such that $f_m \uparrow f$.*
3. *If f is any measurable function, define $\int f(\omega) d\mu(\omega)$ by $\int f^+(\omega) d\mu(\omega) - \int f^-(\omega) d\mu(\omega)$, provided that this expression is not of the form $+\infty - \infty$. Otherwise, $\int f(\omega) d\mu(\omega)$ is undefined.*

We illustrate the usefulness of Definition 5.41 by giving another proof of Proposition 5.29, namely that if X and Y are independent with finite means, then $\mathrm{E}(XY) = \mathrm{E}(X)\mathrm{E}(Y)$.

Assume first that X and Y are independent, nonnegative simple random variables, $X = \sum_{i=1}^{m} a_i I(A_i)$ and $Y = \sum_{j=1}^{n} b_j I(B_j)$, all $a_i \geq 0$, $b_j \geq 0$. Then each (A_i, B_j) pair of events are independent. Also, XY is a nonnegative simple random variable taking values $a_i b_j$ with probability $P(A_i \cap B_j)$, $i = 1, \ldots, m$, $j = 1, \ldots, n$. Therefore, by definition of expectation for nonnegative simple functions,

$$
\begin{aligned}
\mathrm{E}(XY) &= \sum_{i=1}^{m} \sum_{j=1}^{n} a_i b_j P(A_i \cap B_j) \\
&= \sum_{i=1}^{m} \sum_{j=1}^{n} a_i b_j P(A_i) P(B_j) \quad \text{(independence of } A_i, B_j\text{)} \\
&= \sum_{i=1}^{m} a_i P(A_i) \sum_{j=1}^{n} b_j P(B_j) \\
&= \mathrm{E}(X)\mathrm{E}(Y). \qquad (5.38)
\end{aligned}
$$

This proves the result for independent, nonnegative simple random variables X and Y.

Now suppose that X and Y are arbitrary nonnegative, independent random variables. There exist sequences X_n and Y_n of nonnegative simple random variables increasing to X and Y respectively such that X_n and Y_n are independent. This follows from the fact that X_n can be constructed solely from X and Y_n solely from Y, and X and Y are independent. Note that $Z_n = X_n Y_n$ are nonnegative simple functions increasing to $Z = XY$. If follows that

$$\begin{aligned}
\mathrm{E}(XY) &= \mathrm{E}\left(\lim_{n\to\infty} X_nY_n\right) \\
&= \lim_{n\to\infty} \mathrm{E}(X_nY_n) && \text{(MCT)} \\
&= \lim_{n\to\infty} \mathrm{E}(X_n)\mathrm{E}(Y_n) && \text{(result for nonnegative simple functions)} \\
&= \lim_{n\to\infty} \mathrm{E}(X_n) \lim_{n\to\infty} \mathrm{E}(Y_n) \\
&= \mathrm{E}(X)\mathrm{E}(Y). && \text{(MCT)} \qquad (5.39)
\end{aligned}$$

This proves the result for arbitrary nonnegative, independent random variables X and Y with finite means.

The last step is to prove that the result holds for arbitrary independent random variables X and Y with finite mean. Write X as $X^+ - X^-$ and Y as $Y^+ - Y^-$. The pair (X^-, X^+) are Borel functions of X and (Y^-, Y^+) are Borel functions of Y. It follows that (X^-, X^+) are independent of (Y^-, Y^+), so

$$\begin{aligned}
\mathrm{E}(XY) &= \mathrm{E}(X^+Y^+ - X^+Y^- - X^-Y^+ + X^-Y^-) \\
&= \mathrm{E}(X^+)\mathrm{E}(Y^+) - \mathrm{E}(X^+)\mathrm{E}(Y^-) - \mathrm{E}(X^-)\mathrm{E}(Y^+) + \mathrm{E}(X^-)\mathrm{E}(Y^-) \\
&= \mathrm{E}(X^+)\{\mathrm{E}(Y^+) - \mathrm{E}(Y^-)\} - \mathrm{E}(X^-)\{\mathrm{E}(Y^+) - \mathrm{E}(Y^-)\} \\
&= \mathrm{E}(X^+)\mathrm{E}(Y) - \mathrm{E}(X^-)\mathrm{E}(Y) \\
&= \{\mathrm{E}(X^+) - \mathrm{E}(X^-)\}\mathrm{E}(Y) \\
&= \mathrm{E}(X)\mathrm{E}(Y), \qquad (5.40)
\end{aligned}$$

completing the proof.

5.8 Summary

1. Comparison between Lebesgue and Riemann integration:
   (a) Lebesgue partitions the y-axis, Riemann the x-axis.
   (b) Poorly behaved functions like $f(\omega) = I(\omega \text{ is rational})$ are Lebesgue-integrable but not Riemann-integrable.

2. Law of the Unconscious Statistician: If random variable X on $(\Omega, \mathcal{F}, P)$ induces probability measure P' on $(R, \mathcal{B}, P')$, then

$$\int_\Omega X(\omega)dP(\omega) = \int_R x dP'(x).$$

3. *** If $f_n(\omega) \to f(\omega)$ pointwise, then $\int f_n(\omega)d\mu(\omega) \to \int f(\omega)d\mu(\omega)$ if:
   (a) MCT: f_n is nonnegative and $f_n \uparrow f$.
   (b) DCT: $|f_n| \le F$, where $\int F(\omega)d\mu(\omega) < \infty$.

4. Under measurability and completeness assumptions,

$$\int\int_{R\times R} g(x,y)d(\mu_X \times \mu_Y) = \int \left[\int g(x,y)d\mu_X(x)\right] d\mu_Y(y).$$

   is valid for sigma-finite measures μ_X and μ_Y if g is nonnegative (Tonelli's theorem) or if $\int\int_{R\times R} |g(x,y)|d(\mu_X \times \mu_Y) < \infty$ (Fubini's theorem).

5. Important inequalities:

   (a) Jensen: If f is convex, $\mathrm{E}\{f(X)\} \geq f\{\mathrm{E}(X)\}$.

   (b) Markov: $P(|X| \geq c) \leq \mathrm{E}(|X|)/c$.

   (c) *** Chebychev: $P(|X-\mu| \geq c) \leq \sigma^2/c^2$
   (apply Markov to $(X-\mu)^2$).

   (d) Hölder: If $p > 0$, $q > 0$, and $1/p + 1/q = 1$, then

   $$\mathrm{E}(|XY|) \leq \{\mathrm{E}(|X|^p)\}^{1/p}\{\mathrm{E}(|X|^q)\}^{1/q}.$$

   (e) Schwarz: $\mathrm{E}(|XY|) \leq \sqrt{\mathrm{E}(X^2)\mathrm{E}(Y^2)}$
   (apply Hölder to $p = q = 2$).

   (f) Minkowski: If $p \geq 1$, then

   $$\{\mathrm{E}(|X+Y|^p)\}^{1/p} \leq \{\mathrm{E}(|X|^p)\}^{1/p} + \{\mathrm{E}(|Y|^p)\}^{1/p}$$

6. A density function $f(x)$ for a probability measure μ_X with distribution function $F(x)$ satisfies $\int_B f(x)dx = \mu_X(B)$ for all Borel sets B.

   (a) **Radon-Nikodym theorem** A density function exists if and only if μ_X is absolutely continuous with respect to Lebesgue measure (i.e., $\mu_X(A) = 0$ for all Borel sets A with Lebesgue measure 0).

   (b) $F'(x)$ exists and equals $f(x)$ except on a set of Lebesgue measure 0.

   (c) **Fundamental theorem of calculus** $F'(x) = f(x)$ if f is continuous at x.

7. *** Important technique for proving a property concerning integrals:

   (a) Prove it first for nonnegative simple functions.

   (b) Extend the proof to arbitrary nonnegative functions.

   (c) Prove it for an arbitrary function $f = f^+ - f^-$.

Chapter 6

Modes of Convergence

An important part of the theory of statistical inference is understanding what happens to an estimator $\hat{\theta}_n$ like the mean or median as the sample size n tends to ∞. Several different questions might come to mind. First, does $\hat{\theta}_n$ get close to the true parameter value θ as $n \to \infty$? There are several alternative ways to define "get close to," namely:

1. $\hat{\theta}_n$ converges to θ with probability 1.
2. For any given $\epsilon > 0$, the probability that $\hat{\theta}_n$ is within ϵ of θ tends to 1.
3. The expected value of $|\hat{\theta}_n - \theta|^p$ tends to 0.

These are called almost sure convergence, convergence in probability, and convergence in L^p, respectively.

We might also be interested in conducting a hypothesis test about a parameter. For instance, suppose that X_i are iid $\mathrm{N}(\mu, \sigma^2)$, and we want to test the null hypothesis H_0 that $\mu = 0$ against the alternative H_1 that $\mu > 0$. Let $s^2 = (n-1)^{-1}\sum_{i=1}^n (X_i - \bar{X})^2$ be the usual variance estimate. Under H_0, $Z_n = n^{1/2}\bar{X}_n/s$ has a t-distribution with $n-1$ degrees of freedom, which is close to the standard normal if n is large. Even if the X_i are not normally distributed, Z_n is approximately standard normal under the null when n is large. Therefore, an approximate level α test rejects the null hypothesis when $Z_n > c$, where c is the $100(1-\alpha)$th percentile of a standard normal distribution. For this application, we needed only the **distribution** of Z_n to get close to something, namely $\mathrm{N}(0,1)$. We did not need the sequence Z_n to get close to another random variable in one of the senses 1,2, or 3 above. This weaker condition is called convergence in distribution.

Now suppose that Z_n is the usual t-statistic defined above and Z_n' is the same thing except that the usual variance estimate s^2 is replaced by the maximum likelihood estimator (MLE) $\hat{\sigma}^2 = (1/n)\sum(X_i - \bar{X})^2$, so that $\hat{\sigma}^2$ is just $(n-1)/n$ times s^2. To show that we reach essentially the same reject/do not reject conclusion whether we use Z_n or Z_n', we must show more than just that the **distributions** of Z_n and Z_n' are close. For example, if X_i are iid $\mathrm{N}(0,1)$ under the null hypothesis, X_1 and $n^{1/2}\bar{X}_n$ have identical null distributions, namely standard normal, but the reject/do not reject decisions based on these two test statistics can be completely different in any given sample. To conclude that Z_n and Z_n' reach essentially the same conclusion **in any given sample**, we need to show that $Z_n - Z_n'$ is close to 0 in the sense of 1,2, or 3 above.

We hope this brief introduction gives an understanding of the different modes of convergence integral to the theory of statistical inference. The rest of the chapter expands on these four modes: convergence almost surely, convergence in probability, convergence in L^p, and convergence in distribution and shows which of these imply others. We sprinkle in applied problems where possible, but many examples are mathematical constructs selected to highlight important properties. Future chapters will build the tools necessary to go more in depth into applications.

6.1 Convergence of Random Variables

6.1.1 Almost Sure Convergence

Let $X_1, X_2, \ldots$ be a sequence of random variables on a probability space $(\Omega, \mathcal{F}, P)$. For fixed ω, $X_n(\omega) = x_n$, $n = 1, 2, \ldots$ is just a sequence of numbers, so we understand what it means for the sequence of numbers $X_n(\omega) = x_n$ to converge to the number $X(\omega) = x$. It might happen that for each ω, $X_n(\omega)$ converges, but the limit $X(\omega)$ is different for different ω. For instance, suppose $(\Omega, \mathcal{F}, P) = ([0,1], \mathcal{B}_{[0,1]}, \mu_L)$ and

$$X_n(\omega) = \omega^n. \tag{6.1}$$

Then $X_n(\omega) \to I(\omega = 1)$ as $n \to \infty$. It might also happen that for some values of ω, $X_n(\omega)$ has no limit or has an infinite limit. For example, modify $X_n(\omega)$:

$$X_n(\omega) = (-\omega)^n. \tag{6.2}$$

Then $X_n(\omega)$ still converges to 0 as $n \to \infty$ for $\omega < 1$, but has no limit when $\omega = 1$ because the sequence $X_n(1)$ alternates between -1 and $+1$.

Even though the behavior of $X_n(\omega)$ for $\omega = 1$ was different in Equations (6.1) and (6.2), this is irrelevant in the sense that $\{\omega = 1\}$ has probability 0 of occurring anyway. It makes sense to ignore sets of probability 0, and this motivates the following definition.

Definition 6.1. Almost sure convergence

1. *The sequence of random variables $X_1(\omega), X_2(\omega), \ldots$ is said to converge almost surely (or with probability 1) to $X(\omega)$ if, for each fixed ω outside a set of probability 0, the sequence of numbers $X_n(\omega)$ converges to $X(\omega)$ as $n \to \infty$. We write $X_n \to X$ a.s. or $X_n \stackrel{a.s.}{\to} X$.*

2. *The sequence of functions $f_1(\omega), f_2(\omega), \ldots$ on an arbitrary measure space $(\Omega, \mathcal{F}, \mu)$ is said to converge almost everywhere (a.e.) to $f(\omega)$ if, for each ω outside a set of μ-measure 0, the sequence of numbers $f_n(\omega)$ converges to $f(\omega)$.*

The only difference between convergence almost everywhere and convergence almost surely is whether the measure μ is a probability measure. We focus almost exclusively on probability measures, though some results that extend to general measures are very helpful.

Example 6.2. Again let $(\Omega, \mathcal{F}, P) = ([0,1], \mathcal{B}_{[0,1]}, \mu_L)$. Whether we define $X_n(\omega)$ by Equation (6.1) or (6.2), $X_n \stackrel{a.s.}{\to} 0$ because $X_n(\omega) \to 0$ outside the set $\{\omega = 1\}$ of probability 0. □

Example 6.2 shows how helpful it is when assessing almost sure convergence to fix ω and think about the sequence of numbers $X_1(\omega) = x_1, \ldots, X_n(\omega) = x_n, \ldots$ Nonetheless, we need not necessarily know the underlying probability space to know that X_n converges almost surely, as the following example illustrates.

Example 6.3. Suppose that $Y(\omega)$ is a finite random variable on any probability space, and let $X_n(\omega) = Y(\omega)/n$. For each ω, $Y(\omega)$ is just a finite number, so $Y(\omega)/n \to 0$ as $n \to \infty$. This holds for every ω, so $X_n \to 0$ almost surely.

Now let us relax the assumption that Y is finite for every ω. Assume instead that $Y(\omega)$ is finite with probability 1. Because $X_n(\omega) = Y(\omega)/n \to 0$ for every ω such that $Y(\omega)$ is finite, the exceptional set on which X_n does not converge to 0 has probability 0. In other words, it is still true that $X_n \stackrel{a.s.}{\to} 0$. □

Example 6.4. In the preceding examples, the limiting random variable $X(\omega)$ was a constant (0) with probability 1, but there are many examples in which the limiting random variable is not constant. For example, let X be any random variable, and let $X_1 \equiv X_2 \equiv \ldots \equiv X_n \ldots \equiv X$. Then $X_n \stackrel{a.s.}{\to} X$ trivially. Similarly, if Y is any random variable that is finite with probability 1, then $X_n = \{1 + Y(\omega)/n\}^n \to X(\omega) = \exp\{Y(\omega)\}$ for each ω for which $Y(\omega)$ is finite because $(1 + a/n)^n \to \exp(a)$ for each finite constant a. Therefore, $X_n \stackrel{a.s.}{\to} X = \exp(Y)$. □

Proposition 6.5. Basic facts of almost sure convergence *Suppose that* $X_n \stackrel{a.s.}{\to} X$ *and* $Y_n \stackrel{a.s.}{\to} Y$. *Then*

1. *If* $X_n \stackrel{a.s.}{\to} X'$, *then* $P(X = X') = 1$.
2. $f(X_n) \stackrel{a.s.}{\to} f(X)$ *for any continuous function* f. *In fact,* $f : R \longmapsto R$ *need only be a Borel function whose set* D *of discontinuities is such that* $\{\omega : X(\omega) \in D\} \in \mathcal{F}$ *and* $P(X \in D) = 0$.
3. $X_n \pm Y_n \stackrel{a.s.}{\to} X \pm Y$.
4. $X_n Y_n \stackrel{a.s.}{\to} XY$.
5. *If* $P(Y = 0) = 0$, *then* $X_n/Y_n \stackrel{a.s.}{\to} X/Y$.

Proof. These all follow almost immediately from the corresponding properties of convergence of sequences of numbers. For instance, for part 3, $X_n(\omega) \to X(\omega)$ except on a set N_1 of probability 0, and $Y_n(\omega) \to Y(\omega)$ except on a set N_2 of probability 0. Therefore, outside the set $N = N_1 \cup N_2$ of probability $P(N_1 \cup N_2) \le P(N_1) + P(N_2) = 0$, $X_n(\omega) + Y_n(\omega) \to X(\omega) + Y(\omega)$. The proofs of the other items are left as an exercise. □

Example 6.6. A.s. convergence of 2-sample estimators from a.s. convergence of 1-sample estimators One implication of Proposition 6.5 concerns one- and two-sample estimators. For instance, in a one-sample setting, we may be interested in estimating a (finite) mean μ using the sample mean $\bar{X}$ of iid random variables. We will show in Chapter 7 that $\bar{X}$ converges almost surely to μ. In a two-sample setting, we estimate the difference in means $\mu_1 - \mu_2$ by the difference $\bar{X}_1 - \bar{X}_2$ in sample means. By part 3 of Proposition 6.5, this estimator converges almost surely to $\mu_1 - \mu_2$. Similarly, in a one-sample binary outcome setting, we are interested in estimating the probability p of an event such as death by 28 days. Chapter 7 shows that the sample proportion $\hat{p}$ of people dying by 28 days converges almost surely to p. In a two-sample setting, we may be interested in estimating the relative risk p_1/p_2, the ratio of event probabilities in the two groups. By part 5 of Proposition 6.5, the sample relative risk $\hat{p}_1/\hat{p}_2$ converges almost surely to p_1/p_2 if $p_2 > 0$.

In some cases it is helpful to use a transformation to improve estimation. For example, it is helpful to estimate the logarithm of p_1/p_2 using $\ln(\hat{p}_1/\hat{p}_2)$. Part 2 of Proposition 6.5 assures us that as long as neither p_1 nor p_2 is 0, $\ln(\hat{p}_1/\hat{p}_2)$ converges almost surely to $\ln(p_1/p_2)$. □

Example 6.7. Non-degenerate iid Bernoullis cannot converge a.s. Suppose that $X_1, X_2, \ldots$ are iid Bernoulli random variables with parameter $0 < p < 1$. Is there a random variable X such that $X_n \to X$ almost surely? Again fix ω and ask yourself under what conditions will the sequence of numbers $X_1(\omega) = x_1, X_2(\omega) = x_2, \ldots$ converge? Because each $X_n(\omega)$ is 0 or 1, $\lim_{n\to\infty} X_n(\omega)$ exists if and only if $X_n(\omega)$ consists of all zeroes or all ones from some point on. To see this, suppose that, on the contrary, no matter how far we go in the sequence, $x_N, x_{N+1}, \ldots$ has both zeroes and ones. Then x_n cannot converge to 0 or 1 or anything else for that matter because the sequence oscillates between 0 and 1 indefinitely. Thus, if we can argue that the set of ω such that $X_N(\omega), X_{N+1}(\omega), \ldots$ are all zeroes or all ones for some N has probability zero, then we have shown that X_n diverges almost surely.

For any given N, let A_N be the event that $X_N = 0, X_{N+1} = 0, X_{N+2} = 0, \ldots$, and B_N be the event that $X_N = 1, X_{N+1} = 1, X_{N+2} = 1, \ldots$ Because the X_n are independent, $P(A_N) = (1-p)\cdot(1-p)\cdot(1-p)\ldots = 0$, and $P(B_N) = p\cdot p\cdot p\ldots = 0$ by Proposition 4.35. Therefore, $P(A_N \cup B_N) \le P(A_N) + P(B_N) = 0 + 0 = 0$. The probability that there is some N for which the sequence terminates with all zeroes or all ones is $P\{\cup_{N=1}^{\infty}(A_N \cup B_N)\} \le \sum_{N=1}^{\infty} P(A_N \cup B_N) = \sum_{N=1}^{\infty} 0 = 0$. We have shown that the set of ω such that the sequence of numbers $X_1(\omega) = x_1, X_2(\omega) = x_2, \ldots$ terminates with all zeroes or all ones has probability 0. Therefore, the probability that $X_n(\omega)$ converges is 0; X_n cannot converge almost surely to any random variable.

It makes sense that the X_n cannot converge almost surely. If X_n converged almost surely, then knowledge of the value of X_n should give us a lot of information about the value of X_{n+1}. But if the X_n are iid, then X_n gives us no information about X_{n+1}. These two views are consistent only if the X_ns are constant. □

Example 6.7 can be generalized to any iid random variables X_n. Almost sure convergence of X_n precludes them from being iid unless they are constants (exercise).

Exercises

For Problems 1–4, let $(\Omega, \mathcal{F}, P)$ be $([0,1], \mathcal{B}_{[0,1]}, \mu_L)$, where μ_L is Lebesgue measure.

1. Let $X_n(\omega) \equiv 1$ and $Y_n(\omega) = I(\omega > 1/n)$, where I denotes an indicator function. Does X_n/Y_n converge for every $\omega \in [0,1]$? Does it converge almost surely to a random variable? If so, what random variable?

2. Let
$$Y_n = \begin{cases} (-1)^n & \text{if } \omega \text{ is rational} \\ \omega & \text{if } \omega \text{ is irrational.} \end{cases}$$
Does Y_n converge almost surely to a random variable? If so, specify a random variable on $(\Omega, \mathcal{F}, P)$ that Y_n converges almost surely to.

3. In the preceding problem, reverse the words "rational" and "irrational." Does Y_n converge almost surely to a random variable? If so, specify a random variable on $(\Omega, \mathcal{F}, P)$ that Y_n converges almost surely to.

4. For each n, divide $[0,1]$ into $[0,1/n),[1/n,2/n),\ldots,[(n-1)/n,1]$, and let X_n be the left endpoint of the interval containing ω. Prove that X_n converges almost surely to a random variable, and identify the random variable.

5. Prove that if $X_n \stackrel{a.s.}{\to} 0$, then $Y_n = |X_n|/(1+|X_n|) \stackrel{a.s.}{\to} 0$. Does $\ln(Y_n)$ converge almost surely to a finite random variable?

6. Suppose that X_n converges almost surely to X. Suppose further that the distribution and density functions of X are $F(x)$ and $f(x)$, respectively. Let

$$G(x,y) = \begin{cases} \frac{F(y)-F(x)}{y-x} & \text{if } x \neq y \\ f(y) & \text{if } x = y. \end{cases}$$

Prove that $G(X, X_n)$ converges almost surely to $f(X)$.

7. Prove parts 1 and 2 of Proposition 6.5.

8. Prove parts 4 and 5 of Proposition 6.5.

9. Extend Example 6.7 to show that if $X_1, X_2, \ldots$ are iid random variables with any non-degenerate distribution, then they cannot converge almost surely.

10. Using the same reasoning as in Example 6.7, one can show the following. For each fixed subsequence $n_1, n_2, \ldots$, $P(X_{n_1} = 1, X_{n_2} = 1, \ldots, X_{n_k} = 1, \ldots) = 0$. Does this imply that, with probability 1, there is no subsequence $m_1, m_2, \ldots$ such that $X_{m_1} = 1, X_{m_2} = 1, \ldots, X_{m_k} = 1, \ldots$? Explain.

6.1.2 Convergence in Probability

Many statistical applications involve estimation of a parameter θ using a statistic $\hat{\theta}_n$ based on a sample of n observations. How can we formulate the idea that $\hat{\theta}_n$ should be close to θ if n is large? We could insist on using an estimator $\hat{\theta}_n(\omega)$ that converges almost surely to θ, but that may be too strong a requirement. Almost sure convergence involves the behavior of the infinite sequence $\hat{\theta}_1(\omega), \hat{\theta}_2(\omega), \ldots$ corresponding to sample sizes of $n = 1, 2, \ldots$, but we have only one sample of size n. The real question is, if n is large enough, is there high probability that $\hat{\theta}_n$ is within a small error band $\pm\epsilon$ of θ? This motivates the following definition.

Definition 6.8. Convergence in probability or measure

1. *A sequence of random variables $X_1(\omega), X_2(\omega), \ldots$ is said to converge in probability to $X(\omega)$, written $X_n \stackrel{p}{\to} X$ if, for each $\epsilon > 0$, $P(|X_n(\omega) - X(\omega)| \geq \epsilon) \to 0$ as $n \to \infty$.*

2. *More generally, a sequence of measurable functions $f_1(\omega), f_2(\omega), \ldots$ on a measure space $(\Omega, \mathcal{F}, \mu)$ is said to converge in measure to f if, for each $\epsilon > 0$, $\mu\{\omega : |f_n(\omega) - f(\omega)| \geq \epsilon\} \to 0$ as $n \to \infty$.*

Example 6.9. Consistency of empirical distribution function In a study of patients with very advanced cancer, the time X_i from study entry to death is observed for all n patients. Assume that X_i are iid with distribution function $F(x)$. We are trying to estimate $F(x)$ for different x. For a fixed x, the number of patients dying by time x is binomial with n trials and success probability $F(x)$. If $\hat{F}_n(x)$ denotes the proportion of patients dying by time x, then

$$
\begin{aligned}
P(|\hat{F}_n(x) - F(x)| \geq \epsilon) &\leq \operatorname{var}\{\hat{F}_n(x)\}/\epsilon^2 \text{ (Chebychev's inequality)} \\
&= \frac{F(x)\{1 - F(x)\}}{n\epsilon^2} \to 0. \qquad (6.3)
\end{aligned}
$$

That is the $\hat{F}_n(x) = (1/n)\sum_{i=1}^n I(X_i \leq x)$ converges in probability to the true distribution function $F(x)$ for each value of x. □

Example 6.10. Consistency of sample median Let X_i be as in Example 6.9, but now assume that $F(x)$ has a unique median θ. That is, there is a unique number θ such that $P(X \leq \theta) \geq 1/2$ and $P(X \geq \theta) \geq 1/2$. Let $\hat{\theta}_n$ be the median from a sample of size n. That is, $\hat{\theta}_n$ is the middle observation if n is odd, and the average of the middle two observations if n is even. The event $\hat{\theta}_n \leq \theta - \epsilon$ implies that $\hat{p}_n \geq 1/2$, where $\hat{p}_n$ is the proportion of X_is that are less than or equal to $\theta - \epsilon$. Therefore, if $p = F(\theta - \epsilon)$, then $p < 1/2$ and

$$
\begin{aligned}
P(\hat{\theta}_n \leq \theta - \epsilon) &\leq P(\hat{p}_n \geq 1/2) \\
&= P(\hat{p}_n - p \geq 1/2 - p) \\
&\leq P(|\hat{p}_n - p| \geq 1/2 - p) \\
&\leq \frac{\operatorname{var}(\hat{p}_n)}{(1/2 - p)^2} \text{ (Chebychev's inequality)} \\
&= \frac{p(1-p)}{n(1/2 - p)^2} \to 0. \qquad (6.4)
\end{aligned}
$$

A similar argument shows that $P(\hat{\theta}_n \geq \theta + \epsilon) \to 0$ (exercise). Therefore, $\hat{\theta}_n \xrightarrow{p} \theta$. In other words, there is high probability that the sample median will be close to the true median if the sample size is large. □

Notice that there is nothing in the definition of convergence in probability about pointwise convergence of $X_n(\omega)$ to $X(\omega)$. In fact, $X_n(\omega)$ need not converge to $X(\omega)$ for any ω, as the following celebrated example demonstrates.

Example 6.11. Convergence in probability, but not almost surely Let $(\Omega, \mathcal{F}, P)$ again be $([0,1], \mathcal{B}_{[0,1]}, \mu_L)$. Think of the experiment as selecting a point ω randomly on the unit interval. The following stepwise algorithm pins down the location of ω arbitrarily closely.

Step 0: ω must reside somewhere in $[0,1]$. If $X_0(\omega)$ is the indicator that $\omega \in [0,1]$, then $X_0(\omega) \equiv 1$ (Row 0 of Figure 6.1).

Step 1: Divide $[0,1]$ in half, creating two bins $[0,1/2)$ and $[1/2,1]$ of size $1/2^1 = 1/2$. ω must be in exactly one of the two bins. If $X_1(\omega)$ and $X_2(\omega)$ are the indicator functions that ω is in the first and second bins, respectively, then exactly one of $X_1(\omega)$ and $X_2(\omega)$ is 1 (Row 1 of Figure 6.1).

Step 2: Divide each of the bins in Step 1 in half, creating 4 mutually exclusive bins: $[0,1/4), [1/4,1/2), [1/2,3/4)$, and $[3/4,1]$, each of length $1/2^2 = 1/4$. Let $X_3(\omega), \ldots,$ $X_6(\omega)$ be the indicator functions that ω lies in the first, second, third, and fourth bins, respectively. For each ω, exactly one of $X_3(\omega), X_4(\omega), X_5(\omega)$, and $X_6(\omega)$ is 1 because

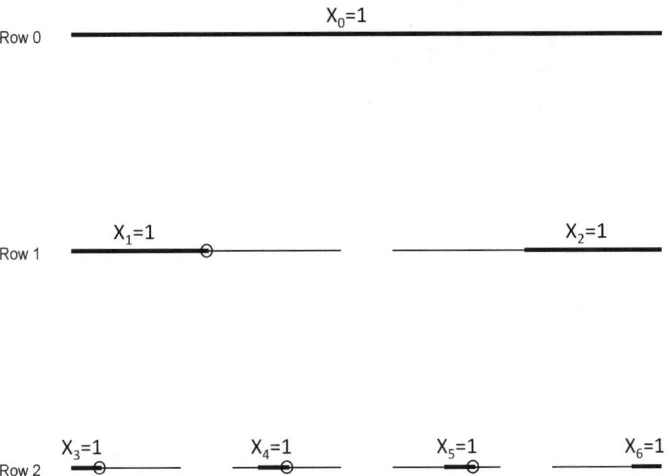

Figure 6.1: The darkened parts of $[0, 1]$ show where the Bernoulli random variables $X_i(\omega)$ in Example 6.11 take the value 1.

ω is in exactly one of the mutually exclusive and exhaustive bins (Row 2 of Figure 6.1).

⋮

Step k: Divide each bin in Step k-1 in half, creating 2^k bins of length $1/2^k$ and define 2^k random variables $X_i(\omega)$ accordingly. More formally, $n = 2^0 + 2^1 + \ldots + 2^{k-1} + j$ for some $j = 0, \ldots, 2^k - 1$ corresponds to a specific bin at step k, and $X_n(\omega)$ is the indicator that ω is in that bin. Exactly one of the 2^k random variables X_i created at step k is 1.

As $n \to \infty$, so does k, and the bin length tends to 0. Because ω is in only one of the 2^k bins, $|X_n(\omega) - 0| = 0$ except when ω is in that bin of length $1/2^k$. Therefore, $P(|X_n(\omega) - 0| > \epsilon) = 1/2^k \to 0$, so $X_n \xrightarrow{p} 0$. But for each ω, exactly one of the 2^k random variables $X_i(\omega)$ in Row k of Figure 6.1 is 1. Therefore, for each ω, $X_n(\omega) = 0$ for infinitely many n and $X_n(\omega) = 1$ for infinitely many n. It follows that for no ω can the sequence of numbers $X_n(\omega)$ converge. This shows that $X_n \xrightarrow{p} 0$, but with probability 1, $X_n(\omega)$ fails to converge. □

We have just seen that we can have convergence in probability but not almost surely, but what about the other way around? It is by no means obvious, but almost sure convergence implies convergence in probability. We prove this in Section 6.2.3.

Proposition 6.12. Basic facts about convergence in probability *If $X_n \xrightarrow{p} X$ and $Y_n \xrightarrow{p} Y$, then:*

1. *If $X_n \xrightarrow{p} X'$, then $P(X = X') = 1$.*

2. *$f(X_n) \xrightarrow{p} f(X)$ for any continuous function f. In fact, $f : R \longmapsto R$ need only be a Borel function whose set D of discontinuities is such that $\{\omega : X(\omega) \in D\} \in \mathcal{F}$ and $P(X \in D) = 0$.*

3. *$X_n \pm Y_n \xrightarrow{p} X \pm Y$.*

4. *$X_n Y_n \xrightarrow{p} XY$.*

5. *$X_n/Y_n \xrightarrow{p} X/Y$, provided that $P(Y = 0) = 0$.*

Proof of part 3. $|X_n + Y_n - (X + Y)| \geq \epsilon \Rightarrow |X_n - X| \geq \epsilon/2$ or $|Y_n - Y| \geq \epsilon/2)$. Therefore,

$$\begin{aligned} P(|X_n + Y_n - (X+Y)| \geq \epsilon) &\leq P(|X_n - X| \geq \epsilon/2 \cup |Y_n - Y| \geq \epsilon/2) \\ &\leq P(|X_n - X| \geq \epsilon/2) + P(|Y_n - Y| \geq \epsilon/2) \end{aligned}$$

by the Bonferroni inequality. Each probability tends to 0 as $n \to \infty$, so $P(|X_n + Y_n - (X + Y)| \geq \epsilon) \to 0$, proving that $X_n + Y_n \xrightarrow{p} X + Y$. The other parts are left as exercises. □

Exercises

1. Let $(\Omega, \mathcal{F}, P)$ be $([0,1], \mathcal{B}_{[0,1]}, \mu_L)$, where μ_L is Lebesgue measure. Let $X_n(\omega) = \omega I(\omega \leq 1 - 1/n)$. Prove that X_n converges in probability to a random variable, and identify that random variable.

2. Sometimes we transform an estimator using a continuous function $f(\hat{\theta}_n)$. What can we conclude about the transformed estimator if $\hat{\theta}_n \xrightarrow{p} \theta$?

3. Explain the relevance of part 3 of Proposition 6.12 in terms of one- and two-sample estimators.

4. Give an example to show that convergence of X_n to X in probability is not sufficient to conclude that $f(X_n)$ converges in probability to $f(X)$ for an arbitrary function f.

5. Prove that the following are equivalent.
   (a) $X_n \xrightarrow{p} X$.
   (b) For each $\epsilon > 0$, $P(|X_n - X| > \epsilon) \to 0$ (that is, the $\geq$ symbol in Definition 6.8 can be replaced by $>$).
   (c) For each $\epsilon > 0$, $P(X_n - X > \epsilon) \to 0$ and $P(X_n - X < -\epsilon) \to 0$ as $n \to \infty$.

6. Prove that if $X_n \xrightarrow{p} X$, then there exists an N such that $P(|X_n - X| \geq \epsilon) \leq \epsilon$ for $n \geq N$.

7. Prove parts 1 and 2 of Proposition 6.12.

8. Prove parts 4 and 5 of Proposition 6.12.

9. Prove that $P(\hat{\theta}_n \geq \theta + \epsilon) \to 0$ in Example 6.10.

10. If $X_1, X_2, \ldots$ are iid random variables with a non-degenerate distribution function, can X_n converge in probability to a constant c? If so, give an example. If not, prove that it cannot happen.

11. If $X_1, \ldots, X_n$ are identically distributed (not necessarily independent) with $\mathrm{E}(|X_i|) < \infty$, then $Y_n = \max(|X_1|, \ldots, |X_n|)/n \xrightarrow{p} 0$. Hint: $P(Y_n \geq \epsilon) = P(\cup_{i=1}^n A_i)$, where $A_i = \{|X_i(\omega)| \geq n\epsilon\}$.

6.1.3 Convergence in L^p

Another common measure of the closeness of an estimator $\hat{\theta}_n$ to a parameter θ is the mean squared error (MSE), $\mathrm{E}(\hat{\theta}_n - \theta)^2$. This is analogous to using $\sum_{i=1}^k (x_i - y_i)^2$ as a measure of how close vectors $\mathbf{x}$ and $\mathbf{y}$ are. But just as we take the square root of $\sum_{i=1}^k (x_i - y_i)^2$ to get the Euclidean distance between two vectors $\mathbf{x}$ and $\mathbf{y}$, we take the square root of the MSE to get a distance between $\hat{\theta}_n$ and θ. This so-called L^2 distance is a special case the L^p distance defined as follows.

Definition 6.13. $\mathbf{L^p}$ distance *If X and Y are two random variables with finite pth moment, $p > 0$, the L^p distance between X and Y is defined to be* $\{\mathrm{E}(|X - Y|^p)\}^{1/p}$.

It can be seen that for $p \geq 1$, the L^p distance satisfies the criteria of a distance measure d, namely that 1) $d(X,Y) \geq 0$, with $d(X,Y) = 0$ if and only if $X = Y$, 2) $d(X,Y) = d(Y,X)$, and 3) $d(X,Z) \leq d(X,Y) + d(Y,Z)$. The first two properties are obvious, and the third follows easily from Minkowski's inequality (Proposition 5.24).

Proposition 6.14. L^p distance increases in p for $p \geq 1$ *The L^p distance between X and Y is, for $p \geq 1$, an increasing function of p.*

Proof. To see this, let $Z = |X-Y|^p$. By Corollary 5.20, for $1 \leq p \leq q$, $\mathrm{E}(Z) \leq \{\mathrm{E}(Z^{q/p})\}^{p/q}$. That is, $\mathrm{E}(Z) \leq \{\mathrm{E}(|X - Y|^q)\}^{p/q}$. Taking the pth root of both sides yields $\{E(|X - Y|^p)\}^{1/p} \leq \{\mathrm{E}(|X - Y|^q)\}^{1/q}$. □

Closely related to the L^2 distance between two random variables is the mean squared error (MSE), as noted above. Let X be an estimator of a parameter θ, and Y be the constant θ. Assuming X has mean μ and finite variance, the MSE is:

$$\begin{aligned} \mathrm{MSE} &= \mathrm{E}(X-\theta)^2 = \mathrm{E}(X-\mu+\mu-\theta)^2 \\ &= \mathrm{E}(X-\mu)^2 + (\mu-\theta)^2 + 2(\mu-\theta)\mathrm{E}(X-\mu) \\ &= \mathrm{var}(X) + \mathrm{bias}^2 + 0 \\ &= \mathrm{var}(X) + \mathrm{bias}^2, \end{aligned} \tag{6.5}$$

where bias is $\mathrm{E}(X - \theta)$. If X is unbiased (i.e., the bias is 0), then its variance, the square of the L^2 distance between X and θ, indicates how close X tends to be to θ. The L^1 distance $\mathrm{E}(|X - Y|)$ is also commonly used in statistics.

Definition 6.15. Convergence in L^p *A sequence of random variables $X_1, X_2, \ldots$ is said to converge to the random variable X in L^p if* $\mathrm{E}(|X_n - X|^p) \to 0$ *as* $n \to \infty$.

Proposition 6.16. L^q *convergence implies* L^p *convergence for* $1 \leq p \leq q$ *Convergence in L^q implies convergence in L^p for $1 \leq p \leq q$.*

Proof. This follows immediately from Proposition 6.14. □

Proposition 6.17. Bias→ 0 and variance→ 0 implies $\hat{\theta}_n$ is consistent *Let $\hat{\theta}_n$ be an estimator of a parameter θ such that* $\mathrm{bias}(\hat{\theta}_n) \to 0$ *and* $\mathrm{var}(\hat{\theta}_n) \to 0$. *Then $\hat{\theta}_n \to \theta$ in L^2.*

Proof. This follows immediately from Expression (6.5). □

Corollary 6.18. Sample mean of iid observations with finite variance is consistent *If $X_1, X_2, \ldots$ are iid with mean μ and finite variance σ^2, then the sample mean $\bar{X}_n$ converges to μ in L^2.*

Proof. This follows from Proposition 6.17 and the fact that $\bar{X}_n$ has bias 0 and variance $\sigma^2/n \to 0$. □

Proposition 6.19.

1. *If X_n converges to X in L^p and X_n converges to X' in L^p, then $P(X = X') = 1$.*

2. *If X_n converges to X in L^p and Y_n converges to Y in L^p, then $X_n \pm Y_n$ converges to $X \pm Y$ in L^p.*

Proof of part 2. This follows immediately from either Minkowski's inequality (Proposition 5.24) or Proposition 5.23. □

Exercises.

1. Let
$$X_n = \begin{cases} \sqrt{n} & \text{with probability } 1/n \\ 0 & \text{with probability } 1 - 1/n. \end{cases}$$
Show that $X_n \to 0$ in L^1, but not in L^2. Are there any examples in which $X_n \to 0$ in L^2, but not in L^1?

2. What does Proposition 6.19 imply about the MSE of two-sample estimators such as the difference in means or proportions when their one-sample MSEs converge to 0?

3. Prove part 1 of Proposition 6.19.

4. Show by a counterexample that X_n converging to X in L^p and Y_n converging to Y in L^p does not necessarily imply that X_nY_n converges to XY in L^p.

5. Show by counterexample that X_n converging to X in L^p does not necessarily imply that $f(X_n)$ converges to $f(X)$ in L^p for a continuous function f.

6. Prove that if $X_n \xrightarrow{p} 0$, then $\mathrm{E}\{|X_n|/(1 + |X_n|)\} \to 0$.

6.1.4 Convergence in Distribution

A completely different type of convergence from the ones we have considered so far concerns not the random variables themselves, but their distribution functions. We could consider X_n to be close to X in distribution if the distribution function $F_n(x)$ for X_n converges to the distribution function $F(x)$ for X. But consideration of the simplest possible case where X_n is the constant $1 + 1/n$ and X is the constant 1 reveals potential problems in trying to define this concept. It certainly seems that the distribution function $F_n(x)$ for $X_n \equiv 1 + 1/n$ should converge to the distribution function $F(x)$ for $X \equiv 1$, namely

$$F(x) = \begin{cases} 0 & \text{if } x < 1 \\ 1 & \text{if } x \geq 1. \end{cases} \tag{6.6}$$

But $F_n(x) = P(X_n \le x)$ is

$$F_n(x) = \begin{cases} 0 & \text{if } x < 1 + 1/n \\ 1 & \text{if } x \ge 1 + 1/n, \end{cases}$$

which converges to

$$\begin{cases} 0 & \text{if } x \le 1 \\ 1 & \text{if } x > 1. \end{cases}$$

Not only does this not coincide with the distribution function in Equation (6.6), but it is not even a distribution function because it is not right continuous at $x = 1$. Therefore, requiring that $F_n(x)$ converge to $F(x)$ for every x seems too strong. The problem is at the lone point $x = 1$ where $F(x)$ is discontinuous. At all continuity points x of $F(x)$, $F_n(x) \to F(x)$. This leads us to the following definition.

Definition 6.20. Convergence in distribution *Let X_n and X have distribution functions $F_n(x)$ and $F(x)$, respectively. Then X_n is said to converge in distribution to X if $F_n(x) \to F(x)$ for every continuity point x of $F(x)$. We write $X_n \stackrel{D}{\to} X$ or $F_n \stackrel{D}{\to} F$.*

Example 6.21. Let X_n be normal with mean μ_n and standard deviation σ_n, where $\mu_n \to \mu$ and $\sigma_n \to \sigma > 0$. Then $P(X_n \le x) = P\{(X_n - \mu_n)/\sigma_n \le (x - \mu_n)/\sigma_n\} = \Phi\{(x - \mu_n)/\sigma_n\}$. Because Φ is continuous and $(x - \mu_n)/\sigma_n \to (x - \mu)/\sigma$, $P(X_n \le x) \to \Phi\{(x - \mu)/\sigma\}$. That is, $X_n \stackrel{D}{\to} \mathrm{N}(\mu, \sigma^2)$. □

Example 6.22. Discrete uniform converging in distribution to continuous uniform Many biological and other applications involve testing whether data are uniformly distributed over time or space. For example, we might be interested in whether heart attacks are equally likely to occur any time of day. However, we usually do not know the exact times of patients' heart attacks. Therefore, we may have to settle for identifying the hour during which it occurred. The question is, if we can pin down the time to a sufficiently small interval, can we ignore the fact that the data are actually from a discrete, rather than continuous, uniform distribution? To be more concrete, consider a setting in which the original interval of time or space has been scaled to have length 1. Suppose we first identify which half of the interval contains the observation. Having identified the correct half, we then identify which half of that half contains the observation, etc. After n steps, the observation should have a discrete uniform distribution on the dyadic rationals of order n. We will show that this discrete uniform converges in distribution to a continuous uniform.

Let X_n have a discrete uniform distribution on the dyadic rationals of order n. That is, $P(X_n = i/2^n) = 1/2^n$, $i = 1, 2, \ldots, 2^n$. We show that $X_n \stackrel{D}{\to} X$, where X is uniform $(0, 1)$. Let x_d be a dyadic rational of some order. That is, $x_d = i/2^m$ for some m and some $i = 1, \ldots, 2^m$. Then for $n \ge m$, $F_n(x_d) = P(X_n \le x_d) = i/2^m = x_d$. Therefore, $F_n(x_d) \to x_d$ for every dyadic rational number. Now suppose that x is an arbitrary number in the interval $(0, 1)$. Let x_1 and x_2 be arbitrary dyadic rationals with $x_1 < x < x_2$. Then $F_n(x_1) \le F_n(x) \le F_n(x_2)$. It follows that $x_1 = \underline{\lim}\, F_n(x_1) \le \underline{\lim}\, F_n(x)$. Similarly, $\overline{\lim}\, F_n(x) \le \overline{\lim}\, F_n(x_2) = x_2$. Because $x_1 < x$ and $x_2 > x$ are arbitrary dyadic rationals and we can find dyadic rationals arbitrarily close to x (either less than or greater than), $x \le \underline{\lim}\, F_n(x)$ and $\overline{\lim}\, F_n(x) \le x$; i.e., $\underline{\lim}\, F_n(x) = \overline{\lim}\, F_n(x) = x$. Therefore, $F_n(x) \to x$. This being true for every $x \in (0, 1)$, $X_n \stackrel{D}{\to} X$, where X is uniform $(0, 1)$. □

The same technique used in Example 6.22 can be used to prove the following result.

Proposition 6.23. Equivalent condition for weak convergence: convergence on a dense set $X_n \stackrel{D}{\to} X$ *if and only if $F_n(x) \to F(x)$ for all x in a dense set D of reals* *(meaning that if $y \in R$, every nonempty interval containing y contains at least one element of D).*

Proof. Exercise.

Proposition 6.24. Law of small numbers *Let X_n be binomial (n, p_n), where $np_n \to \lambda$ as $n \to \infty$. Then $P(X_n = k) \to \exp(-\lambda)\lambda^k/k!$ and X_n converges in distribution to a Poisson with parameter λ.*

Proof.

$$\begin{aligned}
P(X_n = k) &= \binom{n}{k} p_n^k (1-p_n)^{n-k} \\
&= \frac{n(n-1)\ldots,(n-k+1)p_n^k(1-p_n)^n}{k!(1-p_n)^k.} \\
&= \frac{nn\left(1-\frac{1}{n}\right)n\left(1-\frac{2}{n}\right)\ldots n\left(1-\frac{k-1}{n}\right)p_n^k\left(1-\frac{np_n}{n}\right)^n}{k!(1-p_n)^k} \\
&= \frac{(np_n)^k\left\{\left(1-\frac{1}{n}\right)\left(1-\frac{2}{n}\right)\ldots\left(1-\frac{k-1}{n}\right)\right\}\left(1-\frac{np_n}{n}\right)^n}{k!(1-p_n)^k} \\
&\to \frac{\lambda^k \cdot 1 \cdot \exp(-\lambda)}{k! \cdot 1} \\
&= \exp(-\lambda)\lambda^k/k! \qquad (6.7)
\end{aligned}$$

For each positive x, $F_n(x) = \sum_{k=0}^{\lfloor x \rfloor} P(X_n = k)$, where $\lfloor x \rfloor$ is the greatest integer less than or equal to x. Each term of the finite sum converges to the corresponding Poisson probability, so $P(X_n \le x) \to \sum_{k=0}^{\lfloor x \rfloor} \exp(-\lambda)\lambda^k/k! = P(X \le x)$, where X is Poisson with parameter λ, completing the proof. □

Example 6.25. Application of law of small numbers: Bonferroni adjustment is asymptotically not very conservative for independent comparisons Researchers conducting multiple statistical tests in the same study recognize that the probability of at least one error—called the familywise error rate (FWE)—is inflated if each test uses level α. The Bonferroni correction using α/n for each of the n tests ensures that the FWE $P(\cup_{i=1}^n E_i)$ is no greater than $\sum_{i=1}^n P(E_i) \le \sum_{i=1}^n (\alpha/n) = \alpha$. The Bonferroni correction tends to become quite conservative if the test statistics are highly correlated, especially if the number of comparisons is large. An interesting question is: what happens to the FWE for a large number of **independent** comparisons?

The probability of exactly k errors assuming the null hypothesis for each test is computed as follows. For each test, whether an error is made is Bernoulli with probability $p_n = \alpha/n$. Therefore, the total number of errors X_n is binomial with parameters n and p_n. By the law of small numbers (Proposition 6.24), X_n converges in distribution to an exponential with parameter $\lambda = \lim_{n\to\infty} n\alpha/n = \alpha$. The probability of no errors tends to $\exp(-\alpha)\alpha^0/0! = \exp(-\alpha)$. Typically, α is .05, in which case the probability $\exp(-.05)$ of no errors is approximately .951. Accordingly, the FWE is approximately $1 - .951 = .049$. Because the FWE is very close to the intended value of .05, the Bonferroni method is only slightly conservative for a large number of independent comparisons. □

Example 6.26. Relative error in estimating tail probabilities may not tend to 0 In Example 6.9, we showed that the empirical distribution function $\hat{F}(x)$ converges in probability to the actual distribution function $F(x)$. Equivalently, $\hat{S}(x)/S(x)$ converges in probability to 1, where $\hat{S}(x) = 1 - \hat{F}(x)$ and $S(x) = 1 - F(x)$. That is, the relative error in estimating a survival probability tends to 0 as $n \to \infty$ for each x. But suppose now that we are estimating a survival probability that is close to 0. We can formulate the problem as estimating $S(t_n)$ for a sequence of numbers $t_n \to \infty$ such that $nS(t_n) \to \lambda$. Does the relative error still tend to 0 in probability? The number N_n of $X_1, \ldots, X_n$ exceeding t_n is binomial with n trials and success probability $p_n = S(t_n)$, with $np_n \to \lambda$. The law of small numbers (Proposition 6.24) implies that N_n converges in distribution to a Poisson (λ). Then

$$P\{\hat{S}_n(t_n)/S(t_n) > (1+\epsilon)\} = P\{N_n > (1+\epsilon)nS(t_n)\}. \tag{6.8}$$

As $n \to \infty$, $(1+\epsilon)nS(t_n)$ tends to $\lambda(1+\epsilon)$. If $\lambda(1+\epsilon)$ is not an integer, then $m < \lambda(1+\epsilon) < m+1$ for some integer m. For n sufficiently large, the probability in Equation (6.8) equals $P(N_n > m)$, which converges to $\sum_{k=m+1}^{\infty} \exp(-\lambda)\lambda^k/k!$. This is nonzero, so the relative error in estimating a survival probability tending to 0 at a rate of $1/n$ does not tend to 0. This is usually inconsequential because the absolute error (i.e., the difference between estimated and actual survival probabilities) is small. It can be consequential if the goal is to prove that survival beyond a given time is less than some tiny number like 10^{-4}. □

Remark 6.27. Convergence in distribution is "weak" convergence *Convergence in distribution is also called weak convergence, and for good reason. The random variables X_n need not be close to X in any conventional sense. In fact, because convergence in distribution depends only on the distributions of the random variables, there is no requirement for $X_1, \ldots,$ $X_n, \ldots$ to be defined on the same probability space. In contrast, convergence almost surely, in probability, and in L^p do require the random variables to be defined on the same probability space. Fortunately, the following very useful result shows that if X_n converges in distribution to X, we can define random variables on a common probability space that have the same marginal distributions and converge almost surely.*

Theorem 6.28. Skorokhod representation theorem *If $X_n \stackrel{D}{\to} X$, then there exist random variables X_n' and X' defined on $(\Omega, \mathcal{F}, P) = ((0,1), \mathcal{B}_{(0,1)}, \mu_L)$ such that X_n' has the same distribution as X_n, X' has the same distribution as X, and X_n' converges almost surely to X'.*

Proof. We proved in Section 4.6 that the random variables $X_n'(\omega) = F_n^{-1}(\omega)$ and $X'(\omega) = F^{-1}(\omega)$ defined on $((0,1), \mathcal{B}_{(0,1)}, \mu_L)$ have respective distribution functions F_n and F. The proof that $X_n'(\omega) \stackrel{a.s.}{\to} X'(\omega)$ for all but countably many ω using Proposition 4.54 is straightforward, though tedious (see Serfling, 1980, page 21). □

The Skorokhod representation theorem greatly facilitates proofs, as we see in the next result.

Proposition 6.29. Alternative definition of convergence in distribution involving bounded, continuous functions $X_n \stackrel{D}{\to} X \Leftrightarrow \mathrm{E}\{f(X_n)\} \to \mathrm{E}\{f(X)\}$ *for every bounded continuous function.*

Proof of $\Rightarrow$. Notice that whether $\mathrm{E}\{f(X_n)\} \to \mathrm{E}\{f(X)\}$ depends only on the marginal distributions of X_n and X, not on joint distributions. Therefore, it suffices to prove that $\mathrm{E}\{f(X_n')\} \to \mathrm{E}\{f(X')\}$, where X_n' and X' are as stated in Theorem 6.28. Because $X_n' \to X'$ almost surely and f is continuous, $f(X_n)$ converges almost surely to $f(X)$ by part 2 of Proposition 6.5. Because f is bounded, the bounded convergence theorem (Theorem 5.13) implies that $\mathrm{E}\{f(X_n)\} \to \mathrm{E}\{f(X)\}$, completing the proof. □

Proposition 6.30. *If $X_n \stackrel{D}{\to} X$ and f is a continuous function, then $f(X_n) \stackrel{D}{\to} f(X)$. The same is true if the continuity assumption is replaced by the following weaker version: $f : R \longmapsto R$ is a Borel function whose set D of discontinuities is a Borel set with $P(X \in D) = 0$.*

Proof. Exercise.

Skorokhod's representation theorem can be used to strengthen the bounded convergence theorem presented in Section 5.3.

Theorem 6.31. A more general bounded convergence theorem *If $X_n \stackrel{D}{\to} X$ and c is a finite constant such that $P(|X_n| \le c) = 1$ for each n, then $\mathrm{E}(X_n) \to \mathrm{E}(X)$.*

Proof. Let $X_n \sim F_n$ and $X \sim F$. By the Skorokhod representation theorem, we can find random variables $X_n' \sim F_n$ and $X' \sim F$ on the same probability space such that $X_n' \stackrel{a.s.}{\to} X'$. Because X_n' has the same distribution as X_n and $P(|X_n| \le c) = 1$, $P(|X_n'| \le c) = 1$ as well. By the usual BCT (Theorem 5.13), $\mathrm{E}(X_n') \to \mathrm{E}(X')$. But this implies that $\mathrm{E}(X_n) \to \mathrm{E}(X)$ because X_n and X_n' have the same distribution, as do X and X'. □

Exercises

1. Suppose that $X_n \sim N(\mu_n, 1)$. Prove that if $X_n \stackrel{D}{\to} N(\mu, 1)$, then $\mu_n \to \mu$ as $n \to \infty$.

2. Suppose that $X_n \sim N(0, \sigma_n^2)$. Prove that if $X_n \stackrel{D}{\to} N(0, \sigma^2)$, then $\sigma_n \to \sigma$ as $n \to \infty$.

3. Prove that if X_n has a discrete uniform distribution on $\{1/n, 2/n, \ldots, n/n = 1\}$, $X_n \stackrel{D}{\to} X$, where X is uniform $(0, 1)$.

4. Prove Proposition 6.23 using the same technique as in Example 6.22.

5. Prove Proposition 6.30.

6. Let $U_1, U_2, \ldots$ be iid uniform $[0, 1]$ random variables, and λ be a fixed number in $(0, 1)$. Let X_n be the indicator that $U_i \in [\lambda/n, 1]$, and Y_n be the indicator that $X_1 = 1, X_2 = 1, \ldots, X_n = 1$. Does Y_n converge in distribution? If so, what is its limiting distribution?

7. Let X be a random variable with distribution function $F(x)$ and strictly positive density $f(x)$ that is continuous at $x = 0$. Let G_n be the conditional distribution of nX given that $nX \in [a, b]$. Show that G_n converges in distribution to a uniform on $[a, b]$.

6.2 Connections between Modes of Convergence

Figure 6.2 shows which modes of convergence imply other modes. An arrow leading from mode 1 to mode 2 means that convergence in mode 1 implies convergence in mode 2. If there is no arrow connecting mode 1 to mode 2, then mode 1 convergence does not necessarily imply mode 2 convergence without additional conditions.

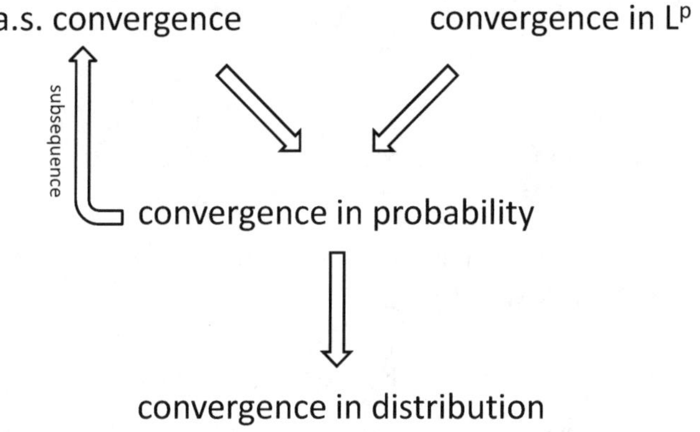

Figure 6.2: Connections between modes of convergence.

6.2.1 Convergence in L^p Implies Convergence in Probability, but Not Vice Versa

The easiest implication to prove in Figure 6.2 is the following.

Proposition 6.32. Convergence in L^p implies convergence in probability *If $X_n \rightarrow X$ in L^p, then $X_n \xrightarrow{p} X$.*

Proof. By Markov's inequality (Proposition 5.17),

$$\begin{aligned} P(|X_n - X| \geq \epsilon) &= P(|X_n - X|^p \geq \epsilon^p) \\ &\leq \mathrm{E}(|X_n - X|^p)/\epsilon^p \rightarrow 0. \end{aligned} \tag{6.9}$$

□

Proposition 6.33. Bias and variance both tending to 0 implies convergence in probability *An estimator $\hat{\theta}_n$ of θ whose bias and variance both tend to 0 as $n \rightarrow \infty$ converges to θ in probability. In particular, the sample mean from a sample of n iid observations with mean μ and finite variance converges in probability to μ.*

Proof. By Proposition 6.17, $\hat{\theta}_n \rightarrow \theta$ in L^2. By Proposition 6.32, $\hat{\theta}_n \xrightarrow{p} \theta$. □

The statement that the sample mean converges in probability to μ is known as a law of large numbers, or in the common vernacular, a law of averages. In the next chapter, we will strengthen this result.

Note that convergence in probability does not imply convergence in L^p, as the following example shows.

Example 6.34. counterexample to convergence in probability implies convergence in L^p Let

$$X_n = \begin{cases} 0 & \text{with probability } 1 - 1/n \\ \exp(n) & \text{with probability } 1/n \end{cases}$$

Then $X_n \xrightarrow{p} 0$ because, for $\epsilon \leq \exp(n)$, $P(|X_n - 0| \geq \epsilon) = 1/n \to 0$. On the other hand, $\mathrm{E}(|X_n - 0|^p) = \exp(np)/n \to \infty$ as $n \to \infty$ for every $p > 0$. □

6.2.2 Convergence in Probability Implies Convergence in Distribution, but Not Vice Versa

The next easiest implication to prove in Figure 6.2 is the following.

Proposition 6.35. Convergence in probability implies convergence in distribution *If $X_n \xrightarrow{p} X$, then $X_n \xrightarrow{D} X$.*

Proof. The easiest way to prove this uses Proposition 6.29 (exercise). However, the following proof using only the definition of convergence in distribution is instructive. Let F_n and F be the distributions of X_n and X, and let x be a continuity point of F. Let ϵ be such that $x - \epsilon$ and $x + \epsilon$ are also continuity points of F. Notice that

$$\begin{aligned} \{X \leq x - \epsilon\} \cap \{|X_n - X| < \epsilon\} &\Rightarrow X_n \leq x \\ P(X \leq x - \epsilon \cap |X_n - X| < \epsilon) &\leq F_n(x) \\ \underline{\lim}\, \{P(X \leq x - \epsilon \cap |X_n - X| < \epsilon)\} &\leq \underline{\lim}\, F_n(x) \\ F(x - \epsilon) &\leq \underline{\lim}\, F_n(x). \end{aligned}$$

The last step follows from writing $P(X \leq x - \epsilon \cap |X_n - X| < \epsilon)$ as $P(X \leq x - \epsilon) - P(X \leq x - \epsilon \cap |X_n - X| \geq \epsilon)$ and noting that $P(X \leq x - \epsilon \cap |X_n - X| \geq \epsilon) \leq P(|X_n - X| \geq \epsilon) \to 0$.

Similarly,

$$\begin{aligned} X_n \leq x \cap |X_n - X| < \epsilon &\Rightarrow X \leq x + \epsilon \\ P(X_n \leq x) - P(X_n \leq x \cap |X_n - X| \geq \epsilon) &\leq F(x + \epsilon) \\ P(X_n \leq x) &\leq P(X_n \leq x \cap |X_n - X| \geq \epsilon) + F(x + \epsilon) \\ \overline{\lim}\, P(X_n \leq x) &\leq \overline{\lim}\, \{P(X_n \leq x \cap |X_n - X| \geq \epsilon) + F(x + \epsilon)\} \\ \overline{\lim}\, F_n(x) &\leq F(x + \epsilon). \end{aligned}$$

We have shown that $\underline{\lim}\, F_n(x) \geq F(x - \epsilon)$ and $\overline{\lim}\, F_n(x) \leq F(x + \epsilon)$. But ϵ was arbitrary subject only to $x - \epsilon$ and $x + \epsilon$ being continuity points of F. We can find a sequence $\epsilon_m \to 0$ such that $x - \epsilon_m$ and $x + \epsilon_m$ are continuity points of F (exercise). Therefore, $\underline{\lim}\, F_n(x) = \overline{\lim}\, F_n(x) = F(x)$, proving that $F_n(x) \to F(x)$ for each continuity point x of F. □

The simplest example showing that the reverse direction does not hold is when the X_n have the same distribution, but are on different probability spaces (see Remark 6.27). Then X_n converges in distribution to X_1, but X_n cannot converge in probability if they are on different probability spaces. Another example is the following.

Example 6.36. Counterexample to convergence in distribution implies convergence in probability Let $X_1, X_2, \ldots$ be iid Bernoulli (1/2) random variables. Then X_n converges in distribution to X_1, but X_n does not converge in probability to X_1 because $P(|X_n - X_1| > 1/2) = P(X_n = 1, X_1 = 0) + P(X_n = 0, X_1 = 1) = 1/4 + 1/4 = 1/2$. □

6.2.3 Almost Sure Convergence Implies Convergence in Probability, but Not Vice Versa

Proposition 6.37. Almost sure convergence implies convergence in probability: *If $X_n \stackrel{a.s.}{\to} X$, then $X_n \stackrel{p}{\to} X$.*

Proof. We must prove that for each $\epsilon > 0$, $P(|X_n - X| \geq \epsilon) \to 0$ as $n \to \infty$. Let $B_n = \{|X_n - X| \geq \epsilon\}$ and $B = \{|X_n - X| \geq \epsilon \text{ i.o.}\}$. The almost sure convergence of X_n to X implies that $P(B) = 0$. Also, B occurs if and only if $\omega \in \cup_{k=n}^{\infty} B_k$ for every $n = 1, 2, \ldots$. Equivalently,

$$B = \cap_{n=1}^{\infty} A_n, \text{ where } A_n = \cup_{k=n}^{\infty} B_k$$

is a decreasing sequence of sets. It follows from the continuity property of probability (Proposition 3.30) that

$$\begin{aligned} 0 &= P(B) = P\left(\cap_{n=1}^{\infty} A_n\right) \\ &= \lim_{n\to\infty} P(A_n) = \lim_{n\to\infty} P\left(\cup_{k=n}^{\infty} B_k\right). \end{aligned} \tag{6.10}$$

We have shown that $P\left(\cup_{k=n}^{\infty} B_k\right) \to 0$ as $n \to \infty$. But $0 \leq P(B_n) \leq P\left(\cup_{k=n}^{\infty} B_k\right)$, so $P(B_n) \to 0$. Thus, $P(|X_n - X| \geq \epsilon) \to 0$ as $n \to \infty$. Because this is true for every $\epsilon > 0$, $X_n \stackrel{p}{\to} X$. Therefore, almost sure convergence implies convergence in probability. □

Examples 6.7 and 6.11 and Proposition 6.37 illustrate that the concept of events occurring infinitely often in n is very useful in convergence settings. The following is an indispensable tool for determining the probability of events occurring infinitely often.

Lemma 6.38. Borel-Cantelli lemma

1. *If A_n, $n = 1, 2, \ldots$ is an infinite sequence of events and $\sum P(A_n) < \infty$, then $P(A_n \text{ i.o.}) = 0$. This remains true if P is replaced by an arbitrary measure μ.*

2. *If the A_n are independent and $\sum_n P(A_n) = \infty$, then $P(A_n \text{ i.o.}) = 1$.*

Proof. For part 1, define the nonnegative random variable X by $X(\omega) = \sum_n I(\omega \in A_n)$. Then $\omega \in A_n$ for infinitely many n if and only if $X(\omega) = \infty$. If $X = \infty$ with nonzero probability, then clearly $\mathrm{E}(X) = \infty$, so

$$\infty = \mathrm{E}(X) = \mathrm{E}\left\{\sum_n I(A_n)\right\} = \sum_n \mathrm{E}\{I(A_n)\} = \sum_n P(A_n) \tag{6.11}$$

(interchange of expectation and summation is justified by the MCT because $\sum_{n=1}^{N} I(A_n) \uparrow \sum_{n=1}^{\infty} I(A_n)$). Therefore, if $\sum_n P(A_n) < \infty$, X must be finite a.s., implying that $P(A_n \text{ i.o.}) = 0$. Each step of the proof holds when P is replaced by an arbitrary measure μ and expectation is replaced by integration with respect to μ.

For part 2, recall that

$$P(A_n \text{ i.o.}) = P(\cap_{N=1}^{\infty} \cup_{n=N}^{\infty} A_n) = P(\cap_{N=1}^{\infty} B_N),$$

where $B_N = \cup_{n=N}^{\infty} A_n$ is a decreasing sequence of sets. Therefore, by the continuity property of probability (Proposition 3.30),

$$P(\cap_{N=1}^{\infty} \cup_{n=N}^{\infty} A_n) = \lim_{N\to\infty} P(\cup_{n=N}^{\infty} A_n).$$

It suffices to prove that $P(\cup_{n=N}^{\infty} A_n) = 1$ for each N. Because

$$P(\cup_{n=N}^{\infty} A_n) = 1 - \prod_{n=N}^{\infty} (1 - p_n),$$

where $p_n = P(A_n)$, we need only show that $\prod_{n=N}^{\infty}(1-p_n) = 0$, or equivalently, $\sum_{n=N}^{\infty} \ln(1-p_n) = -\infty$. Use a one term Taylor series of $f(x) = \ln(1-x)$ expanded about $x = 0$ to see that

$$\ln(1-x) = -x - (1-c)^{-2}x^2/2,$$

where $0 \le c \le x$ (Proposition A.49). Therefore, $\ln(1-x) \le -x$ for $x > 0$. This shows that $\sum_{n=N}^{\infty} \ln(1-p_n) \le -\sum_{n=N}^{\infty} p_n = -\infty$, which completes the proof. □

The Borel-Cantelli lemma (Lemma 6.38) allows us to construct new examples involving almost sure convergence or convergence in probability.

Example 6.39. Let X_n be independent Bernoulli random variables with $P(X_n = 1) = p_n = 1 - P(X_n = 0)$, and suppose that $p_n = 1/n^2$. Does X_n converge to 0 almost surely, in probability or neither? Notice that the probability that $X_n = 1$ for infinitely many n is 0 by part 1 of the Borel-Cantelli lemma because $\sum_n P(X_n = 1) = \sum_n 1/n^2 < \infty$. This means that for each ω outside a set of probability 0, there is an N such that $X_n(\omega) = 0$ for all $n \ge N$. For each such ω, $X_n(\omega) \to 0$. Therefore, $X_n \overset{a.s.}{\to} 0$.

Suppose that the Bernoulli parameter p_n had been $1/n$ instead of $1/n^2$. Because $\sum_n 1/n = \infty$ and the events $\{X_n = 1\}$ are independent, part 2 of the Borel-Cantelli lemma implies that $P(X_n = 1 \text{ i.o.}) = 1$. Of course $P(X_n = 0 \text{ i.o.}) = 1$ as well, so $P(X_n = 0 \text{ i.o.} \cap X_n = 1 \text{ i.o.}) = 1$. But each ω for which $X_n(\omega) = 0$ i.o. and $X_n(\omega) = 1$ i.o. is an ω for which $X_n(\omega)$ does not converge. Therefore, if p_n had been $1/n$, X_n would not have converged almost surely to any random variable, although X_n would still have converged to 0 in probability.

Even though X_n does not converge almost surely to 0 if $p_n = 1/n$, the subsequence $X_{n^2} = (X_1, X_4, X_9, \ldots)$ does converge almost surely to 0 by part 1 of the Borel-Cantelli lemma: $P(|X_{n^2} - 0| > 0 \text{ i.o.}) = 0$ because $\sum_n 1/n^2 < \infty$. We will soon see that this example is typical in that it is always possible to find a subsequence converging almost surely to X whenever $X_n \overset{p}{\to} X$. □

Exercises

1. **Let** $(\Omega, \mathcal{F}, P) = ([0,1], \mathcal{B}_{[0,1]}, \mu_L)$ **and define** $X_1 = I(\omega \in [0,1/2])$, $X_2 = I(\omega \in (1/2, 3/4])$, $X_3 = I(\omega \in (3/4, 7/8])$, **etc. (the intervals have widths** $1/2, 1/4, 1/8, 1/16, \ldots$**) Prove that** $X_n \stackrel{a.s.}{\to} 0$**.**

2. Let $X_1, X_2, \ldots$ be iid with a continuous distribution function, and let Y_1 be the indicator that X_1 is the larger of (X_1, X_2), Y_2 be the indicator that X_3 is the largest of (X_3, X_4, X_5), Y_3 be the indicator that X_6 is the largest of (X_6, X_7, X_8, X_9), etc. Prove that $Y_n \stackrel{p}{\to} 0$ but not almost surely.

3. In the proof of Proposition 6.35, we asserted that it is possible to find a sequence $\epsilon_m \to 0$ such that $x - \epsilon_m$ and $x + \epsilon_m$ are both continuity points of the distribution of F. Prove this fact.

4. It can be shown that the bounded convergence theorem applies to convergence in probability as well as almost sure convergence. Use this fact and Proposition 6.29 to supply a much simpler proof of Proposition 6.35.

5. Prove the converse of Exercise 6 in Section 6.1.3, namely that if $\mathrm{E}\{|X_n|/(1+|X_n|)\} \to 0$, then $X_n \stackrel{p}{\to} 0$.

6. Using the same technique as in the proof of the first part of the Borel-Cantelli lemma (namely, using a sum of indicator random variables) compute the following expected numbers:

   (a) One method of testing whether basketball players have "hot" and "cold" streaks is as follows. Let X_i be 1 if shot i is made and -1 if it is missed. The number of sign changes in consecutive shots measures how "streaky" the player is: a very small number of sign changes means the player had long streaks of made or missed shots. For example, $-1, -1, -1, +1, +1, +1, +1, +1, +1, +1$ contains only one sign change and has a streak of 3 missed shots followed by 7 made shots. Under the null hypothesis that the X_i are iid Bernoulli p (i.e., streaks occur randomly), what is the expected number of sign changes in n shots? Hint: let $Y_i = I(X_i \neq X_{i-1})$.

   (b) What is the expected number of different values appearing in a bootstrap sample (a sample drawn with replacement) of size n from $\{x_1, \ldots, x_n\}$? Show that the expected proportion of values not appearing in the bootstrap sample is approximately $\exp(-1)$ if n is large.

7. Regression analysis assumes that errors from different observations are independent. One way to test this assumption is to count the numbers n_+ and n_- of positive and negative residuals, and the number n_R of runs of the same sign (see pages 198-200 of Chatterjee and Hadi, 2006). For example, if the sequence of signs of the residuals is $+ + - - - + - - - -$, then $n_+ = 3$, $n_- = 7$, and $n_R = 4$. Assume that the residuals are exchangeable (each permutation has the same joint distribution). Using indicator functions, prove that the expected number of runs, given n_+ and n_-, is $2n_+n_-/(n_+ + n_-) + 1$.

8. Suppose that $Y_1, Y_2, \ldots$ are independent. Show that $P[\sup_n\{Y_n\} < \infty] = 1$ if and only if $\sum_{n=1}^{\infty} P(Y_n > B) < \infty$ for some constant B.

9. Let $A_1, A_2, \ldots$ be a countable sequence of independent events with $P(A_i) < 1$ for each i. Then $P(\cup_{i=1}^{\infty} A_i) = 1$ if and only if $P(A_i \text{ i.o.}) = 1$.

10. Prove that if $A_1, \ldots, A_n$ are events such that $\sum_{i=1}^n P(A_i) > n-1$, then $P(\cap_{i=1}^n A_i) > 0$.

11. Give an example to show that independence is required in the second part of the Borel-Cantelli lemma.

12. * Show that almost sure convergence and convergence in L^p do not imply each other without further conditions. Specifically, use the Borel-Cantelli lemma to construct an example in which X_n takes only two possible values, one of which is 0, and:

    (a) X_n converges to 0 in L^p for every $p > 0$, but not almost surely.

    (b) X_n converges almost surely to 0, but not in L^p for any $p > 0$ (hint: modify Example 6.34).

13. Construct an example such that X_n are independent and $X_n \xrightarrow{p} 0$, yet S_n/n does not converge almost surely to 0. Hint, let X_n take values 0 or 2^n with certain probabilities.

14. Let $\hat{\theta}_n$ be an estimator with finite mean.

    (a) If $\hat{\theta}_n$ is consistent for θ (i.e., $\hat{\theta}_n \xrightarrow{p} \theta$), is $\hat{\theta}_n$ asymptotically unbiased (i.e., $\mathrm{E}(\hat{\theta}_n) \to \theta$)?

    (b) If $\hat{\theta}_n$ is asymptotically unbiased, is $\hat{\theta}_n$ consistent?

    (c) If $\hat{\theta}_n$ is asymptotically unbiased and $\mathrm{var}(\hat{\theta}_n)$ exists and tends to 0, is $\hat{\theta}_n$ consistent?

15. Let X_i be iid random variables, and let $Y_n = X_n/n$. Prove that $Y_n \xrightarrow{a.s.} 0$ if and only if $\mathrm{E}(|X|) < \infty$. Hint: use Proposition 5.9 in conjunction with the Borel-Cantelli lemma.

6.2.4 Subsequence Arguments

The technique at the end of Example 6.39 of examining subsequences of X_n is very useful. Because we will be considering subsequences of subsequences, the notation can get cumbersome. The following simplifies things.

Definition 6.40. Subsequence notation *Suppose $(n_1, n_2, \ldots)$ is a sequence and $(i_1, i_2, \ldots)$ is a subsequence. That is, $i_1 = n_{k_1}, i_2 = n_{k_2}, \ldots$ for some $k_1 < k_2 < \ldots$ Letting (N) denote $(n_1, n_2 \ldots)$ and (I) denote $(i_1, i_2, \ldots)$, we write $(I) \subset (N)$. We use the term subsequence to refer to either (I) or x_i for $i \in (I)$.*

Subsequence arguments can be helpful even in nonrandom settings, as the following result shows.

Proposition 6.41. Subsequence argument for convergence of numbers *The sequence of numbers x_n converges to x if and only if every subsequence $(M) \subset (N)$ has a further subsequence $(J) \subset (M)$ such that $x_j \to x$ along $j \in (J)$.*

Proof. It is clear that if $x_n \to x$, then every subsequence x_m, $m \in (M)$ also converges to x. To prove the other direction, we use the contrapositive. If x_n does not converge to x, then we can find an $\epsilon > 0$ and a subsequence (K) such that $|x_k - x| \geq \epsilon$ for all $k \in (K)$. But then clearly no subsequence x_j $j \in (J) \subset (K)$ can converge to x. Therefore, if x_n does not converge to x, then it cannot be true that every subsequence contains a further subsequence converging to x. By the contrapositive, this means that if every subsequence contains a further subsequence converging to x, then $x_n \to x$. □

We can use a subsequence argument to prove convergence in probability by piggybacking on almost sure convergence.

Proposition 6.42. Subsequence argument for convergence in probability or measure

1. *$X_n \xrightarrow{p} X$ if and only if every subsequence $(M) \subset (N)$ has a further subsequence $(K) \subset (M)$ such that $X_k \xrightarrow{a.s.} X$ along (K).*

2. *If f_n and f are functions on an arbitrary measure space $(\Omega, \mathcal{F}, \mu)$ and f_n converges to f in measure, then every subsequence $(M) \subset (N)$ contains a further subsequence $(K) \subset (M)$ such that $f_k \to f$ a.e. with respect to μ (although the reverse direction is not necessarily true).*

Proof. Suppose first that $X_n \xrightarrow{p} X$, and let $(M) \subset (1, 2, \ldots)$ be a subsequence. For a fixed number k, $\epsilon = 1/k$ is a positive number and $P(|X_m - X| \geq \epsilon) \to 0$ as $m \to \infty$ along (M). This means we can find $m_k \in (M)$ such that

$$P(|X_{m_k} - X| \geq \epsilon = 1/k) \leq 1/k^2.$$

Because $\sum_k 1/k^2 < \infty$, part 1 of the Borel-Cantelli lemma implies that $P(|X_{m_k} - X| \geq 1/k \text{ i.o. in } k) = 0$. But each ω for which $|X_{m_k}(\omega) - X(\omega)| \geq 1/k$ for only finitely many k is an ω for which $X_{m_k}(\omega) \to X(\omega)$ as $k \to \infty$. Therefore, $X_{m_k} \xrightarrow{a.s.} X$ as $k \to \infty$. This shows that $X_n \xrightarrow{p} X$ implies that the subsequence X_m, $m \in M$ contains a further subsequence X_{m_k} converging almost surely to X. The same argument works for f_n converging in measure to f with respect to a more general measure μ.

The proof of the reverse direction for part 1 is very similar to the proof of Proposition 6.41, and is left as an exercise. □

Subsequence arguments can be used to prove many useful results. For instance, consider part 3 of Proposition 6.12. Let (M) be an arbitrary subsequence of $(N) = (1, 2, \ldots)$. We will prove that there is a further subsequence $(I) \subset (M)$ such that $X_i + Y_i \xrightarrow{a.s.} X + Y$ along (I). Because $X_m \xrightarrow{p} X$ along (M), there exists a subsequence $(J) \subset (M)$ such that $X_j \xrightarrow{a.s.} X$ along (J) by Proposition 6.42. Because $Y_j \xrightarrow{p} Y$ along (J), Proposition 6.42 implies that there exists a further subsequence $(I) \subset (J)$ such that $Y_i \xrightarrow{a.s.} Y$ along (I). Along the subsequence I, $X_i \xrightarrow{a.s.} X$ and $Y_i \xrightarrow{a.s.} Y$. By Proposition 6.5, $X_i + Y_i \xrightarrow{a.s.} X + Y$ along (I). We have shown that an arbitrary subsequence (M) has a further subsequence $(I) \subset (M)$ such that $X_i + Y_i \xrightarrow{a.s.} X + Y$ along (I). By Proposition 6.42, $X_n + Y_n \xrightarrow{p} X + Y$, completing the proof. The same type of subsequence argument can be used for the other parts of Proposition 6.12 as well. We leave this as an exercise.

Proposition 6.43. More general DCT for arbitrary measures *If $f_n \to f$ in measure with respect to μ and $|f_n| \leq g$ a.e. (μ), where $\int g(\omega)d\mu(\omega) < \infty$, then $\int f_n(\omega)d\mu(\omega) \to \int f(\omega)d\mu(\omega)$.*

Proof. We will prove that each subsequence of $\nu_n = \int f_n(\omega)d\mu(\omega)$ has a further subsequence that converges to $\nu = \int f(\omega)d\mu(\omega)$. Let $(M) \subset (1, 2, \ldots)$ be a subsequence. Because $f_n \to f$ in measure, there is a further subsequence (K) such that $f_k \to f$ a.e. (μ) as $k \to \infty$ along (K) by Proposition 6.42. Also, $|f_k| \leq g$ and $\int g\, d\mu(\omega) < \infty$. By the usual DCT (Theorem 5.12), $\int f_k d\mu(\omega) \to \int f d\mu(\omega)$ as $k \to \infty$ along (K). Therefore, each subsequence (M)

contains a further subsequence $(K) \subset (M)$ such that $\nu_k \to \nu$ along (K). By Proposition 6.41, $\int f_n(\omega)d\mu(\omega) \to \int f(\omega)d\mu(\omega)$. □

Subsequence arguments can also be used for convergence in distribution. The proof of the following result is very similar to the proof of Proposition 6.41 and is left as an exercise.

Proposition 6.44. Subsequence argument for convergence in distribution *$X_n \xrightarrow{D} X$ if and only if every subsequence $(M) \subset (N)$ has a further subsequence $(K) \subset (M)$ such that $X_k \xrightarrow{D} X$ along (K).*

Notice that is not sufficient for every subsequence to contain a further subsequence that converges in distribution; it must be the **same** distribution, as the following example shows.

Example 6.45. Let X_n be N(0, 1) if n is odd and N(0, 2) if n is even. Then every subsequence contains a further subsequence converging to a distribution function, but X_n does not converge in distribution because $F_{2n+1} \xrightarrow{D} N(0,1)$ but $F_{2n} \xrightarrow{D} N(0,2)$. □

In Example 6.45, different subsequences converge to different distribution functions. A very interesting question about a sequence of distribution functions is whether there is **any** subsequence that converges to a distribution function. The investigation of this question produces a very useful technique called the diagonal technique, illustrated in the following theorem.

Theorem 6.46. Helley Selection Theorem *If $F_1(x), F_2(x), \ldots$ is a sequence of distribution functions, there is a subsequence (M) and an increasing, right-continuous function $F(x)$ such that $F_m(x) \to F(x)$ along (M) at all continuity points.*

Sketch of Proof. Let $\{r_i\}_i$ be an enumeration of the rational numbers. The ith row of Table 6.1 shows the set $\{F_1(r_i), F_2(r_i), \ldots\}$, a bounded infinite set of numbers. Consider the first row. By the Bolzano-Weierstrass theorem (Theorem A.17), there is a limit point, so there is a subsequence (M_1) such that $F_m(r_1) \to y_1 = F(r_1)$ as $m \to \infty$ along (M_1). For instance, that subsequence might be the bolded elements $1, 3, 4, 6, 7, \ldots$ shown in row 1. In the second row, consider the bounded infinite set $(F_m(r_2))_{m \in (M_1)}$. Again by the Bolzano-Weierstrass theorem, there must be a further subsequence $(M_2) \subset (M_1)$ such that $F_m(r_2) \to y_2 = F(r_2)$ as $m \to \infty$ along (M_2). For instance, that subsequence of $1, 3, 4, 6, 7, \ldots$ might be $1, 4, 6, 7 \ldots$, the bolded sequence in row 2. Continuing in this fashion, we can find in row $i + 1$ a subsequence (M_{i+1}) of (M_i) such that $F_m(r_{i+1}) \to y_{i+1} = F(r_{i+1})$ as $m \to \infty$ along (M_{i+1}). Now define a subsequence (M) by taking the first bolded element of row 1, the second bolded element of row 2, the third bolded element of row 3, etc. In Table 6.1, this corresponds to the subsequence $(M) = (1, 4, 7, \ldots)$. Along this subsequence, $F_m(r_i) \to y_i = F(r_i)$ as $m \to \infty$ along (M) for each rational r_i.

Now define $F(x)$ for irrational x by $\inf\{F(r),\ r > x,\ r \text{ rational}\}$. It is not difficult to show that F is right continuous and increasing. □

Notice that the function $F(x)$ in Helley's selection theorem satisfies two of the three conditions of a distribution function. Conspicuously absent is the condition that $\lim_{x\to-\infty} F(x) = 0 = 1 - \lim_{x\to\infty} F(x)$. The next example shows that this condition is not necessarily satisfied.

Table 6.1:

$\boldsymbol{F_1(r_1)}$	$F_2(r_1)$	$\boldsymbol{F_3(r_1)}$	$\boldsymbol{F_4(r_1)}$	$F_5(r_1)$	$\boldsymbol{F_6(r_1)}$	$\boldsymbol{F_7(r_1)}$	$\ldots$
$\boldsymbol{F_1(r_2)}$	$F_2(r_2)$	$F_3(r_2)$	$\boldsymbol{F_4(r_2)}$	$F_5(r_2)$	$\boldsymbol{F_6(r_2)}$	$\boldsymbol{F_7(r_2)}$	$\ldots$
$F_1(r_3)$	$F_2(r_3)$	$F_3(r_3)$	$\boldsymbol{F_4(r_3)}$	$F_5(r_3)$	$\boldsymbol{F_6(r_3)}$	$\boldsymbol{F_7(r_3)}$	$\ldots$
$\vdots$					$\vdots$		

Example 6.47. Limit of d.f.s need not be a d.f Let

$$X_n = \begin{cases} -n & \text{w.p. } 1/2 \\ +n & \text{w.p. } 1/2. \end{cases}$$

Then for any subsequence (M), the distribution function F_m for X_m satisfies $F_m(x) \to 1/2$ as $m \to \infty$ along (M). Therefore, there is no subsequence (M) such that $F_m(x)$ converges to distribution function $F(x)$ as $m \to \infty$ along (M). □

To ensure that the limiting $F(x)$ is a distribution function, we need another condition called tightness.

Definition 6.48. Tight sequence of d.f.s *The sequence of distribution functions $F_1(x)$, $F_2(x), \ldots$ is said to be tight if for each $\epsilon > 0$ there is a number M such that $F_n(M) < \epsilon$ and $1 - F_n(M) < \epsilon$ for all n.*

It is clear that the sequence of distribution functions in Example 6.47 is not tight because for any M, $1 - F_n(M) = 1/2$ for all $n > M$. Thus, tightness precludes examples such as Example 6.47.

Proposition 6.49. Connection between convergence in distribution and tightness *The subsequence $F_1(x), F_2(x), \ldots$ has a further subsequence converging to a distribution function if and only if $F_1, F_2, \ldots$ is tight.*

Corollary 6.50. *If $F_n \stackrel{D}{\to} F$, then F_n is tight.*

We close this section by collecting part 2 of Proposition 6.5, part 2 of Proposition 6.12, and Proposition 6.30 into a single "continuous mapping theorem", also called the Mann-Wald theorem. The idea is that if X_n converges to X in some manner, then a continuous function $g(X_n)$ ought to converge to $g(X)$ in the same manner. We proved these results earlier, but we are including them here because they can also be proven using subsequence arguments (see Exercise 4, for example).

Theorem 6.51. Mann-Wald mapping theorem *Suppose that $X_1, X_2, \ldots$ is a sequence of random variables and $f : R \longmapsto R$ is a Borel function whose set D of discontinuities is such that $\{\omega : X(\omega) \in D\} \in \mathcal{F}$ and $P(X \in D) = 0$. If X_n converges to X either*

1. *almost surely,*
2. *in probability, or*
3. *in distribution,*

Then $g(X_n)$ converges to $g(X)$ in the same sense.

Exercises

1. Prove the reverse direction of part 1 of Proposition 6.42.

2. State whether the following is true: X_n converges in probability if and only if every subsequence contains a further subsequence that converges almost surely. If true, prove it. If not, give a counterexample.

3. Determine and demonstrate whether the distribution functions associated with the following sequences of random variables are tight.

   (a)
   $$X_n = \begin{cases} 0 & \text{with probability } 1/2 \\ n^{1/2} & \text{with probability } 1/2. \end{cases}$$

   (b) $X_n = \ln(U_n)$, where U_n is uniform $[1/n, 1]$.

   (c) $X_n = \cos(nU)$, where U is uniform $[0, 1]$.

   (d) $X_n \sim \mathrm{N}(n, 1)$.

   (e) $X_n \sim \mathrm{N}(0, 1 + 1/n)$.

4. Use subsequence arguments to prove items 1 and 2 of Proposition 6.12.

5. Use a subsequence argument to prove items 4 and 5 of Proposition 6.12.

6. Prove Proposition 6.44 using Proposition 6.29 and a subsequence argument.

7. Provide an alternative proof of Proposition 6.44 by contradiction using the definition of convergence in distribution and a subsequence argument.

8. Prove that $X_n \to X$ in L^p if and only if each subsequence (M) contains a further subsequence $(K) \subset (M)$ such that $X_k \to X$ in L^p along (K).

6.2.5 Melding Modes: Slutsky's Theorem

A very useful result melding convergence in distribution and convergence in probability is Slutsky's theorem:

Theorem 6.52. Slutsky's theorem *Suppose that* $X_n \stackrel{D}{\to} X$, $A_n \stackrel{p}{\to} A$, *and* $B_n \stackrel{p}{\to} B$, *where* A *and* B *are constants. Then* $A_n X_n + B_n \stackrel{D}{\to} AX + B$.

Proof. It is elementary to show that $AX_n + B \stackrel{D}{\to} AX + B$, and we can write $A_n X_n + B_n$ as $AX_n + B + C_n$, where $C_n = (A_n - A)X_n + (B_n - B)$ converges in probability to 0. Therefore, it suffices to prove that if $Y_n \stackrel{D}{\to} Y$ and $C_n \stackrel{p}{\to} 0$, then $Y_n + C_n \stackrel{D}{\to} Y$.

Let y, $y - \epsilon$, and $y + \epsilon$ be continuity points of the distribution function $F(y)$ for Y. This is always possible because there are only countably many discontinuity points of F.

$$\begin{aligned} Y_n \le y - \epsilon \cap |C_n| \le \epsilon \quad &\Rightarrow \quad Y_n + C_n \le y \\ P(Y_n \le y - \epsilon \cap |C_n| \le \epsilon) \quad &\le \quad P(Y_n + C_n \le y) \end{aligned}$$

$$\begin{aligned}
P(Y_n \le y-\epsilon) - P(Y_n \le y-\epsilon \cap |C_n| > \epsilon) &\le P(Y_n + C_n \le y) \\
\underline{\lim}\, \{P(Y_n \le y-\epsilon) - P(Y_n \le y-\epsilon \cap |C_n| > \epsilon)\} &\le \underline{\lim}\, P(Y_n + C_n \le y) \\
F(y-\epsilon) &\le \underline{\lim}\, P(Y_n + C_n \le y).
\end{aligned}$$

Similarly,

$$Y_n + C_n \le y \cap |C_n| \le \epsilon \;\Rightarrow\; Y_n \le y+\epsilon$$

$$P(Y_n + C_n \le y) - P(Y_n + C_n \le y \cap |C_n| > \epsilon) \le P(Y_n \le y+\epsilon)$$

$$\begin{aligned}
P(Y_n + C_n \le y) &\le P(Y_n + C_n \le y \cap |C_n| > \epsilon) + P(Y_n \le y+\epsilon) \\
\overline{\lim}\, P(Y_n + C_n \le y) &\le \overline{\lim}\, \{P(Y_n \le y \cap |C_n| > \epsilon) + P(Y_n \le y+\epsilon)\} \\
\overline{\lim}\, P(Y_n + C_n \le y) &\le F(y+\epsilon).
\end{aligned}$$

We have shown that $\underline{\lim}\, P(Y_n+C_n \le y) \ge F(y-\epsilon)$ and $\overline{\lim}\, P(Y_n+C_n \le y) \le F(y+\epsilon)$. Because ϵ is arbitrary and F is continuous at y, $\underline{\lim}\, P(Y_n + C_n \le y) = \overline{\lim}\, P(Y_n + C_n \le y) = F(y)$, completing the proof. □

Example 6.53. Replacing variance by consistent estimator Many estimators $\hat{\theta}_n$ are asymptotically normally distributed with mean θ and some variance v_n, meaning that $Z_n = (\hat{\theta}_n - \theta)/v_n^{1/2} \stackrel{D}{\to} Z$, where Z is N(0, 1). If $\hat{v}_n$ is an estimator of the variance such that $v_n/\hat{v}_n \stackrel{p}{\to} 1$, then

$$\begin{aligned}
\frac{\hat{\theta}_n - \theta}{\sqrt{\hat{v}_n}} &= \sqrt{\frac{v_n}{\hat{v}_n}} \left(\frac{\hat{\theta}_n - \theta}{\sqrt{v_n}} \right) \\
&= A_n Z_n \stackrel{D}{\to} Z \qquad (6.12)
\end{aligned}$$

by Slutsky's theorem because $A_n \stackrel{p}{\to} 1$ and $Z_n \stackrel{D}{\to} Z$. Therefore, asymptotically it does not matter whether we use the actual variance or a consistent estimator of the variance. □

Exercises

1. Give an example to show that the following proposition is false. If $X_n \stackrel{D}{\to} X$ and $Y_n \stackrel{p}{\to} Y$, then $X_n + Y_n \stackrel{D}{\to} X + Y$.

2. In the first step of the proof of Slutsky's theorem, we asserted that $X_n \stackrel{D}{\to} X$ implies that $AX_n + B \stackrel{D}{\to} AX + B$ for constants A and B. Why does this follow?

3. Show that convergence in distribution to a constant is equivalent to convergence in probability to that constant.

4. Let X_n and Y_n be random variables on the same probability space with $X_n \stackrel{p}{\to} 0$ and $Y_n \stackrel{D}{\to} Y$. Prove that $X_n Y_n \stackrel{p}{\to} 0$.

5. ↑ In the preceding problem, why is there a problem if you try to use the Skorokhod representation theorem to prove the result?

6. What is the problem with the following attempted proof of Slutsky's theorem? As noted in the first paragraph of the proof, it suffices to prove that if $Y_n \stackrel{D}{\to} Y$ and $C_n \stackrel{p}{\to} 0$, then $Y_n + C_n \stackrel{D}{\to} Y$.

   (a) By the Skorokhod representation theorem, there exist random variables Y_n' and Y' such that Y_n' has the same distribution as Y_n, Y' has the same distribution as Y, and $Y_n' \stackrel{a.s.}{\to} Y'$.

   (b) Because almost sure convergence implies convergence in probability (Proposition 6.37), $Y_n' \stackrel{p}{\to} Y'$.

   (c) Since $C_n \stackrel{p}{\to} 0$ by assumption, $Y_n' + C_n \stackrel{p}{\to} Y' + 0 = Y'$ by part 3 of Proposition 6.12.

   (d) Also, convergence in probability implies convergence in distribution (Proposition 6.35). Therefore, $Y_n' + C_n \stackrel{D}{\to} Y'$.

   (e) But Y_n' has the same distribution as Y_n and Y' has the same distribution as Y, so $Y_n + C_n \stackrel{D}{\to} Y$ as well, completing the proof.

7. In Chapter 8, we will prove the following. If X_i are iid Bernoulli random variables with parameter $p \in (0,1)$, and $\hat{p}_n$ is the proportion of $X_i = 1$ among $X_1, X_2, \ldots, X_n$, then $Z_n = n^{1/2}(\hat{p}_n - p)/\{p(1-p)\}^{1/2} \stackrel{D}{\to} \mathrm{N}(0,1)$. Taking this as a fact, prove that $\tilde{Z}_n = n^{1/2}(\hat{p}_n - p)/\{\hat{p}_n(1-\hat{p}_n)\}^{1/2} \stackrel{D}{\to} \mathrm{N}(0,1)$.

8. If X_n is asymptotically normal with mean $n\mu$ and variance $n\sigma^2$, does Slutsky's theorem imply that $\{n/(n+1)\}^{1/2}X_n$ is also asymptotically normal with mean $n\mu$ and variance $n\sigma^2$? Explain.

6.3 Convergence of Random Vectors

We can extend definitions for convergence of random variables to convergence of random vectors of dimension k

Definition 6.54. *Let $\mathbf{X}_n$ and $\mathbf{X}$ be k-dimensional random vectors.*

1. **Almost sure convergence for random vectors** *$\mathbf{X}_n$ converges a.s. to $\mathbf{X}$, written $\mathbf{X}_n \stackrel{a.s.}{\to} \mathbf{X}$, if, for each fixed ω outside a set of probability 0, the sequence of vectors $\mathbf{x}_n = \mathbf{X}_n(\omega)$ converges to the vector $\mathbf{x} = \mathbf{X}(\omega)$ as $n \to \infty$.*

2. **Convergence in probability for random vectors** *$\mathbf{X}_n$ converges in probability to $\mathbf{X}$, written $\mathbf{X}_n \stackrel{p}{\to} \mathbf{X}$, if $P(||\mathbf{X}_n - \mathbf{X}|| \geq \epsilon) \to 0$ as $n \to \infty$, where $||\ ||$ denotes Euclidean length of vectors.*

Proposition 6.55. Reduction to random variables *Let $\mathbf{X}_n$ and $\mathbf{X}$ be k-dimensional random vectors.*

1. *$\mathbf{X}_n \stackrel{a.s.}{\to} \mathbf{X}$ if and only if each component X_{ni} converges almost surely to the corresponding component X_i as $n \to \infty$.*

2. *$\mathbf{X}_n \stackrel{p}{\to} \mathbf{X}$ if and only if each component X_{ni} converges in probability to the corresponding component X_i as $n \to \infty$.*

Proposition 6.56. A.s. convergence implies convergence in probability (extension of Proposition 6.37) *If* $\mathbf{X}_n$ *and* $\mathbf{X}$ *are* k*-dimensional random vectors, then* $\mathbf{X}_n \overset{a.s.}{\to} \mathbf{X}$ *implies that* $\mathbf{X}_n \overset{p}{\to} X$.

Proof. This follows immediately from Propositions 6.55 and 6.37. □

Proposition 6.57. Convergence in probability is equivalent to a.s. convergence of a subsequence (extension of Proposition 6.42) *If* $\mathbf{X}_n$ *and* $\mathbf{X}$ *are* k*-dimensional random vectors,* $\mathbf{X}_n \overset{p}{\to} \mathbf{X}$ *if and only if every subsequence* $(M) \subset (1, 2, \ldots)$ *has a further subsequence* $(K) \subset (M)$ *such that* $\mathbf{X}_k \overset{a.s.}{\to} \mathbf{X}$ *along* (K).

The proof is identical to that of the univariate case with $|\ |$ replaced by $||\ ||$.

The most convenient way to extend the definition of convergence in distribution to random vectors is as follows.

Definition 6.58. Convergence in distribution for random vectors *The random vector* $\mathbf{X}_n$ *converges in distribution to the vector* $\mathbf{X}$ *if* $E\{f(\mathbf{X}_n)\} \to \mathrm{E}\{f(\mathbf{X})\}$ *for every bounded continuous function* $f : R^n \longmapsto R$.

Theorem 6.59. Multivariate Mann-Wald mapping theorem *Let* $\mathbf{X}_n$ *be a sequence of* k*-dimensional random vectors and* $f : R^k \longmapsto R^p$ *be a* $\mathcal{B}^k$*-measurable function whose set* D *of discontinuities is such that* $\{\omega : \mathbf{X}(\omega) \in D\} \in \mathcal{F}$ *and* $P(\mathbf{X} \in D) = 0$. *If* $\mathbf{X}_n$ *converges to* $\mathbf{X}$ *either*

1. *almost surely,*
2. *in probability, or*
3. *in distribution,*

then $f(\mathbf{X}_n)$ *converges to* $f(\mathbf{X})$ *in the same sense.*

Proof. We prove part 3 when f is continuous on R^k. Let $\mathbf{Y}_n = f(\mathbf{X}_n)$, and $\mathbf{Y} = f(\mathbf{X})$. If $g : R^p \longmapsto R$ is a bounded continuous function, then $h(\mathbf{X}_n)$ defined as $g\{f(\mathbf{X}_n)\}$ is a bounded continuous function of $\mathbf{X}_n$. Because $\mathbf{X}_n \overset{D}{\to} \mathbf{X}$, $\mathrm{E}\{h(\mathbf{X}_n)\} \to \mathrm{E}\{h(\mathbf{X})\}$ by Definition 6.58. That is, $\mathrm{E}\{g(\mathbf{Y}_n)\} \to \mathrm{E}\{g(\mathbf{Y})\}$. Because this is true for any bounded continuous function $g : R^p \longmapsto R$, $\mathbf{Y}_n \overset{D}{\to} \mathbf{Y}$. □

Theorem 6.60. Extension of the Skorokhod representation theorem *If* $\mathbf{X}_n \overset{D}{\to} \mathbf{X}$, *there exist random vectors* $\mathbf{X}'_n$ *and* $\mathbf{X}$ *defined on* $((0,1), \mathcal{B}_{(0,1)}, \mu_L)$ *such that* $\mathbf{X}'_n$ *has the same joint distribution as* $\mathbf{X}_n$, $\mathbf{X}$ *has the same distribution as* $\mathbf{X}$, *and* $\mathbf{X}_n \overset{a.s.}{\to} X$.

Exercises

1. Show that Proposition 6.55 is not necessarily true for convergence in distribution. For example, let (X_n, Y_n) be bivariate normal with means $(0,0)$, variances $(1,1)$ and correlation
$$\rho_n = \begin{cases} -0.5 & \text{if } n \text{ is odd} \\ +0.5 & \text{if } n \text{ is even.} \end{cases}$$
Then the marginal distributions of X_n and Y_n converge weakly to standard normals, but the joint distribution does not converge weakly.

2. Let $X_{n1},\ldots,X_{nk}$ be independent with respective distribution functions F_{ni} converging weakly to F_i, $i=1,\ldots,k$. Prove that

   (a) $(X_{n1},\ldots,X_{nk}) \stackrel{D}{\to} (X_1,\ldots,X_k)$ as $n\to\infty$, where the X_i are independent and $X_i \sim F_i$.

   (b) For any constants $a_1,\ldots,a_k$, $\sum_{i=1}^k a_i X_{ni} \stackrel{D}{\to} \sum_{i=1}^k a_i X_i$ as $n\to\infty$.

3. Let (X_n, Y_n) be bivariate normal with means $(0,0)$, variances $(1/n,1)$, and correlation $\rho_n = 1/2$. What is the limiting distribution of $X_n + Y_n$? What would the limiting distribution be if the variances of (X_n, Y_n) were $(1+1/n, 1)$? Justify your answers.

4. Suppose that $(X_n, Y_n) \stackrel{D}{\to} (X,Y)$. Prove that $X_n \stackrel{D}{\to} X$.

5. Prove Proposition 6.55.

6.4 Summary

1. X_n can converge to X:

   (a) Almost surely: $P\{\omega : X_n(\omega) \to X(\omega)\} = 1$.
   (b) In L^p: $\mathrm{E}(|X_n - X|^p) \to 0$.
   (c) In probability: $P(|X_n - X| \ge \epsilon) \to 0$ for all $\epsilon > 0$.
   (d) In distribution: $F_n(x) \to F(x)$ for all continuity points x of F.

   (a)⇒(c), (b)⇒(c), (c)⇒(d).

2. *** To compute the expected number of events $A_1, A_2, \ldots$ that occur, form $S = \sum_{i=1}^\infty I(A_i)$ and take the expected value: $\mathrm{E}(S) = \sum_{i=1}^\infty P(A_i)$.

3. *** Borel-Cantelli:

   (a) If $\sum_n P(A_n) < \infty$, then $P(A_n \text{ i.o.}) = 0$.
   (b) If A_n are independent and $\sum_n P(A_n) = \infty$, then $P(A_n \text{ i.o.}) = 1$.

4. *** Subsequence arguments:

   (a) $X_n \stackrel{p}{\to} X$ if and only if every subsequence contains a further subsequence converging to X a.s.

   (b) $F_n \stackrel{D}{\to} F$ if and only if every subsequence contains a further subsequence converging in distribution to F.

   (c) Every tight sequence of distribution functions has a subsequence converging to a distribution function.

   (d) If the sequence of distribution functions F_n converges weakly to a distribution function F, then F_n is tight.

5. *** Mann-Wald theorem: A continuous function f preserves convergence in the sense that if X_n converges to X either almost surely, in probability, or in distribution, then $f(X_n)$ converges to $f(X)$ in the same manner.

6. *** Skorokhod representation theorem: If $X_n \stackrel{D}{\to} X$, there exist X_n', X' on the same probability space such that X_n' and X_n have the same distribution, X' and X have the same distribution, and $X_n' \stackrel{a.s.}{\to} X'$.

Chapter 7

Laws of Large Numbers

7.1 Basic Laws and Applications

Many important results in probability theory concern sums or averages of independent random variables. Some of these are dubbed "laws of averages" among the lay community, but go by the name "laws of large numbers" to probabilists and statisticians. Proposition 6.33 was a very weak version, stating that the sample mean of independent and identically distributed (iid) random variables with finite variance converges in probability to $\mathrm{E}(X)$. It turns out that only a finite mean is required for convergence in probability and the stronger, almost sure convergence.

Proposition 7.1. Laws of large numbers *Let $X_1, X_2, \ldots$ be iid with finite mean μ, and let $\bar{X}_n = (1/n)\sum_{i=1}^n X_i$.*

1. **Weak law of large numbers (WLLN)** *$\bar{X}_n$ converges in probability to μ.*
2. **Strong law of large numbers (SLLN)** *$\bar{X}_n$ converges almost surely to μ.*

We defer proofs to Section 7.2. For now we consider carefully the meaning of the WLLN and SLLN. Note first that the SLLN implies the WLLN because almost sure convergence implies convergence in probability (Proposition 6.37). The sample mean $\bar{X}_n$ is an estimator of $\mu = \mathrm{E}(X)$, so the WLLN says that there is high probability that this estimator will be within ϵ of μ when the sample size is large enough. To understand the SLLN, consider a specific ω, which amounts to considering a specific infinite sequence $x_1, x_2, \ldots$, where $x_i = X_i(\omega)$. For some ω the sequence $x_1, x_2, \ldots$ results in sample means $\bar{x}_1, \bar{x}_2, \ldots$ that converge to μ. For others it might converge to a different number or to $\pm\infty$, or it might fail to converge. For example, suppose that $X_1, X_2, \ldots$ are iid standard normals. One ω might generate the infinite sample $1, 1, \ldots, 1 \ldots$, in which case the sequence of observed sample means $\bar{x}_n$ is also $1, 1, \ldots$, which converges to $1 \neq \mu = 0$. Another ω might generate the sequence $1, -1, 3, -3, 5, -5, \ldots$ In that case the observed sample mean $\bar{x}_n$ is 0 if n is even and 1 if n is odd, so $\bar{x}_n$ fails to converge for that ω. These ωs each have probability 0, as does every other single ω in this experiment. However, the SLLN says that the set of ω, and therefore the set of infinite samples, for which the observed sequence $\bar{x}_n$ converges to μ has probability 1. Therefore, the probability that our infinite sample is one for which $\bar{x}_n$ either fails to converge or converges to something other than μ is 0.

If $X_1, X_2, \ldots$ are iid Bernoulli (p) random variables, $\bar{X}$ is the proportion $\hat{p}_n$ of $X_1, X_2, \ldots,$ X_n that are 1. The WLLN says that $\hat{p}_n$ converges in probability to $\mathrm{E}(X_1) = p$. In other words, the probability that $\hat{p}_n$ differs from p by more than $\pm\epsilon$ tends to 0 as $n \to \infty$, no matter how small ϵ is. The SLLN says that the set of ω such that $\hat{p}_n(\omega)$ converges to p has probability 1. Again it is helpful to think about different infinite samples of outcomes $x_i = X_i(\omega)$, $i = 1, 2, \ldots$ One ω might generate $0, 0, \ldots, 0, \ldots$, in which case $\hat{p}_n \to 0$. Another ω might generate $1, 1, \ldots, 1, \ldots$, in which case $\hat{p}_n \to 1$, etc. The SLLN says that the set of infinitely long strings of 0s and 1s such that $\hat{p}_n$ converges to p has probability 1.

Example 7.2. In a two-armed clinical trial using simple 1:1 randomization, a computer program simulates fair coin flips to assign patients to arm 1 (heads) or arm 2 (tails). The WLLN ensures that the probability that the proportion $\hat{p}_n$ of patients assigned to arm 1 differs from 1/2 by more than a small amount ϵ will be very small if the number of assignments, n, is large enough. This is important because the treatment effect estimator has the smallest variance when the numbers of patients assigned to the two arms are equal. For instance, if the outcome Y is the change in cholesterol or some other continuous measurement from baseline to the end of the trial, the variance of the between-arm difference in sample means $\bar{Y}_T - \bar{Y}_C$ is $\sigma^2(1/n_T + 1/n_C)$. For fixed total sample size $n = n_T + n_C$, the variance of $\bar{Y}_T - \bar{Y}_C$ is minimized when $n_T = n_C$, i.e., when the proportion $\hat{p}_n$ of patients assigned to arm 1 is 1/2. As $|\hat{p}_n - 1/2|$ increases, $\mathrm{var}(\bar{Y}_T - \bar{Y}_C)$ increases. Therefore, it is reassuring that $\hat{p}_n$ is very likely to be close to 1/2. Of course, the SLLN implies that the treatment effect estimate, $\bar{Y}_T - \bar{Y}_C$, converges almost surely to the difference in population means, $\mu_T - \mu_C$. □

We can recast the WLLN and SLLN for Bernoulli (1/2) random variables in terms of Lebesgue measure. Recall that when $(\Omega, \mathcal{F}, P) = ([0,1], \mathcal{B}_{[0,1]}, \mu_L)$, we can write ω in base 2 as $0.X_1, X_2, \ldots$, where the X_i are iid Bernoulli (1/2) random variables. Therefore, the WLLN states that with high probability, the proportion of ones in the first n digits of the base 2 representation of a number drawn randomly from $[0,1]$ will be close to 1/2 if n is large enough. The SLLN states that the proportion of ones in the base 2 representation of ω tends to 1/2 for all ω in a set of Lebesgue measure 1.

Example 7.3. Application of SLLN: A random variable with no univariate density Example 1.1 involved flipping an unfair coin with probability $p \neq 1/2$ infinitely many times and letting X_i be the indicator that the ith flip is heads, $i = 1, 2, \ldots$ We stated without proof that the random variable $Y = 0.X_1X_2\ldots = X_1/2 + X_2/2^2 + X_3/2^3 + \ldots$ has no density function. We are now in a position to prove this. Let ν_Y be the probability measure for the random variable Y, and let A be the set of numbers $\omega \in [0,1]$ whose base 2 representations have the property that the proportion of ones in the first n digits of ω converges to p as $n \to \infty$. Then $\nu_Y(A)$ is the probability that $(X_1 + \ldots + X_n)/n \to p$ as $n \to \infty$, which is 1 by the SLLN. If Y had a density $f(y)$, then $\nu_Y(A) = \int_0^1 f(y)I(y \in A)dy$. But the Lebesgue measure of A is 0 because, under Lebesgue measure, the digits of the base 2 representation of a number y in $[0,1]$ are iid Bernoulli (1/2); with probability 1 the proportion of ones must converge to $1/2 \neq p$ as $n \to \infty$ by the SLLN. Therefore, $f(y)I(y \in A) = 0$ except on a set of Lebesgue measure 0. It follows from Proposition 5.3 that $\int_0^1 f(y)I(y \in A)dy = 0$. Therefore, if Y had a density $f(y)$, then

$$1 = \nu_Y(A) = \int_0^1 f(y)I(y \in A)dy = 0,$$

a contradiction. It follows that Y cannot have a density. See also Example 5.33, which shows that permuted block randomization leads to a base 2 representation that also has no density. It is interesting that too much forced imbalance from using an unfair coin or too much forced balance from using permuted block randomization both lead to random numbers with no density. Simple randomization is "just right" in that it leads to a random number that has a density—the uniform density—on $[0,1]$.

Another way to look at this example is to imagine how the density $f(y)$ of Y would have to appear, if it existed. Suppose $p = 2/3$. The density would have to be twice as high on the second half as on the first half of $[0,1]$. The density's relative appearance on $[0,1/2]$ must be a microcosm of its appearance on $[0,1]$; it is twice as high on the second half of $[0,1/2]$ as it is on the first half, etc. The same is true on the two halves of $[1/2,1]$, and likewise on intervals of length $1/4, 1/8$, etc. This self-similarity property causes no problem when $p = 1/2$ because then the density $f(y)$ is constant. When $p \neq 1/2$, it is impossible to draw a density that has this self-similarity property. The reader can gain an appreciation of the problem by performing the following simulation. For $n = 10$, generate thousands of replications of $\sum_{i=1}^{n} X_i/2^i$, where X_i are iid Bernoulli (p). Make a histogram. Then repeat with $n = 20$ and $n = 50$. □

Example 7.4. Application of SLLN: a.s. convergence of distribution functions Return to Example 6.9, involving estimation of the distribution function $F(x)$. We used the fact that $\hat{F}_n(x) = (1/n)\sum_{i=1}^{n} I(X_i \leq x)$ is binomial $(n, p = F(x))$ to prove that $\hat{F}_n(x)$ converges in probability to $F(x)$. We can strengthen this result because $\hat{F}_n(x)$, being the sample mean of n iid Bernoullis, converges almost surely to $\mathrm{E}\{I(X_i \leq x)\} = F(x)$ by the SLLN. Therefore, at each point x, the empirical distribution function converges almost surely to the true distribution function. Likewise, we can strengthen Example 6.10 to prove that when the distribution function has a unique median θ, the sample median converges not just in probability, but almost surely, to θ (exercise). □

We can say even more than what is stated in Example 7.4. Recall that in Example 6.9, we showed that $P\{|\hat{F}_n(x) - F(x)| > \epsilon\} \leq F(x)\{1 - F(x)\}/(n\epsilon^2)\}$ for all x. Also, $F(x)\{1 - F(x)\} \leq (1/2)(1 - 1/2) = 1/4$. Therefore, $P\{|\hat{F}_n(x) - F(x)| > \epsilon\} \leq (1/4)/(n\epsilon^2)$ for all x. This tells us that the convergence in probability of $\hat{F}_n(x)$ to $F(x)$ is uniform in x: $\sup_x P\{|\hat{F}_n(x) - F(x)| > \epsilon\} \to 0$. The Glivenko-Cantelli theorem stated below strengthens this to uniform almost sure convergence of $\hat{F}_n(x)$ to $F(x)$.

Theorem 7.5. Glivenko-Cantelli theorem: uniform a.s. convergence of distribution functions *The sample distribution function $\hat{F}_n(x)$ for iid random variables $X_1, \ldots, X_n$ satisfies $\sup_x |\hat{F}_n(x) - F(x)| \stackrel{a.s.}{\to} 0$.*

We do not prove the Glivenko-Cantelli theorem, but note that Example 7.4 shows that for each fixed x, $\hat{F}_n(x,\omega) \stackrel{a.s.}{\to} F(x)$. It follows that $P(\hat{F}_n(r,\omega) \to F(r)$ for all rational $r) = 1$. Because the set of rational numbers is a dense set, $\hat{F}_n(x,\omega) \stackrel{D}{\to} F(x)$ for ω outside a null set. This shows a small part of what is needed, but does not show that the convergence is for all x, not just continuity points of F, and does not show that the convergence is uniform in x. The proof is simplified considerably under the additional assumption that F is continuous. See Polya's theorem (Theorem 9.2) for details.

Example 7.6. Application of SLLN: a.s. convergence of sample variance if $\sigma^2 < \infty$ Let X_i be iid with finite variance σ^2. Then the sample variance $s_n^2 = (n-1)^{-1}\sum_{i=1}^{n}(X_i - \bar{X}_n)^2$ converges almost surely to σ^2 as $n \to \infty$. To see this, note that

$$\begin{aligned}
\sum_{i=1}^{n}(X_i - \bar{X}_n)^2 &= \sum_{i=1}^{n}(X_i - \mu + \mu - \bar{X}_n)^2 \\
&= \sum_{i=1}^{n}\{(X_i - \mu)^2 + (\mu - \bar{X}_n)^2 + 2(X_i - \mu)(\mu - \bar{X}_n)\} \\
&= \sum_{i=1}^{n}(X_i - \mu)^2 + n(\mu - \bar{X}_n)^2 + 2(\mu - \bar{X}_n)n(\bar{X}_n - \mu) \\
&= \sum_{i=1}^{n}(X_i - \mu)^2 - n(\bar{X}_n - \mu)^2. \qquad (7.1)
\end{aligned}$$

It follows that

$$\frac{1}{n-1}\sum_{i=1}^{n}(X_i - \bar{X}_n)^2 = \left(\frac{n}{n-1}\right)\left\{(1/n)\sum_{i=1}^{n}(X_i - \mu)^2 - (\bar{X}_n - \mu)^2\right\}. \qquad (7.2)$$

By the SLLN applied to the random variables $Y_i = (X_i - \mu)^2$, $(1/n)\sum_{i=1}^{n}(X_i - \mu)^2 \stackrel{a.s.}{\to} \mathrm{E}(X_i - \mu)^2 = \sigma^2$. Furthermore, $(\bar{X}_n - \mu)^2 \stackrel{a.s.}{\to} 0^2 = 0$, so Expression (7.2) converges almost surely to $(1)(\sigma^2 - 0^2) = \sigma^2$. This shows that the sample variance converges almost surely to the population variance σ^2. Likewise, the pooled variance estimate in a t-test converges almost surely to the population variance, as does the pooled variance estimate in analysis of variance with k groups (exercise). □

Example 7.7. Application of SLLN: convergence of correlation coefficient and regression estimator In some applications we are interested in studying the association between two variables X and Y. For example, X and Y might be systolic and diastolic blood pressure, HIV (Human immunodeficiency virus) viral load and CD4+ T cell counts, etc. Pairs (X_i, Y_i) corresponding to different patients are independent, though observations within a pair—i.e., within a patient—are dependent. Assume that the variances σ_X^2 and σ_Y^2 of X and Y are finite. We estimate the population correlation coefficient $\rho = \mathrm{E}\{(X - \mu_X)(Y - \mu_Y)\}/(\sigma_X\sigma_Y)$ using the sample correlation coefficient

$$R_n = \frac{\sum_{i=1}^{n}(X_i - \bar{X}_n)(Y_i - \bar{Y}_n)}{\sqrt{\sum_{i=1}^{n}(X_i - \bar{X}_n)^2\sum_{i=1}^{n}(Y_i - \bar{Y}_n)^2}}.$$

By arguments similar to those of Example 7.6, we can show that R_n converges almost surely to ρ as $n \to \infty$ (exercise). Notice that no assumption of bivariate normality is required: the joint distribution of (X, Y) is arbitrary. Because almost sure convergence implies convergence in probability, there is high probability that R_n will be close to ρ if the sample size n is large.

Another way to study the association between X and Y is through the regression equation

$$Y_i = \beta_0 + \beta_1 x_i + \epsilon_i,\ i = 1, \ldots, n \qquad (7.3)$$

where the errors ϵ_i are assumed iid from a distribution with mean 0. Notice that here we are treating the X_i as fixed constants x_i. That is, we assume that the relationship (7.3) holds for fixed x_i. We study conditioning in much greater detail in a subsequent chapter, but the reader is assumed to have some familiarity with this concept. The least squares estimate of β_1 is

$$\hat{\beta}_1 = \frac{\sum_{i=1}^{n}(x_i - \bar{x})(Y_i - \bar{Y})}{\sum_{i=1}^{n}(x_i - \bar{x})^2} = \frac{\sum_{i=1}^{n}(x_i - \bar{x})Y_i}{\sum_{i=1}^{n}(x_i - \bar{x})^2}. \qquad (7.4)$$

Conditioned on the xs, $\hat{\beta}_1$ has mean

$$\begin{aligned} E(\hat{\beta}_1) &= \frac{\sum_{i=1}^n (x_i - \bar{x})E(Y_i)}{\sum_{i=1}^n (x_i - \bar{x})^2} = \frac{\sum_{i=1}^n (x_i - \bar{x})(\beta_0 + \beta_1 x_i)}{\sum (x_i - \bar{x})^2} \\ &= \frac{\beta_0 \sum_{i=1}^n (x_i - \bar{x}) + \beta_1 \sum (x_i - \bar{x})^2}{\sum (x_i - \bar{x})^2} = \beta_0 \cdot 0 + \beta_1 = \beta_1 \end{aligned} \tag{7.5}$$

and variance

$$\mathrm{var}(\hat{\beta}_1) = \frac{\sum_{i=1}^n (x_i - \bar{x})^2 \sigma^2}{\{\sum_{i=1}^n (x_i - \bar{x})^2\}^2} = \frac{\sigma^2}{\sum_{i=1}^n (x_i - \bar{x})^2}.$$

If $\sum_{i=1}^n (x_i - \bar{x})^2 \to \infty$ as $n \to 0$, then $\mathrm{var}(\hat{\beta}_1) \to 0$ and $\hat{\beta}_1$ converges in probability to β_1 by Chebychev's inequality. Furthermore, if the x_i are realizations from iid random variables X_i with positive variance σ_X^2, then the set of ω such that $\sum_{i=1}^n \{X_i(\omega) - \bar{X}(\omega)\}^2 \to \infty$ has probability 1 by Example 7.6 because $(n-1)^{-1} \sum_{i=1}^n (X_i - \bar{X})^2 \overset{a.s.}{\to} \sigma_X^2$. Thus, we are certain to encounter an infinite sample $x_1, x_2, \ldots$ such that the conditional probability $P(|\hat{\beta}_1 - \beta_1| > \epsilon \,|\, X_1 = x_1, \ldots, X_n = x_n)$ converges to 0 as $n \to \infty$ for every $\epsilon > 0$.

It is important to understand the distinction between the two approaches to studying association between two variables. The first used the sample correlation coefficient R_n and involved treating both X and Y as random. We imagine the so-called *repeated samples behavior* of the experiment. If we repeat the experiment, we draw a new ω, which generates new values of $(X_1(\omega), Y_1(\omega)), \ldots, (X_n(\omega), Y_n(\omega)), \ldots$, from which we compute the sample correlation coefficients R_n. But now imagine performing the experiment in two steps. In the first step, generate an infinite string $x_1, x_2, \ldots$ of x values as realizations of $X_1(\omega), X_2(\omega), \ldots$ In the second step, generate independent observations $Y_1, Y_2, \ldots$ according to the regression equation (7.3). Suppose that instead of repeating the entire experiment, we repeat only the second part. That is, $x_1, x_2, \ldots$ remain fixed when we replicate the process of generating $Y_1, Y_2, \ldots$ and computing $\hat{\beta}_1$. This is the conditional inference approach we adopted when considering the distribution of $\hat{\beta}_1$. We were then interested in which fixed infinite sequences $x_1, x_2, \ldots$ have the property that $\hat{\beta}_1 \overset{p}{\to} \beta$. We found that the answer was "almost all infinite sequences $x_1, x_2, \ldots$ have this property." □

The next application of the WLLN involves approximating a function by polynomials. We have seen one such approximation by polynomials, namely a Taylor series involving higher order derivatives at a given point. In the next application, we approximate a continuous function f at the point x using values of f on a grid of other points.

Theorem 7.8. Bernstein approximation theorem: an application of the WLLN *Let $f(x) : [0,1] \longmapsto R$ be a continuous function, and let*

$$B_n(x) = \sum_{k=0}^n f(k/n) \binom{n}{k} x^k (1-x)^{n-k}. \tag{7.6}$$

Then $B_n(x) \to f(x)$ uniformly in $x \in [0,1]$.

Proof. Notice that $\binom{n}{k} x^k (1-x)^{n-k}$ in Equation (7.6) is a binomial probability mass function with probability parameter x. Let S_n be binomial with parameters (n, x). Then

$$E\{f(S_n/n)\} = \sum_{k=0}^n f(k/n) P(S_n = k) = B_n(x).$$

Also, $S_n/n \xrightarrow{p} x$ by either Proposition 6.33 or the WLLN. Because f is continuous on the compact set $[0,1]$, f is uniformly continuous on $[0,1]$ (Proposition A.62). It follows that f is bounded on $x \in [0,1]$. By the DCT for convergence in probability (Proposition 6.43),

$$B_n(x) = \mathrm{E}\{f(S_n/n)\} \to f(x). \tag{7.7}$$

To demonstrate that the convergence is uniform in $x \in [0,1]$, let $\epsilon > 0$ be given. We must show that there exists an N such that $|\mathrm{E}\{f(S_n/n)\} - f(x)| < \epsilon$ for all $n \geq N$ and $x \in [0,1]$. Because f is uniformly continuous, there exists a $\delta > 0$ such that $|f(y) - f(x)| < \epsilon/2$ whenever $|x - y| < \delta$. Let $A_{n,\delta}$ be the event that $|S_n/n - x| < \delta$, and let $B = 2\sup_{x\in[0,1]} |f(x)|$. Then

$$\begin{aligned}
&|\mathrm{E}\{f(S_n/n)\} - f(x)| \leq \mathrm{E}\{|f(S_n/n) - f(x)|\} \\
= \;& \mathrm{E}\{|f(S_n/n) - f(x)|I(A_{n,\delta})\} + \mathrm{E}\{|f(S_n/n) - f(x)|I(A_{n,\delta}^C)\} \\
< \;& (\epsilon/2)P(A_{n,\delta}) + BP(A_{n,\delta}^C) \\
\leq \;& \epsilon/2 + B\,\mathrm{var}(S_n/n)/\delta^2 \text{ (Chebychev)} \\
= \;& \epsilon/2 + Bx(1-x)/(n\delta^2) \leq \epsilon/2 + B(1/2)(1-1/2)/(n\delta^2) \\
= \;& \epsilon/2 + B/(4n\delta^2).
\end{aligned} \tag{7.8}$$

Now take N to be the smallest integer such that $B/(4N\delta^2) < \epsilon/2$. Then $n \geq N \Rightarrow |\mathrm{E}\{f(S_n/n)\} - f(x)| < \epsilon$. This completes the proof that the convergence is uniform. □

The B_n are known as Bernstein polynomials . We have used results from probability theory to prove that the B_n uniformly approximate f.

Example 7.9. Connection between Bernstein polynomials and cubic splines In data analysis, we often fit simple curves such as lines or quadratics to data. But sometimes these simple curves do not fit too well. We can use higher degree terms such as cubics and higher. One problem is that with higher order terms, the y values for some xs strongly influence the curve fitting in other x locations. For example, Figure 7.1 shows the best fitting fifth degree polynomial applied to data whose actual model is $Y = \cos(x) + \epsilon$ over the interval $[0, 6\pi]$. The fit is quite poor. It can be improved by using a tenth order polynomial, but it is undesirable to have to continue fitting higher order polynomials until we find one that looks acceptable. An alternative that avoids this problem is to use spline fitting. This involves estimating $y(x)$ locally by averaging the y values of nearby x values and forming different polynomials in different regions of the curve. The resulting fit is quite good (Figure 7.2), and the process obviates the need to check the plot to make sure the curve fits the data.

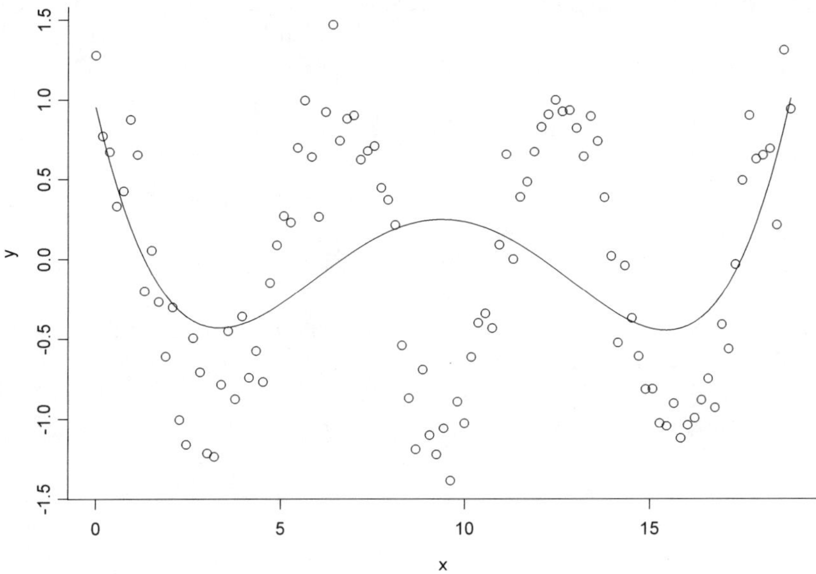

Figure 7.1: The best fitting fifth degree polynomial fit to data generated from the model $y = \cos(x) + \epsilon$.

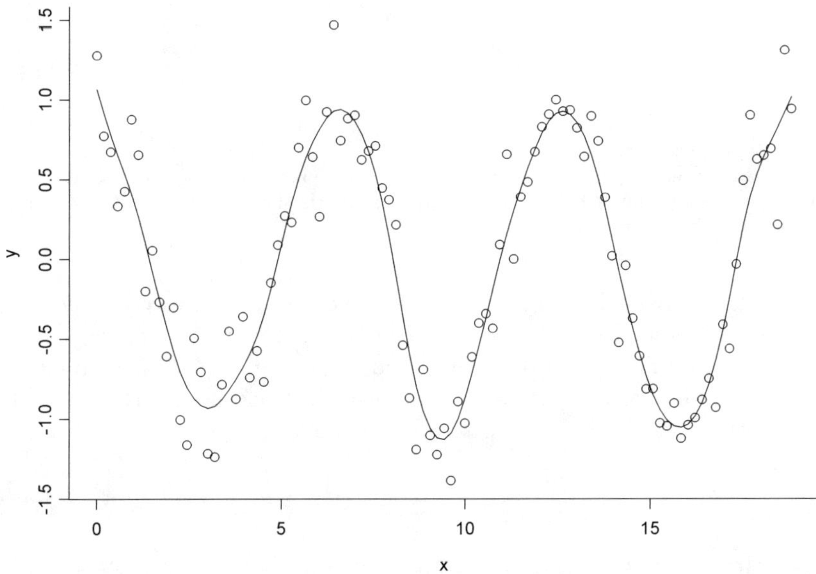

Figure 7.2: Spline curve fit to the data of Figure 7.1.

One cubic spline method uses Bezier curves . Let $(x_0, y_0), \ldots, (x_3, y_3)$ be 4 points whose x values are equally spaced. We can always reparameterize to make the xs equally spaced on $[0, 1]$. For instance, if x'_k are not on $[0, 1]$, take $x_k = x'_0 + (x'_3 - x'_0)x_k$, where $x_k = k/3$, $k = 0, 1, 2, 3$, is between 0 and 1. Therefore, assume without loss of generality that $x_k = k/3$, $k = 0, 1, 2, 3$. Define $y_k = f(x_k)$ and, for $0 \le x \le 1$, define

$$f(x) = \sum_{k=0}^{3} f(k/3)\binom{3}{k}x^k(1-x)^{3-k} = \sum_{k=0}^{3} w_k y_k,$$

where $w_k = \binom{3}{k}x^k(1-x)^{3-k}$. The fitted curve $f(x)$ (Figure 7.3) is a weighted combination of the 4 nearby y_ks. The weight w_k applied to y_k is the binomial probability mass function with $n = 3$ and probability parameter x, evaluated at k. This binomial mass function is largest when its probability parameter x equals $k/3$ (the maximum likelihood estimate of x), and diminishes as x moves away from $k/3$. That is, as x approaches $k/3 = x_k$, the weight applied to y_k grows, while the weights applied to y_is with x_is far from x diminish. The Bezier spline then links contiguous Bezier curves in a smooth way.

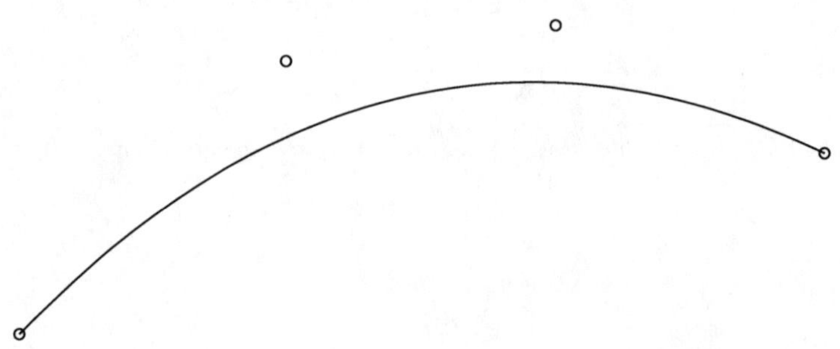

Figure 7.3: Bezier polynomial fit to four points. The curve goes through the first and fourth points, the derivative at the first point equals the slope of the secant line joining the first two points, and the derivative at the last point equals the slope of the secant line joining the third and fourth points. These properties facilitate the joining of contiguous Bezier curves.

The connection between Bezier spline fitting and Bernstein polynomials is that the Bezier curves that are joined together are Bernstein polynomials of degree 3. Bernstein's approximation theorem (Theorem 7.8) involves convergence of $B_n(x)$ for n large; the Bezier curves in cubic spline fitting use the small value $n = 3$. Nonetheless, the method of local approximation is the same as that of Bernstein polynomials.

Exercises

1. Prove that if $X_1, \ldots, X_n$ are iid Bernoulli (1/2) random variables, then

   (a) $S_n/n^r \stackrel{a.s.}{\to} 0$ if $r > 1$.

   (b) With probability 1, infinitely many X_i must be 0.

   (c) $P(A) = 0$, where $A = \{$fewer than 40 percent of $X_1, \ldots, X_n$ are 1 for infinitely many $n\}$.

2. The following statements about iid random variables $X_1, X_2 \ldots$ with mean μ are all consequences of the WLLN or SLLN. For each, say whether the result follows from the WLLN, or it requires the SLLN.

   (a) There is an N such that $P(\bar{X}_n > \mu - .001) \geq .999$ for all $n \geq N$.

   (b) With probability at least .999, there is an $N = N(\omega)$ such that $\bar{X}_n > \mu - .001$ for all $n \geq N$.

   (c) $P\{\underline{\lim}\,(\bar{X}_n) > \mu - .001\} = 1$.

   (d) $\underline{\lim}\,\{P(\bar{X}_n > \mu - .001)\} = 1$.

3. Let $X_1, X_2, \ldots$ be iid with distribution function F having a unique median θ. Prove that the sample median $\hat{\theta}$ converges almost surely to θ.

4. In analysis of variance with k groups and n_i observations in group i, the pooled variance is $\sum_{i=1}^{k}\sum_{j=1}^{n_i}(Y_{ij} - \bar{Y}_i)^2 / \sum_{i=1}^{k}(n_i - 1)$. Assume that the Y_{ij} are independent, and Y_{ij}, $j = 1, \ldots n_i$, are identically distributed with mean μ_i and variance σ^2. Prove that $\hat{\sigma}^2$ converges almost surely to the common within-group variance σ^2 as k remains fixed and each $n_i \to \infty$.

5. Pick countably infinitely many letters randomly and independently from the 26-letter alphabet. Let X_1 be the indicator that letters 1–10 spell "ridiculous," X_2 be the indicator that letters 2–11 spell "ridiculous," X_3 be the indicator that letters 3–12 spell "ridiculous," etc.

   (a) Can you apply the SLLN to $X_1, X_2, \ldots$?

   (b) What if X_1 had been the indicator that letters 1–10 spell "ridiculous," X_2 had been the indicator that letters 11–20 spell "ridiculous," X_3 had been the indicator that letters 21–30 spell "ridiculous," etc.

   (c) Prove that the probability that some set of 10 consecutive letters spells "ridiculous" is 1.

6. Let $(X_1, Y_1), (X_2, Y_2), \ldots$ be independent, though (X_i, Y_i) may be correlated. Assume that the X_i are identically distributed with finite mean $\mathrm{E}(X_1) = \mu_X$ and the Y_i are identically distributed with finite mean $\mathrm{E}(Y) = \mu_Y$. Which of the following are necessarily true?

   (a) $(1/n)\sum(X_i + Y_i) \stackrel{a.s.}{\to} \mu_X + \mu_Y$.

   (b) You flip a single coin. If it is heads, set $Z_n = (1/n)\sum_{i=1}^{n} X_i$, $n = 1, 2, \ldots$ If it is tails, set $Z_n = (1/n)\sum_{i=1}^{n} Y_i$, $n = 1, 2, \ldots$. Then $Z_n \stackrel{a.s.}{\to} (\mu_X + \mu_Y)/2$.

   (c) You flip a coin for each i. Set $U_i = X_i$ if the ith flip is heads, and $U_i = Y_i$ if the ith flip is tails. Then $(1/n)\sum_{i=1}^{n} U_i \stackrel{a.s.}{\to} (\mu_X + \mu_Y)/2$.

   (d) You generate an infinite sequence $U_1, U_2, \ldots$ as described in part c, and let $Z_1(\omega) = \lim_{n\to\infty}(1/n)\sum_{i=1}^{n} U_i(\omega)$ if this limit exists and is finite, and $Z_1(\omega) = 0$ if the limit either does not exist or is infinite. Repeat the entire experiment countably infinitely many times, and let $Z_1, Z_2, \ldots$ be the results. Then $P\{\omega : Z_i(\omega) = (\mu_X + \mu_Y)/2 \text{ for all } i\} = 1$.

   (e) Form **all** possible combinations obtained by selecting either X_i or Y_i, $i = 1, 2, \ldots$ Let s index the different combinations; for the sth combination, let Z_s denote the corresponding limit in part d. Then $P\{\omega : Z_s(\omega) \to (\mu_X + \mu_Y)/2 \text{ for all } s\} = 1$.

7. A gambler begins with an amount A_0 of money. Each time he bets, he bets half of his money, and each time he has a 60% chance of winning.

   (a) Show that the amount of money A_n that the gambler has after n bets may be written as $A_0 \prod_{i=1}^n X_i$ for iid random variables X_i. Show that $\mathrm{E}(A_n) \to \infty$ as $n \to \infty$.

   (b) Show that $A_n \to 0$ a.s.

8. Prove that if $(X_1, Y_1), \ldots, (X_n, Y_n)$ are iid pairs with $0 < \mathrm{var}(X) < \infty$ and $0 < \mathrm{var}(Y) < \infty$, then the sample correlation coefficient R_n converges almost surely to $\rho = \mathrm{E}\{(X - \mu_X)(Y - \mu_Y)\}/(\sigma_X \sigma_Y)$. Prove that R_n converges to ρ in L^1 as well.

9. Consider the analysis of covariance (ANCOVA) model $Y_i = \beta_0 + \beta_1 X_i + \beta_2 z_i + \epsilon_i$, where X_i and Y_i are the baseline and end of study values of a continuous variable like cholesterol, z_i is a treatment indicator (1 for treatment, 0 for control), and the ϵ_i are iid from a distribution with mean 0 and variance σ_ϵ^2. Though we sometimes consider the x_i as fixed constants, regard the X_i now as iid random variables with finite, nonzero variance σ_X^2, and assume that the vector of Xs is independent of the vector of ϵs. The estimated treatment effect in this model is

$$\hat{\beta}_2 = \bar{Y}_T - \bar{Y}_C - \hat{\beta}_1(\bar{X}_T - \bar{X}_C),$$

   where

$$\hat{\beta}_1 = \frac{\sum_T (X_i - \bar{X}_T)(Y_i - \bar{Y}_T) + \sum_C (X_i - \bar{X}_C)(Y_i - \bar{Y}_C)}{\sum_T (X_i - \bar{X}_T)^2 + \sum_C (X_i - \bar{X}_C)^2}.$$

   Let $n = n_T + n_C$, and suppose that $n_T/n_C \to \lambda$ as $n \to \infty$. Prove that $\hat{\beta}_1$ converges almost surely to β_1 and $\hat{\beta}_2$ converges almost surely to β_2.

10. Use the weak law of large numbers to prove that if Y_n is Poisson with parameter n, $n = 1, 2, \ldots$, then $P(|Y_n/n - 1| > \epsilon) \to 0$ as $n \to \infty$ for every $\epsilon > 0$ (hint: what is the distribution of the sum of independent Poisson(1)s?). Can the strong law of large numbers be used to conclude that $Y_n/n \stackrel{a.s.}{\to} 1$?

11. $\uparrow$ Prove that if f is a bounded continuous function on $x \geq 0$, then $f(x) = \lim_{n\to\infty} \sum_{k=0}^{\infty} f(k/n) \exp(-nx)(nx)^k/k!$ (see the proof of Theorem 7.8).

7.2 Proofs and Extensions

Rather than proving the strongest law of large numbers (see Billingsley, 2012 for proof), we prove other versions that illustrate useful techniques in probability theory. One such technique is a subsequence argument, which we use next to prove the following result.

Theorem 7.10. Weak law for uncorrelated random variables *If X_i are uncorrelated random variables with variance $\sigma^2 < 0$, then $\bar{X}_n - \mathrm{E}(\bar{X}_n) \stackrel{a.s.}{\to} 0$.*

Proof. Without loss of generality, we can take X_n to have mean 0 because we can consider $X_n' = X_n - \mathrm{E}(X_n)$. We first argue that $\bar{X}_{n^2} \stackrel{a.s.}{\to} 0$ as $n \to \infty$. This follows from Chebychev's inequality coupled with the Borel-Cantelli lemma, as follows.

$$\sum_{n=1}^{\infty} P(|\bar{X}_{n^2}| > \epsilon) \leq \sum_{n=1}^{\infty} \frac{\mathrm{var}(\bar{X}_{n^2})}{\epsilon^2} = \sum_{n=1}^{\infty} \frac{\sigma^2}{n^2 \epsilon} < \infty.$$

By the Borel-Cantelli lemma, $P(|\bar{X}_{n^2}| > \epsilon \text{ i.o.}) = 0$. Replacing ϵ by $1/k$ and defining A_k to be the event that $|\bar{X}_{n^2}| > 1/k$ for infinitely many n, we have $P(A_k) = 0$ for each k. Then $P(\cup_{k=1}^{\infty} A_k) \le \sum_{k=1}^{\infty} P(A_k) = 0$. That is, with probability 1, for each k, $|\bar{X}_{n^2}| \le 1/k$ for all but finitely many n. This implies that $\bar{X}_{n^2} \overset{a.s.}{\to} 0$.

The next step is to prove that $M_n \overset{a.s.}{\to} 0$, where $M_n = \max_{n^2 < k \le (n+1)^2} |\bar{X}_k - \bar{X}_{n^2}|$.

$$
\begin{aligned}
P(M_n > \epsilon) &= P\left[\bigcup_{k=n^2+1}^{(n+1)^2} \{(\bar{X}_k - \bar{X}_{n^2})^2 > \epsilon^2\}\right] \le \sum_{k=n^2+1}^{(n+1)^2} \frac{\text{var}(\bar{X}_k - \bar{X}_{n^2})}{\epsilon^2} \\
&= \frac{1}{\epsilon^2}\sum_{k=n^2+1}^{(n+1)^2}\left\{\frac{\sigma^2}{k} + \frac{\sigma^2}{n^2} - 2\,\text{cov}(\bar{X}_k, \bar{X}_{n^2})\right\} = \frac{1}{\epsilon^2}\sum_{k=n^2+1}^{(n+1)^2}\left\{\frac{\sigma^2}{k} + \frac{\sigma^2}{n^2} - \frac{2}{kn^2}\text{cov}(S_k, S_{n^2})\right\} \\
&= \frac{1}{\epsilon^2}\sum_{k=n^2+1}^{(n+1)^2}\left\{\frac{\sigma^2}{k} + \frac{\sigma^2}{n^2} - \frac{2}{kn^2}n^2\sigma^2\right\} = \frac{\sigma^2}{\epsilon^2}\sum_{k=n^2+1}^{(n+1)^2}\left(\frac{k-n^2}{kn^2}\right) \\
&\le \frac{\sigma^2}{\epsilon^2}\sum_{k=n^2+1}^{(n+1)^2}\frac{\{(n+1)^2 - n^2\}}{n^2 n^2} = \frac{(2n+1)^2\sigma^2}{\epsilon^2 n^4} \le \frac{(2n+n)^2\sigma^2}{\epsilon^2 n^4} \\
&= \frac{9\sigma^2}{\epsilon^2 n^2}. \qquad (7.9)
\end{aligned}
$$

Because $\sum_n (1/n^2) < \infty$, the Borel-Cantelli lemma implies that $P(M_n > \epsilon \text{ i.o.}) = 0$. By the same argument as in the preceding paragraph, this implies that $M_n \overset{a.s.}{\to} 0$. Putting these results together, we get

$$|\bar{X}_k| = |\bar{X}_k - \bar{X}_{n^2} + \bar{X}_{n^2}| \le M_n + |\bar{X}_{n^2}| \overset{a.s.}{\to} 0 + 0 = 0,$$

completing the proof. □

Notice that Theorem 7.10 requires a finite variance, but does not require the X_i to be identically distributed or even independent. It is clear that without some condition akin to independence, we could easily concoct counterexamples to laws of large numbers. For example, let X_1 be any non-degenerate random variable with finite mean μ, and define $X_2 = X_1, X_3 = X_1, \ldots$ Then $(X_1 + X_2 + \ldots + X_n)/n = X_1$ does not converge to μ almost surely or in probability. Theorem 7.10 eliminates such a counterexample by requiring the X_i to be uncorrelated.

Another useful technique in the proofs of laws of large numbers is truncation. Truncated random variables behave very similarly to their untruncated counterparts, but have the advantage of possessing finite moments of all orders. Let X_i be identically distributed (not necessarily independent) with finite mean μ. Let

$$Y_i = \begin{cases} X_i & \text{if } |X_i| < i \\ i & \text{if } |X_i| \ge i. \end{cases} \qquad (7.10)$$

The truncated random variables Y_i have finite variance regardless of whether the X_i do. Also, $\{Y_i \ne X_i\}$ implies that $|X_i| \ge i$, so $P(Y_i \ne X_i) \le P(|X_i| \ge i)$. Therefore, if X_i are iid with finite mean,

$$\sum_i P(Y_i \neq X_i) \leq \sum_i P(|X_i| \geq i) = \sum_i P(|X_1| \geq i) < \infty$$

by Proposition 5.9 because $\mathrm{E}(|X_1|) < \infty$. By part 1 of the Borel-Cantelli lemma, $P(Y_i \neq X_i \text{ i.o.}) = 0$. In other words, with probability 1, only finitely many Y_i differ from the corresponding X_i. This implies that $\bar{Y}_n - \bar{X}_n \stackrel{a.s.}{\to} 0$.

Lemma 7.11. Truncated r.v.s are asymptotically equivalent to original variables *Let X_i be identically distributed (not necessarily independent) with finite mean μ, and Y_i be the truncated random variables defined by Equation (7.10). Then*

1. $\bar{X}_n \stackrel{a.s.}{\to} \mu$ *if and only if* $\bar{Y}_n \stackrel{a.s.}{\to} \mu$.
2. $\bar{X}_n \stackrel{p}{\to} \mu$ *if and only if* $\bar{Y}_n \stackrel{p}{\to} \mu$.

Proof. These results follow from the fact that $\bar{Y}_n - \bar{X}_n \stackrel{a.s.}{\to} 0$. For example, suppose $\bar{X}_n \stackrel{p}{\to} \mu$. Write $\bar{Y}_n$ as $\bar{X}_n + \bar{Y}_n - \bar{X}_n$ and note that $\bar{Y}_n - \bar{X}_n \stackrel{p}{\to} 0$ because $\bar{Y}_n - \bar{X}_n \stackrel{a.s.}{\to} 0$. Therefore, $\bar{Y}_n \stackrel{p}{\to} \mu + 0 = \mu$. □

We are now in a position to prove a version of the WLLN requiring only finite mean and pairwise independence.

Theorem 7.12. Weak law for pairwise independent variables *If $X_1, \ldots, X_n, \ldots$ are identically distributed random variables with mean μ, and all pairs (X_i, X_j), $i \neq j$, are independent, then $\bar{X}_n \stackrel{p}{\to} \mu$.*

Proof. By Lemma 7.11, it suffices to prove that $(1/n)\sum_{i=1}^n Y_i \stackrel{p}{\to} \mu$. Notice that

$$\begin{aligned} \mathrm{E}(Y_i) &= \mathrm{E}\{X_i I(|X_i| < i) + i I(|X_i| \geq i)\} \\ &= \mathrm{E}\{X_i I(|X_i| < i)\} + i P(|X_i| \geq i) \\ &= \mathrm{E}\{X_1 I(|X_1| < i)\} + i P(|X_1| \geq i). \end{aligned}$$

By the DCT, $\mathrm{E}\{X_1 I(|X_1| < i)\} \to \mathrm{E}(X_1)$ as $i \to \infty$ because $X_1 I(|X_1| < i)$ converges a.s. to X_1 as $i \to \infty$ and is dominated by the integrable random variable $|X_1|$. Furthermore, $iP(|X_1| \geq i) \leq \mathrm{E}\{|X_1| I(|X_1| \geq i)\} \to 0$ by the DCT. Thus, $\mathrm{E}(Y_i) \to \mathrm{E}(X_1)$ as $i \to \infty$. It follows that $(1/n)\sum_{i=1}^n \mathrm{E}(Y_i) \to \mathrm{E}(X_1)$ (Exercise 1).

We will establish that $\bar{Y}_n \stackrel{p}{\to} 0$ by proving that $\mathrm{var}\{(1/n)\sum_{i=1}^n Y_i\} \to 0$. Notice that

$$\begin{aligned} \mathrm{var}\left\{(1/n)\sum_{i=1}^n Y_i\right\} &= (1/n^2)\sum_{i=1}^n \mathrm{var}(Y_i) \\ &\leq (1/n^2)\sum_{i=1}^n \mathrm{E}(Y_i^2). \end{aligned} \tag{7.11}$$

Therefore, we need only show that $(1/n^2)\sum_{i=1}^n \mathrm{E}(Y_i^2) \to 0$. The following technique almost works.

$$Y_i^2 = |Y_i||Y_i| \leq i|Y_i| \leq i|X_i|, \text{ so } \mathrm{E}(Y_i^2) \leq i\,\mathrm{E}(|X_1|). \tag{7.12}$$

Therefore,

$$\begin{aligned}(1/n^2)\sum_{i=1}^n \mathrm{E}(Y_i^2) &\le \frac{\mathrm{E}(|X_1|)}{n^2}\sum_{i=1}^n i \\ &= \frac{\mathrm{E}(|X_1|)n(n+1)/2}{n^2} \rightarrow \mathrm{E}(|X_1|)/2. \end{aligned} \tag{7.13}$$

This shows that $(1/n^2)\sum_{i=1}^n \mathrm{E}(Y_i^2)$ is bounded, but does not show that it converges to 0. We can improve this technique as follows. For any sequence $a_n < n$,

$$(1/n^2)\sum_{i=1}^n \mathrm{E}(Y_i^2) = (1/n^2)\sum_{i=1}^{a_n} \mathrm{E}(Y_i^2) + (1/n^2)\sum_{i=a_n+1}^{n} \mathrm{E}(Y_i^2). \tag{7.14}$$

If we choose $a_n = \lfloor \ln(n) \rfloor$ and use inequality (7.12), then the first term on the right side of Equation (7.14) is no greater than

$$\begin{aligned}(1/n^2)\sum_{i=1}^{a_n} i\mathrm{E}(|X_1|) &= (1/n^2)\mathrm{E}(|X_1|)a_n(a_n+1)/2 \\ &\le \frac{\lfloor \ln(n)\rfloor(\lfloor \ln(n)\rfloor + 1)\mathrm{E}(|X_1|)}{2n^2} \rightarrow 0. \end{aligned} \tag{7.15}$$

Write the second term on the right of Equation (7.14) as

$$\begin{aligned} & (1/n^2)\sum_{i=a_n+1}^{n} \mathrm{E}\left\{Y_i^2 I(|Y_i| \le a_n)\right\} + (1/n^2)\sum_{i=a_n+1}^{n} \mathrm{E}\left\{Y_i^2 I(|Y_i| > a_n)\right\} \\ \le\ & (1/n^2)\sum_{i=a_n+1}^{n} a_n^2 + (1/n^2)\sum_{i=a_n+1}^{n} i\mathrm{E}\left\{|Y_i| I(|Y_i| > a_n)\right\} \\ \le\ & \frac{(n-a_n)a_n^2}{n^2} + (1/n^2)\sum_{i=a_n+1}^{n} i\mathrm{E}\left\{|X_i| I(|X_i| > a_n)\right\} \\ \le\ & \frac{a_n^2}{n} + (1/n^2)\sum_{i=a_n+1}^{n} i\mathrm{E}\left\{|X_1| I(|X_1| > a_n)\right\} \\ \le\ & \frac{a_n^2}{n} + \frac{\mathrm{E}\left\{|X_1| I(|X_1| > a_n)\right\} n(n+1)/2}{n^2} \\ =\ & \frac{\lfloor \ln(n)\rfloor^2}{n} + \frac{(1+1/n)\mathrm{E}\left\{|X_1| I(|X_1| > \lfloor \ln(n)\rfloor)\right\}}{2} \rightarrow 0, \end{aligned} \tag{7.16}$$

because $\mathrm{E}(|X_1|) < \infty$. This completes the proof. □

It is interesting to ask what happens to the sample mean of independent and identically distributed random variables $X_1, X_2, \ldots$ such that $\mathrm{E}(X)$ does not exist or is not finite. Does $\bar{X}_n$ still converge to a constant with probability 1? For simplicity, assume that the X_n are nonnegative iid random variables with mean ∞. If we let $Y_n(A) = X_n I(X_n \le A)$, then the $Y_n(A)$ are iid with finite mean, so $\sum_{n=1}^{\infty} Y_n(A)/n \rightarrow \mathrm{E}\{Y_1(A)\}$ as $n \rightarrow \infty$ by the SLLN. It follows that

$$\underline{\lim}\,(1/n)\sum_{i=1}^n X_i \ge \underline{\lim}\,(1/n)\sum_{i=1}^n Y_i(A) = \mathrm{E}\{Y_1(A)\}.$$

Therefore, $\underline{\lim}\,\bar{X}_n \geq \mathrm{E}\{Y_1(A)\}$ for every positive number A. But $\mathrm{E}\{Y_1(A)\} \to \mathrm{E}(X_1) = \infty$ as $A \to \infty$ by the MCT because $Y_1(A) \uparrow X_1$ as $A \to \infty$. Therefore, $\underline{\lim}\,(1/n)\sum_{i=1}^n X_i = \infty$. It follows that $\bar{X}_n \to \infty$ with probability 1.

Now lift the restriction that the X_n be nonnegative. If X_n has mean $\pm\infty$, then exactly one of the nonnegative random variables X_n^- and X_n^+ has infinite mean. The above argument can be used to show that $\bar{X}_n$ converges a.s. to $\pm\infty$. For instance, if $\mathrm{E}(X_n^+) = \mu^+ < \infty$ and $\mathrm{E}(X_n^-) = \infty$, then $\bar{X}_n = (1/n)\sum_{i=1}^n X_i^+ - (1/n)\sum_{i=1}^n X_i^- \to \mu^+ - \infty = -\infty$. We have shown that $\bar{X}_n$ cannot converge almost surely to a finite constant if $\mathrm{E}(X_i) = \pm\infty$. It can also be shown that if $\mathrm{E}(X_n^+) = \mathrm{E}(X_n^-) = \infty$, $\bar{X}_n$ does not converge almost surely to a constant. The following theorem summarizes these conclusions.

Theorem 7.13. Kolmogorov *If $X_1, X_2, \ldots$ are iid, then $\bar{X}_n$ converges almost surely to some finite constant c if and only if $\mathrm{E}(X_1) = \mu$ is finite, in which case $c = \mu$.*

A good illustration of the necessity of a finite mean is the following example.

Example 7.14. Need for finite mean Let $X_1, \ldots, X_n$ be iid from a Cauchy distribution with respective distribution and density functions

$$F_\theta(x) = 1/2 + (1/\pi)\arctan(x - \theta) \text{ and } f_\theta(x) = \frac{1}{\pi\{1 + (x - \theta)^2\}}.$$

The mean of a Cauchy random variable does not exist, although the median is θ. We will see in the next chapter that $\bar{X}_n$ has the same Cauchy distribution F_θ, so that $P(|\bar{X}_n - \theta| > \epsilon) = P(|X_1 - \theta| > \epsilon)$. This means that the sample mean $\bar{X}$ is no better than a single observation at estimating θ. In particular, $\bar{X}_n$ does not converge to θ almost surely as the sample size tends to ∞. Of course, $\bar{X}_n$ does not converge almost surely to any other constant either, which is consistent with Theorem 7.13. □

Exercises

1. Let a_n be a sequence of numbers such that $a_n \to a$ as $n \to \infty$. Prove that $(1/n)\sum_{i=1}^n a_i \to a$.

2. Prove the SLLN for the special case that X_i are iid Bernoulli with parameter p. Hint: use the Borel-Cantelli lemma (Lemma 6.38) together with the fact that the binomial random variable $S_n = \sum_{i=1}^n X_i$ satisfies $\mathrm{E}(S_n - np)^4 = 3n^2p^2(1-p)^2 + np(1-p)\{1 - 6p(1-p)\}$.

3. What is wrong with the following reasoning? If X_n are iid with mean 0, then $\bar{X}_n \overset{a.s.}{\to} 0$ by the SLLN. That means there exists an N such that $|\bar{X}_n - 0| \leq 1$ for all $n \geq N$. The DCT then implies that $\mathrm{E}(|\bar{X}_n|) \to 0$.

7.3 Random Walks

Imagine starting at the origin in the plane and stepping one unit to the right and either one unit up or one unit down with probability p or $1-p$, respectively. Wherever we end up after the first step, we again step one unit to the right and then randomly step up or down

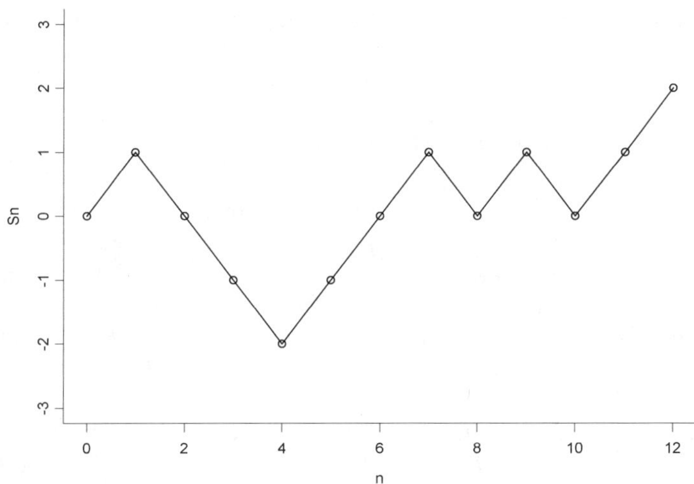

Figure 7.4: A random walk.

one unit with probability p or $1-p$, respectively. We continue this process ad infinitum. If $X_1, X_2, \ldots$ are iid random variables representing the steps up or down, then

$$X_i = \begin{cases} -1 & \text{w.p. } 1-p \\ +1 & \text{w.p. } p. \end{cases} \tag{7.17}$$

The sum $S_n = \sum_{i=1}^n X_i$ is the vertical position after n steps.

Definition 7.15. Random walk *The sequence $S_n = \sum_{i=1}^n X_i$, where X_i are iid with distribution (7.17), is called a random walk. When $p = 1/2$, it is called a symmetric random walk.*

We represent a random walk with a graph of the points (n, S_n) in the plane, with linear interpolation between points, as shown in Figure 7.4. The behavior of random walks has been studied extensively (Feller, 1968), first in connection with gambling, but with many other applications as well. We give only a few examples and results for symmetric random walks that can be proven using laws of large numbers.

Example 7.16. Consider a study in which everyone receives both treatments A and B. Let X_i be -1 or $+1$ depending on whether patient i does better on treatment A or B, respectively. Patients enter the study sequentially, and we observe results after each is evaluated. The sum $S_n = \sum_{i=1}^n X_i$ measures whether treatment A or B is doing better after n patients. If the two treatments are equally effective, then $\mu = \mathrm{E}(X_i) = (1/2)(1) + (1/2)(-1) = 0$, so $\mathrm{E}(S_n) = 0$ as well. That is, we "expect" S_n to be 0. The SLLN says that $S_n/n \stackrel{a.s.}{\to} 0$, so for each $\epsilon > 0$, there is an $N = N(\omega)$ such that $-\epsilon < S_n(\omega)/n < \epsilon$ for all $n \geq N(\omega)$. Still, does this preclude S_n from wandering far from 0? How large do we expect $\max_n S_n$ to be if $\mu = 0$? Knowing the answer to this question involving fluctuations of S_n is crucial when monitoring results of the study after every new observation. □

Proposition 7.17. *If S_n is a symmetric random walk, then with probability 1, $|S_n|$ eventually reaches every level $0, 1, 2 \ldots$*

Proof. $|S_0| = 0$, so level 0 is reached immediately. Consider a fixed positive integer k. Note that $|S_n|$ will reach k (in fact, it will reach $2k$) if the first $2k$ X_i are all -1 or all $+1$. If $|S_n|$ does not reach level k within the first $2k$ steps, then regardless of the value of S_{2k}, $|S_n|$ will reach level k by step $4k$ if $X_{2k+1}, \ldots, X_{4k}$ are all -1 or all $+1$. Likewise, if $|S_n|$ does not reach level k by step $4k$, then regardless of the value of S_{4k}, $|S_n|$ will reach level k by step $6k$ if $X_{4k+1}, \ldots, X_{6k}$ are all -1 or all $+1$, etc. Denote by A_i the event that $X_{2k(i-1)+1}, \ldots, X_{2ki}$ are all -1 or all $+1$, and let I_i be the indicator of A_i. Then $|S_n|$ eventually reaches level k if any of $I_1, I_2, \ldots$ is 1. These indicator variables are iid with mean $(1/2)^{2k} + (1/2)^{2k} = (1/2)^{2k-1} > 0$. By the SLLN, the proportion of I_i equaling 1 converges almost surely to $(1/2)^{2k-1}$, which certainly implies that at least one of $I_1, I_2, \ldots$ is 1. In other words, with probability 1, $|S_n|$ will reach level k. We have shown that $P(B_k) = 1$ for each natural number k, where B_k is the event that $|S_n|$ eventually reaches level k. If B denotes $\cap_k B_k$, then $P(B) = 1$ because $P(B^C) = P(\cup_k B_k^C) \leq \sum_k P(B_k^C) = 0$. Therefore, the probability is 1 that $|S_n|$ will eventually reach every level $0, 1, 2, \ldots$ □

Proposition 7.18. Symmetric random walks eventually hit every level *With probability* 1, *the symmetric random walk eventually reaches every level* $0, -1, +1, -2, +2, \ldots$ *In fact, with probability* 1 *it reaches each level infinitely often, including infinitely many returns to the origin.*

Sketch of proof. First we prove that with probability 1 the random walk eventually reaches level $+1$. After one step, S_n reaches either -1 or $+1$ with probability $1/2$ each. Let A_1 be the event $S_1 = -1$. Now imagine a random walk beginning at level -1 instead of 0. Proposition 7.17 ensures that it will reach either -3 or $+1$ eventually, and by symmetry, it is equally likely to reach either of these first (see Figure 7.5). Let A_2 be the event that it reaches -3 before reaching $+1$. Now imagine a random walk beginning at level -3. By Proposition 7.17, the random walk eventually reaches either -7 or $+1$, and by symmetry, it is equally likely to reach either first. Let A_3 denote the event that it reaches -7 first, etc. The probability that S_n never reaches level 1 is $P(\cap_i A_i) = (1/2)(1/2)\ldots = 0$. A similar proof can be used for other k, or one can proceed by induction; the new random walk starting at level 1 must eventually end up one level higher, etc. Also, the same proof works for $-k$ by symmetry.

To see that S_n must reach each level infinitely often, note that if $S_n(\omega)$ reached level k only finitely many times, then there would be an $N = N(\omega)$ such that either S_N, $S_{N+1}, \ldots$ are all less than k or all larger than k. Either way, infinitely many levels (either all levels $> k$ or all levels $< k$) would never be reached at or after observation N. Only finitely many of those levels could have been reached before observation N, so for that ω, some levels would never be reached. Such ω have probability 0 by what we proved in the preceding paragraph. Therefore, each level is reached infinitely often. This also implies that S_n returns to the origin infinitely often. □

Consider the implications of Proposition 7.18 on Example 7.16. Suppose that the treatments are equally effective and we monitor the study after every new patient is evaluated. The WLLN ensures high probability that the **proportion** of patients for whom treatment A does better than treatment B will be close to 1/2. Nonetheless, the difference between the **number** of patients doing better on treatment A and its expected number, $n/2$, can be very substantial.

The number of returns to the origin of a random walk has many important applications. The example offered below relates to the potential for selection bias in certain types of clinical trials.

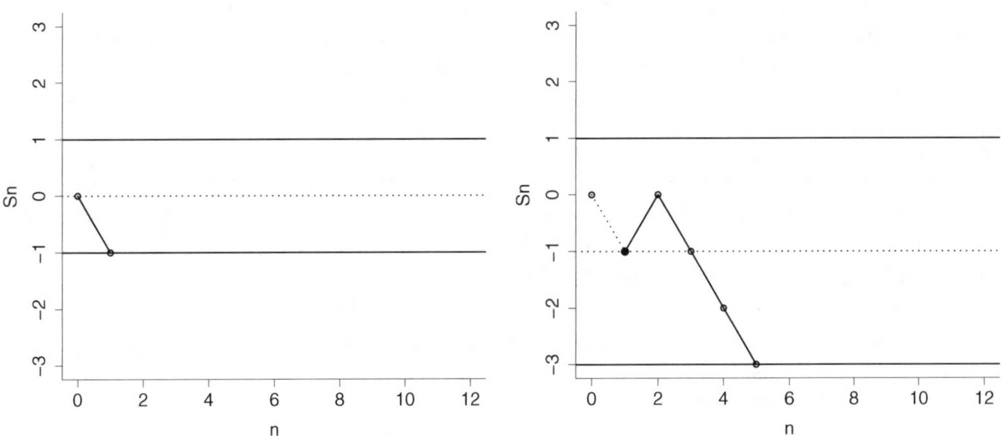

Figure 7.5: For a random walk never to reach level $+1$, it must reach level -1 before level $+1$ (left panel). It begins anew at $(1, -1)$, and then it must reach level -3 before reaching level $+1$ (right panel), etc.

Example 7.19. Trying to guess treatment assignments in a clinical trial Well-designed clinical trials keep the patient and investigators blinded to the treatment given. In some cases, maintaining the blind is virtually impossible. For example, suppose that one treatment is surgery and the other is a pill. Once the patient receives treatment, he or she and the investigator know the treatment received. Therefore, an investigator could keep track of all previous treatment assignments. Let X_i be -1 or $+1$ if patient i is assigned pill or surgery, respectively. An investigator keeping track of all previous assignments knows S_n, the difference between the numbers of patients assigned to surgery versus pill. If simple randomization is used, keeping track of past assignments will not help the investigator predict the next assignment. But most methods of randomization try to force more balance in the numbers of patients assigned to the two treatments than simple randomization would. For example, permuted block randomization with blocks of size $2k$ is equivalent to placing k Cs (denoting control) and k Ts (denoting treatment) in a box and drawing without replacement for the next $2k$ patients. Balance is forced after every $2k$ patients. Forced balance means that the investigator usually has a better than 50 percent chance of correctly guessing the next assignment. With blocks of size 2, the investigator is sure to correctly guess the second patient in every block.

Having a better than 50 percent chance of correctly guessing the next assignment could lead to *selection bias*, a phenomenon whereby patients assigned to treatment are systematically different from those assigned to control. This could happen if the investigator vetoes potential new patients until the "right" one comes along. For instance, an investigator may believe that treatment helps. If the investigators knows that the next patient will receive the new treatment, he or she may unwittingly veto healthy looking patients until a particularly sick one who "really needs the treatment" arrives. The end result is that patients receiving treatment are sicker than control patients. Likewise, an investigator who is confident that the treatment works may introduce bias such that those receiving treatment are healthier than the controls to ensure that the trial produces the "right" answer that treatment works. In either case the investigator's bias could influence the results of the trial. Such bias is usually subtle, but to understand its potential influence, we consider an extreme scenario whereby an investigator intentionally tries to influence the results of a trial with permuted block randomization and a continuous outcome Y. The truth is that the treatment has no effect on Y. An investigator who believes that the next patient will be assigned control (respectively, treatment) selects one with mean μ (respectively, $\mu+\Delta$). The probability of a false positive result depends on the number of correct guesses of the investigator (Proschan, 1994).

The guessing strategy used by the investigator is as follows. The first guess is determined by the flip of a fair coin. After the first patient is assigned, the investigator continues guessing the opposite treatment until the numbers assigned to the two treatments balance again. Let ν_1 be the index of the observation when things balance again. That is, ν_1 is the first index $n \geq 1$ such that $S_n = 0$, i.e., the first return to the origin. The investigator will be correct exactly one time more than he or she is incorrect on patients $2, 3, \ldots, \nu_1$ (see Figure 7.6). Moreover, he or she has probability 1/2 of being correct on patient 1. The same thing happens over the next interval $\nu_1 + 1, \ldots, \nu_1 + \nu_2$, where $\nu_1 + \nu_2$ is the index of the second time of balance—i.e., the index of the second return of the random walk to the origin. The same thing happens between all successive returns to the origin. If the randomization forces balance at the end of the trial, then the number of correct treatment assignment guesses depends crucially on the total number of returns to the origin (forced or unforced).

Note that simple randomization avoids this problem because we can correctly predict only half the assignments, on average. Keeping track of previous treatment assignments offers no advantage because the conditional probability that the next assignment is to treatment A is 1/2 regardless of previous assignments. □

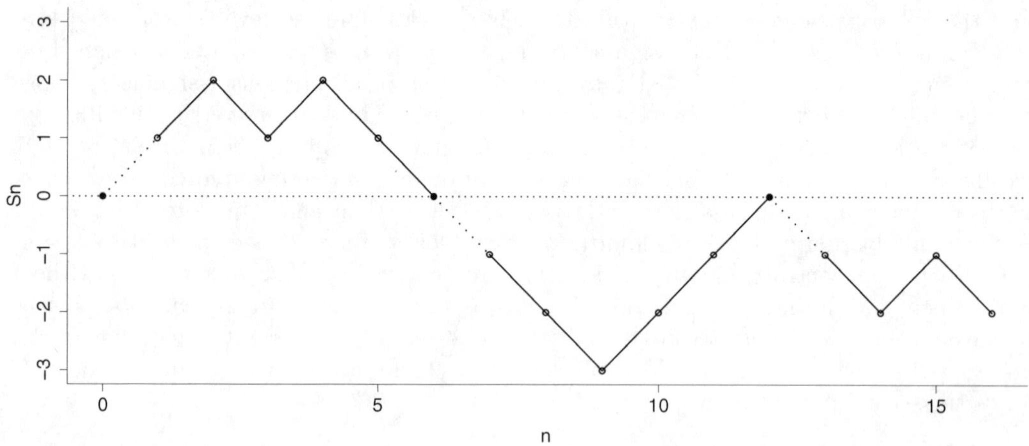

Figure 7.6: The first step is +1, so we predict −1 for each of steps 2-6, at which time the random walk returns to the origin. Steps 2-6 contain one more correct than incorrect predictions, while step 1 has probability 1/2 of being correctly predicted. Thus, over steps 1-6, the number correct minus number incorrect is $1 + Y_1$, where $Y_1 = -1$ or 1 with probability 1/2. The same thing happens between any two consecutive returns to the origin.

It turns out that thinking in terms of the investigator in Example 7.19 leads to the following useful result for symmetric random walks.

Theorem 7.20. *The expected number of returns to the origin of a symmetric random walk S_k before time $2n$ (not including any return at time $2n$) is* $\mathrm{E}(|S_{2n}|) - 1$.

Proof. Consider the investigator trying to predict treatment assignments in Example 7.19 with $2n$ patients when simple randomization is used. The difference between the number of correct and incorrect predictions over the interval between returns i and $i+1$ to the origin is $1+Y_i$, where $Y_i = \pm 1$ with probability 1/2 (see Figure 7.6). Let L be the time of the last return to the origin before $2n$, with $L = 0$ if there is no return before time $2n$. The difference between the numbers of correct and incorrect predictions from step $L+1$ to step $2n$ is either $-|S_{2n}|$ or $-|S_{2n}|+2$, depending on whether the prediction on step $L+1$ is incorrect or correct, respectively. Thus, the number correct minus number incorrect from step $L+1$ to step $2n$ is $-|S_{2n}|+U$, where U is 0 or 2 with probability 1/2. Over all $2n$ steps, the number correct minus number incorrect is $-|S_{2n}|+U+\sum_{i=1}^{N}(1+Y_i)$, where N is the number of returns to the origin before time $2n$. It is not difficult to show that $\mathrm{E}\{\sum_{i=1}^{N}(1+Y_i)\} = \mathrm{E}(N)$ (see Exercise 1), so

$$\mathrm{E}(\#\text{correct} - \#\text{incorrect}) = -\mathrm{E}(|S_{2n}|) + 1 + \mathrm{E}(N). \tag{7.18}$$

But for any guessing strategy, the number correct minus number incorrect must have mean 0. This is because the steps of the random walk are iid; the conditional probability of being correct on step k, given the results on steps $1, \ldots, k-1$, is still 1/2. Equating Expression (7.18) to 0 gives the stated result. □

What is the expected amount of time until a random walk returns to the origin? If this expectation were finite, then the number of returns before time $2n$ would be of order $2n$. To see this, let ν_1 be the time to the first return to the origin (note: ν_1 could be $> 2n$), ν_2 be the time between the first and second return, etc. Then $\nu_1, \nu_2, \ldots$ are iid random variables. If the mean time between returns $\mu_\nu = \mathrm{E}(\nu_i)$ were finite, then $\bar{\nu} = (1/k)\sum_{i=1}^{k}\nu_i$ would converge almost surely to μ_ν as $k \to \infty$ by the SLLN. But each ω for which $\bar{\nu}(\omega) \to \mu_\nu$ is an ω for which $\hat{p}_{2n} = (1/2n)(\#\text{ returns before } 2n) \to 1/\mu_\nu$. Therefore, if $\mu_\nu < \infty$, then $\hat{p}_{2n}$ would converge almost surely to $1/\mu_\nu$. Moreover, $\hat{p}_{2n}$ is bounded by 1, so the BCT would imply that $\mathrm{E}(\hat{p}_{2n})$ also converges to $1/\mu_\nu$. In other words, the expected number of returns by time $2n$ would be a constant times $2n$, i.e., it would be of order $2n$. However, this is all predicated on the mean time between returns being finite. If it is not finite, then the mean number of returns need not be of order $2n$. One way to show that the mean time between returns is **not** finite is to show that the mean number of returns is not of order $2n$. To that end, use Theorem 7.20 and note that

$$\begin{aligned} \mathrm{E}(|S_{2n}|) &\leq \sqrt{E(S_{2n}^2)} \\ &= \sqrt{\mathrm{var}(S_{2n})} = \sqrt{2n}. \end{aligned} \tag{7.19}$$

The inequality in the first line is from Corollary 5.20 with $p = 2$. We have shown that the expected number of returns before $2n$ is of order at most $(2n)^{1/2}$. Because the expected number of returns would have been of order $2n$ if the expected return time were finite, this proves that the expected return time must be infinite. We have proven the following theorem.

Theorem 7.21. Infinite expected time to return to origin for symmetric random walks *The expected time for a symmetric random walk to return to the origin is* ∞.

What we have shown is that although a symmetric random walk is guaranteed to return to the origin infinitely often, the time it takes to return can be extremely long, such that its expectation is infinite. Very long times to return mean that the random walk could spend a high proportion of time above the x-axis or a high proportion of time below the x-axis. See Feller (1968) for a formalization of this fact known as an *arcsin law.*

Exercises

1. Let $X_1, X_2, \ldots$ be iid random variables independent of the positive integer-valued random variable N with $\mathrm{E}(N) < \infty$. If $\mathrm{E}(|X_1|) < \infty$, then $\mathrm{E}(S_N) = \mathrm{E}(X_1)\mathrm{E}(N)$.

2. Let S_n be a non-symmetric random walk (i.e., $p \neq 1/2$) and k be any integer. Prove that with probability 1, S_n reaches level k only finitely many times. Prove also that $P(S_n$ reaches level k infinitely many times for some integer $k)$ is 0.

7.4 Summary

1. The sample mean of iid random variables with mean μ converges to μ in two ways:
   (a) SLLN: $\bar{X}_n \stackrel{a.s.}{\to} \mu$.
   (b) WLLN: $\bar{X}_n \stackrel{p}{\to} \mu$.

2. (Kolmogorov) If $X_1, X_2, \ldots$ are iid, then $\bar{X}_n$ converges almost surely to some finite constant c if and only if $\mathrm{E}(X_1) = \mu$ is finite, in which case $c = \mu$.

3. A useful technique for identically distributed (not necessarily independent) random variables $X_1, \ldots, X_n$ with finite mean μ is truncation: $Y_i = X_i I(|X_i| < i) + iI(|X_i| \geq i)$.
   (a) The Y_i are bounded, so they have finite moments of all orders.
   (b) $\bar{X} \stackrel{a.s.}{\to} \mu$ if and only if $\bar{Y} \stackrel{a.s.}{\to} \mu$ and $\bar{X} \stackrel{p}{\to} \mu$ if and only if $\bar{Y} \stackrel{p}{\to} \mu$.

4. If X_i are iid Bernoulli (p), $S_n = \sum_{i=1}^n X_i$ is called a random walk. If $p = 1/2$, S_n reaches every level infinitely often, but the expected time to return to the origin is infinite.

Chapter 8

Central Limit Theorems

Part 1: CLT Applications

8.1 CLT for iid Random Variables and Applications

Many test statistics and estimators involve sample means of iid random variables $X_1, \ldots, X_n$. If we make assumptions about the X_i, we can get the exact distribution of the sample mean $\bar{X}_n$. For example, if the X_i are normally distributed, so is $\bar{X}_n$. But in reality, we do not know the distribution of most data. It would be nice if the distribution of $\bar{X}_n$, properly standardized, did not depend on the distribution of the X_i. In this chapter we show that if n is large and $\text{var}(X_i) = \sigma^2 < \infty$, the distribution of a properly standardized version $(\bar{X}_n - a_n)/b_n$ is approximately the same—standard normal—regardless of the distribution of $X_1, \ldots, X_n$. Later in the chapter we show that, under certain conditions, a similar result holds if the X_i are independent but not identically distributed.

Imagine that we know nothing about the central limit theorem, but we know only that some standardized version $(\bar{X}_n - a_n)/b_n$ converges in distribution to some non-constant random variable Z. How can we deduce the proper standardization of $\bar{X}_n$ and the distribution of Z? It is natural to subtract the mean of $\bar{X}_n$ to give the standardized version mean 0. Therefore, it makes sense to take $a_n = \text{E}(\bar{X}) = \mu$. The question then becomes: what sequence b_n makes $(\bar{X}_n - \mu)/b_n$ converge in distribution to a non-constant random variable Z? The variance of

$$Z_n = (\bar{X}_n - \mu)/b_n$$

is $\sigma^2/(nb_n^2)$. If we want Z_n to converge in distribution to a non-constant, then $\text{var}(Z_n)$ must not tend to 0; otherwise, Z_n would converge in probability to 0 by Chebychev's inequality. We could choose b_n such that $\text{var}(Z_n)$ is a nonzero constant. For example, $b_n = \sigma/n^{1/2}$ makes $\text{var}(Z_n) = 1$. Therefore, it is natural to postulate that $n^{1/2}(\bar{X}_n - \mu)/\sigma$ converges in distribution to a non-constant random variable.

The next question is: what is the distribution of the limiting random variable Z? If the X_i are normally distributed, then $Z_n = n^{1/2}(\bar{X}_n - \mu)/\sigma$ has an exact standard normal distribution, so the limiting random variable Z must be standard normal in that case. But we are assuming that Z_n converges in distribution to the same Z regardless of the distribution of the X_i. Therefore, that Z must be standard normal. We have surmised, but not proven, the most famous theorem in probability, stated in Theorem 8.1.

Theorem 8.1. Standard CLT *Let X_i be iid random variables with mean μ and variance $\sigma^2 < \infty$. Then $Y_n = n^{1/2}(\bar{X}_n - \mu)/\sigma \xrightarrow{D} N(0,1)$ (equivalently, $(S_n - n\mu)/(n\sigma^2)^{1/2} \xrightarrow{D}$ $\mathrm{N}(0,1)$).*

We defer the proof to Section 8.5.

Remark 8.2. Terminology: asymptotic normality *Sometimes people use the somewhat less formal terminology "$\bar{X}_n$ is asymptotically normal with mean μ and variance σ^2/n" to mean that $(\bar{X}_n - \mu)/(\sigma^2/n)^{1/2} \xrightarrow{D} \mathrm{N}(0,1)$. Likewise, "$S_n$ is asymptotically normal with mean $n\mu$ and variance $n\sigma^2$" means that $(S_n - n\mu)/(n\sigma^2)^{1/2} \xrightarrow{D} \mathrm{N}(0,1)$.*

We illustrate several applications of the CLT.

Example 8.3. Application of CLT: asymptotic distribution of one-sample t-statistic In some settings we have continuous measurements of the same people at two time points or under two different conditions (e.g., treatment and placebo). We are interested in the paired differences $D_1, \ldots, D_n$ for the n people in the study. Assume that $D_1, \ldots, D_n$ are iid from a distribution with mean μ and finite variance σ^2. The one-sample t-statistic for testing the null hypothesis that $\mu = 0$ is

$$T_n = \frac{\bar{D}}{\sqrt{s^2/n}},$$

where $\bar{D} = \bar{D}_n$ is the sample mean and $s^2 = s_n^2 = (n-1)^{-1}\sum(D_i - \bar{D})^2$ is the sample variance. If the underlying distribution of the data D_i is normal with mean 0 and variance σ^2, then T_n has a central t-distribution with $n-1$ degrees of freedom. If the alternative hypothesis is that $\mu > 0$, we reject the null hypothesis if $T_n > t_{n-1,\alpha}$, where $t_{n-1,\alpha}$ is the $(1-\alpha)$th quantile of a t-distribution with $n-1$ degrees of freedom. If the paired differences do come from a normal distribution, the type I error rate is exactly α, as desired.

Suppose that our assumption that the data are normal is wrong. Suppose that the correct distribution is $F(d)$, which has mean 0 and variance σ^2. What happens to the type I error rate of the t-test for large sample sizes? Notice that $T_n = Z_n(\sigma^2/s^2)^{1/2}$ where

$$Z_n = \frac{\bar{D}}{\sqrt{\sigma^2/n}} = \frac{\sqrt{n}\bar{D}}{\sigma}.$$

The CLT implies that $Z_n \xrightarrow{D} \mathrm{N}(0,1)$ as $n \to \infty$. Furthermore, Example 7.6 showed that $s^2 \xrightarrow{a.s.} \sigma^2$, implying that $(\sigma^2/s^2)^{1/2} \xrightarrow{p} 1$. Slutsky's theorem (Theorem 6.52) implies that $T_n = Z_n(\sigma^2/s^2)^{1/2} \xrightarrow{D} \mathrm{N}(0,1)$. In other words, regardless of the underlying distribution $F(d)$ for D, the limiting distribution of the t-statistic as $n \to \infty$ is $\mathrm{N}(0,1)$ under the null hypothesis. In particular, this holds if the D_i are iid normals with mean 0 and variance $\sigma^2 < \infty$, in which case T_n has an exact t-distribution with $n-1$ degrees of freedom. This shows that a t-distribution with $n-1$ degrees of freedom tends to a $\mathrm{N}(0,1)$ as its degrees of freedom tends to ∞. This implies that the critical value $t_{n-1,\alpha}$ converges to z_α, the $(1-\alpha)$th quantile of a standard normal distribution, as $n \to \infty$ (exercise). More importantly, the t-test has approximately correct type I error rate for large sample sizes regardless of the underlying distribution of the data. □

Example 8.4. Application of CLT: asymptotic distribution of two-sample t-statistic Let $X_1, \ldots, X_m$ and $Y_1, \ldots, Y_n$ be values of a continuous variable in two different groups. Assume that the X_i are iid with mean μ_X, the Y_i are iid with mean μ_Y and independent of the X_i, and $\text{var}(X_i) = \text{var}(Y_i) = \sigma^2$. We wish to test the null hypothesis that $\mu_X = \mu_Y$. The two-sample t-statistic is

$$T_{m,n} = \frac{\bar{X}_m - \bar{Y}_n}{\sqrt{s^2(1/m + 1/n)}},$$

where $s^2 = \{(m-1)s_X^2 + (n-1)s_Y^2\}/(m+n-2)$ is the pooled variance of the individual sample variances s_X^2 and s_Y^2. If the X_i and Y_i are normally distributed, $T_{m,n}$ has a t-distribution with $m + n - 2$ degrees of freedom under the null hypothesis that $\mu_X = \mu_Y$.

Suppose that X_i and Y_i have arbitrary distributions F and G with the same mean μ and same variance σ^2. Suppose that $m = m_n$ is such that $m_n/(m_n + n) \to \lambda$ as $n \to \infty$. The CLT implies that

$$Z_{m_n} = \frac{\sqrt{m_n}\,(\bar{X}_{m_n} - \mu)}{\sigma} \xrightarrow{D} \text{N}(0,1) \text{ and } Z_n = \frac{\sqrt{n}\,(\bar{Y}_n - \mu)}{\sigma} \xrightarrow{D} \text{N}(0,1) \text{ as } n \to \infty.$$

Then

$$T_{m_n,n} = \frac{\sigma\left(\frac{Z_{m_n}}{\sqrt{m_n}} - \frac{Z_n}{\sqrt{n}}\right)}{\sqrt{s^2\left(\frac{1}{m_n} + \frac{1}{n}\right)}} = \left\{\sqrt{\frac{n}{m_n + n}}\, Z_{m_n} - \sqrt{\frac{m_n}{m_n + n}}\, Z_n\right\}\sqrt{\frac{\sigma^2}{s^2}}. \tag{8.1}$$

Also, $n/(m_n + n) \to 1 - \lambda$ as $n \to \infty$, so $\{n/(m_n + n)\}^{1/2} Z_{m_n}$ converges in distribution to $(1-\lambda)^{1/2}\text{N}(0,1)$ as $n \to \infty$ by Slutsky's theorem. Similarly, $\{m_n/(m_n+n)\}^{1/2} Z_n$ converges in distribution to $\lambda^{1/2}\text{N}(0,1)$ as $n \to \infty$. Moreover, $\{n/(m_n + n)\}^{1/2} Z_{m_n}$ and $\{m_n/(m_n + n)\}^{1/2} Z_n$ are independent, so their difference converges in distribution to $(1-\lambda)^{1/2}Z - \lambda^{1/2}Z'$ as $n \to \infty$, where Z and Z' are iid standard normals. But $(1 - \lambda)^{1/2}Z - \lambda^{1/2}Z'$ has a standard normal distribution. Also, $(\sigma^2/s^2)^{1/2} \xrightarrow{p} 1$ (see Exercise 4 of Section 7.3), so Slutsky's theorem implies that Expression (8.1) converges in distribution to $\text{N}(0, 1)$.

We have shown that if the proportion of total study participants who are in the first group tends to a constant, then the two sample t-statistic has the correct type I error rate asymptotically when the variances in the two groups are equal. Notice that nothing in our derivation required the distributions in the two groups to be the same; as long as the means and variances in the two groups are equal, the two-sample t-test has the correct asymptotic type I error rate.

Example 8.5. Application of CLT: asymptotic distribution of sample variance In some cases we are interested in tests or confidence intervals on the variance σ^2 of iid random variables $X_1, \ldots, X_n$. For example, we may have such a huge amount of data from one population that we are confident that its variance is 1; we have a smaller number of observations from another population, and we want to test whether its variance is also 1. We estimate σ^2 using the sample variance $s_n^2 = (n-1)^{-1}\sum_{i=1}^n (X_i - \bar{X})^2$. Without loss of generality, assume that $\text{E}(X_i) = 0$ because subtracting $\text{E}(X)$ from every observation does not change the sample or population variance. The usual test and confidence interval on σ^2 assume the X_i are normally distributed, in which case $(n-1)s^2/\sigma^2$ has a chi-squared distribution with $n - 1$ degrees of freedom. For example, to test whether $\sigma^2 = 1$ at level α, we reject the null hypothesis if $(n - 1)s^2/1$ exceeds the $(1 - \alpha)$th quantile of a chi-squared distribution with $n - 1$ degrees of freedom.

Now suppose we have a large sample size and do not want to assume that the data are normal. How can we determine the asymptotic distribution of the sample variance? Assume that $\mathrm{var}(X_i^2) < \infty$. Apply the CLT to the iid random variables $Y_i = X_i^2$:

$$\frac{\sum_{i=1}^n (X_i^2 - \sigma^2)}{\sqrt{n\,\mathrm{var}(X_1^2)}} = \frac{\sum_{i=1}^n \{Y_i - \mathrm{E}(Y_i)\}}{\sqrt{n\,\mathrm{var}(Y_1)}} \xrightarrow{D} N(0,1).$$

Also,

$$\frac{\sum_{i=1}^n \{(X_i - \bar{X})^2 - \sigma^2\}}{\sqrt{n\,\mathrm{var}(X_1^2)}} = \frac{\sum_{i=1}^n (X_i^2 - \sigma^2)}{\sqrt{n\,\mathrm{var}(X_1^2)}} + R_n,$$

where

$$\begin{aligned} R_n &= \frac{\sum_{i=1}^n \{(X_i - \bar{X})^2 - \sigma^2\}}{\sqrt{n\,\mathrm{var}(X_1^2)}} - \frac{\sum_{i=1}^n (X_i^2 - \sigma^2)}{\sqrt{n\,\mathrm{var}(X_1^2)}} \\ &= \frac{-n\bar{X}^2}{\sqrt{n\,\mathrm{var}(X_1^2)}} = \frac{-\sigma(\sqrt{n}\bar{X}/\sigma)(\bar{X})}{\sqrt{\mathrm{var}(X_1^2)}}. \end{aligned} \tag{8.2}$$

By the CLT, $n^{1/2}\bar{X}/\sigma \xrightarrow{D} \mathrm{N}(0,1)$, and by the WLLN, $\bar{X} \xrightarrow{p} 0$. Therefore, $R_n \xrightarrow{p} 0$ (see Exercise 4 of Subsection 6.2.5). By Slutsky's theorem,

$$\frac{\sum_{i=1}^n \{(X_i - \bar{X})^2 - \sigma^2\}}{\sqrt{n\,\mathrm{var}(X_1^2)}} \xrightarrow{D} \mathrm{N}(0,1)$$

Write the left side as

$$\frac{(n-1)s^2/\sigma^2 - (n-1)}{\sqrt{n\,\mathrm{var}(X_1^2)/\sigma^4}} - \frac{1}{\sqrt{n\,\mathrm{var}(X_1^2)/\sigma^4}},$$

and note that the last term tends to 0. By Slutsky's theorem, we can ignore it. We have shown that $(n-1)s^2/\sigma^2$ is asymptotically normal with mean $n-1$ and variance $n\,\mathrm{var}(X_1^2)/\sigma^4$. On the other hand, if we had assumed normality, then as noted previously, $(n-1)s^2/\sigma^2$ follows a chi-squared distribution with $n-1$ degrees of freedom. We can write a chi-squared random variable with $n-1$ degrees of freedom as $\sum_{i=1}^{n-1} U_i$, where the U_i are iid chi-squared random variables with 1 degree of freedom. By the CLT, this sum is asymptotically normal with mean $(n-1)$ and variance $2(n-1)$.

In summary, we get the following asymptotic distributions assuming normality versus not:

$$\frac{(n-1)s^2/\sigma^2 - (n-1)}{\sqrt{2(n-1)}} \xrightarrow{D} \mathrm{N}(0,1) \text{ and } \frac{(n-1)s^2/\sigma^2 - (n-1)}{\sqrt{n\,\mathrm{var}(X_1^2)/\sigma^4}} \xrightarrow{D} \mathrm{N}(0,1).$$

The numerators are identical, and the ratio of the denominators tends to 1 if and only if $\mathrm{var}(X_1^2) = 2\sigma^4$. This is a peculiar condition holding for the normal distribution, but not necessarily for other distributions. Therefore, using the test based on normality gives the wrong answer, **even asymptotically**, if that normality assumption is wrong. This is in sharp contrast to the t-test of means. As we saw in Example 8.3, the t-test is asymptotically correct regardless of the underlying distribution of the data. □

Example 8.6. Application of CLT: asymptotic distribution of sample covariance The sample covariance between X and Y is

$$\hat{\sigma}_{XY} = \frac{1}{n-1}\sum_{i=1}^n (X_i - \bar{X})(Y_i - \bar{Y}). \tag{8.3}$$

What is the asymptotic distribution of $\hat{\sigma}_{XY}$? Without loss of generality, we may assume that X and Y have mean 0 because subtracting $\mathrm{E}(X)$ from every X observation and $\mathrm{E}(Y)$ from every Y observation does not change the sample or population covariance.

Because $\bar{X}$ and $\bar{Y}$ should be close to 0, we should be able to approximate $\sum_{i=1}^n (X_i - \bar{X})(Y_i - \bar{Y})$ by $\sum_{i=1}^n X_i Y_i$. Apply the CLT to the random variables $U_i = X_i Y_i$ with mean $\mu = \mathrm{E}(XY) = \mathrm{cov}(X, Y)$.

$$\frac{\sum_{i=1}^n X_i Y_i - n\,\mathrm{cov}(X,Y)}{\sqrt{n\,\mathrm{var}(X_1Y_1)}} = \frac{\sum U_i - n\,\mathrm{E}(U_1)}{\sqrt{n\,\mathrm{var}(U_1)}} \xrightarrow{D} \mathrm{N}(0,1).$$

It is an exercise to justify the replacement of $\sum_{i=1}^n (X_i - \bar{X})(Y_i - \bar{Y})$ by $\sum_{i=1}^n X_i Y_i$ using Slutsky's theorem. Therefore,

$$\frac{\sum_{i=1}^n (X_i - \bar{X})(Y_i - \bar{Y}) - n\,\mathrm{cov}(X,Y)}{\sqrt{n\,\mathrm{var}(X_1Y_1)}} \xrightarrow{D} \mathrm{N}(0,1).$$

□

Example 8.7. Application of CLT: asymptotic distribution of ANCOVA test statistic Consider the ANCOVA model

$$Y_i = \beta_0 + \beta_1 X_i + \beta_2 z_i + \epsilon_i,$$

where X_i and Y_i are the baseline and end of study values of the continuous outcome, z_i is the treatment indicator (1 for treatment, 0 for control), and the ϵ_i are iid errors from a distribution with mean 0 and variance σ_ϵ^2. Example 7.7 treated the x_i as fixed constants, whereas Exercise 9 of Section 7.3 treated them as random variables. Here we regard them as random variables. The treatment effect estimator using this ANCOVA model is

$$\hat{\beta}_2 = \bar{Y}_T - \bar{Y}_C - \hat{\beta}_1(\bar{X}_T - \bar{X}_C), \tag{8.4}$$

where bars denote sample means, T and C denote treatment and control groups, and $\hat{\beta}_1$ is the slope estimator. We saw in Exercise 9 of Section 7.3 that $\hat{\beta}_1 \xrightarrow{p} \beta_1$. Replace $\hat{\beta}_1$ with β_1. Then (8.4) is $\bar{U}_T - \bar{U}_C$, where $U_i = Y_i - \beta_1 X_i$. In the treatment arm, the U_i are iid with mean $\beta_0 + \beta_2$ and variance σ_ϵ^2. By the CLT, $\bar{U}_T$ is asymptotically normal with mean $\beta_0 + \beta_2$ and variance σ_ϵ^2/n_T. Similarly, in the control arm, the U_i are iid with mean β_0 and variance σ_ϵ^2; $\bar{U}_C$ is asymptotically normal with mean β_0 and variance σ_ϵ^2/n_C. Also, because the treatment and control observations are independent, $\bar{U}_T$ and $\bar{U}_C$ are independent. Therefore,

$$\frac{\bar{Y}_T - \bar{Y}_C - \beta_1(\bar{X}_T - \bar{X}_c) - \beta_2}{\sqrt{\sigma_\epsilon^2(1/n_T + 1/n_C)}} = \frac{\bar{U}_T - \bar{U}_C - \beta_2}{\sqrt{\sigma_\epsilon^2(1/n_T + 1/n_C)}} \xrightarrow{D} \mathrm{N}(0,1), \tag{8.5}$$

It is an exercise to show that the same thing holds if we replace β_1 on the left side of (8.5) by its estimator $\hat{\beta}_1$. Therefore, if the ϵ_i are iid mean 0 random variables from **any** distribution with finite variance, the ANCOVA test statistically has asymptotically the correct coverage probability. □

We next apply the CLT to iid Bernoulli (p) random variables $X_1, \ldots, X_n$. The sum $S_n = \sum_{i=1}^n X_i$ has a binomial distribution with parameters n and p. The CLT says that $Y_n = (S_n - np)/\{np(1-p)\}^{1/2}$ converges in distribution to a $\mathrm{N}(0,1)$.

Example 8.8. In an actual case of suspected cheating on a multiple choice exam at a university in Florida, a student was observed to be looking at another student's test (Boland and Proschan, 1990). We refer to the suspect and her neighbor as S and N, respectively. There were 5 choices for each question, and there were 16 questions that both students missed. S and N matched answers on 13 of these 16. Under the professor's dubious assumption that each incorrect answer is equally likely to be selected, the probability of a match on any given question that both S and N missed, assuming no cheating, is 1/4. The binomial probability of matching on 13 or more questions is $\sum_{i=13}^{16} \binom{16}{13}(1/4)^i(3/4)^{16-i} \approx 4 \times 10^{-6}$. Because this probability is very low, the professor argued that S must have cheated.

The actual analysis was flawed on many levels: (1) different incorrect answers on multiple choice exams are almost never equally likely to be selected, (2) whether two students match on one answer may not be independent of whether they match on another question, and (3) the professor erroneously used a match probability of 1/16 instead of 1/4.

Our focus is on how well the CLT approximates the binomial probability of 13 or more correct answers out of 16 when $p = 1/4$. Under the professor's assumptions, the number of matches S_{16} is the sum of 16 independent Bernoulli trials with $p = 1/4$. Therefore, $\mathrm{E}(S_{16}) = 16(1/4) = 4$ and $\mathrm{var}(S_n) = 16(1/4)(3/4) = 3$. The CLT approximation is

$$\begin{aligned} P(S_{16} \geq 13) &= P\left(\frac{S_{16} - 16(1/4)}{\sqrt{16(1/4)(3/4)}} \geq \frac{13 - 16(1/4)}{\sqrt{16(1/4)(3/4)}}\right) \\ &= 1 - \Phi(5.2) \approx 10^{-9}. \end{aligned} \tag{8.6}$$

Is this close to the binomial probability of 4×10^{-6}? That depends on whether we are talking about "close" in an absolute or relative sense. The difference between 4×10^{-6} and 10^{-9} is tiny, yet 10^{-9} is 4,000 times smaller than 4×10^{-6}. When one is attempting to demonstrate a very remote probability, 4×10^{-6} and 10^{-9} are very different, though either would be considered strong evidence against S. □

In general, the CLT approximation to the binomial works better when p is closer to 1/2. For example, suppose $p = 1/2$. Even if n is as small as 10, the probability of 7 or more successes is 0.17 under the binomial and 0.16 under the CLT approximation.

Exercises

1. We have two different asymptotic approximations to a binomial random variable X_n with parameters n and p_n: the law of small numbers when $np_n \to \lambda$ and the CLT for fixed p. In each of the following scenarios, compute the exact binomial probability and the two approximations. Which approximation is closer?

   (a) $n = 100$ and $p = 1/100$; compute $P(X_n \leq 1)$.

   (b) $n = 100$ and $p = 40/100$; compute $P(X_n \leq 38)$.

2. Prove that if X_n has a chi-squared distribution with n degrees of freedom, then $(X_n - n)/(2n)^{1/2}$ converges in distribution to a standard normal deviate. Hint: what is the distribution of the sum of n iid chi-squared random variables with parameter 1?

3. Prove that if X_n has a Poisson distribution with parameter n, then $(X_n - n)/n^{1/2} \xrightarrow{D} X$, where $X \sim \mathrm{N}(0, 1)$. Hint: what is the distribution of the sum of n iid Poisson random variables with parameter 1?

4. ↑ In the preceding problem, let us approximate $P(X_n = n)$ by $P(n - 1/2 \le X_n \le n + 1/2)$ using the CLT.

   (a) Show that the CLT approximation is asymptotic to $\phi(0)/n^{1/2}$ as $n \to 0$, where $\phi(x)$ is the standard normal density function.

   (b) Equate $P(X_n = n)$ to $\phi(0)/n^{1/2}$ and solve for $n!$ to obtain Stirling's formula.

   (c) Is the above a formal proof of Stirling's formula? Why or why not?

5. Suppose that $X_n \sim F_n$ and $X \sim F$, where F_n and F are continuous and F is strictly increasing. Let $x_{n,\alpha}$ be a $(1-\alpha)$th quantile of F_n, and suppose that $X_n \xrightarrow{D} X$. Prove that $x_{n,\alpha} \to x_\alpha$, the $(1-\alpha)$th quantile of X.

6. In Example 8.6, use Slutsky's theorem to justify the replacement of $\sum_{i=1}^n (X_i - \bar{X})(Y_i - \bar{Y})$ by $\sum_{i=1}^n X_i Y_i$ when showing that

$$\frac{\sum_{i=1}^n (X_i - \bar{X})(Y_i - \bar{Y}) - n\,\mathrm{cov}(X,Y)}{\sqrt{n\,\mathrm{var}(X_1 Y_1)}} \xrightarrow{D} \mathrm{N}(0,1).$$

7. In Example 8.7, use Slutsky's theorem to justify rigorously the replacement of $\hat{\beta}_1$ by β_1 when showing that

$$\frac{\bar{Y}_T - \bar{Y}_C - \hat{\beta}_1(\bar{X}_T - \bar{X}_C) - \beta_2}{\sqrt{\sigma_\epsilon^2(1/n_T + 1/n_C)}} \xrightarrow{D} \mathrm{N}(0,1). \tag{8.7}$$

8.2 CLT for Non iid Random Variables

There are some settings in which observations are independent but not identically distributed. For instance, a more realistic model in Example 8.8 allows different probabilities for the different incorrect choices on a given question. For example, suppose that the proportions of students in a large class choosing incorrect options A,B,C, and D among those who missed that question are .50, .20, .20, and .10, respectively. If these were actual probabilities instead of estimates, we would calculate the probability of a match on that question as $(.50)^2 + (.20)^2 + (.20)^2 + (.10)^2 = .34$. The indicators of matches on different questions would be Bernoullis with question-specific match probabilities p_i. It is arguably reasonable to assume that these Bernoullis are independent if different questions cover different material. Under this assumption, the number of matches is the sum of independent Bernoullis with different p_is. This prompts us to investigate central limit theorems for independent, but not identically distributed summands. It makes sense that if the random variables are nearly iid, a central limit theorem should still hold. For instance, the following example shows that if we start with an iid sequence and replace one observation with an arbitrary random variable with finite variance, a CLT still holds.

Example 8.9. Let $X_1, X_2, \ldots$ be iid with mean μ and finite variance σ^2, and let Y have an arbitrary distribution with mean μ_Y and variance σ_Y^2. Replace X_1 with Y; that is, define $Y_1 = Y$, $Y_i = X_i$ for $i \ge 2$. The following argument shows that $\sum_{i=1}^n Y_i$ is asymptotically normal with mean $\mu_Y + (n-1)\mu$ and variance $\sigma_Y^2 + (n-1)\sigma^2$.

By the CLT,

$$Z_n = \frac{\sum_{i=2}^n Y_i - (n-1)\mu}{\sqrt{(n-1)\sigma^2}} \xrightarrow{D} \mathrm{N}(0,1).$$

Let

$$U_n = \frac{\sum_{i=1}^n Y_i - \{\mu_Y + (n-1)\mu\}}{\sqrt{\sigma_Y^2 + (n-1)\sigma^2}}.$$

Then $U_n = A_n Z_n + D_n$, where $A_n = \{(n-1)\sigma^2\}^{1/2}/\{\sigma_Y^2 + (n-1)\sigma^2\}^{1/2} \to 1$ and

$$D_n = \frac{Y - \mu_Y}{\sqrt{\sigma_Y^2 + (n-1)\sigma^2}} \xrightarrow{p} 0.$$

By Slutsky's theorem, $U_n \xrightarrow{D} \mathrm{N}(0,1)$.

The conclusion from this example is that the CLT still holds if we replace one of n iid random variables with an arbitrary random variable with finite variance σ_Y^2. The same is true if we replace any finite and fixed set of the X_i by other random variables Y_i. □

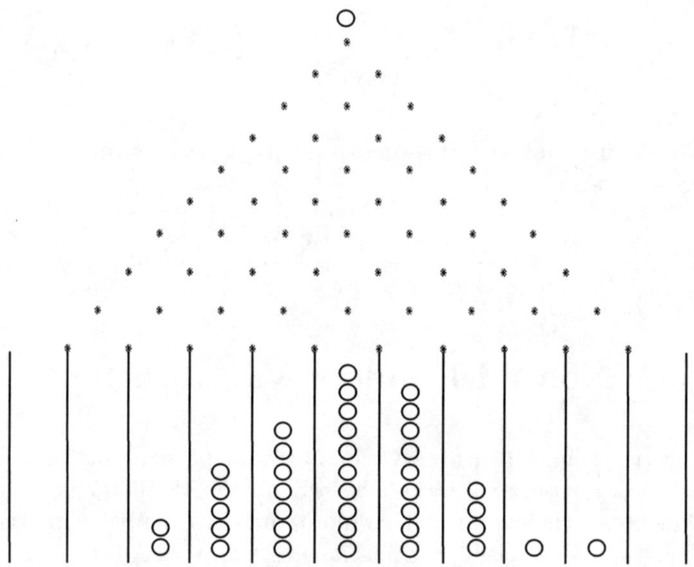

Figure 8.1: Quincunx.

A very useful device for developing intuition about central limit theorems is the quincunx (Figure 8.1). The standard quincunx simulates the sum of iid random variables X_i taking values -1 or $+1$ with probability 1/2 each. Balls roll down a board toward a triangular array of nails. When a ball hits the nail in the first row, it is equally likely to bounce left ($X_1 = -1$) or right ($X_1 = +1$). Whichever way it bounces, it then strikes a nail in the second row and bounces left ($X_2 = -1$) or right ($X_2 = +1$) with equal probability, etc. Each left or right deflection in a given row represents one binary ± 1 random variable. Bins at the end of the board collect the balls after they pass through the n rows of nails. If the numbers of left and right deflections are equal, the ball will collect in the middle bin ($\sum_{i=1}^n X_i = 0$), whereas if all deflections are to the right, the ball will collect in the rightmost bin ($\sum_{i=1}^n X_i = n$), etc. The ball's location at the end is the sum of n iid ± 1 deflections, and therefore has the distribution of $2Y - n$, where Y is binomial with parameters n and 1/2. The numbers of balls in the different bins is a frequency histogram estimating the distribution of $\sum_{i=1}^n X_i$. With a large number of balls, this empirical distribution is a good approximation to the asymptotic distribution of S_n, which is $N(0, n)$ by the CLT.

We can modify the quincunx to allow different sized deflections in different rows. This helps us envision scenarios when a central limit theorem might hold even if the random variables are independent but not identically distributed. As long as the deflection sizes in the different rows are not wildly different, the distribution of the balls at the bottom is approximately normal.

Example 8.10. One-sample permutation test Suppose data consist of iid paired differences D_i from a distribution symmetric about μ and with finite variance, and we wish to test the null hypothesis that $\mu = 0$. A permutation test in this setting corresponds to treating the data d_i as fixed numbers, and regarding $-d_i$ and $+d_i$ as equally likely. For instance, if $d_1 = 8$, we treat -8 or $+8$ as equally likely. The different observations are still independent, binary observations, but are not identically distributed because the d_i have different magnitudes. A quincunx with deflection size $|d_i|$ in row i represents the permutation distribution. Intuitively, because the d_i arose as iid observations from some distribution with finite variance, they will not differ so radically that they cause the distribution of balls at the bottom to be bimodal or have some other non-normal shape. Later we prove that a central limit theorem holds in this setting. □

If the size of the deflection in one row dominates the sizes in other rows, the distribution of the sum may not be asymptotically normal. We illustrate how to use a quincunx to construct an example for which the sum is not asymptotically normal.

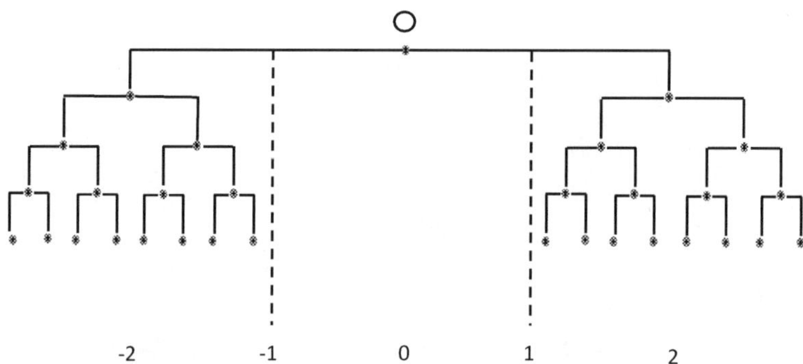

Figure 8.2: A quincunx whose first deflection has magnitude 2, and the sum of magnitudes of all subsequent deflections is 1. Then $P(-1 < S_n < 1) = 0$ for all n.

Example 8.11. CLT may not hold if one variable dominates Suppose that $U_1, U_2, \ldots$ are iid Bernoulli (1/2). To create a quincunx whose first row deflection has size 2, let X_1 be $4U_1 - 2$. We want row $i \geq 2$ to have deflection size $1/2^{i-1}$, so let $X_i = (2U_i - 1)/2^{i-1}$ for $i \geq 2$. Then the sum of magnitudes of all deflections from row 2 onward is $\sum_{i=2}^{\infty} |X_i| = \sum_{i=2}^{\infty} 1/2^{i-1} = 1$ (Figure 8.2). Because this is only half as large as the deflection in the first row, balls can never end up between -1 and 1. That is, $P\{S_n \in (-1, 1)\} = 0$ for every n. In this example, $\mathrm{E}(S_n) = 0$ and $\mathrm{var}(S_n) = v_n = 4 + \sum_{i=2}^{n} 1/2^{2i-2} \to 4 + 1/3 = 13/3$ as $n \to \infty$. If the CLT held, then $v_n^{-1/2} S_n$ would converge in distribution to $\mathrm{N}(0, 1)$. Slutsky's theorem would then imply that $S_n \xrightarrow{D} (13/3)^{1/2}\mathrm{N}(0, 1) = \mathrm{N}(0, 13/3)$. Clearly this cannot be the case when $P\{S_n \in (-1, 1)\}$ is exactly 0 for every n. Therefore, the CLT cannot hold; S_n is not asymptotically normal with mean 0 and variance $\mathrm{var}(S_n)$. □

In Example 8.11, we used intuition from a quincunx to concoct an example, but then we needed to show rigorously that $S_n/\{\mathrm{var}(S_n)\}^{1/2}$ was not asymptotically $\mathrm{N}(0,1)$. Quincunxes can provide the key idea, but rigor is required to prove normality or lack thereof.

In some applications of interest, the distribution of $X_1,\ldots,X_n$ changes with n. Consider Example 6.25 of the Bonferroni correction applied to n independent test statistics with continuous distributions. The indicators $X_1,\ldots,X_n$ of rejection for the n test statistics are iid Bernoullis with parameter $p_n=\alpha/n$. The Bernoulli probability parameter changes with n. This means that X_1 has one distribution when $n=5$ and another when $n=10$, for example. Therefore, we use a second subscript, writing $X_{n1},\ldots,X_{nn}$. The rest of this section considers a triangular array of random variables:

$$\begin{pmatrix} X_{11} & & & & \\ X_{21} & X_{22} & & & \\ \vdots & \vdots & & \vdots & \vdots \\ X_{n1} & X_{n2} & X_{n3} & \ldots & X_{nn} \end{pmatrix}. \tag{8.8}$$

Row 1 corresponds to $n=1$, row 2 to $n=2$, etc. The distribution of X_{nj} can change with n (i.e., can differ by row). Within a row, random variables are independent with $\mathrm{var}(X_{nj})=\sigma^2_{nj}$, $j=1,\ldots,n$. Without loss of generality, we assume that X_{nj} has mean 0 because we can always consider the distribution of $\sum_{i=1}^n\{X_{nj}-\mathrm{E}(X_{nj})\}$. We seek conditions under which $S_n/\{\mathrm{var}(S_n)\}^{1/2}\stackrel{D}{\rightarrow}N(0,1)$.

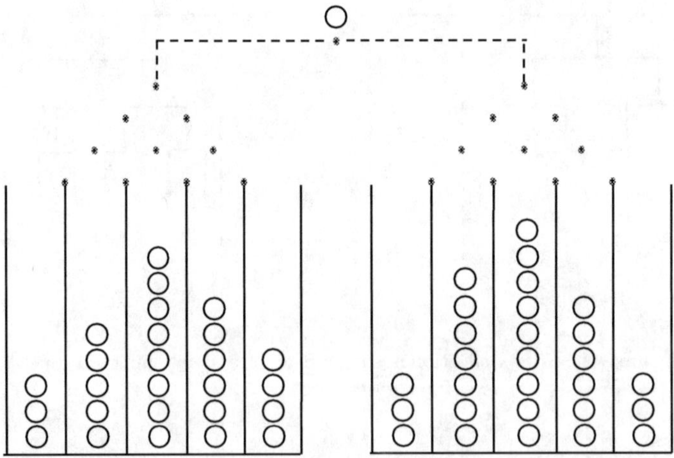

Figure 8.3: A quincunx whose first deflection is much larger than the subsequent 4, creating a bimodal distribution.

Allowing distributions to change with n makes it even easier to construct examples such that S_n is not asymptotically normal with mean 0 and variance $\mathrm{var}(S_n)$. Figure 8.3 shows a quincunx whose first deflection is much larger than the subsequent 4 deflections. The resulting distribution is bimodal. If we simultaneously add more rows and increase the size of the first deflection, the distribution of the sum will be a mixture of normals. If the deflection size in the first row is so large that there is complete separation between the two humps, the probability of being in a small interval about 0 is 0. The limiting random variable will also have probability 0 of being in a sufficiently small interval about 0, as in Example 8.11 and the following example.

Example 8.12. Another example of CLT not holding if one variable dominates Let $U_1, U_2, \ldots$ be iid Bernoulli random variables with parameter 1/2. We create a quincunx whose first row has a deflection size of n by defining $X_{n1} = (2n)U_1 - n$. Make subsequent deflections have size 1 by defining $X_{nj} = 2U_j - 1$, $j = 2, \ldots, n$. The variance of the sum $S_n = \sum_{j=1}^n X_{nj}$ is $n^2 + n - 1$. It follows that

$$\begin{aligned} \frac{S_n}{\sqrt{\text{var}(S_n)}} &= \frac{X_{n1}}{\sqrt{n^2+n-1}} + \frac{\sum_{j=2}^n X_{nj}}{\sqrt{n^2+n-1}} \\ &= Y_n + \left(\frac{\sum_{j=2}^n X_{nj}}{\sqrt{n-1}}\right)\sqrt{\frac{n-1}{n^2+n-1}}, \end{aligned} \tag{8.9}$$

where $Y_n = X_{n1}/(n^2+n-1)^{1/2}$ converges in distribution to a random variable Y taking values ± 1 with probability 1/2 each. Also, $\sum_{j=2}^n X_{nj}/(n-1)^{1/2} \xrightarrow{D} \text{N}(0,1)$ by the CLT, and $\{(n-1)/(n^2+n-1)\}^{1/2} \to 0$. It follows that the term to the right of Y_n in Expression (8.9) tends to 0 in probability. By Slutsky's theorem, $S_n/\{\text{var}(S_n)\}^{1/2} \xrightarrow{D} Y$. That is, the asymptotic distribution of $S_n/\{\text{var}(S_n)\}^{1/2}$ is a discrete distribution on only two values. Therefore, the CLT does not hold for X_{nj}. □

By making the deflection size in the first row of order $n^{1/2}$ instead of n, we can create a sum whose asymptotic distribution is a mixture of normals, as we see in the following example.

Example 8.13. When the limiting distribution of S_n is a mixture Again let $U_1, U_2, \ldots$ be iid Bernoulli random variables with parameter 1/2, but now define $X_{n1} = (2n^{1/2})U_1 - n^{1/2}$, $X_{nj} = 2U_j - 1$, $j = 2, \ldots, n$. This corresponds to a quincunx whose first deflection is only of size $n^{1/2}$ instead of n. Unlike in Example 8.12, the two humps of the distribution of S_n are no longer completely separated. Therefore, the asymptotic distribution of $S_n/\{\text{var}(S_n)\}^{1/2}$ no longer puts probability 0 on an interval about 0. Instead, it is a mixture of normals. More specifically, $S_n/\{\text{var}(S_n)\}^{1/2} \xrightarrow{D} Y$, where $F(y) = (1/2)\Phi(\sqrt{2}\,y + 1) + (1/2)\Phi(\sqrt{2}\,y - 1)$ (exercise). Therefore, the CLT does not hold for X_{nj}. □

One condition preventing scenarios like Examples 8.12 and 8.13 is the Lindeberg condition:

$$\left(\frac{1}{\text{var}(S_n)}\right)\sum_{i=1}^n \text{E}\left[X_{ni}^2 I\left\{X_{ni}^2 \geq \epsilon\,\text{var}(S_n)\right\}\right] \to 0 \quad \text{(Lindeberg condition)} \tag{8.10}$$

for each $\epsilon > 0$. The term $\text{E}[X_{ni}^2 I\{X_{ni}^2 \geq \epsilon\,\text{var}(S_n)\}]$ is essentially a tail variance for the random variable X_{ni}. Therefore, the Lindeberg condition says that the sum of tail variances must be a negligible fraction of the total variance.

We now show that the Lindeberg condition is not satisfied in Example 8.12, where $\text{var}(S_n) = n^2 + n - 1$. Take $\epsilon = 1/2$. The first term of the sum of Expression (8.10) is $\text{E}[X_{n1}^2 I\{X_{n1}^2 \geq (1/2)(n^2+n-1)\}]$. But $X_{n1}^2 = n^2$, which exceeds $(1/2)(n^2+n-1)$ for all $n = 1, 2, \ldots$ Therefore, the first term of the sum of Expression (8.10) is n^2. Each subsequent term $\text{E}[X_{ni}^2 I\{X_{ni}^2 \geq (1/2)(n^2+n-1)\}]$ is 0 for $n \geq 2$ because $X_{ni}^2 = 1 < (1/2)(n^2+n-1)$ for $n \geq 2$ and $i = 2, 3, \ldots, n$. Thus, the left side of Expression (8.10) is $n^2/(n^2+n-1) \to 1$. Therefore, the Lindeberg condition is not satisfied. In fact, if the Lindeberg condition were satisfied, then a CLT would hold:

Theorem 8.14. Lindeberg's CLT *If the Lindeberg condition (8.10) holds for the triangular array (8.8) of independent random variables with mean 0 and variances σ_{ni}^2, then $S_n/\{\text{var}(S_n)\}^{1/2} \xrightarrow{D} N(0,1)$ as $n \to \infty$.*

The proof may be found in Billingsley (2012). We illustrate through several examples how to apply the theorem.

Example 8.15. CLT for Bernoulli random variables For each n, let $U_{n1}, \ldots, U_{nn}$ be iid Bernoulli random variables with parameter p_n such that $np_n(1-p_n) \to \infty$. Let $X_{ni} = U_{ni} - p_n$, so that the X_{ni} have mean 0. We will show that the Lindeberg condition is satisfied. Note that $\text{var}(S_n) = n\,\text{var}(X_{n1}) = n\,\text{var}(U_{n1}) = np_n(1-p_n) \to \infty$. Consequently, $\epsilon\,\text{var}(S_n) \to \infty$ for each $\epsilon > 0$. Whenever n is large enough that $\epsilon\,\text{var}(S_n) > 1$, $X_{ni}^2 I\{X_{ni}^2 \geq \epsilon\,\text{var}(S_n)\}$ is 0 because $X_{ni}^2 \leq 1$ for all n and i. Therefore, there exists an N such that each term of the sum in Expression (8.10) is 0 for $n \geq N$. This implies that the sum is 0 for $n \geq N$, so the limit of Expression (8.10) is 0. That is, the Lindeberg condition is satisfied. By Theorem 8.14, $\sum_{i=1}^n X_{ni}/\{\text{var}(S_n)\}^{1/2} \xrightarrow{D} \text{N}(0,1)$. □

Example 8.16. Bernoulli random variables with S_n asymptotically Poisson For each n, let $U_{n1}, \ldots, U_{nn}$ be iid Bernoulli random variables with parameter p_n such that $np_n \to \lambda < \infty$. By the law of small numbers (Proposition 6.24), the binomial (n, p_n) distribution of $S_n = \sum_{i=1}^n U_{ni}$ is asymptotically Poisson (λ). If we center the Bernoullis by subtracting p_n from each, then Slutsky's theorem implies that $(S_n - np_n)/\{np_n(1-p_n)\}^{1/2} \xrightarrow{D} (X-\lambda)/\lambda^{1/2}$, where X is Poisson (λ). Given that this is not a normal distribution, the Lindeberg condition cannot be satisfied. It is an exercise to verify directly that the Lindeberg condition is not satisfied. □

Example 8.16 is noteworthy because no single random variable dominates the others the way the first random variable did in Example 8.12. One condition ensuring that no variable dominates is the *uniform asymptotic negligibility condition*

$$\max_{1 \leq i \leq n} P\left(|X_{ni}|/\{\text{var}(S_n)\}^{1/2} > \epsilon\right) \to 0 \text{ as } n \to \infty. \tag{8.11}$$

We have seen that the Lindeberg condition is sufficient for S_n to be asymptotically $\text{N}(0, \text{var}(S_n))$. When (8.11) holds, it is necessary as well:

Theorem 8.17. Feller's theorem *Suppose that X_{ni} are independent with mean 0 and satisfy the uniform asymptotic negigibility condition (8.11). If $S_n/\{\text{var}(S_n)\}^{1/2} \xrightarrow{D} \text{N}(0,1)$, then the Lindeberg condition (8.10) is satisfied.*

Remark 8.18. Lindeberg-Feller *Combining the Lindeberg and Feller theorems, we see that under the uniform asymptotic negligibility condition (8.11), $S_n/\{\text{var}(S_n)\} \xrightarrow{D} \text{N}(0,1)$ if and only if the Lindeberg condition (8.10) is satisfied. This is sometimes called the Lindeberg-Feller theorem.*

In Example 8.15, it was easy to verify the Lindeberg condition because the sum in Expression (8.10) became 0 for n sufficiently large. Another setting in which the Lindeberg condition is easy to verify is when higher moments are available and Lyapounov's condition is satisfied:

$$\frac{\sum_{i=1}^n \text{E}(|X_{ni}|^r)}{\left(\sqrt{\text{var}(S_n)}\right)^r} \to 0 \text{ for some } r > 2 \text{ (\textit{Lyapounov Condition})} \tag{8.12}$$

Theorem 8.19. *Lyapounov's condition (8.12) implies the Lindeberg condition (8.10).*

Proof (exercise).

Our final example requires slightly more work to verify the Lindeberg condition.

Example 8.20. Asymptotic normality of permutation distribution in paired setting Consider the permutation test setting of Example 8.10. The permutation distribution of $\sum_{i=1}^n D_i$ is obtained by fixing d_i and defining $X_i = \pm d_i$, with probability 1/2 each; the distribution of $\sum_{i=1}^n X_i$ is the permutation distribution of $\sum_{i=1}^n D_i$. We will show that, for almost all ω, the permutation distribution is asymptotically normal by verifying that the Lindeberg condition holds with probability 1.

Note that $\mathrm{var}(X_i) = d_i^2$ and $\Gamma_n = \mathrm{var}(S_n) = \sum_{i=1}^n d_i^2$. Also,

$$(1/\Gamma_n)\sum_{j=1}^n \mathrm{E}\{X_j^2 I(X_j^2 \geq \epsilon\Gamma_n)\} = (1/\Gamma_n)\sum_{j=1}^n d_j^2 I\left(d_j^2 \geq \epsilon\sum_{i=1}^n d_i^2\right). \tag{8.13}$$

Let L_n denote the expression on the right side of Equation (8.13). We must prove that $L_n \to 0$ as $n \to \infty$.

Before we fixed $D_1 = d_1, D_2 = d_2, \ldots$, they were iid random variables. By the SLLN, $(1/n)\sum_{i=1}^n D_i^2 \to \mathrm{E}(D_i^2) > 0$ for almost all ω as $n \to \infty$, so $\sum_{i=1}^n D_i^2 \to \infty$. Thus, for the fixed numbers $d_1, d_2, \ldots$, $\sum_{i=1}^n d_i^2 \to \infty$ as $n \to \infty$. This means that for any A we can determine an integer N_A such that

$$\sum_{i=1}^n d_i^2 \geq A \text{ for } n \geq N_A. \tag{8.14}$$

It follows that

$$\begin{aligned} L_n &\leq (1/\Gamma_n)\sum_{j=1}^n d_j^2 I(d_j^2 \geq \epsilon A) \\ &= \frac{(1/n)\sum_{j=1}^n d_j^2 I(d_j^2 \geq \epsilon A)}{(1/n)\sum_{j=1}^n d_j^2} \text{ for } n \geq N_A. \end{aligned} \tag{8.15}$$

Apply the SLLN separately to the numerator and denominator: for ω in a set of probability 1, the observed sequence $d_j = D_j(\omega)$ is such that Expression (8.15) converges to $\mathrm{E}\{D_1^2 I(D_1^2 \geq \epsilon A)\}/\mathrm{E}(D_1^2)$ as $n \to \infty$. We conclude that $\overline{\lim}_{n\to\infty} L_n \leq \mathrm{E}\{D_1^2 I(D_1^2 \geq \epsilon A)\}/\mathrm{E}(D_1^2)$. This holds for arbitrarily large A, and this expression tends to 0 as $A \to \infty$. Because $\mathrm{E}(D^2) < \infty$, $\overline{\lim}_{n\to\infty} L_n = 0$ and the Lindeberg condition is satisfied.

We conclude that, for almost all ω (equivalently, for almost all sequences $d_1, d_2, \ldots$), the permutation distribution of $\sum_{j=1}^n D_j$ is asymptotically normal with mean 0 and variance $\sum_{j=1}^n d_j^2$. In other words, the one-sided permutation test is asymptotically equivalent to rejecting the null hypothesis if $Z_n > z_\alpha$, where

$$Z_n = \frac{\sum_{j=1}^n D_j}{\sqrt{\sum_{j=1}^n D_j^2}} \tag{8.16}$$

and z_α is the $(1-\alpha)$th quantile of the standard normal distribution. Note that Z_n is very closely related to the usual t-statistic, except that the variance estimate is $n^{-1}\sum_{j=1}^n D_j^2$ instead of $(n-1)\sum_{j=1}^n (D_j - \bar{D})^2$. It is an exercise to show that these two variance estimates are asymptotically equivalent under the null hypothesis. □

We end this section with an important theorem on the rate of convergence of the normalized statistic $(S_n - n\mu)/(n\sigma^2)^{1/2}$ to the standard normal distribution.

Theorem 8.21. Berry-Esseen: rate of convergence to normal *Let X_i be iid with mean μ and variance σ^2, and set $S_n = \sum_{i=1}^n X_i$. Let $F_n(z)$ be the distribution function for $Z_n = (S_n - n\mu)/(n\sigma^2)^{1/2}$, and $\Phi(z)$ be the standard normal distribution function. There is a universal constant C such that*

$$\sup_z |F_n(z) - \Phi(z)| \leq \frac{C\,\mathrm{E}(|X_1 - \mu|^3}{\sigma^3 n^{1/2}}$$

for all $n = 1, 2, \ldots$. Here, "universal" means that the same C can be used regardless of the distribution of the X_i (subject only to having mean μ, variance σ^2, and third absolute moment $\mathrm{E}(|X_1 - \mu|^3)$.

Note that the inequality in the Berry-Esseen Theorem holds vacuously if $\mathrm{E}(|X_1 - \mu|^3) = \infty$.

Exercises

1. Imagine infinitely many quincunxes, one with a single row, another with two rows, another with three rows, etc. Roll one ball on each quincunx. What is the probability that the ball is in the rightmost bin of infinitely many of the quincunxes?

2. Let X_1 have distribution F with mean 0 and variance 1, and $X_2, X_3, \ldots$ be iid with point mass distributions at 0. That is, $X_i \equiv 0$ with probability 1 for $i \geq 2$. What is the asymptotic distribution of $S_n = \sum_{i=1}^n X_i$? Does the CLT hold?

3. In Example 8.16, prove directly that the Lindeberg condition does not hold.

4. Let τ_{ni} be independent Bernoullis with probability p_n, $1 \leq i \leq n$, and let $X_{ni} = \tau_{ni} - p_n$. Prove that if $\epsilon < p_n < 1 - \epsilon$ for all n, where $\epsilon > 0$, then X_{ni} satisfies Lyapounov's condition with $r = 3$.

5. Let X_{ni}, $1 \leq i \leq n$, be independent and uniformly distributed on $(-a_n, a_n)$, $a_n > 0$. Prove that Lyapounov's condition holds for $r = 3$.

6. Let Y_i be iid random variables taking values ± 1 with probability 1/2 each. Prove that the random variables $X_{ni} = (i/n)Y_i$ satisfy Lyapounov's condition with $r = 3$.

7. Prove that the Lindeberg CLT (Theorem 8.14) implies the standard CLT (Theorem 8.1).

8. What is the asymptotic distribution of S_n in Example 8.11? Hint: recall that $\sum_{i=2}^\infty U_i/2^i$ is the base 2 representation of a number picked randomly from $[0, 1]$.

9. Let D_n be iid from a distribution with mean 0 and finite variance σ^2, and let T_n be the usual one-sample t-statistic

$$T_n = \frac{\sum_{i=1}^n D_i}{\sqrt{\left(\frac{n}{n-1}\right)\{\sum_{i=1}^n D_i^2 - (\sum_{i=1}^n D_i)^2/n\}}}.$$

With Z_n defined by Equation (8.16), prove that, $T_n - Z_n \xrightarrow{p} 0$ under the null hypothesis that $\mathrm{E}(D_i) = 0$. What does this say about how the t-test and permutation test compare under the null hypothesis if n is large?

10. ↑ Consider the context of the preceding problem. To simulate what might happen to the t-test if there is an outlier, replace the nth observation by $n^{1/2}$.

    (a) Prove that $T_n \stackrel{D}{\rightarrow} \mathrm{N}(1/\sqrt{2}, 1/2)$.

    (b) If n is very large, what is the approximate type I error rate for a one-tailed test rejecting for T_n exceeding the $(1-\alpha)$th quantile of a t-distribution with $n-1$ degrees of freedom?

Part 2: CLT Proof

The rest of the chapter is devoted to proving the standard CLT using a very helpful tool called the characteristic function. We motivate the characteristic function by first presenting harmonic regression, a useful technique for analyzing periodic data. The reader will then see that characteristic functions may be viewed as an extension of harmonic regression. Section 8.3 gives an informal, heuristic motivation. Section 8.4 gives a more rigorous presentation.

8.3 Harmonic Regression

Medical researchers sometimes study circadian rhythms of continuous biological phenomena like blood pressure. The word "circadian" comes from "circa dia," meaning "around the day," and describes the tendency of the biological measurement $Y = Y(x)$ to vary systematically as a function of time of day, x. For instance, Figure 8.4 shows the average systolic blood pressure $Y(x)$ of 40 patients at hour x, where $x = 0$ is midnight. Note that $Y(x)$ is lowest in the middle of the night when participants are sleeping. Time is circular rather than linear in the sense that $x + 24$ is the same time of day as x. Therefore, when we model Y as a function of x, we want to use a function that is periodic with a period of 24 hours. We can re-parameterize so that x represents values in the unit circle $[0, 2\pi]$. Then time $x + 2\pi$ is the same as time x. Assume that data are equally spaced in time. This is reasonable because ambulatory blood pressure monitoring (ABPM) machines can take blood pressure readings every hour, every half hour, every quarter hour, etc. Assume there are n equally-spaced measurements over the course of the day, where n is a multiple of 24. Then $\mathbf{x} = (x_1, \ldots, x_n)'$, where $x_j = 2\pi j/n$.

The simplest possible model for $\mathrm{E}\{Y(x)\}$ is $\mathrm{E}\{Y(x)\} = \mu$, meaning that the expected blood pressure is constant over time. The best estimate of μ is just $\bar{Y} = 105.052$, the sample mean blood pressure. A cursory glance at Figure 8.4 shows that this model is not appropriate. It is clear that blood pressure is not constant over time.

A logical next attempt adds the first harmonic to the model:

$$\mathrm{E}\{Y(x)\} = \mu + \lambda \cos(x - \theta) \tag{8.17}$$

for constants μ, λ and θ. If $\lambda \geq 0$, then θ corresponds to the time when blood pressure is highest. Equation (8.17) is linear in μ and λ, but not θ. We can re-parameterize Equation (8.17) to make it linear in all coefficients using the formula for the cosine of a difference. This leads to

$$\mathrm{E}\{Y(x)\} = \mu + \alpha_1 \cos(x) + \beta_1 \sin(x), \tag{8.18}$$

where $\alpha_1 = \lambda \cos(\theta)$ and $\beta_1 = \lambda \sin(\theta)$. Thus, the first harmonic includes both the $\cos(x)$ and $\sin(x)$ term. Equation (8.18) is linear in the coefficients μ, α_1 and β_1. This simple

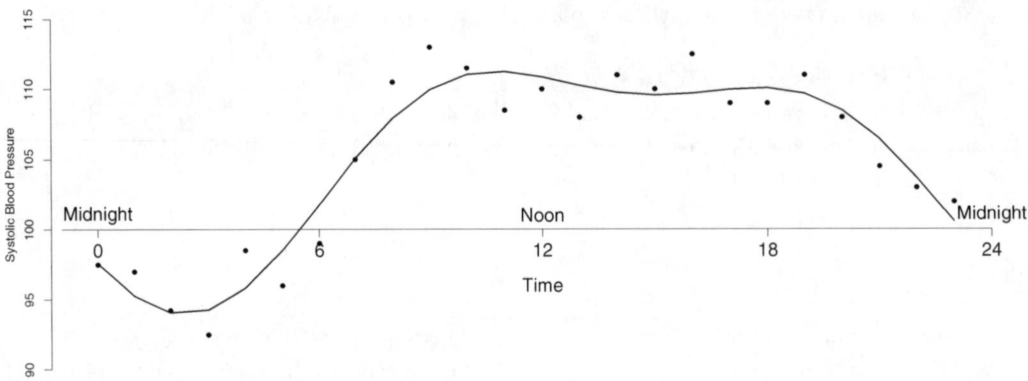

Figure 8.4: Circadian variation in systolic blood pressure.

model accommodates a single peak and single trough in the interval $[0, 2\pi]$. Figure 8.5 shows that the first harmonic model applied to the data in Figure 8.4 does not seem to fit. For example, it does not seem to represent what is happening between hours 0 and 6.

To better fit the data, we can add a second harmonic consisting of $\cos(2x)$ and $\sin(2x)$:

$$\mathrm{E}\{Y(x)\} = \mu + \alpha_1 \cos(x) + \beta_1 \sin(x) + \alpha_2 \cos(2x) + \beta_2 \sin(2x). \tag{8.19}$$

In matrix notation,

$$\begin{pmatrix} \mathrm{E}\{Y(x_1)\} \\ \mathrm{E}\{Y(x_2)\} \\ \vdots \\ \mathrm{E}\{Y(x_n)\} \end{pmatrix} = \begin{pmatrix} 1 & \cos(x_1) & \sin(x_1) & \cos(2x_1) & \sin(2x_1) \\ 1 & \cos(x_2) & \sin(x_2) & \cos(2x_2) & \sin(2x_2) \\ \vdots & \vdots & \vdots & \vdots & \vdots \\ 1 & \cos(x_n) & \sin(x_n) & \cos(2x_n) & \sin(2x_n) \end{pmatrix} \begin{pmatrix} \mu \\ \alpha_1 \\ \beta_1 \\ \alpha_2 \\ \beta_2 \end{pmatrix} \tag{8.20}$$

This model allows up to two peaks and two troughs, and fits the data quite well. In fact, the curve in Figure 8.4 uses a two harmonic model.

Table 8.1: Summary of harmonic regression applied to systolic blood pressure.

Terms	0 Harmonic Model	1 Harmonic Model	2 Harmonic Model
Constant	105.0521	105.0521	105.0521
$\cos(x)$		-6.6431	-6.6431
$\sin(x)$		-4.1638	-4.1638
$\cos(2x)$			-0.8448
$\sin(2x)$			-3.1463

Table 8.1 summarizes the results of our harmonic regression. Notice that the coefficients for a given term do not depend on whether other terms are included in the model. For example, the constant term remains 105.0521 whether 0, 1, or 2 harmonics are included in

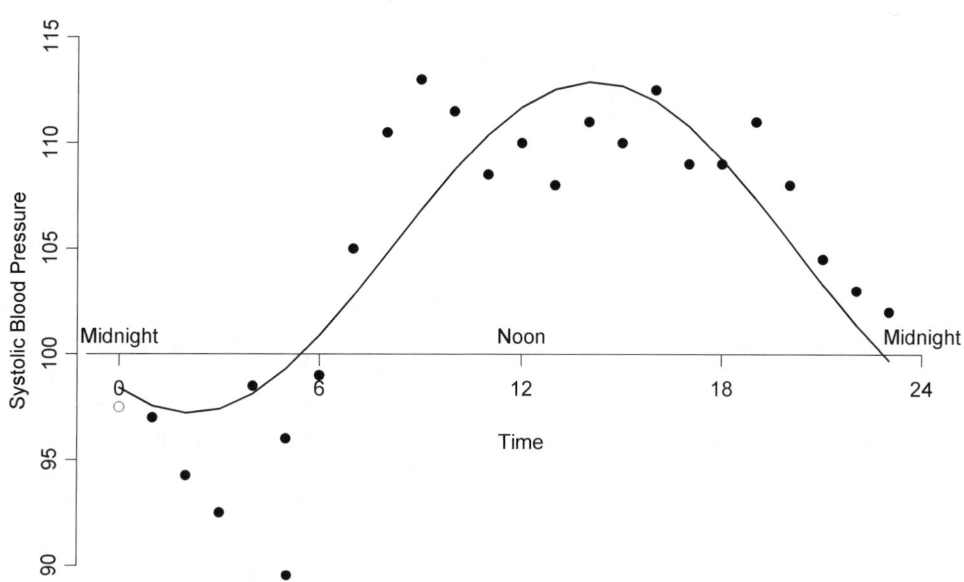

Figure 8.5: First harmonic does not fit the blood pressure data well.

the model; the coefficient for $\cos(x)$ is -6.6431 whether 1 or 2 harmonics are included in the model. This is because the predictor vectors—the columns of the design matrix in the middle of Equation (8.20)—are all orthogonal.

If the two harmonic model does not fit, we can add more harmonics. In fact, we can overfit the data using a saturated model with as many parameters as there are observations:

$$\mathrm{E}\{Y(x_i)\} = \mu + \sum_{k=1}^{n/2} \alpha_k \cos(kx_i) + \sum_{k=1}^{n/2-1} \beta_k \sin(kx_i). \tag{8.21}$$

The regression coefficient for each term of the saturated model is the same as it would be in univariate regression. For $0 < k < n/2$,

$$\begin{aligned} \hat{\alpha}_k &= \frac{\sum_{i=1}^n Y_i \cos(kx_i)}{\sum_{i=1}^n \cos^2(kx_i)} = \frac{\sum_{i=1}^n Y_i \cos(kx_i)}{n/2} \\ \hat{\beta}_k &= \frac{\sum_{i=1}^n Y_i \sin(kx_i)}{\sum_{i=1}^n \sin^2(kx_i)} = \frac{\sum_{i=1}^n Y_i \sin(kx_i)}{n/2} \end{aligned} \tag{8.22}$$

Note that the denominators are $n/2$ because $\sum_{i=1}^n \cos^2(kx_i) = \sum_{i=1}^n \sin^2(kx_i)$ and $\sum_{i=1}^n \cos^2(kx_i) + \sum_{i=1}^n \sin^2(kx_i) = \sum_{i=1}^n \{\cos^2(kx_i) + \sin^2(kx_i)\} = \sum_{i=1}^n (1) = n$.

When we write the regression model in matrix notation as in Equation (8.20), we see that the n vectors of the design matrix form an orthogonal basis for R^n, so Equation (8.21) reproduces the Ys without error. But the Ys are completely arbitrary; they could represent values of an arbitrary function $f(x)$ with domain the n equally spaced values $x_i = 2\pi i/n$. In other words, for any such function $f(x)$, we have

$$f(x_i) = \hat{\mu} + \sum_{k=1}^{n/2} \hat{\alpha}_k \cos(kx_i) + \sum_{k=1}^{n/2-1} \hat{\beta}_k \sin(kx_i) \tag{8.23}$$

$$\hat{\mu} = (1/n)\sum_{i=1}^{n} f(x_i); \quad \hat{\alpha}_k = (2/n)\sum_{i=1}^{n} f(x_i)\cos(kx_i); \quad \hat{\beta}_k = (2/n)\sum_{i=1}^{n} f(x_i)\sin(kx_i) \tag{8.24}$$

for $k < n/2$. For $k = n/2$, replace $(2/n)$ of $\hat{\alpha}_k$ with $(1/n)$.

Now suppose that $f(x)$ is the probability mass function of a random variable X taking on n possible, equally-spaced values $x_i = 2\pi i/n$, $i = 1, \ldots, n$. Then $\hat{\mu} = 1/n$, $\hat{\alpha}_k = (2/n)\mathrm{E}\{\cos(kX)\}$ and $\hat{\beta}_k = (2/n)\mathrm{E}\{\sin(kX)\}$. Once we know $\mathrm{E}\{\cos(kX)\}$ and $\mathrm{E}\{\sin(kX)\}$ for all $k = 1, \ldots, n/2$, we can reproduce the probability mass function f, and therefore the distribution function of X. The same argument shows that for any discrete random variable taking on n equally-spaced values (not necessarily restricted to the interval $[-2\pi, 2\pi]$), knowledge of $\mathrm{E}\{\cos(tX)\}$ and $\mathrm{E}\{\sin(tX)\}$ for sufficiently many values of t reproduces the distribution function $F(x)$ of X. As n increases, the number of t for which we need to know $\mathrm{E}\{\cos(tX)\}$ and $\mathrm{E}\{\sin(tX)\}$ to reproduce F also increases.

8.4 Characteristic Functions

The previous section motivates the following definition of a characteristic function.

Definition 8.22. Characteristic function *The characteristic function $\psi(t)$ for an arbitrary random variable X is defined to be*

$$\psi(t) = \mathrm{E}\{\cos(tX)\} + i\,\mathrm{E}\{\sin(tX)\},$$

where i is the imaginary number $\sqrt{-1}$.

The key result that we have motivated through harmonic regression is that the characteristic function, as its name implies, characterizes a distribution function.

Theorem 8.23. The characteristic function characterizes a distribution function *The characteristic function $\psi(t)$ uniquely determines the distribution function $F(x)$ of the random variable X. That is, if F_1 and F_2 are distribution functions with the same characteristic function $\psi(t)$, then $F_1 \equiv F_2$.*

Before studying other properties of characteristic functions, we briefly review some important properties of complex numbers. Topologically, the complex plane is identical to the plane R^2: each complex number $z = a + bi$ may be viewed as a vector (a, b) with *length* (also called *magnitude* or *norm*) defined by Euclidean length $||z|| = (a^2 + b^2)^{1/2}$. Immediate consequences of this geometric way of viewing complex numbers are the following.

Proposition 8.24. Elementary inequalities for complex numbers

1. *If $z = a + bi$, then $\max(|a|, |b|) \leq ||z|| \leq |a| + |b|$.*

2. *Triangle inequality for complex numbers: If z_j, $j = 1, \ldots, n$ are complex numbers, then $||\sum_{j=1}^{n} z_j|| \leq \sum_{j=1}^{n} ||z_j||$.*

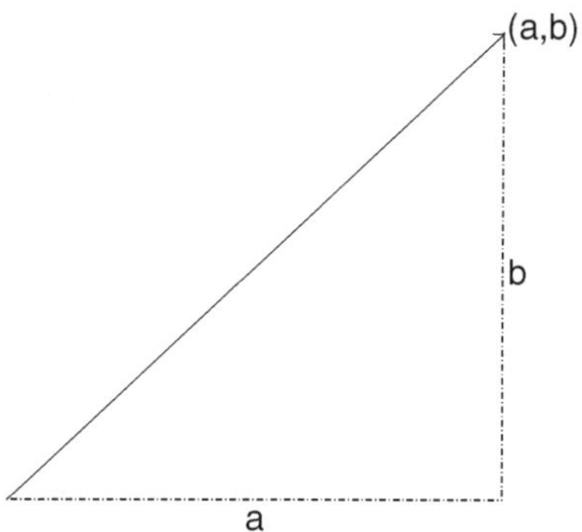

Figure 8.6: Part 1 of Proposition 8.24 says that the length of the hypotenuse in a right triangle is between the length of the longer leg and the sum of lengths of the two legs.

It is expedient to express the characteristic function in an alternative way using the exponential function for complex arguments. This involves a power series, so we first define the infinite sum $\sum_{n=0}^{\infty} z_n$ of complex numbers $z_n = a_n + b_n i$ in the obvious way as $\sum_{n=0}^{\infty} a_n + i\sum_{n=0}^{\infty} b_n$, provided that both of these sums exist. The following is the complex analog of Proposition A.44 for real sums.

Proposition 8.25. Infinite series with complex coefficients: convergence and rearrangement of terms *If* $\sum_{n=0}^{\infty} ||z_n|| < \infty$, *then* $\sum_{n=0}^{\infty} z_n$ *exists, is finite, and has the same value for any rearrangement of terms.*

Proof. If $z_n = a_n + b_n i$, then $\sum_{n=0}^{\infty} |a_n| \le \sum_{n=0}^{\infty} ||z_n||$ and $\sum_{n=0}^{\infty} |b_n| \le \sum_{n=0}^{\infty} ||z_n||$ by part 1 of Proposition 8.24, so the real sums $\sum_{n=0}^{\infty} a_n$ and $\sum_{n=0}^{\infty} b_n$ converge absolutely. By Propositions A.43 and A.44, these sums converge and are invariant to rearrangement of terms. Therefore, $\sum_{n=0}^{\infty} z_n$ is finite and invariant to rearrangement of terms. □

We are now in a position to extend power series to complex arguments. Any real power series $\sum_{n=0}^{\infty} a_n x^n$ converging absolutely for $|x| < r$ has a complex analog $\sum_{n=0}^{\infty} a_n z^n$ converging for $||z|| < r$. This follows from part 1 of Proposition 8.24 because, with Re denoting the real part of a complex number, $\sum_{n=0}^{\infty} |Re(a_n z^n)| \le \sum_{n=0}^{\infty} ||a_n z^n|| = \sum_{n=0}^{\infty} |a_n| \, ||z||^n < \infty$ by assumption if $||z|| < r$, and similarly for the imaginary part.

Definition 8.26. Power series for exponential functions *For complex argument* z, *the exponential function is defined by* $\exp(z) = \sum_{n=0}^{\infty} z^n/n!$.

The series in Definition 8.26 converges for all complex numbers z by Proposition 8.25 because $\sum_{n=0}^{\infty} ||z||^n/n!$ is the (convergent) Taylor series for the function $\exp(||z||)$ of the real-valued argument $||z||$.

We can now express the characteristic function in terms of the exponential function:

Proposition 8.27. *If t is a real number, then*

$$\begin{aligned}\exp(it) &= \sum_{k=0}^{\infty}(it)^k/k! \\ &= (1 - t^2/2! + t^4/4! \ldots) + i(t^1/1! - t^3/3! + t^5/5! \ldots) \\ &= \cos(t) + i\sin(t).\end{aligned}$$

Therefore, the characteristic function may also be written as $\mathrm{E}\{\exp(itX)\}$.

One reason that it is convenient to write the characteristic function as an exponential function is that the property $\exp(x_1 + x_2) = \exp(x_1)\exp(x_2)$ for real arguments holds for complex arguments as well:

Proposition 8.28. Product rule extends to exponential functions with complex arguments *For complex numbers z_1 and z_2, $\exp(z_1 + z_2) = \exp(z_1)\exp(z_2)$.*

Proof.

$$\begin{aligned}\exp(z_1 + z_2) &= \sum_{k=0}^{\infty}\frac{(z_1+z_2)^k}{k!} = \sum_{k=0}^{\infty}\left\{\sum_{j=0}^{k}\binom{k}{j}\frac{z_1^j z_2^{k-j}}{k!}\right\} \\ &= \sum_{k=0}^{\infty}\left\{\sum_{j=0}^{k}\frac{z_1^j z_2^{k-j}}{j!(k-j)!}\right\} = \sum_{j=0}^{\infty}\left\{\sum_{k=j}^{\infty}\frac{z_1^j z_2^{k-j}}{j!(k-j)!}\right\} \\ &= \sum_{j=0}^{\infty}\frac{z_1^j}{j!}\left\{\sum_{m=0}^{\infty}\frac{z_2^m}{m!}\right\} = \exp(z_1)\exp(z_2). \qquad (8.25)\end{aligned}$$

The reversal of order of summation in the second line is justified either using Tonelli's theorem or as follows. The double sum may be viewed as a single sum of countably many terms. The norm of the summand is $||z_1^j z_2^{k-j}/\{j!(k-j)!\}|| = ||z_1||^j||z_2||^{k-j}/\{j!(k-j)!\}$, a nonnegative sequence of real numbers whose sum is invariant to order of summation. Going backwards from line 2 to line 1 but replacing the summand by its norm, we find that $\sum_{k=0}^{\infty}\sum_{j=0}^{k}||z_1||^j||z_2||^{k-j}/\{j!(k-j)!\}|| = \exp(||z_1|| + ||z_2||) < \infty$. The reversal of order of summation in line 2 is now justified by Proposition 8.25. □

Proposition 8.29. Basic Properties of ch.f.s *Suppose that $\psi(t)$ is the ch.f. of a random variable X. Then:*

1. $||\psi(t)|| \leq 1 = \psi(0)$.
2. *If a and b are constants, the ch.f. of $aX + b$ is $e^{itb}\psi(at)$.*
3. *If $X \equiv c$ for some constant c, then $\psi(t) = e^{itc}$.*
4. *$\psi_{-X}(t)$ is the complex conjugate of $\psi_X(t)$.*
5. *X is symmetric about 0 (i.e., $-X$ and X have the same distribution functions) if and only if $\psi_X(t)$ is real.*
6. *$\psi(t)$ is a uniformly continuous function of t.*

Items 2–4 are immediate consequences of the definition of characteristic functions. For example, for item 1,

$$\begin{aligned}
||\psi(t)|| &= \left|\left|\mathrm{E}\{\cos(tX)\} + i\mathrm{E}\{\sin(tX)\}\right|\right| \\
&= \sqrt{[\mathrm{E}\{\cos(tX)\}]^2 + [\mathrm{E}\{\sin(tX)\}]^2} \\
&\leq \sqrt{\mathrm{E}\{\cos^2(tX)\} + \mathrm{E}\{\sin^2(tX)\}} = \sqrt{\mathrm{E}\{\cos^2(tX) + \sin^2(tX)\}} \\
&= \sqrt{\mathrm{E}(1)} = 1 = \mathrm{E}\{\exp(it0)\} = \psi(0).
\end{aligned}$$

Going from step 2 to step 3 follows from the fact that $\mathrm{E}(Y^2) \geq \{\mathrm{E}(Y)\}^2$ because $\mathrm{E}(Y^2) - \{\mathrm{E}(Y)\}^2 = \mathrm{var}(Y) \geq 0$ (alternatively, we could apply Jensen's inequality).

Item 5 is proven as follows. If $-X$ and X have the same distribution functions, then $\mathrm{E}\{\sin(tX)\} = \mathrm{E}[\sin\{t(-X)\}] = -\mathrm{E}\{\sin(tX)\}$, which implies that $\mathrm{E}\{\sin(tX)\}$ must be 0. Thus, if $-X$ and X have the same distribution functions, then the imaginary part of $\psi_X(t)$ is 0 (i.e., $\psi_X(t)$ is real). To go the opposite direction, suppose that $\psi_X(t)$ is real. Note that $\mathrm{E}[\cos\{t(-X)\}] = \mathrm{E}\{\cos(tX)\}$ because $\cos(-\theta) = \cos(\theta)$ for all θ. Therefore, the real parts of the ch.f.s of X and $-X$ coincide. The fact that $\psi_X(t)$ is real means that $\mathrm{E}\{\sin(tX)\} = 0$, so $\mathrm{E}[\sin\{t(-X)\}] = -\mathrm{E}\{\sin(tX)\} = 0$. Therefore, the imaginary parts of the ch.f.s of $-X$ and X coincide as well. We have shown that $-X$ and X have the same ch.f.s. By Theorem 8.23, $-X$ and X have the same distribution functions.

Item 6 is proven as follows:

$$\begin{aligned}
||\psi(t+\Delta) - \psi(t)|| &= \left|\left|\int [\exp\{i(t+\Delta)x\} - \exp(itx)]dF(x)\right|\right| \\
&= \left|\left|\int \exp(itx)[\exp(i\Delta x) - 1]dF(x)\right|\right| \\
&\leq \int \left|\left|\exp(itx)\right|\right|\left|\left|\exp(i\Delta x) - 1\right|\right| dF(x) \\
&= \int \left|\left|\exp(i\Delta x) - 1\right|\right| dF(x) \\
&= \int \left|\left|\cos(\Delta x) - 1 + i\sin(\Delta x)\right|\right| dF(x) \\
&\leq \int \left|\cos(\Delta x) - 1\right| dF(x) + \int \left|\sin(\Delta x)\right| dF(x). \qquad (8.26)
\end{aligned}$$

As $\Delta \to 0$, $|\cos(\Delta x) - 1| \to 0$, and is bounded by 2. Therefore, by the BCT (Theorem 5.13), the left term of Expression (8.26) approaches 0. A similar argument shows that the right term also approaches 0. Also, as Expression (8.26) does not depend on t, the convergence of $\psi(t+\Delta)$ to $\psi(t)$ as $\Delta \to 0$ is uniform in t.

Table 8.2 shows the characteristic function for some common distributions.

A very important property of characteristic functions is the following.

Proposition 8.30. Product rule for the characteristic function of the sum of independent random variables *If X_1 and X_2 are independent random variables with respective characteristic functions ψ_1 and ψ_2, then the characteristic function for $X_1 + X_2$*

Table 8.2: The characteristic functions of random variables with different density or mass functions.

Distribution	Density/mass function $f(x)$	Characteristic function
Binomial	$\binom{n}{x}p^x(1-p)^{n-x}$	$(1-p+pe^{it})^n$
Bernoulli	$p^x(1-p)^{1-x}$	$1-p+pe^{it}$
Poisson	$\exp(-\lambda)\lambda^x/x!$	$\exp\{\lambda(e^{it}-1)\}$
Geometric	$p(1-p)^x,\ x=0,1,2\ldots$	$\frac{p}{1-(1-p)e^{it}}$
Uniform	$(b-a)^{-1}I(a\le x\le b)$	$\frac{e^{ibt}-e^{iat}}{(b-a)it}$
Normal	$(2\sigma^2)^{-1/2}\exp\left\{\frac{-(x-\mu)^2}{2\sigma^2}\right\}$	$\exp(i\mu t-\sigma^2t^2/2)$
Cauchy	$\frac{1}{\pi\lambda[1+\{(x-\theta)/\lambda\}^2]}$	$\exp(it\theta-\lvert t\rvert\lambda)$
Gamma	$\frac{\theta^r x^{r-1}\exp(-\theta x)}{\Gamma(r)}$	$\frac{1}{(1-it/\theta)^r}$
Chi-squared	$\frac{x^{k/2-1}\exp(-x/2)}{\{2^{k/2}\Gamma(k/2)\}}$	$\frac{1}{(1-2it)^{k/2}}$
Exponential	$\theta\exp(-\theta x)$	$\frac{1}{1-it/\theta}$

is $\psi_1(t)\psi_2(t)$. Likewise, the characteristic function of the sum of n independent random variables with respective ch.f.s $\psi_1,\ldots,\psi_n$ is $\prod_{i=1}^n \psi_i(t)$.

Proof. The ch.f. of X_1+X_2 is

$$
\begin{aligned}
& \mathrm{E}\{\cos(tX_1)\cos(tX_2)-\sin(tX_1)\sin(tX_2)\} \\
& +i\mathrm{E}\{\sin(tX_1)\cos(tX_2)+\cos(tX_1)\sin(tX_2)\} \\
= {} & \mathrm{E}\{\cos(tX_1)\}\mathrm{E}\{\cos(tX_2)\}-\mathrm{E}\{\sin(tX_1)\}\mathrm{E}\{\sin(tX_2)\} \\
& +i\Big[\mathrm{E}\{\sin(tX_1)\}\mathrm{E}\{\cos(tX_2)\}+\mathrm{E}\{\cos(tX_1)\}\mathrm{E}\{\sin(tX_2)\}\Big] \\
= {} & \mathrm{E}\{\cos(tX_2)\}\Big[\mathrm{E}\{\cos(tX_1)\}+i\mathrm{E}\{\sin(tX_1)\}\Big] \\
& +i\mathrm{E}\{\sin(tX_2)\}\Big[\mathrm{E}\{\cos(tX_1)\}+i\mathrm{E}\{\sin(tX_1)\}\Big] \\
= {} & \mathrm{E}\{\cos(tX_2)\}\psi_{X_1}(t)+i\mathrm{E}\{\sin(tX_2)\}\psi_{X_1}(t) \\
= {} & \psi_{X_1}(t)\Big[\mathrm{E}\cos(tX_2)+i\mathrm{E}\{\sin(tX_2)\}\Big] \\
= {} & \psi_{X_1}(t)\psi_{X_2}(t).
\end{aligned} \tag{8.27}
$$

The result for n random variables follows by induction. □

Proposition 8.31. Linear combinations of independent normals are normal *Let $X_1\sim \mathrm{N}(\mu_1,\sigma_1^2)$ and $X_2\sim \mathrm{N}(\mu_2,\sigma_2^2)$ be independent, where $\sigma_1^2<\infty$, $\sigma_2^2<\infty$. If a_1 and a_2 are finite constants, then $a_1X_1+a_2X_2\sim \mathrm{N}(a_1\mu_1+a_2\mu_2, a_1^2\sigma_1^2+a_2^2\sigma_2^2)$.*

Proof. The ch.f. of X_1 is $\psi_{X_1}(t)=\exp(it\mu_1-\sigma_1^2t^2/2)$ (Table 8.2). The ch.f. of a_1X_1 is $\mathrm{E}\{\exp(ita_1X_1)\}=\psi_{X_1}(ta_1)=\exp(ita_1\mu_1-t^2a_1^2\sigma_1^2/2)$. That is, a_1X_1 has the ch.f. of a $\mathrm{N}(a_1\mu_1,a_1^2\sigma_1^2)$ random variable. By Theorem 8.23, $a_1X_1\sim\mathrm{N}(a_1\mu_1,a_1^2\sigma_1^2)$. Similarly, $a_2X_2\sim\mathrm{N}(a_2\mu_2,a_2^2\sigma_2^2)$. Moreover, a_1X_1 and a_2X_2 are independent, so Proposition 8.30 implies that the ch.f. of $a_1X_1+a_2X_2$ is the product $\exp\{it(a_1\mu_1+a_2\mu_2)-t^2(a_1^2\sigma_1^2+a_2^2\sigma_2^2)/2\}$.

This is the ch.f. of a $\mathrm{N}(a_1\mu_1 + a_2\mu_2, a_1^2\sigma_1^2 + a_2^2\sigma_2^2)$ random variable. The result now follows from Theorem 8.23. □

Proposition 8.32. Decomposition of independent sums *Let X_i be iid with characteristic function $\psi_X(t)$ with no real roots, and let $F_j(x)$ be the distribution of $\sum_{i=1}^j X_i$, $j = 1, \ldots$ Suppose that Y_1 and Y_2 are independent with $Y_1 \sim F_k$ and $Y_1 + Y_2 \sim F_n$ for some $n > k$. Then $Y_2 \sim F_{n-k}$.*

Proof. Let $\psi_X(t)$ be the ch.f. of X. By Proposition 8.30, the ch.f.s of Y_1 and $Y_1 + Y_2$ are $\psi_X^k(t)$ and $\psi_X^n(t)$. Also, the ch.f. of $Y_1 + Y_2$ is $\psi_X^k(t)\psi_{Y_2}(t)$ by Proposition 8.30. Thus, $\psi_X^k(t)\psi_{Y_2}(t) = \psi_X^n(t)$. Because $\psi_X(t)$ has no real roots, we can divide both sides of this equation by $\psi_X^k(t)$, yielding $\psi_{Y_2}(t) = \psi_X^{n-k}(t)$. By Theorem 8.23, Y_2 has distribution function F_{n-k}. □

Proposition 8.33. Reproducing moments from differentiation of characteristic function *Let $\psi(t)$ be the characteristic function of a random variable X. If $\mathrm{E}(|X^k|) < \infty$ for a positive integer k, then $\psi(t)$ has a finite kth derivative obtained by differentiating inside the integral: $\psi^{(k)}(t) = \int_{-\infty}^{\infty} (ix)^k \exp(itx)dF(x)$. Therefore, $\psi^{(k)}(0) = i^k\mathrm{E}(X^k)$.*

Proof. It is helpful to consider the real and imaginary parts $\psi_R(t)$ and $\psi_I(t)$ of $\psi(t)$ separately. By the mean value theorem from calculus,

$$\begin{aligned} \frac{\psi_R(t+\Delta) - \psi_R(t)}{\Delta} &= \int \frac{\cos\{(t+\Delta)x\} - \cos(tx)}{\Delta} dF(x) \\ &= \int \frac{-\sin(\eta)(x\Delta)}{\Delta} dF(x) \\ &= \int -x\sin(\eta)dF(x), \qquad (8.28) \end{aligned}$$

where η is between tx and $tx + x\Delta$. Now take the limit as $\Delta \to 0$. The magnitude of the integrand is $|-x\sin(\eta)| \le |x|$, which is integrable by assumption. Moreover, the integrand tends to $-x\sin(tx)$ as $\Delta \to 0$ because $\eta \to tx$. The DCT implies that the integral tends to $-\int_{-\infty}^{\infty} x\sin(tx)dF(x)$ as $\Delta \to 0$. This proves that $\psi_R(t)$ is differentiable and $\psi_R'(t) = -\int_{-\infty}^{\infty} x\sin(tx)dF(x)$. A similar argument shows that the derivative of $\psi_I(t) = \int \sin(tx)dF(x)$ is $\psi_I'(t) = \int x\cos(tx)dF(x)$. It follows that the derivative of $\psi(t)$ is $\int\{-x\sin(tx)\}dF(x) + i\int x\cos(tx)dF(x) = \int(ix)\exp(itx)dF(x)$. This proves that the result is true for $k = 1$. The rest of the proof using induction is similar and is left as an exercise. □

The ability of the characteristic function $\mathrm{E}\{\exp(itX)\}$ to recover moments of the random variable X via Proposition 8.33 is reminiscent of the moment generating function (m.g.f.), defined as follows.

Definition 8.34. Moment generating function *The moment generating function (m.g.f.) of a random variable X is defined as $\mathrm{E}\{\exp(tX)\}$, provided that this expectation is finite.*

The moment generating function was so-named because it also recovers moments, as seen by the following result.

Theorem 8.35. Moment generating functions: uniqueness, power series representation, and ability to reproduce moments *If the m.g.f. $E\{\exp(tX)\}$ exists for $t \in [-r, r]$ for some $r > 0$, then:*

1. *All moments of X are finite.*

2. *The m.g.f. for X has the power series representation $E\{\exp(tX)\} = \sum_{n=0}^{\infty} a_n t^n$ for $|t| < r$, where $a_n = \mu_n/n!$ and $\mu_n = \mathrm{E}(X^n)$.*

3. *The distribution function for X is uniquely determined by its m.g.f. I.e., two random variables with the same m.g.f. for $t \in [-r, r]$ and some $r > 0$ have the same distribution function.*

Theorem 8.35 seems to suggest that any random variable with finite moments of all orders must be uniquely determined by those moments. After all, the power series representation in part 2 depends only on those moments. However, a key assumption is that the m.g.f. exists for all t in an interval $[-r, r]$ with $r > 0$. In other words, the radius of convergence of the power series for the m.g.f. is not 0. It turns out that there **are** distinct distribution functions with the same moments μ_n if $\sum_{n=0}^{\infty} r^n \mu_n/n!$ does not converge for any $r > 0$.

Exercises

1. Use ch.f.s to prove that if X_1 and X_2 are independent Poissons with respective parameters λ_1 and λ_2, then $X_1 + X_2$ is Poisson with parameter $\lambda_1 + \lambda_2$.

2. Use ch.f.s to prove that if X_1 and X_2 are independent exponentials with parameter θ, then $X_1 + X_2$ is gamma with parameters 2 and θ.

3. Use ch.f.s to prove that if X_1 and X_2 are iid random variables, then the distribution function for $X_1 - X_2$ is symmetric about 0.

4. Use the representation $\cos(t) = \{\exp(it) + \exp(-it)\}/2$ to read out the probability mass function that corresponds to the characteristic function $\cos(t)$. Describe the distribution corresponding to the ch.f. $\{\cos(t)\}^n$.

5. $\uparrow$ Use the CLT in conjunction with the preceding problem to deduce that $\{\cos(t/n^{1/2})\}^n$ converges to $\exp(-t^2/2)$ as $n \to \infty$. Then verify this fact directly. Hint: write the log of $\{\cos(t/n^{1/2})\}^n$ as $\ln\{\cos(t/n^{1/2})\}/(1/n)$ (this is not problematic because $\cos(t/n^{1/2})$ is nonnegative for n sufficiently large) and use L'Hospital's rule as many times as needed.

6. Let Y be a mixture of two normal random variables: $Y = X_1$ or $Y = X_2$ with probability λ and $1 - \lambda$, respectively, where $X_i \sim \mathrm{N}(\mu_i, \sigma_i^2)$. Show that Y has ch.f. $\psi_Y(t) = \lambda \exp(i\mu_1 t - \sigma_1^2 t^2/2) + (1 - \lambda)\exp(i\mu_2 t - \sigma_2^2 t^2/2)$.

7. Use ch.f.s to prove that the distribution of the sample mean of n iid observations from the Cauchy distribution with parameters θ and λ is Cauchy with parameters θ and λ.

8. Let Y_1, Y_2 be iid Cauchy with parameters θ and λ, and let $\tau \in (0, 1)$. Use ch.f.s to deduce the distribution of $\tau Y_1 + (1 - \tau)Y_2$.

9. The geometric distribution is the distribution of the number of failures before the first success in iid Bernoulli trials with success probability p. Given its ch.f. in Table 8.2, determine the ch.f. of the number of failures before the sth success.

10. Suppose that Y_1 and Y_2 are independent, Y_1 has a chi-squared distribution with k degrees of freedom, and $Y_1 + Y_2$ has a chi-squared distribution with n degrees of freedom. Prove that Y_2 has a chi-squared distribution with $n - k$ degrees of freedom.

11. Suppose that Z_1 and Z_2 are independent, $Z_1 \sim \mathrm{N}(0,1)$, and $Z_1 + Z_2 \sim N(0,2)$. Prove that $Z_2 \sim \mathrm{N}(0,1)$.

12. Show that the following are NOT ch.f.s.

    (a) $\psi(t) = \cos(t) + \sin(t)$.

    (b) $\psi(t) = (1/2)\cos(t)$

    (c) $\psi(t) =$
    $$\begin{cases} \sin(1/t) & \text{for } t \neq 0 \\ 1 & \text{for } t = 0. \end{cases}$$

13. Use induction to finish the proof of Proposition 8.33.

8.5 Proof of Standard CLT

We are now in a position to use ch.f.s to prove the CLT. The key theorem is the following.

Theorem 8.36. Continuity property for characteristic functions *Let X_n be a sequence of random variables with ch.f. $\psi_n(t)$, and let X be a random variable with ch.f. $\psi(t)$. Then $X_n \stackrel{D}{\to} X$ if and only if $\psi_n(t) \to \psi(t)$ as $n \to \infty$ for every t.*

Proof. Suppose first that $X_n \stackrel{D}{\to} X$. Then $\mathrm{E}\{f(X_n)\} \to \mathrm{E}\{f(X)\}$ for every bounded continuous function f by Proposition 6.29. Since $\cos(tX)$ and $\sin(tX)$ are bounded continuous functions, $\mathrm{E}\{\cos(tX_n)\} \to \mathrm{E}\{\cos(tX)\}$ and $\mathrm{E}\{\sin(tX_n)\} \to \mathrm{E}\{\sin(tX)\}$. It follows that $\psi_n(t) \to \psi(t)$. Thus, $X_n \stackrel{D}{\to} X$ implies that $\psi_n(t) \to \psi(t)$.

Now suppose that $\psi_n(t) \to \psi(t)$ for all t. It can be shown that the sequence of distribution functions $F_n(x)$ for X_n is tight. We will prove that every subsequence contains a further subsequence converging to the distribution function F for X. Proposition 6.44 will then imply that $X_n \stackrel{D}{\to} X$.

To prove that every subsequence contains a further subsequence converging to F, let $\{m\} \subset \{n\}$ be any subsequence. Because F_n is tight, so is F_m. Thus, there is a subsequence $\{k\} \subset \{m\}$ such that F_k converges weakly to some distribution function G. By what we proved in the preceding paragraph, $\psi_k(t)$ must converge to the ch.f. of $G(x)$. But $\psi_k(t)$ must also converge to $\psi(t)$, the ch.f. of $F(x)$, because $\psi_k(t)$ is just a subsequence of $\psi_n(t)$ and $\psi_n(t) \to \psi(t)$. But two distribution functions with the same ch.f. must be identical by Theorem 8.23, so $F = G$. This proves that every subsequence $\{m\}$ contains a further subsequence $\{k\}$ such that $X_k \stackrel{D}{\to} X$. By Proposition 6.44, $X_n \stackrel{D}{\to} X$. □

Let X_i be iid random variables with mean 0 and variance 1, $S_n = \sum_{i=1}^n X_i$, and $Y_n = S_n/n^{1/2}$. Theorem 8.36 implies that to prove the CLT, we need only prove that the ch.f. of Y_n converges to the ch.f. $\exp(-t^2/2)$ of a $\mathrm{N}(0,1)$. The first step is the following.

Lemma 8.37. *If X is a random variable with mean 0 and variance 1, then its characteristic function $\psi(t)$ satisfies*

$$\psi(t) = 1 - t^2/2 + o(t^2) \text{ as } t \to 0.$$

Proof. Taylor's theorem for a function of a real variable (Theorem A.49) and Proposition 8.33 imply that the real and imaginary parts of ψ satisfy

$$\begin{aligned}\psi_R(t) &= \psi_R(0) + \psi_R'(0)t + \psi_R''(0)t^2/2 + o(t^2) = 1 - t^2/2 + o(t^2) \\ \psi_I(t) &= \psi_I(0) + \psi_I'(0)t + \psi_I''(0)t^2/2 + o(t^2) = 0 + o(t^2).\end{aligned} \tag{8.29}$$

It follows that the ch.f. $\psi(t)$ is $1 - t^2/2 + o(t^2)$ as $t \to 0$. □

8.5.1 Proof of CLT for Symmetric Random Variables

It is helpful to start with symmetric random variables because the characteristic function is real (Proposition 8.29, part 5). The ch.f. of $Y_n = n^{-1/2}\sum_{i=1}^n X_i$ is also real and is given by

$$\begin{aligned}\psi_{Y_n}(t) &= \mathrm{E}[\exp\{i(t/n^{1/2})(X_1 + \ldots + X_n)\}] \\ &= \psi_X^n(t/n^{1/2}) \text{ (Proposition 8.30)} \\ &= [1 - t^2/(2n) + r_n]^n, \text{ (Proposition 8.37)}\end{aligned} \tag{8.30}$$

where the remainder term r_n is $o\{t^2/(2n)\}$. Now take the natural logarithm of both sides to get

$$\begin{aligned}\ln\{\psi_{Y_n}(t)\} &= n\ln\{1 - t^2/(2n) + r_n\} \\ &= n[-t^2/(2n) + r_n + o\{-t^2/(2n) + r_n\}] \\ &\to -t^2/2 \text{ as } n \to \infty.\end{aligned} \tag{8.31}$$

Therefore, the ch.f. of Y_n tends to $\exp(-t^2/2)$, the ch.f. of a $N(0,1)$ random variable. By Theorem 8.36, Y_n converges in distribution to $N(0,1)$. This completes the proof when the X_i are symmetric about 0.

8.5.2 Proof of CLT for Arbitrary Random Variables

The proof in the preceding subsection was seamless because the ch.f.s for X_i and Y_n are real when the X_i are symmetric about 0. When the X_i are arbitrary iid random variables with mean 0 and variance 1, the ch.f. of Y_n is complex-valued. Taking its logarithm is problematic because the logarithm of a complex variable is not unique. To avoid this problem, we must extend a result from calculus to complex numbers.

Lemma 8.38. The exponential function as a limit of $(1 + z_n/n)^n$ *Let z_n be a sequence of complex constants such that $z_n \to z$. Then $x_n = (1 + z_n/n)^n \to \exp(z)$.*

Proof.

$$x_n = \sum_{k=0}^n \binom{n}{k}(z_n/n)^k(1)^{n-k} = \sum_{k=0}^m \binom{n}{k}(z_n/n)^k + \sum_{k=m+1}^n \binom{n}{k}(z_n/n)^k \tag{8.32}$$

for any integer m between 0 and n.

Because $z_n \to z$, z_n is bounded: $||z_n|| \le B$ for all n. Therefore,

$$\left|\left| x_n - \sum_{k=0}^m \binom{n}{k}\left(\frac{z_n}{n}\right)^k \right|\right| \le \sum_{k=m+1}^n \left|\left| \left(\frac{n^k}{k!}\right)\left(\frac{z_n}{n}\right)^k \right|\right|$$

$$= \sum_{k=m+1}^{n} \frac{||z_n||^k}{k!} \le \sum_{k=m+1}^{\infty} \frac{B^k}{k!}. \text{ I.e.,}$$

$$\left\| x_n - \sum_{k=0}^{m} \binom{n}{k} \left(\frac{z_n}{n}\right)^k \right\| \le \sum_{k=m+1}^{\infty} \frac{B^k}{k!}. \tag{8.33}$$

Assume first that $\lim_{n\to\infty} x_n = x$, where x is finite. Take the limit as $n \to \infty$ of Equation (8.33). The limit of the left side is $||x - \sum_{k=0}^{m} z^k/k!||$, while the limit of the right side is $\sum_{k=m+1}^{\infty} B^k/k!$. Thus, we see that

$$\left\| x - \sum_{k=0}^{m} \frac{z^k}{k!} \right\| \le \sum_{k=m+1}^{\infty} \frac{B^k}{k!}. \tag{8.34}$$

Now take the limit as $m \to \infty$ of Equation (8.34). The limit of the left side is $||x - \sum_{k=0}^{\infty} z^k/k!||$, while the limit of the right side is 0. We conclude that $||x - \sum_{k=0}^{\infty} z^k/k!|| = 0$, so $x = \sum_{k=0}^{\infty} z^k/k! = \exp(z)$.

Even without assuming that x_n has a finite limit, the same argument shows that any convergent subsequence x_{n_k} must converge to $\exp(z)$. Furthermore, no subsequence x_{n_k} can converge to an infinite limit because $||z_n|| \le B$ implies that $x_n = (1 + z_n/n)^n$ satisfies

$$(1 - B/n)^n \le ||x_n|| \le (1 + B/n)^n,$$

and $(1 - B/n)^n \to \exp(-B)$, $(1 + B/n)^n \to \exp(B)$. Because every convergent subsequence must converge to the same limit, $\exp(z)$, and no subsequence can converge to an infinite limit, $x_n \to \exp(z)$. □

To use this lemma to prove the CLT, note that Expression (8.30) remains valid even if the distribution of X_i is not symmetric about 0. The only difference is that the remainder term r_n is complex instead of real. Therefore, the ch.f. of $Y_n = S_n/n^{1/2}$ is of the form $1 + z_n/n$, where $z_n = -t^2/2 + nr_n \to -t^2/2$ as $n \to \infty$ because $r_n = o\{t^2/(2n)\}$. By Lemma 8.38, the ch.f. of Y_n converges to $\exp(-t^2/2)$, the ch.f. of a standard normal deviate. By Theorem 8.36, $Y_n \xrightarrow{D} \mathrm{N}(0,1)$. This completes the proof of the standard CLT.

Exercises

1. Use ch.f.s to give another proof of the fact that if $X_n \xrightarrow{D} X$, $Y_n \xrightarrow{D} Y$, and X_n and Y_n are independent, then $X_n + Y_n \xrightarrow{D} X + Y$, where X and Y are independent (see also Problem 2 of Section 6.3).

2. Use ch.f.s to give another proof of the fact that, in Example 8.13, the asymptotic distribution of $\sum_{j=1}^{n} X_{nj}/\{\mathrm{var}(S_n)\}^{1/2}$ is a mixture of two normals (see Problem 6 of Section 8.4), where $S_n = \sum_{j=1}^{n} X_{nj}$.

3. Modify Example 8.13 so that the first random variable is $-n^{1/2}$, 0, or $+n^{1/2}$ with probability 1/3. Show that the asymptotic distribution of $S_n/\{\mathrm{var}(S_n)\}^{1/2}$ is a mixture of three normals.

4. Use characteristic functions to prove the law of small numbers (Proposition 6.24).

8.6 Multivariate Ch.f.s and CLT

The characteristic function of a random vector $\mathbf{X} = (X_1, \ldots, X_k)'$ is defined analogously to that of a random variable. Because the distribution of $\mathbf{X}$ is k-dimensional, its characteristic function is a function of $(t_1, \ldots, t_k)$.

Definition 8.39. Multivariate characteristic function *The characteristic function of the random vector* $\mathbf{X} = (X_1, \ldots, X_k)'$ *is*

$$\begin{aligned} \psi(t_1, \ldots, t_k) &= \mathrm{E}\{i(\mathbf{t}'\mathbf{X})\} = \mathrm{E}\left(i \sum_{j=1}^{k} t_j X_j\right) \\ &= \int \cdots \int \exp\left(i \sum_{j=1}^{k} t_j x_j\right) dF(x_1, \ldots, x_k). \end{aligned}$$

The key results for ch.f.s of random variables carry over to ch.f.s of random vectors as well. For example, the following is a generalization of Proposition 8.23.

Proposition 8.40. Extension of Proposition 8.23: the multivariate ch.f. characterizes the distribution of X *A ch.f.* $\psi(\mathbf{t})$ *uniquely determines the distribution function* $F(\mathbf{x})$ *of the random vector* $\mathbf{X}$*. That is, two random vectors with the same ch.f.s have the same distribution functions as well.*

Proposition 8.41. Extension of Proposition 8.30: product rule for the ch.f. of a sum of independent random vectors *Let* $\mathbf{X}_1$ *and* $\mathbf{X}_2$ *be* k*-dimensional random vectors with respective ch.f.s* $\psi_{\mathbf{X}_1}(\mathbf{t})$ *and* $\psi_{\mathbf{X}_2}(\mathbf{t})$*. If* $\mathbf{X}_1$ *and* $\mathbf{X}_2$ *are independent, then the ch.f.* $\psi_{\mathbf{X}_1+\mathbf{X}_2}(\mathbf{t})$ *of* $\mathbf{X}_1 + \mathbf{X}_2$ *is* $\psi_{\mathbf{X}_1}(\mathbf{t})\psi_{\mathbf{X}_2}(\mathbf{t})$*. Similarly, if* $\mathbf{X}_1, \ldots, \mathbf{X}_n$ *are independent* k*-dimensional random vectors with respective ch.f.s* $\psi_{\mathbf{X}_1}(\mathbf{t}), \ldots, \psi_{\mathbf{X}_n}(\mathbf{t})$*, the ch.f.* $\psi_{\mathbf{X}_1+\ldots\mathbf{X}_n}(\mathbf{t})$ *of* $\sum_{i=1}^{n} \mathbf{X}_i$ *is* $\prod_{i=1}^{n} \psi_{\mathbf{X}_i}(\mathbf{t})$*.*

Proof. For $\mathbf{t} \in R^k$, let $Y_1 = \mathbf{t}'\mathbf{X}_1$ and $Y_2 = \mathbf{t}'\mathbf{X}_2$. Then Y_1 and Y_2 are independent 1-dimensional random variables. By Proposition 8.30, the ch.f. $\psi_{Y_1+Y_2}(s) = \mathrm{E}[\exp\{is(Y_1 + Y_2)\}]$ of $Y_1 + Y_2$ is $\psi_{Y_1}(s)\psi_{Y_2}(s)$. Take $s = 1$ to deduce the result for two random vectors. The proof for n random vectors can be proven by induction. □

We have seen that convergence in distribution of a sequence X_n to a random variable X is equivalent to convergence of the corresponding characteristic functions. This holds true in the multivariate case as well:

Proposition 8.42. Extension of Proposition 8.36: continuity property of characteristic functions *Let* $\mathbf{X}_n$ *be a sequence of* k*-dimensional random vectors with ch.f.* $\psi_n(\mathbf{t})$*, and let* $\mathbf{X}$ *be a* k*-dimensional random vector with ch.f.* $\psi(\mathbf{t})$*. Then* $\mathbf{X}_n \xrightarrow{D} \mathbf{X}$ *if and only if* $\psi_n(\mathbf{t}) \to \psi(\mathbf{t})$ *as* $n \to \infty$ *for all* $\mathbf{t} \in R^k$*.*

Notice that we can obtain the ch.f. of a random vector from knowledge of the distribution of $\mathbf{t}'\mathbf{X}$ for each $\mathbf{t} \in R^k$. Moreover, the ch.f. completely determines the distribution function of $\mathbf{X}$. Thus, we can deduce the multivariate distribution function of $\mathbf{X}$ from the distribution functions of $\mathbf{t}'\mathbf{X}$ for all $\mathbf{t} \in R^k$. This suggests the possibility of deducing the limiting distribution function of a sequence $\mathbf{X}_n$ from the limiting distribution of $\mathbf{t}'X_n$, $\mathbf{t} \in R^k$. This important reduction technique, known as the Cramer-Wold device, can be proven using Proposition 8.42.

Corollary 8.43. Cramer-Wold device: reducing convergence in distribution of random vectors to convergence in distribution of random variables *The k-dimensional random vector* $\mathbf{X}_n$ *converges in distribution to the k-dimensional random vector* $\mathbf{X}$ *if and only if* $\mathbf{a}'\mathbf{X}_n$ *converges in distribution to* $\mathbf{a}'\mathbf{X}$ *for all vectors* $\mathbf{a} = (a_1, \ldots, a_k)' \in R^k$.

Proof. Exercise.

A very important consequence of the Cramer-Wold device is a multivariate generalization of the central limit theorem (CLT).

We begin by defining the multivariate normal distribution function, starting with the bivariate normal. Its density was given in Section 5.6, but here we provide an alternative definition in terms of linear combinations of independent standard normals. Let Z_1, Z_2 be iid standard normals, and consider the joint distribution of two linear combinations, $Y_1 = a_{11}Z_1 + a_{12}Z_2$, $Y_2 = a_{21}Z_1 + a_{22}Z_2$. We can write $\mathbf{Y}$ in matrix notation as $\mathbf{Y} = A\mathbf{Z}$, where $\mathbf{Y}' = (Y_1, Y_2)$, $\mathbf{Z}' = (Z_1, Z_2)$, and

$$A = \begin{pmatrix} a_{11} & a_{12} \\ a_{21} & a_{22} \end{pmatrix}.$$

By Proposition 8.31, the marginal distributions of Y_1 and Y_2 are $Y_1 \sim \mathrm{N}(0, a_{11}^2 + a_{12}^2)$ and $Y_2 \sim \mathrm{N}(0, a_{21}^2 + a_{22}^2)$. The covariance between Y_1 and Y_2 is $\mathrm{cov}(a_{11}Z_1 + a_{12}Z_2, a_{21}Z_1 + a_{22}Z_2) = a_{11}a_{21} + a_{12}a_{22}$. The key question is whether, for arbitrary variances σ_1^2 and σ_2^2 and correlation ρ, we can find a matrix A such that $\mathbf{Y} = A\mathbf{Z}$ has these variances and correlation. Set

$$a_{11} = \sigma_1,\ a_{12} = 0,\ a_{21} = \rho\sigma_2,\ a_{22} = \sigma_2(1 - \rho^2)^{1/2}. \tag{8.35}$$

It is easy to see that $\mathbf{Y} = A\mathbf{Z}$ has variances σ_1^2 and σ_2^2 and correlation ρ. We can add the constant vector $\boldsymbol{\mu}$ to $\mathbf{Y}$ to make the mean of $\mathbf{Y}$ equal $\boldsymbol{\mu}$. A random vector $\mathbf{Y} = A\mathbf{Z} + \boldsymbol{\mu}$, where A is a 2×2 matrix and $\mathbf{Z}$ are two iid standard normals, is said to have a *bivariate normal distribution*.

More generally, let $\mathbf{Z} = (Z_1, \ldots, Z_k)'$ be iid standard normals, and let $\mathbf{Y} = A\mathbf{Z}$, where A is n $k \times k$ matrix. Each Y_i is normal with mean 0. The covariance matrix of $\mathbf{Y}$, defined by $\mathrm{E}(\mathbf{Y}\mathbf{Y}')$, is

$$\begin{aligned} \mathrm{cov}(A\mathbf{Z}) &= \mathrm{E}\{(A\mathbf{Z})(A\mathbf{Z})'\} = \mathrm{E}(AZZ'A') \\ &= A\mathrm{E}(\mathbf{Z}\mathbf{Z}')A' = A\,\mathrm{cov}(\mathbf{Z})A' = AA'. \end{aligned} \tag{8.36}$$

Any covariance matrix is *positive definite*, meaning that for any k-dimensional vector a, $\mathbf{a}'\Sigma\mathbf{a} \geq 0$ (we say *strictly positive definite* if this quantity is always strictly positive) Again the key question is whether, given an arbitrary covariance matrix Σ, we can find a matrix A such that $\mathrm{cov}(A\mathbf{Z}) = \Sigma$. The following result from linear algebra provides an affirmative answer.

Proposition 8.44. Square root of a covariance matrix *Let* Σ *be an arbitrary* $k \times k$ *covariance matrix (i.e.,* Σ *is symmetric and* $\mathbf{a}'\Sigma\mathbf{a} \geq 0$ *for each real vector* $\mathbf{a}$*). There exists a "square root" matrix* A *such that* $AA' = \Sigma$.

Definition 8.45. Multivariate normal distribution *The k-dimensional random vector* $\mathbf{Y}$ *is said to have a multivariate normal distribution if* $\mathbf{Y} = A\mathbf{Z} + \boldsymbol{\mu}$, *where* A *is a* $k \times k$ *matrix and* $\mathbf{Z}$ *are* k *iid standard normals.*

Proposition 8.46. The characteristic function of the multivariate normal distribution *The characteristic function of the normal distribution with mean vector* $\mathbf{0}$ *and covariance matrix* Σ *is* $\exp\{-(1/2)\mathbf{t}'\Sigma\mathbf{t})\}$.

Proof. The ch.f. for each Z is $\exp(-t^2/2)$. By result 8.30, the ch.f. for $\mathbf{Z}$ is $\psi_{\mathbf{Z}}(\mathbf{t}) = \exp(-\sum_{i=1}^k t_i^2/2) = \exp(-||\mathbf{t}||^2/2)$. The ch.f. for $A\mathbf{Z}$ is $\mathrm{E}\{\exp(\mathbf{t}'A\mathbf{z})\} = \psi_{\mathbf{Z}}(A'\mathbf{t})\} = \exp(-||A'\mathbf{t}||^2/2) = \exp(-\mathbf{t}'AA'\mathbf{t}) = \exp(-\mathbf{t}'\Sigma\mathbf{t}/2)$. □

Theorem 8.47. Multivariate CLT *Let* $\mathbf{X}_1, \mathbf{X}_2, \ldots$ *be iid k-dimensional random vectors with mean vector* $\boldsymbol{\mu}$ *and finite covariance matrix* Σ, *and let* $\mathbf{S}_n = \sum_{i=1}^n \mathbf{X}_i$. *Then* $n^{-1/2}(\mathbf{S}_n - n\boldsymbol{\mu})$ *converges in distribution to a multivariate normal with mean vector* $\mathbf{0}$ *and covariance matrix* Σ.

Proof. By the Cramer-Wold device (Corollary 8.43), it suffices to prove that $\mathbf{t}'n^{-1/2}(\mathbf{S}_n - n\boldsymbol{\mu})$ is asymptotically normal with mean 0 and variance $\mathbf{t}'\Sigma\mathbf{t}$.

Let $Y_i = \mathbf{t}'(\mathbf{X}_i - \boldsymbol{\mu})$. Then the Y_i are iid random variables with mean 0 and variance $\mathbf{t}'\Sigma\mathbf{t}$. By the CLT for random variables, $n^{-1/2}\sum_{i=1}^n Y_i$ converges in distribution to $\mathrm{N}(0, \mathbf{t}'\Sigma\mathbf{t})$. By the Cramer-Wold device (Theorem 8.43), $n^{-1/2}(\mathbf{S}_n - n\boldsymbol{\mu})$ converges in distribution to a multivariate normal with mean vector $\mathbf{0}$ and covariance matrix Σ. □

Example 8.48. Application of multivariate CLT: multiple endpoints Consider a medical study with two outcomes, say 30-day mortality and 30-day cardiovascular mortality. Let X_i and Y_i be the indicators of 30-day death and cardiovascular death for patient i. Then (X_i, Y_i), $i = 1, \ldots, n$ are independent pairs, though of course X_i and Y_i are dependent. The covariance matrix Σ for (X_i, Y_i) is given by $\mathrm{var}(X_i) = \sigma_X^2 = p_X(1 - p_X)$, $\mathrm{var}(Y_i) = \sigma_Y^2 = p_Y(1 - p_Y)$, and $\mathrm{cov}(X_i, Y_i) = \rho\sigma_X\sigma_Y$, where ρ is the correlation between X_i and Y_i. With n patients, the sample proportions $(\hat{p}_X, \hat{p}_Y)$ of patients with the respective events are sample means $(1/n)\sum_{i=1}^n (X_i, Y_i)'$. The multivariate CLT implies that $(\hat{p}_X, \hat{p}_Y)$ is asymptotically normal with mean (p_X, p_Y) and covariance matrix $(1/n)\Sigma$. That is, $n^{1/2}\{(\hat{p}_X, \hat{p}_Y) - (p_X, p_Y)\}$ converges in distribution to a bivariate normal random vector with mean vector $(0, 0)'$ and covariance matrix Σ. □

Example 8.49. Application of multivariate CLT: chi-squared goodness of fit tests One application of the multivariate CLT involves chi-squared statistics and goodness of fit tests. For example, we may want to know whether the flu is equally likely to occur in the 4 different seasons, spring, summer, fall, and winter. We have data from a large number, n, of patients with flu. Each patient's data $\mathbf{Y}$ is either $(1, 0, 0, 0)'$, $(0, 1, 0, 0)'$, $(0, 0, 1, 0)'$ or $(0, 0, 0, 1)'$ depending on whether flu occurred in spring, summer, fall, or winter, respectively. The total numbers of patients with flu in the different seasons is the sum, $\mathbf{S}_n = \sum_{i=1}^n \mathbf{Y}_i$ of n independent vectors. Under the hypothesis that flu is equally likely to occur in any season, the expected number of observations in each season is $n/4$. The chi-squared statistic is of the form $\sum_{j=1}^4 (\mathrm{Observed}_j - \mathrm{Expected}_j)^2/(\mathrm{Expected}_j) = \sum_{j=1}^4 (S_{nj} - n/4)^2/(n/4)$. More generally, with k categories, the chi-squared statistic for testing uniformity is $\sum_{j=1}^k (S_{nj} - n/k)^2/(n/k)$.

We prove that under the null hypothesis, the above goodness of fit statistic converges in distribution to a chi-squared random variable with $k-1$ degrees of freedom as $n \to \infty$. Note that $\mathbf{S}_n$ is the sum of n iid vectors; a generic vector $\mathbf{Y}$ has mean $\mathrm{E}(\mathbf{Y}) = (1/k, \ldots, 1/k)'$. Each component Y_i of $\mathbf{Y}$ is Bernoulli $1/k$, so its variance is $(1/k)(1 - 1/k)$. Also, if $i \neq j$, then $\mathrm{cov}(Y_i, Y_j) = \mathrm{E}(Y_iY_j) - (1/k)^2$. But $Y_iY_j = 0$ because only one of $Y_1, \ldots, Y_k$ is nonzero. Thus, $\mathrm{cov}(Y_i, Y_j) = -(1/k)^2$. Therefore, the covariance matrix $\Sigma = \mathrm{cov}(\mathbf{Y})$ has diagonal element $(1/k)(1-1/k)$ and off diagonal elements $-(1/k)^2$. By the multivariate CLT, $n^{-1/2}(\mathbf{S}_n - n(1/k,\ldots,1/k)')$ converges in distribution to a multivariate normal vector $\mathbf{U}$ with mean vector $\mathbf{0}$ and covariance matrix Σ. By the Mann-Wald mapping theorem (Theorem 6.59), the goodness of fit statistic converges in distribution to $\sum_{j=1}^k U_j^2/(1/k) = k\sum_{j=1}^k U_j^2$.

It remains to show that $k\sum_{j=1}^k U_j^2$ has a chi-squared distribution with $k-1$ degrees of freedom. Let $W_j = k^{1/2}U_j$, so that the goodness of fit statistic is $\sum_{j=1}^k W_j^2$. Each W_j is marginally $\mathrm{N}(0, 1-1/k)$, and the covariance between W_i and W_j is $-1/k$ for each pair (i,j), $i \neq j$. Notice that $(W_1,\ldots,W_k)$ has the same joint distribution as $(X_1-\bar{X},\ldots,X_k-\bar{X})$, where the X_i are iid standard normal. Therefore, the goodness of fit statistic has the same distribution as $\sum_{j=1}^k (X_j-\bar{X})^2$, which is $k-1$ times the sample variance of k iid observations from a standard normal distribution. To see that this has a chi-squared distribution with $k-1$ degrees of freedom, note that $\sum_{j=1}^k X_j^2 - k\bar{X}^2 = \sum_{j=1}^k (X_j-\bar{X})^2$ is independent of the sample mean $\bar{X}$, and therefore of $k\bar{X}^2$. Also, $\sum_{j=1}^k X_j^2 - k\bar{X}^2 + k\bar{X}^2 = \sum_{j=1}^k X_j^2$ has a chi-squared distribution with k degrees of freedom, and $k\bar{X}_k^2$ has a chi-squared distribution with 1 degree of freedom. The result now follows from Proposition 8.32 (see Exercise 10 of Section 8.4). □

Theorem 8.50. Multivariate Slutsky *Let $\mathbf{X}_n$ and $\mathbf{Y}_n$ be k-dimensional random vectors with $\mathbf{X}_n \xrightarrow{D} \mathbf{X}$ and $\mathbf{Y}_n \xrightarrow{p} \mathbf{0}$. Then $\mathbf{X}_n + \mathbf{Y}_n \xrightarrow{D} \mathbf{X}$. Also, if $\mathbf{C}_n$ is a k-dimensional random vector converging in probability to a constant vector $\mathbf{c}$, then $(C_{n1}X_{n1},\ldots,C_{nk}X_{nk}) \xrightarrow{D} (c_1X_1,\ldots,c_kX_k)$.*

Proof. Exercise.

Example 8.51. Clinical trials are monitored several times to ensure patient safety and determine whether efficacy of treatment has been established. Many test statistics involve sums of iid random variables with mean 0 under the null hypothesis. Therefore, if we monitor for efficacy k times, we are examining overlapping sums $S_{n_1},\ldots,S_{n_k}$, where $S_{n_i} = \sum_{j=1}^{n_i} X_j$, $\mathrm{E}(X_j)=0$, $\mathrm{var}(X_j)=\sigma^2$. Let $N=\sum_{i=1}^k n_i$. To protect against falsely declaring a treatment benefit, we must determine the joint distribution of $S_{n_1}/(N\sigma^2)^{1/2},\ldots,S_{n_k}/(N\sigma^2)^{1/2}$ and construct boundaries $b_1,\ldots,b_k$ such that $P\{\cup_{i=1}^k S_{n_i}/(N\sigma^2)^{1/2} \geq b_i\} = \alpha$.

Note that

$$Y_{Ni} = \frac{S_{n_i}-S_{n_{i-1}}}{\sqrt{N\sigma^2}} = \left(\frac{S_{n_i}-S_{n_{i-1}}}{\sqrt{(n_i-n_{i-1})\sigma^2}}\right)\sqrt{t_{Ni}-t_{N\,i-1}} = Z_{Ni}\sqrt{t_{Ni}-t_{N\,i-1}},$$

where $t_{Ni}=n_i/N$ and $Z_{Ni} = (S_{n_i}-S_{n_{i-1}})/\{(n_i-n_{i-1})\sigma^2\}^{1/2}$.

Let $N\to\infty$, and assume that $t_{Ni}\to t_i$, $i=1,\ldots,k$. Then $(t_{Ni}-t_{N\,i-1})^{1/2}\to(t_i-t_{i-1})^{1/2}$. By the CLT for iid random variables (Theorem 8.1), $Z_{Ni}\xrightarrow{D} Z_i \sim \mathrm{N}(0,1)$ as $N\to\infty$. By Slutsky's theorem (Theorem 6.52),

$$\text{as } N\to\infty,\ Y_{Ni}\xrightarrow{D} Y_i = \sqrt{t_i-t_{i-1}}\,Z_i \sim \mathrm{N}(0, t_i-t_{i-1}).$$

Also, the Y_{Ni} are independent because they involve non-overlapping sums. Therefore, the Y_i are independent (see Exercise 2 of Section 6.3).

Let $g : R^k \longmapsto R^k$ be the continuous function $g(\mathbf{y}) = (y_1, y_1 + y_2, \ldots, y_1 + \ldots + y_k)$. By the Mann-Wald mapping theorem (Theorem 6.59),

$$\begin{aligned}\left(\frac{S_{n_1}}{\sqrt{N\sigma^2}}, \ldots, \frac{S_{n_k}}{\sqrt{N\sigma^2}}\right) &= \left(\frac{S_{n_1}}{\sqrt{N\sigma^2}}, \frac{S_{n_1}}{\sqrt{N\sigma^2}} + \frac{S_{n_2} - S_{n_1}}{\sqrt{N\sigma^2}}, \ldots, \frac{\sum_{i=1}^k (S_{n_i} - S_{n_{i-1}})}{\sqrt{N\sigma^2}}\right) \\ &= g(Y_{N1}, \ldots, Y_{Nk}) \xrightarrow{D} g(Y_1, \ldots, Y_k) \sim \text{MN}(\mathbf{0}, \Sigma), \qquad (8.37)\end{aligned}$$

where the (p, q)th element of Σ is, for $p \leq q$,

$$\Sigma_{ij} = \text{cov}\left(\sum_{i=1}^p Y_i, \sum_{j=1}^q Y_j\right) = \sum_{i=1}^p \text{var}(Y_i) = \sum_{i=1}^p (t_i - t_{i-1}) = t_p.$$

The statistic $S_{n_i}/(N\sigma^2)^{1/2}$ is known as the "B-value" and n_i/N is known as the "information fraction." What we have shown is that, under the null hypothesis, the B-values at different information fractions are asymptotically normal with $\text{E}\{B(t_i)\} = 0$ and $\text{cov}\{B(t_i), B(t_j)\} = t_i$ for $t_i \leq t_j$. This can be used to construct monitoring boundaries.

Although we assumed known variance, the same result holds if we have an estimator $\hat{\sigma}^2_{Ni}$ of σ^2 available at information fraction t_{Ni}. In that case

$$\frac{S_{n_i}}{\sqrt{N\hat{\sigma}^2_{Ni}}} = \frac{S_{n_i}}{\sqrt{N\sigma^2}}\sqrt{\frac{\sigma^2}{\hat{\sigma}^2_{Ni}}}, \quad i = 1, \ldots, k.$$

By Slutsky's theorem for random vectors, the asymptotic joint distribution of $S_{n_i}/(N\hat{\sigma}^2_{Ni})^{1/2}$ is also asymptotically multivariate normal with mean vector $\mathbf{0}$ and the same covariance matrix as above. □

Exercises

1. Use ch.f.s to prove Corollary 8.43, the Cramer-Wold device.
2. Prove that two k-dimensional random vectors $\mathbf{X}_1$ and $\mathbf{X}_2$ with respective ch.f.s $\psi_1(\mathbf{t})$ and $\psi_2(\mathbf{t})$ are independent if and only if the joint ch.f. $\text{E}\{i(\mathbf{t}_1'\mathbf{X}_1 + \mathbf{t}_2'\mathbf{X}_2)\}$ of $(\mathbf{X}_1, \mathbf{X}_2)$ is $\psi_1(\mathbf{t}_1)\psi_2(\mathbf{t}_2)$ for all $\mathbf{t}_1 \in R^k$, $\mathbf{t}_2 \in R^k$.
3. Let $\mathbf{Y}$ have a trivariate normal distribution with zero means, unit variances, and pairwise correlations ρ_{12}, ρ_{13}, and ρ_{23}. Show that $\mathbf{Y}$ has the same distribution as $A\mathbf{Z}$, where $\mathbf{Z}$ are iid standard normals and
$$A = \begin{pmatrix} 1 & 0 & 0 \\ \rho_{12} & \sqrt{1-\rho_{12}^2} & 0 \\ \rho_{13} & \frac{\rho_{23}-\rho_{12}\rho_{13}}{\sqrt{1-\rho_{12}^2}} & \sqrt{\frac{(1-\rho_{12}^2)(1-\rho_{13}^2)-(\rho_{23}-\rho_{12}\rho_{13})^2}{1-\rho_{12}^2}} \end{pmatrix}.$$
4. Prove the multivariate version of Slutsky's theorem (Theorem 8.50).
5. Use bivariate ch.f.s to prove that if (X, Y) are nonsingular bivariate normal with correlation ρ, then X and Y are independent if and only if $\rho = 0$.

6. ↑ Let X have a standard normal distribution. Flip a fair coin and define Y by:

$$Y = \begin{cases} -X & \text{if tails} \\ +X & \text{if heads.} \end{cases}$$

Show that X and Y each have a standard normal distribution and $\text{cov}(X, Y) = 0$ but X and Y are not independent. Why does this not contradict the preceding problem?

7. Suppose that X_1 and X_2 are iid $\text{N}(\mu, \sigma^2)$ random variables. Use bivariate ch.f.s to prove that $X_1 - X_2$ and $X_1 + X_2$ are independent.

8. Let (X, Y, Z) be independent with respective (finite) means μ_X, μ_Y, μ_Z and respective (finite) variances $\sigma_X^2, \sigma_Y^2, \sigma_Z^2$. Let (X_i, Y_i, Z_i), $i = 1, \ldots, n$, be independent replications of (X, Y, Z), Show that the asymptotic distribution of $\sum_{i=1}^n (X_i + Y_i, Y_i + Z_i)'$ as $n \to \infty$ is bivariate normal, and find its asymptotic mean and covariance vector.

9. Let $\mathbf{Y}$ be multivariate normal with mean vector $\mathbf{0}$ and strictly positive definite covariance matrix Σ. Let $\Sigma^{1/2}$ be a symmetric square root of Σ; i.e., $\left(\Sigma^{1/2}\right)' = \Sigma^{1/2}$ and $\Sigma^{1/2}\Sigma^{1/2} = \Sigma$. Define $\mathbf{Z} = \left(\Sigma^{1/2}\right)^{-1}\mathbf{Y}$. What is the distribution of $\mathbf{Z}$?

10. Let Γ be an orthogonal matrix (i.e., Γ is $k \times k$ and $\Gamma'\Gamma = \Gamma\Gamma' = I_k$, where I_k is the k-dimensional identity matrix).

    (a) Prove that $||\Gamma\mathbf{y}|| = ||\mathbf{y}||$ for all k-dimensional vectors $\mathbf{y}$. That is, orthogonal transformations preserve length.

    (b) Prove that if $Y_1, \ldots, Y_k$ are iid normals, then the components of $\mathbf{Z} = \Gamma Y$ are also independent. That is, orthogonal transformations of iid normal random variables preserve independence.

11. **Helmert transformation** The *Helmert transformation* for iid $\text{N}(\mu, \sigma^2)$ random variables $Y_1, \ldots, Y_n$ is $\mathbf{Z} = H\mathbf{Y}$, where

$$H = \begin{pmatrix} \frac{1}{\sqrt{2}} & \frac{-1}{\sqrt{2}} & 0 & 0 & 0 & \cdots & 0 \\ \frac{1}{\sqrt{6}} & \frac{1}{\sqrt{6}} & -\sqrt{\frac{2}{3}} & 0 & 0 & \cdots & 0 \\ \vdots & \vdots & \vdots & \vdots & \vdots & \cdots & \vdots \\ \frac{1}{\sqrt{i(i+1)}} & \frac{1}{\sqrt{i(i+1)}} & \cdots & \frac{1}{\sqrt{i(i+1)}} & -\sqrt{\frac{i}{i+1}} & \cdots & 0 \\ \vdots & \vdots & \vdots & \vdots & \vdots & \cdots & \vdots \\ \frac{1}{\sqrt{n}} & \frac{1}{\sqrt{n}} & \frac{1}{\sqrt{n}} & \frac{1}{\sqrt{n}} & \frac{1}{\sqrt{n}} & \cdots & \frac{1}{\sqrt{n}} \end{pmatrix}$$

In row i, the number of $\{i(i+1)\}^{-1/2}$ terms is i.

    (a) Show that H is orthogonal.

    (b) Using the facts that $Z_n = n^{1/2}\bar{Y}$ and orthogonal transformations preserve length, prove that $\sum_{i=1}^{n-1} Z_i^2 = \sum_{i=1}^n Y_i^2 - n\bar{Y}^2 = \sum_{i=1}^n (Y_i - \bar{Y})^2$.

    (c) Using the representation in part b, prove that the sample mean and variance of a sample of size n from a $\text{N}(\mu, \sigma^2)$ distribution are independent.

    (d) Where does this argument break down if the Y_i are iid from a non-normal distribution?

8.7 Summary

1. The ch.f. of a random variable X is $\psi(t) = \psi_X(t) = \mathrm{E}\{\exp(itX)\}$.

   (a) $\psi(t)$ uniquely determines the distribution function of X.

   (b) $X_n \xrightarrow{D} X$ if and only if $\psi_{X_n}(t) \to \psi_X(t)$ for all $t \in R$.

2. The ch.f. of a k-dimensional random vector $\mathbf{X}$ is $\psi(\mathbf{t}) = \mathrm{E}\{\exp(i\mathbf{t}'\mathbf{X})\}$.

   (a) $\psi(\mathbf{t})$ uniquely determines the distribution function of X.

   (b) $\mathbf{X}_n \xrightarrow{D} \mathbf{X}$ if and only if $\psi_{\mathbf{X}_n}(\mathbf{t}) \to \psi_{\mathbf{X}}(\mathbf{t})$ for all $\mathbf{t} \in R^k$.

3. Standard CLT: if X_i are iid with mean μ and variance $\sigma^2 < \infty$, and $S_n = \sum_{i=1}^n X_i$, then $(S_n - n\mu)/(n\sigma^2)^{1/2} \xrightarrow{D} \mathrm{N}(0,1)$.

4. Lindeberg CLT: Let X_{ni}, $i = 1, \ldots, n$ be independent with mean μ_i and variance $\sigma_i^2 < \infty$, and let $S_n = \sum_{i=1}^n X_{ni}$. If the Lindeberg condition

$$\{\mathrm{var}(S_n)\}^{-1} \sum_{i=1}^n \mathrm{E}[(X_{ni} - \mu_i)^2 I\{(X_{ni} - \mu_i)^2 \geq \epsilon\, \mathrm{var}(S_n)\}] \to 0$$

   holds for all $\epsilon > 0$, then $\{S_n - \mathrm{E}(S_n)\}/\{\mathrm{var}(S_n)\}^{1/2} \xrightarrow{D} \mathrm{N}(0,1)$.

5. Multivariate CLT: If $\mathbf{X}_1, \mathbf{X}_2, \ldots$ are independent k-dimensional vectors with mean vector μ and covariance matrix Σ, and $\mathbf{S}_n = \sum_{i=1}^k \mathbf{X}_i$, then $(\mathbf{S}_n - n\boldsymbol{\mu})/n^{1/2} \xrightarrow{D} N(\mathbf{0}, \Sigma)$.

6. Cramer-Wold device: If $\mathbf{X}_n$ and $\mathbf{X}$ are k-dimensional vectors, then $\mathbf{X}_n \xrightarrow{D} \mathbf{X}$ if and only if $\mathbf{a}'\mathbf{X} \xrightarrow{D} \mathbf{a}'\mathbf{X}$ for all $\mathbf{a} \in R^k$.

7. Multivariate Slutsky: If $\mathbf{C}_n$, $\mathbf{D}_n$, and $\mathbf{X}_n$ are k-dimensional random vectors and $\mathbf{c}$ and $\mathbf{d}$ are k-dimensional constants such that $\mathbf{C}_n \xrightarrow{p} \mathbf{c}$, $\mathbf{D}_n \xrightarrow{p} \mathbf{d}$, and $\mathbf{X}_n \xrightarrow{D} \mathbf{X}$, then $(C_{n1}X_{n1}, \ldots, C_{nk}X_{nk})' + \mathbf{D}_n' \xrightarrow{D} (c_1X_1, \ldots, c_kX_k)' + \mathbf{d}'$.

Chapter 9

More on Convergence in Distribution

This chapter gives additional useful results about convergence in distribution, including conditions under which weak convergence of F_n to F is strengthened. The first such condition is when the limiting distribution F is continuous, in which case the convergence of $F_n(x)$ to $F(x)$ is uniform in x (Polya's theorem—Theorem 9.2). An even stronger condition is when F_n and F have density functions f_n and f and $f_n(x) \to f(x)$ for all x. In that case, not only does $P(X_n \in B)$ converge to $P(X \in B)$ for **all** Borel sets B, but the convergence is uniform over $B \in \mathcal{B}$ (Scheffé's theorem—Theorem 9.6). We then cover a very helpful technique, the delta method, for obtaining the asymptotic distribution of a function of random variables that converge in distribution. Next we explore what it means for $Y_n = (X_n - a_n)/b_n$ to converge in distribution to Y. Specifically, this does not necessarily imply that $\mathrm{E}(Y_n) \to \mathrm{E}(Y)$, although we determine additional conditions under which this does hold. We conclude with an investigation of the uniqueness of the normalizing sequences of numbers a_n and b_n for which $(X_n - a_n)/b_n$ converges in distribution to a non-degenerate random variable. We show that if $(X_n - a_n)/b_n$ and $(X_n - a'_n)/b'_n$ converge in distribution to non-degenerate random variables, then there is a very specific relationship between a_n and a'_n, between b_n and b'_n, and between the two asymptotic distribution functions.

9.1 Uniform Convergence of Distribution Functions

Suppose that a one-sided hypothesis test rejects the null hypothesis for large values of a statistic Y_n whose null distribution function $F_n(y)$ converges in distribution to F. Does that imply that the p-value $1 - F_n(y_n)$ will be close to $1 - F(y_n)$? The problem is that although $F_n(y)$ is close to $F(y)$ for **fixed** y if n is large, that does not necessarily mean that $F_n(y_n)$ is close to $F(y_n)$, as the following example shows.

Example 9.1. Test statistic converges in distribution, but p-value does not converge to that of the asymptotic distribution Suppose that $X_1, \ldots, X_n$ are uniformly distributed on the interval $[0, \theta]$, and we test the null hypothesis that $\theta = 1$ versus the alternative hypothesis that $\theta > 1$. It can be shown that the most powerful test rejects the null hypothesis for large values of $Y_n = \max(X_1, \ldots, X_n)$. Suppose that the null hypothesis is true, so $X_i \sim \text{uniform}\,[0, 1]$. The distribution function $F_n(y)$ for Y_n is

$F_n(y) = P(X_1 \le y, \ldots, X_n \le y) = y^n$. The critical value for a test at level α rejects the null hypothesis if $Y_n \ge y_{n,\alpha}$, where $y_{n,\alpha} = (1-\alpha)^{1/n}$. Notice that for each fixed y, $F_n(y) \to F(y)$, where

$$F(y) = \begin{cases} 0 & 0 \le y < 1 \\ 1 & y = 1. \end{cases}$$

Nonetheless, $(1-\alpha) = F_n(y_{n,\alpha})$ is not close to $F(y_{n,\alpha}) = 0$. This demonstrates that the convergence of $F_n(y)$ to $F(y)$ for each y does not necessarily imply that for an arbitrary sequence of numbers y_n, $F_n(y_n)$ is close to $F(y_n)$ for large n. This is another reminder of why convergence in distribution is also called weak convergence.

We can recast this example in terms of p-values. The (random) p-value is $1 - Y_n^n$. For arbitrary p, the probability that this p-value is p or less is $P\{Y_n \ge (1-p)^{1/n}\} = 1 - \{(1-p)^{1/n}\}^n = p$. This is not surprising; because the null distribution of Y_n is continuous, the p-value has a uniform distribution on $[0,1]$. Thus, there is a 50 percent chance that the p-value will be $1/2$ or less. On the other hand, if we approximate the p-value by $1 - F(Y_n)$, then with probability 1 this "approximate" p-value will be 1. □

Example 9.1 is disconcerting because it is quite common to compute an "approximate" p-value using the asymptotic distribution of a test statistic. The example shows the fallacy of the argument that the p-value $1 - F_n(z_n)$ must be close to $1 - F(z_n)$ for large n if F_n converges weakly to F. The reader may feel that Example 9.1 is unfair because we did not standardize the test statistic, and this caused its limiting distribution to be degenerate. When people speak of "asymptotic distributions," they usually mean that the limiting distribution is non-degenerate. Nonetheless, there are non-degenerate counterexamples to the proposition that "approximate" p-values using the asymptotic distribution are close to the p-value using the actual distribution (see Exercise 1). The shared feature among these anomalies is that the limiting distribution is discontinuous. Fortunately, in most statistical applications, the limiting distribution of the standardized estimator or test statistic is continuous (e.g., standard normal). We will see that the following result is instrumental in showing that it **is** valid to approximate a p-value using the asymptotic distribution when that distribution is continuous.

Theorem 9.2. Polya's theorem: convergence in distribution to a continuous distribution implies uniform convergence *Suppose that* $X_n \sim F_n(x) \to F(x)$ *as* $n \to \infty$ *for each* x*, where* $F(x)$ *is a continuous distribution function. Then* $\sup_x |F_n(x) - F(x)| \to 0$.

Proof. Let $\epsilon > 0$. We must show that we can find an N such that $|F_n(x) - F(x)| < \epsilon$ for all x and $n \ge N$.

Because F_n converges in distribution, F_n is a tight sequence of distribution functions by Corollary 6.50. Therefore, we can find a value t_1 such that $P(|X_n| \ge t_1) < \epsilon/2$ for all n. Also, there is a t_2 such that $P(|X| \ge t_2) < \epsilon/2$ (why?). Let $T = \max(t_1, t_2)$. Then for $x < -T$,

$$\begin{aligned} |F_n(x) - F(x)| &\le |F_n(x)| + |F(x)| \le F_n(-T) + F(-T) \\ &< \epsilon/2 + \epsilon/2 = \epsilon. \end{aligned} \tag{9.1}$$

For $x > T$,

$$\begin{aligned} |F_n(x) - F(x)| &= |1 - F(x) - \{1 - F_n(x)\}| \le |1 - F(x)| + |1 - F_n(x)| \\ &\le 1 - F(T) + 1 - F_n(T) < \epsilon/2 + \epsilon/2 = \epsilon. \end{aligned} \tag{9.2}$$

Now consider the closed interval $[-T, T]$. Because the limiting distribution function F is continuous on the compact set $[-T, T]$, F is uniformly continuous on $[-T, T]$ by Proposition A.62. It follows that we can find a δ such that

$$|F(y) - F(x)| < \epsilon/5 \text{ for all } x \text{ and } y \text{ with } |y - x| < \delta.$$

1. Divide the interval $[-T, T]$ into M equal intervals of length $2T/M$, where M is chosen large enough that $2T/M < \delta$: $E_0 = -T, E_1 = -T + 2T/M, \ldots, E_M = T$ (Figure 9.1).

2. Because the left and right endpoints $L_i = E_i$ and $R_i = E_{i+1}$ of each interval satisfy $R_i - L_i < \delta$, $|F(R_i) - F(L_i)| < \epsilon/5$ for each $i = 0, \ldots, M$.

3. Choose N large enough that $n \geq N \Rightarrow \max_{0 \leq i \leq M} |F_n(E_i) - F(E_i)| < \epsilon/5$. This is possible because $F_n(E_i) \to F(E_i)$ as $n \to \infty$ for the finite set $i = 0, \ldots, M$.

Now consider an arbitrary x in the interval $[-T, T]$. Then x must lie within one of the M intervals, say $[L_i, R_i)$. For $n \geq N$,

$$\begin{aligned}
&|F_n(x) - F(x)| \\
&= |F_n(x) - F_n(L_i) + F_n(L_i) - F(L_i) + F(L_i) - F(x)| \\
&\leq |F_n(x) - F_n(L_i)| + |F_n(L_i) - F(L_i)| + |F(L_i) - F(x)| \\
&\leq |F_n(R_i) - F_n(L_i)| + \max_{0 \leq i \leq M} |F_n(L_i) - F(L_i)| + \max_{0 \leq i \leq M} |F(L_i) - F(R_i)| \\
&< |F_n(R_i) - F_n(L_i)| + \epsilon/5 + \epsilon/5 \\
&\leq |F_n(R_i) - F(R_i) + F(R_i) - F(L_i) + F(L_i) - F_n(L_i)| + 2\epsilon/5 \\
&\leq |F_n(R_i) - F(R_i)| + |F(R_i) - F(L_i)| + |F(L_i) - F_n(L_i)| + 2\epsilon/5 \\
&\leq \max_{0 \leq i \leq M} |F_n(R_i) - F(R_i)| + \max_{0 \leq i \leq M} |F(R_i) - F(L_i)| + \max_{0 \leq i \leq M} |F(L_i) - F_n(L_i)| + 2\epsilon/5 \\
&< \epsilon/5 + \epsilon/5 + \epsilon/5 + 2\epsilon/5 = \epsilon,
\end{aligned}$$

completing the proof. □

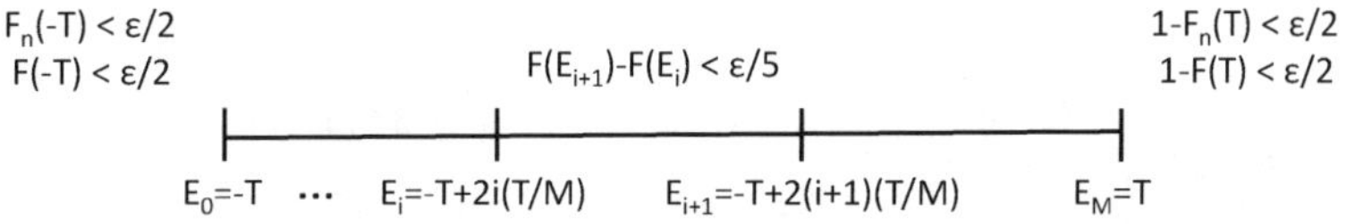

Figure 9.1: Division of $[-T, T]$ into M equal intervals of width $2T/M < \delta$.

Example 9.3. Using continuous asymptotic distributions to approximate p-values is valid The null distribution functions of many standardized test statistics are continuous: standard normal when the CLT applies, chi-squared for Wald tests, etc. If F_n and F denote the actual and asymptotic distribution functions of a standardized test statistic Z_n, the absolute value of the difference between the actual and approximate one-sided p-values is $|F_n(Z_n) - F(Z_n)| \leq \sup_z |F_n(z) - F(z)|$. By Polya's theorem, this difference tends to 0. Therefore, using the asymptotic distribution to approximate a p-value is valid if the asymptotic distribution is continuous. □

Polya's theorem implies uniform convergence of not just $P(X_n \leq x)$ to $P(X \leq x)$, but of $P(X_n \in I)$ to $P(X \in I)$ for all intervals:

Corollary 9.4. Uniform convergence over all intervals *Let $X_n \sim F_n(x) \to F(x)$ as $n \to \infty$ for each x, where $F(x)$ is a continuous distribution function. If I ranges over all intervals, then $\sup_I |P(X_n \in I) - P(X \in I)| \to 0$ as $n \to \infty$.*

Proof. Uniformity of convergence for intervals of the form $(a, b]$, a and b finite, follows from

$$\begin{aligned} \sup_{a,b} |P(X_n \in (a,b]) - P(X \in (a,b])| &= \sup_{a,b} |F_n(b) - F_n(a) - \{F(b) - F(a)\}| \\ &\leq \sup_b |F_n(b) - F(b)| + \sup_a |F_n(a) - F(a)| \end{aligned}$$

and the uniform convergence of F. Also, $\sup_x |P(X_n = x) - P(X = x)| = \sup_x |P(X_n = x)|$

$$\begin{aligned} &\leq \sup_x |F_n(x) - F_n(x - 1/n)| \\ &\leq \sup_x |F_n(x) - F(x) + F(x) - F(x - 1/n) + F(x - 1/n) - F_n(x - 1/n)| \\ &\leq \sup_x |F_n(x) - F(x)| + \sup_x |F(x) - F(x - 1/n)| + \sup_x |F(x - 1/n) - F_n(x - 1/n)|. \end{aligned}$$

The first and third terms tend to 0 as $n \to \infty$ by the uniform convergence of F. Because the limiting distribution F is continuous, F is uniformly continuous (Proposition 4.23). Therefore, the middle term also tends to 0 as $n \to \infty$.

For sets of the form $[a, b]$, a and b finite, use the fact that $P(X_n \in [a, b]) = P(X_n = a) + P(X_n \in (a, b])$ in conjunction with the results we have just proven. The proof for other types of intervals is similar and left as an exercise. □

Polya's theorem can also be extended to higher dimensions. The following is a consequence of a more general result of Ranga Rao (1962).

Theorem 9.5. Polya's theorem in $\boldsymbol{R^k}$: uniform convergence over product sets of intervals *If $\mathbf{X}_n \sim F_n(x_1, \ldots, x_k)$ converges in distribution to $\mathbf{X} \sim F(x_1, \ldots, x_k)$, where F is continuous, then: $F_n(\mathbf{x})$ converges uniformly to $F(x)$. Furthermore, $\sup_I |P(\mathbf{X}_n \in I) - P(\mathbf{X} \in I)| \to 0$, where I ranges over all product sets $I_1 \times I_2 \times \ldots \times I_k$, each I_j being an interval.*

The proof for $k = 2$ is left as an exercise. Ranga Rao (1962) proves a more general result implying Theorem 9.5.

Another useful result concerns convergence of density functions. In general, weak convergence of X_n to X does not imply convergence of all event probabilities; i.e., we cannot conclude that $P(X_n \in B) \to P(X \in B)$ for all Borel sets. For instance, the discrete uniform distribution putting mass $1/n$ on $1/n, \ldots, n/n = 1$ converges in distribution to $U[0, 1]$ as $n \to \infty$, yet the probabilities of the set of rational numbers are 1 and 0 under the discrete and continuous uniform distributions, respectively. In fact, it is not even the case that $F_n(x) \to F(x)$ for all x. Rather, convergence of X_n to X in distribution implies that $F_n(x) \to F(x)$ when x is a continuity point of the limiting distribution F. However, if X_n and X have density functions and $f_n(x) \to f(x)$, a much stronger result holds, as we see from Scheffé's theorem below.

Theorem 9.6. Scheffé's theorem: convergence of densities implies uniform convergence over Borel sets *Let X_n and X be random variables with density functions*

$f_n(x)$ and $f(x)$ with respect to a measure μ. If $f_n(x) \to f(x)$ except on a set of μ-measure 0, then $\int |f_n(x) - f(x)|d\mu(x) \to 0$ and $\sup_{B \in \mathcal{B}} |P(X_n \in B) - P(X \in B)| \to 0$.

Proof.

$$\begin{aligned}
&\int |f_n(x) - f(x)|d\mu(x) \\
&= \int \{f(x) - f_n(x)\}I\{f(x) - f_n(x) \geq 0\}d\mu(x) + \int \{f_n(x) - f(x)\}I\{f_n(x) - f(x) > 0\}d\mu(x) \\
&= \int \{f(x) - f_n(x)\}I\{f(x) - f_n(x) \geq 0\}d\mu(x) \\
&\quad + \int \{f_n(x) - f(x)\}[1 - I\{f_n(x) - f(x) \leq 0\}]d\mu(x) \\
&= \int \{f(x) - f_n(x)\}I\{f(x) - f_n(x) \geq 0\}d\mu(x) + \int f_n(x)d\mu(x) - \int f(x)d\mu(x) \\
&\quad + \int \{f(x) - f_n(x)\}I\{f(x) - f_n(x) \geq 0\}d\mu(x) \\
&= 2\int \{f(x) - f_n(x)\}I\{f(x) - f_n(x) \geq 0\}d\mu(x) + 1 - 1 \\
&= 2\int \{f(x) - f_n(x)\}I\{f(x) - f_n(x) \geq 0\}d\mu(x). \qquad (9.3)
\end{aligned}$$

Note that $\{f(x) - f_n(x)\}I\{f(x) - f_n(x) \geq 0\} \to 0$ except on a set of μ-measure 0. Also, $|\{f(x) - f_n(x)\}I\{f(x) - f_n(x) \geq 0\}| \leq f(x)$ and f is integrable because $\int f(x)d\mu(x) = 1$. By the DCT, $\int \{f(x) - f_n(x)\}I\{f(x) - f_n(x) \geq 0\}d\mu(x) \to 0$. Therefore, $\int |f_n(x) - f(x)|d\mu(x) \to 0$.

For any Borel set B,

$$\begin{aligned}
|P(X_n \in B) - P(X \in B)| &= \left|\int_B f_n(x)d\mu(x) - \int_B f(x)d\mu(x)\right| \\
&\leq \int_B |f_n(x) - f(x)|d\mu(x) \\
&\leq \int_{-\infty}^{\infty} |f_n(x) - f(x)|d\mu(x) \to 0. \qquad (9.4)
\end{aligned}$$

Because $\int_{-\infty}^{\infty} |f_n(x) - f(x)|d\mu(x)$ is not a function of B, $\sup_{B \in \mathcal{B}} |P(X_n \in B) - P(X \in B)| \to 0$. □

Exercises

1. Let X_n have density function $f_n(x) = 1 + \cos(2\pi x)/n$ for $X \in [0, 1]$. Prove that $P(X_n \in B)$ converges to the Lebesgue measure of B for every Borel set B.

2. Consider the t-density

$$f_\nu(x) = \frac{\Gamma\left(\frac{\nu+1}{2}\right)(1 + x^2/\nu)^{-(\nu+1)/2}}{\Gamma(\nu/2)\sqrt{\pi\nu}}.$$

Stirling's formula says that $\Gamma(x)/\{\exp(-x)x^{x-1/2}(2\pi)^{1/2}\} \to 1$ as $x \to \infty$. Use this fact to prove that the t-density converges to the standard normal density as $\nu \to \infty$. If Z and T_ν denote standard normal and T random variables with ν degrees of freedom, for what sets A does $P(T_\nu \in A)$ converge to $P(Z \in A)$ as $\nu \to \infty$? Is the convergence uniform?

3. Prove part of Theorem 9.5 in R^2, namely that if $(X_n, Y_n) \sim F_n(x, y)$ converges in distribution to $(X, Y) \sim F(x, y)$ and F is continuous, then F_n converges uniformly to F. Do this in 3 steps: (1) for given $\epsilon > 0$, prove there is a bound B such that $P(\{|X| > B\} \cup \{|Y| > B\}) < \epsilon/2$ and $P(\{|X_n| > B\} \cup \{|Y_n| > B\}) < \epsilon/2$ for all n; (2) use the fact that F is continuous on the compact set $C = [-B, B] \times [-B, B]$ to divide C into squares such that $|F_n(x, y) - F(x, y)|$ is arbitrarily small for (x, y) corresponding to the corners of the squares; (3) Use the fact that, within each square, $|F_n(x_2, y_2) - F_n(x_1, y_1)|$ and $|F(x_2, y_2) - F(x_1, y_1)|$ are maximized when (x_1, y_1) and (x_2, y_2) are at the "southwest" and "northeast" corners.

4. Recall that in Exercise 11 of Section 3.4, there are n people, each with a different hat. The hats are shuffled and passed back in random order. Let Y_n be the number of people who get their own hat back. You used the inclusion-exclusion formula to see that $P(Y_n \geq 1) \to 1 - \exp(-1)$. Extend this result by proving that $P(Y_n = k) \to P(Y = k)$ as $n \to \infty$, where Y has a Poisson distribution with parameter 1. Conclude that $\sup_A |P(Y_n \in A) - P(Y \in A)| \to 0$, where the supremum is over all subsets of $\Omega = \{0, 1, 2, \ldots\}$. Hint: $P(Y_n = k) = \binom{n}{k} P$(the first k people get their own hat back and none of the remaining $n - k$ people get their own hat back).

9.2 The Delta Method

We often use transformations in statistics. For example, we might apply a logarithmic transformation to reduce skew. If we know that the original statistic is asymptotically normal, what can we conclude about the distribution of the transformed statistic? It seems counterintuitive, but the transformed statistic is also asymptotically normal under mild conditions. The technique for showing this is called the delta method, illustrated in the following example.

Example 9.7. Reducing skew Example 8.5 showed that the distribution of the variance s^2 of a sample of n iid observations is asymptotically normal, but for smaller sample sizes, s^2 has a right-skewed distribution even if the observations are normally distributed. For example, suppose that the X_i are iid N(0, 1). The exact density $f_{n-1}(y)$ for $Y = (n-1)s^2/\sigma^2 = (n-1)s^2$ is chi-squared with $n - 1$ degrees of freedom:

$$f_{n-1}(y) = \frac{y^{(n-1)/2-1}\exp(-y/2)}{2^{(n-1)/2}\Gamma\{(n-1)/2\}}.$$

This density and its normal approximation with the same mean and variance are displayed for $n = 10$ as solid and dotted lines, respectively, in the top panel of Figure 9.2. The normal approximation does not fit well because $f_9(y)$ is right-skewed.

Now consider a logarithmic transformation $U = \ln(Y)$ of $Y = (n-1)s^2$. The density for U is

$$g_{n-1}(u) = \frac{\exp\{(n-1)u/2\}\exp\{-\exp(u)/2\}}{2^{(n-1)/2}\{\Gamma(n-1)/2\}}.$$

The following heuristic argument suggests that we can approximate g_{n-1} by a normal density as well. Use a Taylor series approximation to the function $h(x) = \ln(x)$ expanded about x_0: $h(x) \approx h(x_0) + h'(x_0)(x - x_0) = \ln(x_0) + (1/x_0)(x - x_0)$, which fits well when x is close to x_0. Apply this approximation with $x = s^2$ and $x_0 = \sigma^2 = 1$; s^2 will be close to 1 because s^2 converges almost surely to σ^2 (Example 7.6). Therefore, $\ln(s^2) \approx \ln(1) + (1/1)(s^2 - 1) = s^2 - 1$, which is asymptotically normal with mean 0 and variance $2/(n-1)$. It follows that $\ln\{(n-1)s^2\}$ should be approximately normally distributed with mean $\ln(n-1)$ and variance $2/(n-1)$. The bottom panel of Figure 9.2 shows the exact density $g_{n-1}(y)$ (solid line) and normal approximation for $n = 10$. The normal approximation fits much better to $\ln\{(n-1)s^2\}$ (bottom panel) than to $(n-1)s^2$ (top panel). □

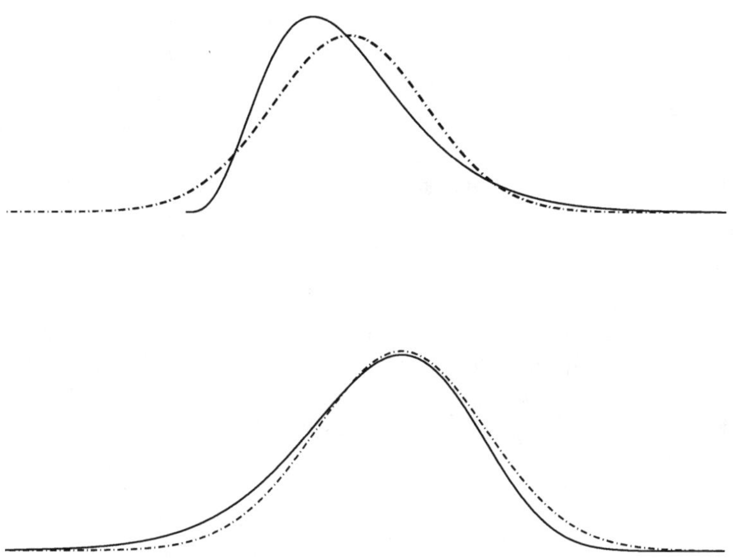

Figure 9.2: Top: The exact density of $(n-1)s^2$ (solid line) and its asymptotic normal approximation (dashed line). Bottom: The exact density of $\ln\{(n-1)s^2\}$ (solid line) and its asymptotic normal approximation (dashed line).

To make the above argument for asymptotic normality rigorous, we must do better than saying that $\ln(x)$ is approximately $\ln(x_0) + (1/x_0)(x - x_0)$. The following result fills in the details.

Proposition 9.8. Delta method: the distribution of a function of an estimator *Suppose that* $X_n = a_n(\hat{\theta}_n - \theta) \xrightarrow{D} X$, *where* a_n *is a sequence of numbers tending to* ∞ *as* $n \to \infty$. *If* $f(x)$ *is a function with derivative* $f'(\theta)$ *at* $x = \theta$, *then*

$$a_n\{f(\hat{\theta}_n) - f(\theta)\} \xrightarrow{D} f'(\theta)X.$$

Proof. Note first that $\hat{\theta}_n \xrightarrow{p} \theta$. This follows from the fact that, for $a_n \neq 0$, $|\hat{\theta}_n - \theta| = |X_n|/|a_n| \xrightarrow{p} 0$ because $X_n = a_n(\hat{\theta}_n - \theta) \xrightarrow{D} X$ and $|a_n| \to \infty$. The caveat that $a_n \neq 0$ is no restriction because $a_n \to \infty$. Furthermore,

$$a_n\left\{f(\hat{\theta}_n) - f(\theta)\right\} = g(\hat{\theta}_n)a_n\left(\hat{\theta}_n - \theta\right) \tag{9.5}$$

where

$$g(x) = \begin{cases} \frac{f(x)-f(\theta)}{x-\theta} & \text{if } x \neq \theta \\ f'(\theta) & \text{if } x = \theta. \end{cases}$$

Note that $g(x)$ is continuous at $x = \theta$. Because $\hat{\theta}_n \xrightarrow{P} \theta$, the Mann-Wald theorem (Theorem 6.51) implies that $g(\hat{\theta}_n) \xrightarrow{P} g(\theta) = f'(\theta)$. By assumption, $a_n(\hat{\theta}_n - \theta) \xrightarrow{D} X$, so the result now follows from Slutsky's theorem applied to Expression (9.5). □

Remark 9.9. Non-degenerate limiting distributions *Although Proposition 9.8 does not require $f'(\theta) \neq 0$, the distribution of $f'(\theta)X$ is degenerate if $f'(\theta) = 0$. The statement that Y_n is "asymptotically normal" or "asymptotically chi-squared," etc., means that there is a sequence of numbers a_n and b_n such that $(Y_n - a_n)/b_n \xrightarrow{D} Y$, where Y is a* **non-degenerate** *normal random variable or* **non-degenerate** *chi-squared random variable, etc. Thus, in the following corollary, we impose the constraint that $f'(\theta)$.*

Corollary 9.10. Asymptotic normality *Suppose that $\hat{\theta}_n$ is asymptotically $N(\theta, \sigma_n^2)$, with $\sigma_n \to 0$. If $f : R \mapsto R$ satisfies $f'(\theta) \neq 0$, then $f(\hat{\theta}_n)$ is asymptotically $N\Big(f(\theta), \{f'(\theta)\}^2\sigma_n^2\Big)$.*

It may seem incongruous that a nonlinear function of an asymptotically normal random variable may also be asymptotically normal. For example, let $\bar{X}$ be the sample mean of n iid normals with mean $\mu \neq 0$ and variance 1. Then $n^{1/2}\bar{X} \sim \text{N}(n^{1/2}\mu, 1)$, so $(n^{1/2}\bar{X})^2$ has a noncentral chi-squared distribution with 1 degree of freedom and noncentrality parameter $(n^{1/2}\mu)^2$. Therefore, how can $(n^{1/2}\bar{X})^2$ be asymptotically normal? This is actually not a contradiction. As the noncentrality parameter tends to infinity, the shape of the noncentral chi-squared density with 1 degree of freedom is approximately normal. However, it is an exercise to show that $\bar{X}^2$ is not asymptotically normal if $\mu = 0$.

Transformations are also used to "stabilize the variance." This happens when the variance of an estimator depends on the parameter it estimates, as in the following example.

Example 9.11. Stabilizing variance In epidemiological studies we are often interested in estimating disease rates, namely the number of cases of disease for a given number of person years of study. Assume that the number of cases N over a total follow-up time of t years has a Poisson distribution with mean λt. Our estimator of the rate λ is simply $\hat{\lambda} = N/t$. Notice that the variance of $\hat{\lambda}$ is $\lambda t/t^2 = \lambda/t$. That is, the variance of $\hat{\lambda}$ depends on λ. Under this circumstance, estimation is usually improved by transforming the estimator via $f(\hat{\lambda})$ for a function f chosen to render the variance of $f(\hat{\lambda})$ free of λ.

As $t \to \infty$, $\hat{\lambda}$ becomes asymptotically normal (see Exercise 3 of Section 8.1). The delta method shows that $f(\hat{\lambda})$ is asymptotically normally distributed with mean $f(\lambda)$ and variance $\{f'(\lambda)\}^2\text{var}(\hat{\lambda}) = \{f'(\lambda)\}^2\lambda/t$. To make this variance free of λ, set $\{f'(\lambda)\}^2\lambda = 1$ and solve for f. This yields $f'(\lambda) = \lambda^{-1/2}$, so $f(\lambda) = 2\lambda^{1/2}$. Clearly, multiplication by 2 is unnecessary, so the square root transformation works in this example. That is, $\hat{\lambda}^{1/2}$ is asymptotically normal with mean $\lambda^{1/2}$ and variance $\{(1/2)\lambda^{-1/2}\}^2(\lambda/t) = 1/(4t)$. For example, a 95% confidence interval for $\lambda^{1/2}$ is $\hat{\lambda}^{1/2} \pm 1.96/(2t^{1/2})$. We then transform this interval to a 95% confidence interval for λ by squaring the limits of the confidence interval for $\lambda^{1/2}$:

$$\left(\left\{\left(\frac{N}{t}\right)^{1/2} - \frac{1.96}{2\sqrt{t}}\right\}^2, \left\{\left(\frac{N}{t}\right)^{1/2} + \frac{1.96}{2\sqrt{t}}\right\}^2\right).$$

For instance, suppose that 100 people have each amassed 2 years of follow-up, and there have been 34 events. Then $t = 100 \times 2 = 200$ years, $\hat{\lambda} = 34/200 = 0.17$ events per year, and the 95% confidence interval for λ is

$$\left(\left\{\left(\frac{34}{200}\right)^{1/2}-\frac{1.96}{2\sqrt{200}}\right\}^2,\left\{\left(\frac{34}{200}\right)^{1/2}+\frac{1.96}{2\sqrt{200}}\right\}^2\right)=(0.12,0.23).$$

Therefore, we can be 95% confident that the true disease rate is between 0.12 per person year and 0.23 per person year.

□

Example 9.12. Asymptotic distribution of relative risk Consider the comparison of two groups with respect to a binary outcome like progression of disease. Let $\hat{p}_1$ and $\hat{p}_2$ denote the sample proportions with disease progression in the two groups, and $p_1 > 0$ and $p_2 > 0$ denote their expectations. The relative risk estimate is $\hat{p}_1/\hat{p}_2$. To determine the asymptotic distribution of the relative risk estimator, we first take logs: $\ln(\hat{p}_1/\hat{p}_2) = \ln(\hat{p}_1) - \ln(\hat{p}_2)$. Apply the delta method to $\ln(\hat{p})$. By the CLT, $\hat{p}$ is AN$(p, p(1-p)/n)$. Then $f(\hat{p}) = \ln(\hat{p})$ satisfies $f'(p) = 1/p \neq 0$. By Corollary 9.10, $\ln(\hat{p})$ is AN$[f(p), \{f'(p)\}^2 p(1-p)/n] = [\ln(p), (1-p)/(np)]$. Also, $\ln(\hat{p}_1)$ and $\ln(\hat{p}_2)$ are independent. Therefore, the asymptotic distribution of $\ln(\hat{p}_1) - \ln(\hat{p}_2)$ is normal with mean $\ln(p_1) - \ln(p_2)$ and variance $(1-p_1)/(n_1p_1) + (1-p_2)/(n_2p_2)$, where n_1 and n_2 are the sample sizes in the two groups. By Slutsky's theorem, we can replace p_1 and p_2 with $\hat{p}_1$ and $\hat{p}_2$. Thus,

$$\frac{\ln(\hat{p}_1/\hat{p}_2)-\ln(p_1/p_2)}{\sqrt{(1-\hat{p}_1)/(n_1\hat{p}_1)+(1-\hat{p}_2)/(n_2\hat{p}_2)}} \stackrel{D}{\to} \mathrm{N}(0,1),$$

which can be used to construct the following asymptotically valid $100(1-\alpha)$ percent confidence intervals for the logarithm of the relative risk:

$$\ln(\hat{p}_1/\hat{p}_2) \pm z_{\alpha/2}\sqrt{(1-\hat{p}_1)/(n_1\hat{p}_1)+(1-\hat{p}_2)/(n_2\hat{p}_2)},$$

where $z_{\alpha/2}$ is such that $1 - \Phi(z_{\alpha/2}) = \alpha/2$. □

Could we compute the asymptotic distribution of $\hat{p}_1/\hat{p}_2$ without using a log transformation? We might hope to use Slutsky's theorem to find the asymptotic distribution of such a ratio, as we did for the t-statistic $n^{1/2}\bar{X}/s$ when the X_i are iid with mean 0 and variance $\sigma^2 < \infty$. The difference is that for the t-statistic, the numerator, $n^{1/2}\bar{X}$, converges in distribution and the denominator, s, converges to a constant, σ. Such is not the case with $\hat{p}_1/\hat{p}_2$. In such a setting, we need a bivariate version of the delta method.

Proposition 9.13. Bivariate delta method: asymptotic distribution of a function of two estimators *Suppose that* $a_n(\hat{\delta}_n - \delta, \hat{\theta}_n - \theta) \stackrel{D}{\to} (X, Y)$, *where* a_n *is a sequence of numbers tending to* ∞ *as* $n \to \infty$. *If* $f : R^2 \longmapsto R$ *is differentiable at* (δ, θ) *and has partial derivatives* $f_\delta(\delta, \theta)$ *and* $f_\theta(\delta, \theta)$, *then*

$$a_n\{f(\hat{\delta}_n, \hat{\theta}_n) - f(\delta, \theta)\} \stackrel{D}{\to} f_\delta(\delta, \theta)X + f_\theta(\delta, \theta)Y.$$

Proof. If a function $g(x, y)$ is differentiable at (x_0, y_0), then the partial derivatives $g_x(x_0, y_0) = \partial g/\partial x|_{(x_0,y_0)}$ and $g_y(x_0, y_0) = \partial g/\partial y|_{(x_0,y_0)}$ exist and comprise the derivative of $g(x, y)$ at (x_0, y_0) (see Section A.6.3). By definition of derivative,

$$\frac{g(x,y)-g(x_0,y_0)-g_x(x_0,y_0)(x-x_0)-g_y(x_0,y_0)(y-y_0)}{||(x-x_0,y-y_0)||} \to 0 \qquad (9.6)$$

as $(x, y) \to (x_0, y_0)$, $(x, y) \neq (x_0, y_0)$.

By the multivariate Skorokhod representation theorem (Theorem 6.60), we can assume that $(\hat{\delta}_n, \hat{\theta}_n, X, Y)$ are defined on the same probability space and $a_n(\hat{\delta}_n - \delta, \hat{\theta}_n - \theta) \stackrel{a.s.}{\to} (X, Y)$. By Equation (9.6) with g replaced by f,

$$a_n\{f(\hat{\delta}_n, \hat{\theta}_n) - f(\delta, \theta)\} = f_\delta(\delta, \theta)a_n(\hat{\delta}_n - \delta) + f_\theta(\delta, \theta)a_n(\hat{\theta}_n - \theta) + a_n R_n, \quad (9.7)$$

where $R_n/||(\hat{\delta}_n - \delta, \hat{\theta}_n - \theta)|| \to 0$ for each ω such that $(\hat{\delta}_n, \hat{\theta}_n) \to (\delta, \theta)$, $(\hat{\delta}_n, \hat{\theta}_n) \neq (\delta, \theta)$. Write $a_n R_n$ as

$$\begin{cases} a_n||(\hat{\delta}_n - \delta, \hat{\theta}_n - \theta)||\frac{R_n}{||(\hat{\delta}_n - \delta, \hat{\theta}_n - \theta)||} & \text{if } (\hat{\delta}_n, \hat{\theta}_n) \neq (\delta, \theta) \\ 0 & \text{if } (\hat{\delta}_n, \hat{\theta}_n) = (\delta, \theta). \end{cases}$$

As $n \to \infty$, $(\hat{\delta}, \hat{\theta}) \stackrel{a.s.}{\to} (\delta, \theta)$ because $a_n(\hat{\delta} - \delta, \hat{\theta} - \theta) \stackrel{a.s.}{\to} (X, Y)$ and $a_n \to \infty$. Therefore, $a_n R_n \stackrel{a.s.}{\to} ||(X, Y)|| \cdot 0 = 0$ as $n \to \infty$. The right side of Equation (9.7) converges almost surely, and therefore in distribution, to $f_\delta(\delta, \theta)X + f_\theta(\delta, \theta)Y$, completing the proof.

□

Remark 9.14. Non-degenerate limiting distribution *As with Proposition 9.8, Proposition 9.13 does not require any additional conditions, but the limiting random variable is degenerate if both $f_x(\delta, \theta)$ and $f_y(\delta, \theta)$ are 0. For this reason, the next corollary assumes that at least one of these partial derivatives is nonzero.*

Corollary 9.15. Asymptotic normality *Suppose that $a_n(\hat{\delta} - \delta, \hat{\theta} - \theta)$ converges in distribution to a bivariate normal random variable with mean vector $(0, 0)'$, finite variances (σ_1^2, σ_2^2), and correlation $\rho \in (-1, 1)$, where a_n is a sequence of numbers converging to ∞ as $n \to \infty$. If f is differentiable at (δ, θ) and at least one of the two partial derivatives $f_\delta(\delta, \theta)$ and $f_\theta(\delta, \theta)$ is nonzero, then $f(\hat{\delta}, \hat{\theta})$ is asymptotically normal with mean $f(\delta, \theta)$ and variance $\{f_\delta(\delta, \theta)\}^2\sigma_1^2 + \{f_\theta(\delta, \theta)\}^2\sigma_2^2$.*

Example 9.16. Asymptotic distribution of relative risk: another proof In the context of Example 9.12, consider the asymptotic distribution of $\hat{p}_1/\hat{p}_2$ using Proposition 9.13. Suppose that $n_2/n_1 \to \lambda$. For instance, in a clinical trial with a k:1 group 2 to group 1 allocation ratio, $\lambda = k$. Then $n_1^{1/2}(\hat{p}_1 - p_1, \hat{p}_2 - p_2) \stackrel{D}{\to} (X, Y)$, where X and Y are independent normals with zero means and variances $p_1(1 - p_1)$ and $p_2(1 - p_2)/\lambda$. Let $f(x, y) = x/y$. Then $(\partial f/\partial x)|_{(p_1, p_2)} = 1/p_2$ and $(\partial f/\partial y)|_{(p_1, p_2)} = -p_1/p_2^2$. By Proposition 9.13 with $a_{n_1} = n_1^{1/2}$,

$$n_1^{1/2}(\hat{p}_1/\hat{p}_2 - p_1/p_2) \stackrel{D}{\to} (1/p_2)X - (p_1/p_2^2)Y.$$

The right side is normal with mean 0 and variance

$$\frac{p_1(1 - p_1)}{p_2^2} + \frac{p_1^2 p_2(1 - p_2)}{\lambda p_2^4} = \frac{p_1(1 - p_1)}{p_2^2} + \frac{p_1^2(1 - p_2)}{\lambda p_2^3}.$$

Using Slutsky's theorem in conjunction with the fact that $n_2/n_1 \to \lambda$, we conclude that

$$\frac{\hat{p}_1/\hat{p}_2 - p_1/p_2}{\sqrt{\frac{p_1(1-p_1)}{n_1 p_2^2} + \frac{p_1^2(1-p_2)}{n_2 p_2^3}}} \stackrel{D}{\to} Z, \quad (9.8)$$

where $Z \sim \mathrm{N}(0, 1)$. □

One arrives at the same asymptotic distribution for $\hat{p}_1/\hat{p}_2$ using the asymptotic distribution of $\hat{\theta} = \ln(\hat{p}_1/\hat{p}_2)$ derived in Example 9.12 and applying the delta method to $f(\hat{\theta}) = \exp(\hat{\theta})$ (exercise).

9.3 Convergence of Moments: Uniform Integrability

It is important to know precisely what a statement like "X_n is asymptotically $\mathrm{N}(\mu_n, \sigma_n^2)$" means, namely that $(X_n - \mu_n)/\sigma_n \stackrel{D}{\to} \mathrm{N}(0,1)$. It does **not** mean that $\mathrm{E}(X_n) = \mu_n$ and $\mathrm{var}(X_n) = \sigma_n^2$. More generally, the fact that $X_n \stackrel{D}{\to} X$ does not mean that $\mathrm{E}(X_n) \to \mathrm{E}(X)$, as the following example demonstrates.

Example 9.17. Estimator has infinite mean, but limiting random variable has finite mean If $\hat{p}_n$ is the sample proportion of ones among iid Bernoulli random variables $X_1, \ldots, X_n$, then by the CLT, $\hat{p}_n$ is asymptotically normal with mean $p = \mathrm{E}(X_1)$ and variance $p(1-p)/n$. It follows by the delta method that $\ln(\hat{p})$ is asymptotically normal with mean $\ln(p)$ and variance $(1-p)/(np)$ (see Example 9.12). That is, $Z_n = \{\ln(\hat{p}_n) - \ln(p)\}/\{(1-p)/(np)\}^{1/2} \stackrel{D}{\to} Z \sim \mathrm{N}(0,1)$. However, $\mathrm{E}\{\ln(\hat{p}_n)\} = -\infty$ for each n because $\hat{p}_n$ has positive probability of being 0. Therefore, $\mathrm{E}(Z_n) = -\infty$, whereas $\mathrm{E}(Z) = 0$.

Even if X_n has finite mean, it is not necessarily the case that if $X_n \stackrel{D}{\to} X$, then $\mathrm{E}(X_n) \to \mathrm{E}(X)$. In Example 6.34, $X_n = \exp(n)$ with probability $1/n$ and 0 with probability $1 - 1/n$. X_n converges in probability (and therefore in distribution) to 0, but $\mathrm{E}(X_n)$ converges to ∞. The limiting random variable is degenerate, but we can easily modify it to make the limiting variable non-degenerate. For example, consider $U_n = X_n + Y$, where Y is any mean 0 random variable independent of X_n. Then U_n converges in distribution to the non-degenerate random variable $U = Y$. Furthermore, $\mathrm{E}(U_n) = \exp(n)/n \to \infty$, whereas $\mathrm{E}(U) = 0$.

We now give a heuristic motivation of an additional condition required to ensure that the convergence of X_n to X in distribution and the finiteness of $\mathrm{E}(X_n)$ imply that $\mathrm{E}(X_n)$ converges to $\mathrm{E}(X)$. Following this informal presentation, we make the arguments rigorous.

Note first that whether $\mathrm{E}(X_n)$ converges to $\mathrm{E}(X)$ depends only on the distribution functions of X_n and X. By the Skorokhod representation theorem, we can assume that the X_n and X are on the same probability space and $X_n \stackrel{a.s.}{\to} X$. For now we assume also that $\mathrm{E}(|X|) < \infty$, though we later show that this is not necessary. Then

$$
\begin{aligned}
&|\mathrm{E}(|X_n|) - \mathrm{E}(|X|)| \\
&\quad = \Big|\mathrm{E}\{(|X_n| - |X|)I(|X_n| \le A)\} + \mathrm{E}\{(|X_n| - |X|)I(|X_n| > A)\}\Big| \\
&\quad = \Big|\mathrm{E}\{(|X_n| - |X|)I(|X_n| \le A)\} + \mathrm{E}\{|X_n| I(|X_n| > A)\} - \mathrm{E}\{|X| I(|X_n| > A)\}\Big|. \qquad (9.9) \\
&\quad = |\text{Part } 1 + \text{Part } 2 - \text{Part } 3|,
\end{aligned}
$$

where the parts under the absolute value sign are numbered from left to right. We can show that Part 3 can be made small for all n by choosing A large enough. Once we choose a large value of A, we can make Part 1 small for sufficiently large n by the DCT because $(|X_n| - |X|)I(|X_n| \le A)$ converges almost surely to 0 and is dominated by the integrable random variable $A+|X|$. Therefore, whether we can make $|\mathrm{E}(|X_n|)-\mathrm{E}(X)|$ small depends entirely on whether we can choose A large enough to make Part 2 of Expression (9.9) small for all n. This leads us to the following definition.

Definition 9.18. Uniform integrability *The sequence X_n is said to be uniformly integrable (UI) if for each $\epsilon > 0$ there is an A such that $\mathrm{E}\{|X_n|I(|X_n| > A\} < \epsilon$ for all n.*

We are now in a position to make rigorous the arguments leading to Definition 9.18.

Theorem 9.19. Connection between uniform integrability and convergence of moments *Suppose that $X_n \stackrel{D}{\to} X$ and $\mathrm{E}(|X_n|) < \infty$ for each n.*

1. *If X_n is UI, then $\mathrm{E}(|X_n|) \to \mathrm{E}(|X|)$ and $\mathrm{E}(X_n) \to \mathrm{E}(X)$.*
2. *If $\mathrm{E}(|X_n|) \to \mathrm{E}(|X|) < \infty$, then X_n is UI.*

Proof. Because uniform integrability and convergence of means depend only on the distribution functions of X_n and X, we may assume, by the Skorokhod representation theorem, that the X_n and X are defined on the same probability space and that $X_n \stackrel{a.s.}{\to} X$.

To prove item 1, assume that X_n is UI. We will show first that $\mathrm{E}(|X|) < \infty$. We can see that $\mathrm{E}(|X_n|)$ is bounded by taking $\epsilon = 1$ in the definition of uniform integrability. That is, there is an A such that $\mathrm{E}\{|X_n|I(|X_n| > A)\} \le 1$ for all n, so

$$\begin{aligned} \mathrm{E}(|X_n|) &= \mathrm{E}\{|X_n|I(|X_n| \le A)\} + \mathrm{E}\{|X_n|I(|X_n| > A)\} \\ &\le A+1. \end{aligned}$$

By Fatou's lemma, $\mathrm{E}(|X|) = \mathrm{E}(\underline{\lim}\,|X_n|) \le \underline{\lim}\,\mathrm{E}(|X_n|) \le A+1 < \infty$. We have completed the first step of showing that $\mathrm{E}(|X|) < \infty$.

We will prove next that $\mathrm{E}(|X_n|) - \mathrm{E}(|X|) \to 0$. Let $\epsilon > 0$. We must find an N such that $|\mathrm{E}(|X_n|) - \mathrm{E}(|X|)| < \epsilon$ for $n \ge N$. Apply the triangle inequality to Expression (9.9) and note that Part 2 and Part 3 are nonnegative to deduce that:

$$|\mathrm{E}(|X_n|) - \mathrm{E}(|X|)| \le |\text{Part 1}| + \text{Part 2} + \text{Part 3}.$$

Write Part 3 as

$$\begin{aligned} \text{Part 3} &= \mathrm{E}\{|X|I(|X_n| > A)\} \\ &= \mathrm{E}\{|X|I(|X_n| > A)I(|X| \le A\} + \mathrm{E}\{|X|I(|X_n| > A)I(|X| > A\} \\ &\le AP(|X_n| > A) + \mathrm{E}\{|X|I(|X| > A\}. \end{aligned}$$

We can choose A large enough to simultaneously make:

1. $AP(|X_n| > A) < \epsilon/6$ because $AP(|X_n| > A) \le \mathrm{E}\{|X_n|I(|X_n| > A)\}$ and X_n is UI.
2. $\mathrm{E}\{|X|I(|X| > A\} < \epsilon/6$ because $\mathrm{E}(|X|) < \infty$.
3. Part 2$< \epsilon/3$ because X_n is UI.

Items 1 and 2 show that Part 3 is less than $\epsilon/3$, while item 3 shows that Part 2 is less than $\epsilon/3$. Thus, we have demonstrated that

$$|\mathrm{E}(|X_n|) - \mathrm{E}(|X|)| < |\text{Part 1}| + 2\epsilon/3. \tag{9.10}$$

We can choose N such that $|\text{Part 1}| < \epsilon/3$ for $n \geq N$ because $|\mathrm{E}\{(|X_n| - |X|)I(|X_n| \leq A)\}| \to 0$ by the DCT (see explanation in paragraph following Equation (9.9)). Thus, $|\mathrm{E}(|X_n|) - \mathrm{E}(|X|)| < \epsilon/3 + 2\epsilon/3 = \epsilon$. This completes the proof that $\mathrm{E}(|X_n|) \to \mathrm{E}(|X|)$.

The same arguments can be applied to X_n^+ and X_n^-; $X_n^+ \stackrel{a.s.}{\to} X^+$ and X_n^+ is UI, so $\mathrm{E}(X_n^+) \to \mathrm{E}(X^+)$, and similarly for X_n^-. Therefore, $\mathrm{E}(X_n) = \mathrm{E}(X_n^+) - \mathrm{E}(X_n^-) \to \mathrm{E}(X^+) - \mathrm{E}(X^-) = \mathrm{E}(X)$. This completes the proof of item 1 of Theorem 9.19.

To prove item 2 of Theorem 9.19, suppose that X_n is not UI. We will show that $\mathrm{E}(|X_n|)$ cannot converge to $\mathrm{E}(|X|)$. Because X_n is not UI, there exists an $\epsilon^* > 0$ such that for any $A > 0$, $\mathrm{E}\{|X_n|I(|X_n| > A)\} \geq \epsilon^*$ for infinitely many n. Choose A large enough that Part 3 of Expression (9.9) is less than $\epsilon^*/3$ for all n. Then choose N large enough that the absolute value of Part 1 of Expression (9.9) is less than $\epsilon^*/3$ for $n \geq N$. Because there are infinitely many $n \geq N$ such that the middle term of Expression (9.9) exceeds ϵ^*, there are infinitely many $n \geq N$ such that $|\mathrm{E}(|X_n|) - \mathrm{E}(|X|)| \geq \epsilon^* - (2/3)\epsilon^* = \epsilon^*/3$. It follows that $\mathrm{E}(|X_n|)$ cannot converge to $\mathrm{E}(|X|)$. □

Proposition 9.20. Sufficient condition for uniform integrability *A sufficient condition for X_n to be uniformly integrable is the existence of numbers $\delta > 0$ and $B > 0$ such that $\mathrm{E}(|X_n|^{1+\delta}) \leq B$ for all n.*

Proof. Exercise.

Corollary 9.21. *Let X_i be iid random variables with mean μ and variance σ^2, and let $Z_n = (\bar{X}_n - \mu)/(\sigma/n^{1/2})$. Then $\mathrm{E}(|Z_n|) \to \{2/(\pi\sigma^2)\}^{1/2}$.*

Proof. Note first that $Z_n \stackrel{D}{\to} Z \sim \mathrm{N}(0,1)$ by the CLT. Also, $\mathrm{E}(Z_n^2) = \mathrm{var}(Z_n) = 1$. By Proposition 9.20 with $\delta = 1$, Z_n is UI. By Theorem 9.19, $\mathrm{E}(|Z_n|) \to (2\pi\sigma^2)^{-1/2} \int_{-\infty}^{\infty} |z| \exp(-z^2/2)$ $= \{2/(\pi\sigma^2)\}^{1/2}$. □

Example 9.22. Theorem 7.20 shows that the expected number of returns of a symmetric random walk before time $2n$ is $\mathrm{E}(|S_{2n}|) - 1$. From this and a crude bound, we argued that the expected number of returns is of order $(2n)^{1/2}$ instead of order $2n$, and used this to conclude that the expected return time must be infinite. We are now able to derive a much better estimate of the expected number of returns before time $2n$. By Corollary 9.21, $\mathrm{E}(|S_{2n}|/(2n)^{1/2}) \to (2/\pi)^{1/2}$, so the expected number of returns to the origin of a symmetric random walk is asymptotic to $(2n)^{1/2}(2/\pi)^{1/2} - 1 = 2(n/\pi)^{1/2} - 1$. □

Proposition 9.23. Sufficient condition for uniform integrability *If $P(|X_n| \leq Y) = 1$, where $\mathrm{E}(Y) < \infty$, then X_n is UI.*

Proof. Exercise.

We can use uniform integrability to strengthen the DCT (Proposition 6.43) by weakening its hypotheses from convergence in probability to convergence in distribution.

Proposition 9.24. Even more general DCT for probability measures *If $X_n \stackrel{D}{\to} X$ and $|X_n| \leq Y$, where $\mathrm{E}(Y) < \infty$, then $\mathrm{E}(X_n) \to \mathrm{E}(X)$.*

Proof. By Proposition 9.23, X_n is UI. The result then follows from Theorem 9.19. □

9.4 Normalizing Sequences

We saw in Section 9.3 that $Y_n = (X_n - a_n)/b_n \overset{D}{\to} \mathrm{N}(0,1)$ for $b_n > 0$ does not necessarily imply that the normalizing sequences of numbers a_n and b_n are the mean and variance of X_n; indeed, X_n need not even have a mean or variance. This raises an interesting question: if $Y_n = (X_n - a_n)/b_n \overset{D}{\to} Y \sim F$, where $b_n > 0$ and F is non-degenerate, could there be other normalizing sequences of numbers α_n and $\beta_n > 0$ and another non-degenerate distribution function G such that $Z_n = (X_n - \alpha_n)/\beta_n \overset{D}{\to} Z \sim G$? The answer is yes, but only in a very limited sense. Suppose that $(a_n - \alpha_n)/\beta_n \to c$ and $b_n/\beta_n \to d$. Then by Slutsky's theorem,

$$\frac{X_n - \alpha_n}{\beta_n} = \left(\frac{X_n - a_n}{b_n}\right)\left(\frac{b_n}{\beta_n}\right) + \frac{a_n - \alpha_n}{\beta_n} \overset{D}{\to} dY + c.$$

Also, $G(z) = P(Z \le z) = P\{dY + c \le z\} = P\{Y \le (z - c)/d\} = F\{(z - c)/d\}$. The following theorem implies that the converse is also true.

Theorem 9.25. Virtual uniqueness of normalizing sequences *Suppose that* $Y_n = (X_n - a_n)/b_n \overset{D}{\to} Y \sim F$, *where* $b_n > 0$ *and* F *is non-degenerate. Then* $Z_n = (X_n - \alpha_n)/\beta_n \overset{D}{\to} Z \sim G$, *where* $\beta_n > 0$ *and* G *non-degenerate, if and only if* $(a_n - \alpha_n)/\beta_n \to c$ *and* $b_n/\beta_n \to d > 0$ *for some constants* c *and* d, *in which case* $G(z) = F\{(z - c)/d\}$.

Proof. Assume that $Y_n = (X_n - a_n)/b_n \overset{D}{\to} Y$, and let X_n' and X_n'' be independent with the same distribution as X_n. Then $Y_n' = (X_n' - a_n)/b_n$ and $Y_n'' = (X_n'' - a_n)/b_n$ are independent with the same distribution as Y_n. It follows that

$$Y_n' - Y_n'' = \frac{X_n' - X_n''}{b_n}$$

converges in distribution to $Y' - Y''$, where Y' and Y'' are independent with the same distribution as Y. Similarly,

$$\frac{X_n' - X_n''}{\beta_n}$$

converges in distribution to $Z' - Z''$. Also,

$$\frac{X_n' - X_n''}{\beta_n} = \left(\frac{X_n' - X_n''}{b_n}\right)\left(\frac{b_n}{\beta_n}\right). \tag{9.11}$$

We claim this implies that b_n/β_n converges to a positive number (remember that $b_n > 0$ and $\beta_n > 0$). Suppose not. Then either $b_n/\beta_n \to 0$ or b_n/β_n does not converge. If $b_n/\beta_n \to 0$, then the right side of Equation (9.11) would converge to 0, so its limiting distribution would be degenerate at 0. Hence, b_n/β_n cannot converge to 0. Also, suppose that b_n/β_n did not converge. Then there would be subsequences $\{j\} \subset \{n\}$ and $\{k\} \subset \{n\}$ such that b_j/β_j converges to r along $\{j\}$, b_k/β_k converges to s along $\{k\}$, and $r < s$ (s could be $+\infty$). If s is finite, Slutsky's theorem implies that the right side of Equation (9.11) converges in distribution to $r(Y' - Y'')$ along $\{j\}$ and to $s(Y' - Y'')$ along $\{k\}$. Unless $Y' - Y''$ is degenerate at 0 (which happens only if Y and Y' and Y'' are degenerate), this contradicts the fact that the left side of (9.11) converges in distribution. We conclude that b_n/β_n converges. We also get a contradiction if $s = +\infty$.

Also,

$$\frac{X_n - \alpha_n}{\beta_n} = \left(\frac{X_n - a_n}{b_n}\right)\left(\frac{b_n}{\beta_n}\right) + \frac{a_n - \alpha_n}{\beta_n}.$$

By an argument similar to that used above, $(a_n - \alpha_n)/\beta_n$ must converge. If $(a_n - \alpha_n)/\beta_n \to c$ and $b_n/\beta_n \to d$, then the argument preceding Theorem 9.25 shows that $G(z) = F\{(z - c)/d\}$. □

9.5 Review of Equivalent Conditions for Weak Convergence

Rather than presenting new material, this section summarizes and presents in one place equivalent conditions for convergence in distribution from different parts of the book or from different sources.

Proposition 9.26. Equivalent conditions for convergence in distribution of random variables *The following are equivalent formulations of a sequence of random variables $X_n \sim F_n$ converging in distribution to $X \sim F$.*

1. *$F_n(x) \to F(x)$ for all continuity points of F (Definition 6.20).*
2. *$F_n(x) \to F(x)$ on a set D of x that is dense in R (Proposition 6.23).*
3. *$\mathrm{E}\{g(X_n)\} \to \mathrm{E}\{g(X)\}$ for all bounded (uniformly) continuous functions $g : R \longmapsto R$ (Proposition 6.29). Note that this formulation is equivalent to the others whether or not we include the word "uniformly."*
4. *The characteristic function $\psi_n(t) = \mathrm{E}\{\exp(itX_n)\}$ of X_n converges to the characteristic function $\psi(t) = \mathrm{E}\{\exp(itX)\}$ of X for all $t \in R$ (Theorem 8.36).*

Proposition 9.27. Equivalent conditions for convergence of random vectors *The following are equivalent formulations of a sequence of k-dimensional random vectors $\mathbf{X}_n \sim F_n(\mathbf{x})$ converging in distribution to $\mathbf{X} \sim F(\mathbf{x})$.*

1. *$F_n(\mathbf{x}) \to F(\mathbf{x})$ for all continuity points of F (Billingsley, 2012).*
2. *$F_n(\mathbf{x}) \to F(\mathbf{x})$ on a set D of $\mathbf{x}$ that is dense in R^k (Billingsley, 2012).*
3. *$\mathrm{E}\{g(\mathbf{X}_n)\} \to \mathrm{E}\{g(\mathbf{X})\}$ for all bounded (uniformly) continuous functions $g : R^k \longmapsto R$. (Definition 6.58)*
4. *The characteristic function $\psi_n(\mathbf{t}) = \mathrm{E}\{\exp(i\mathbf{t}'\mathbf{X}_n)\}$ of $\mathbf{X}_n$ converges to the characteristic function $\psi(\mathbf{t}) = \mathrm{E}\{\exp(i\mathbf{t}'\mathbf{X})\}$ of $\mathbf{X}$ for all $\mathbf{t} \in R^k$ (Proposition 8.42).*

Exercises

1. Suppose that we reject a null hypothesis for small values of a test statistic Y_n that is uniformly distributed on $A_n \cup B_n$ under the null hypothesis, where $A_n = [-1/n, 1/n]$ and $B_n = [1-1/n, 1+1/n]$. Show that the "approximate" p-value using the asymptotic null distribution of Y_n is not necessarily close to the exact p-value.
2. Prove Corollary 9.4 for intervals of the form
   (a) $(-\infty, x)$.
   (b) (x, ∞).

(c) $[x, \infty)$.

(d) $[a, b)$.

3. Let X_n and X have probability mass functions f_n and f on the integers $k = 0, -1, +1, -2, +2, \ldots$, and suppose that $f_n(k) \to f(k)$ for each k. Without using Scheffé's theorem, prove that $X_n \xrightarrow{D} X$. Then prove a stronger result using Scheffé's theorem.

4. Find a sequence of density functions converging to a non-density function.

5. If $X_1, \ldots, X_n$ are iid with mean 0 and variance $\sigma^2 < \infty$, prove that $n\bar{X}^2/\sigma^2$ converges in distribution to a central chi-squared random variable with 1 degree of freedom, which is **not** normal. Why does this not contradict the delta method?

6. Use the asymptotic distribution of $\hat{\theta} = \ln(\hat{p}_1/\hat{p}_2)$ derived in Example 9.12 in conjunction with the one-dimensional delta method to prove that the asymptotic distribution of the relative risk is given by Equation (9.8).

7. Give an example to show that Proposition 9.25 is false if we remove the condition that F and G be non-degenerate.

8. Let S_n be the sum of iid Cauchy random variables with parameters θ and λ (see Table 8.2 and Exercise 7 of Section 8.4). Do there exist normalizing constants a_n and b_n such that $(S_n - a_n)/b_n \xrightarrow{D} \mathrm{N}(0, 1)$? If so, find them. If not, explain why not.

9. Suppose you have two "trick" coins having probabilities 0.20 and 0.80 of heads. Randomly choose a coin, and then flip it ad infinitum. Let X_i be the indicator of heads for flip i, and $\hat{p}_n = (1/n)\sum_{i=1}^{n} X_i$. Does $\hat{p}_n$ converge to a constant (either almost surely or in probability)? If so, what is the constant? Does $\hat{p}_n$ converge in distribution? If so, to what distribution? Is $(\hat{p}_n - a_n)/b_n$ asymptotically normal for some a_n and b_n?

10. Prove Proposition 9.20.

11. Let X_i be iid with $\mathrm{E}(|X_1|) < \infty$. Prove that S_n/n is UI.

12. If $X_n \xrightarrow{D} X$ and $\mathrm{E}(|X_n^r|) \to \mathrm{E}(|X^r|) < \infty$, $r > 0$, then X_n^r is UI.

13. Let X_n be binomial (n, p_n).

    (a) Prove that if $p_n = p$ for all n, then $1/X_n$ is asymptotically normal, and determine its asymptotic mean and variance (i.e., the mean and variance of the asymptotic distribution of $1/X_n$). How do these compare with the exact mean and variance of $1/X_n$? Note that $1/X_n$ is infinite if $X_n = 0$.

    (b) If $p_n = \lambda/n$ for some constant λ, prove that $1/X_n$ does not converge in distribution to a finite-valued random variable.

14. Let μ_n be the binomial probability measure with parameters n and p_n, where $np_n \to \lambda$. If ν is the Poisson probability measure with parameter λ, prove the following improvement of the law of small numbers (Proposition 6.24): $\sup_{B \in \mathcal{B}} |\mu_n(B) - \nu(B)| \to 0$.

15. Consider a permutation test in a paired data setting, as in Examples 8.10 and 8.20. Let $p_n = p_n(Z_n)$ be the exact, one-tailed permutation p-value corresponding to $Z_n = \sum_{i=1}^{n} D_i/\sqrt{\sum D_i^2}$, and let p_n' be the approximate p-value $1 - \Phi(Z_n)$. Using what was shown in Example 8.20, prove that $p_n(Z_n) - p_n'(Z_n) \xrightarrow{a.s.} 0$.

16. Prove Proposition 9.23

9.6 Summary

1. **Polya's theorem** (convergence in distribution to a continuous d.f. implies uniform convergence) If $F_n \stackrel{D}{\to} F$, where F is continuous, then $\sup_x |F_n(x) - F(x)| \to 0$.

2. **Scheffé's theorem** (convergence of densities implies uniform convergence of probability measures over all Borel sets) If X_n and X have densities f_n and f with respect to a measure μ, and $f_n(x) \to f(x)$ except on a set of μ-measure 0, then $\int |f_n(x) - f(x)| d\mu(x) \to 0$ and $\sup_{B \in \mathcal{B}} |P(X_n \in B) - P(X \in B)| \to 0$.

3. **Delta method**

   (a) **Univariate** If $a_n(\hat{\theta}_n - \theta) \stackrel{D}{\to} X$, $a_n \to \infty$, and $f'(\theta)$ exists, then $a_n\{f(\hat{\theta}_n) - f(\theta)\} \stackrel{D}{\to} f'(\theta)X$.

   (b) **Bivariate** If $a_n(\hat{\delta} - \delta, \hat{\theta} - \theta) \stackrel{D}{\to} (X, Y)$, $a_n \to \infty$, and $f(x, y)$ is differentiable at (δ, θ), then

$$a_n\{f(\hat{\delta}, \hat{\theta}) - f(\delta, \theta)\} \stackrel{D}{\to} \left(\left.\frac{\partial f}{\partial x}\right|_{(\delta,\theta)} \right) X + \left(\left.\frac{\partial f}{\partial y}\right|_{(\delta,\theta)} \right) Y.$$

4. **Convergence of moments is nearly equivalent to uniform integrability** If $X_n \stackrel{D}{\to} X$ and $\mathrm{E}(|X_n|) < \infty$ for each n, then

   (a) If X_n is UI, then $\mathrm{E}(|X_n|) \to \mathrm{E}(|X|)$ and $\mathrm{E}(X_n) \to \mathrm{E}(X)$.

   (b) If $\mathrm{E}(|X_n|) \to \mathrm{E}(|X|) < \infty$, then X_n is UI.

5. **"Uniqueness" of normalizing sequences and asymptotic distributions** If $(X_n - a_n)/b_n$ and $(X_n - \alpha_n)/\beta_n$ converge in distribution to non-degenerate r.v.s $Y \sim F$ and $Z \sim G$, then $(a_n - \alpha_n)/b_n \to c$ and $b_n/\beta_n \to d$ for some constants c and d, in which case $G(y) = F\{(y - c)/d\}$.

Chapter 10

Conditional Probability and Expectation

Conditioning is a very important tool in statistics. One application is to eliminate the dependence of the distribution of a test statistic on unknown parameters. For instance, Fisher's exact test, which compares two groups with respect to the proportions with events, conditions on the total number of people with events. Its resulting null hypergeometric distribution does not depend on any unknown parameters. Similarly, clinical trials comparing a vaccine to placebo often use what is called the "conditional binomial procedure," which entails conditioning on the total number of infections across both arms and the amount of follow-up time in each arm. Under the assumption that the numbers of infections in the two arms follow Poisson distributions with rates proportional to the amount of follow-up, the conditional distribution of the number of vaccine infections follows a binomial distribution with probability parameter free of any unknown parameters. Permutation tests condition on even more, namely all data other than the treatment labels. The null distribution of the permutation test statistic depends only on those data, not on parameters.

Another reason for conditioning is to create independence. For example, in the comparison of several treatments to a control with respect to a continuous outcome Y like log viral load or blood pressure, we examine differences of means, $\hat{\delta}_1 = \bar{Y}_1 - \bar{Y}_0, \ldots, \hat{\delta}_k = \bar{Y}_k - \bar{Y}_0$, where $\bar{Y}_i$ is the mean in treatment arm i and $\bar{Y}_0$ is the mean of the control arm. The $\hat{\delta}_i$ are dependent because they share the same control mean $\bar{Y}_0$; once we condition on $\bar{Y}_0$, $\hat{\delta}_1, \ldots, \hat{\delta}_k$ are independent. Likewise, a mixed model might assume person-specific intercepts that make different observations on the same person correlated. Once we condition on the random effect for a given person, these observations become independent.

Conditioning also allows different entities to interpret data from their own perspectives. For instance, ideally, a medical diagnostic test should declare the disease present if the patient truly has it, and absent if the patient does not have it. The probabilities of these conclusions are known as *sensitivity* and *specificity*, respectively. These are conditional probabilities of correct diagnoses given that the patient does or does not truly have the disease. The doctor wants to ensure a small proportion of incorrect diagnoses, hence high sensitivity and specificity. The patient is concerned only about the accuracy of his or her own diagnosis. "Given that the test was positive, what is the probability that I really have the disease," or "Given that the test was negative, what is the probability that I am really disease free?" These are different conditional probabilities, known as *positive predictive value* and *negative predictive value*, respectively.

The importance of conditioning may be matched only by the care required to avoid mistakes while carrying it out. Many paradoxes, including the two envelope paradox of Example 1.4, involve errors in conditional probability or expectation. These generally involve conditioning on sets of probability 0; conditioning on sets of positive probability does not cause problems. If $\mathrm{E}(|Y|) < \infty$ and B is any Borel set with $P(X \in B) > 0$, the expected value of Y given that $X \in B$ is unambiguously defined by

$$\mathrm{E}(Y \mid X \in B) = \frac{\mathrm{E}\{YI(X \in B)\}}{P(X \in B)}. \tag{10.1}$$

But suppose we try to condition on the actual value of X by replacing $X \in B$ with $X = x$. If X is continuous, then $P(X = x) = 0$ for each x, so we would be dividing by 0 in (10.1). The solution is to think more generally: instead of considering the **specific** value x such as $x = 5$, think about the information you gain about ω, and therefore about $Y(\omega)$, from knowledge of $X(\omega)$. This more general thinking leads us to define the random variable $\mathrm{E}(Y \mid X)$, the conditional expectation of Y given X. Once we understand this concept, we generalize even more. Recall that Section 4.1.1 encouraged us to think in terms of the sigma-field generated by a random variable. Knowing X tells us, for each Borel set B, whether $X \in B$ occurred, and therefore whether ω lies in $X^{-1}(B)$. That is, we know whether $\omega \in A$ for each A in $\sigma(X)$, the sigma-field generated by X. Therefore, conditioning on a random variable is really conditioning on the sigma-field generated by that random variable. More generally, we could condition on an arbitrary sigma-field $\mathcal{C} \subset \mathcal{F}$ by knowing, for each $C \in \mathcal{C}$, whether $\omega \in C$. We begin with an elementary setting.

10.1 When There is a Density or Mass Function

This section reviews conditional distribution functions and expectation as usually presented in more elementary probability and statistics courses, under the assumption that the random variables have a joint density or probability mass function. The intent is to motivate the more rigorous definition of conditional expectation and distribution given in subsequent sections.

Let (X, Y) have joint density function or probability mass function $f(x, y)$. The conditional density (or mass function) of Y given $X = x$ is defined as $h(y \mid x) = f(x, y)/g(x)$ if $g(x) \neq 0$. It does not matter how we define $h(y \mid x)$ when $g(x) = 0$. For each x such that $g(x) > 0$, $h(y \mid x)$ is a density function (or mass function) in y, to which there corresponds a conditional distribution function $H(y \mid x) = \int_{-\infty}^{y} h(u \mid x) du$ or $H(y \mid x) = \sum_{u \leq y} h(u \mid x)$; $H(y \mid x)$ has all of the properties of an ordinary distribution function in y.

If $\mathrm{E}(|Y|) < \infty$, we define the conditional expected value of Y given $X = x$ by

$$\mathrm{E}(Y \mid X = x) = \begin{cases} 0 & \text{if } g(x) = 0 \\ \int \frac{yf(x,y)}{g(x)} dy \quad or \quad \sum_y \frac{yf(x,y)}{g(x)} & \text{if } g(x) \neq 0. \end{cases} \tag{10.2}$$

Note that the definition of $\mathrm{E}(Y \mid X = x)$ when $g(x) = 0$ is arbitrary. We could define it to be any fixed number.

Example 10.1. Conditional binomial In a vaccine clinical trial, let N_P and N_V be the numbers of disease events in the placebo and vaccine arms, respectively. A common assumption is that N_P and N_V are independent Poissons with parameters $\lambda_P = \mu_P \sum_{i \in P} T_i$

and $\lambda_V = \mu_V \sum_{i\in V} T_i$, where P and V denote the set of indices for the placebo and vaccine arms, $\sum_{i\in P} T_i$ and $\sum_{i\in V} T_i$ are the total amounts of follow-up time in the two arms, and μ_P and μ_V are the placebo and vaccine rates per unit time. Then $N = N_P + N_V$ is Poisson with parameter $\lambda_P + \lambda_V$. The joint probability mass function of (N, N_P) is

$$\begin{aligned} P(N = n, N_P = n_P) &= P(N_P = n_P \cap N_V = n - n_P) \\ &= \left\{\frac{\exp(-\lambda_P)\lambda_P^{n_P}}{n_P!}\right\}\left\{\frac{\exp(-\lambda_V)\lambda_V^{n-n_P}}{(n-n_P)!}\right\} \\ &= \frac{\exp\{-(\lambda_P+\lambda_V)\}\lambda_P^{n_P}\lambda_V^{n-n_P}}{n_P!(n-n_P)!} \end{aligned} \tag{10.3}$$

whenever n_P and n are nonnegative integers and $n_P \le n$. The conditional probability mass function for N_P given $N = n$, namely $f(n, n_P)/g(n)$ is

$$\frac{\exp\{-(\lambda_P+\lambda_V)\}\lambda_P^{n_P}\lambda_V^{n-n_P} n!}{n_P!(n-n_P)!\exp\{-(\lambda_P+\lambda_V)\}(\lambda_P+\lambda_V)^n} = \binom{n}{n_P}\pi^{n_P}(1-\pi)^{n-n_P}, \tag{10.4}$$

where $\pi = \lambda_P/(\lambda_P + \lambda_V)$ and $n \in \{0, 1, 2, \ldots\}$. When $n = 0$, the conditional probability mass function of N_P given $N = n$ is a point mass at 0. We recognize the right side of Equation (10.4) as the binomial probability mass function with n trials and probability parameter π. We have shown that the conditional distribution of the number of disease events in the placebo arm, given the total number of disease events across both arms and the follow-up times in each arm, is binomial (n, π).

The expected value of N_P given $N = n$ is

$$\sum_{i=0}^{n} i\binom{n}{i}\pi^i(1-\pi)^{n-i} = n\pi = n\left(\frac{\lambda_P}{\lambda_P+\lambda_V}\right). \tag{10.5}$$

We can substitute $n = 3$ or $n = 10$ or any other value of n into Equation (10.5) to get the expected value of N_P given $N = n$. But remember that the ultimate goal is to summarize the information contained not just in a particular value of N, but in the random variable N. We do this by substituting N for n into Equation (10.5). This tells us that the expected value of N_P given the random value N is $\mathrm{E}(N_P \mid N) = N\lambda_P/(\lambda_P + \lambda_V)$. Notice that the expected value of N_P given N is a random variable, namely a linear function of N in this example. □

More generally, if we substitute the random variable $X(\omega)$ for the value x in Equation (10.2), we get the random variable

$$Z(\omega) = \begin{cases} 0 & \text{if } g(X(\omega)) = 0 \\ \int \frac{yf(X(\omega),y)}{g(X(\omega))}dy \ \ or \ \ \sum_y \frac{yf(X(\omega),y)}{g(X(\omega))} & \text{if } g(X(\omega)) \neq 0. \end{cases} \tag{10.6}$$

The random variable Z is a Borel function of X by Fubini's theorem (Theorem 5.28).

We have taken the first big step toward a more rigorous definition of conditional expectation when there is a probability density or mass function, namely conditioning on a random variable rather than on a value of the random variable. The key property of Z defined by (10.6) is that it has the same conditional expectation given $X \in B$ (defined by equation (10.1)) as Y does for all Borel sets B such that $P(X \in B) > 0$. For instance, when (X, Y) are continuous with joint density function $f(x, y)$,

$$\begin{aligned} \mathrm{E}(Z \mid X \in B) &= \frac{1}{P(X \in B)} \int_B \left\{ \int \frac{y f(x,y)}{g(x)} dy \right\} g(x) dx \\ &= \frac{1}{P(X \in B)} \int \int_B y f(x,y) dx dy = \frac{\mathrm{E}\{Y I(X \in B)\}}{P(X \in B)} \\ &= \mathrm{E}(Y \mid X \in B). \end{aligned} \tag{10.7}$$

The interchange of order of integration leading to Equation (10.7) is justified by Fubini's theorem because $\mathrm{E}(|Y|) < \infty$ by assumption. Write Equation (10.7) as

$$\frac{\mathrm{E}\{Z I(X \in B)\}}{P(X \in B)} = \frac{\mathrm{E}\{Y I(X \in B)\}}{P(X \in B)}$$

and multiply both sides by $P(X \in B) > 0$ to deduce the equivalent condition,

$$\mathrm{E}\{Z I(X \in B)\} = \mathrm{E}\{Y I(X \in B)\}. \tag{10.8}$$

In fact, Equation (10.8) holds even if $P(X \in B) = 0$, because in that case both sides are 0.

We have shown that if (X, Y) has joint density function $f(x, y)$, then Equation (10.8) holds. We can also start with Equation (10.8) as the definition of conditional expectation and reproduce Equation (10.2) and therefore (10.6). Equation (10.8) motivates the more rigorous definition of $\mathrm{E}(Y \mid X)$ given in the next section.

Exercises

1. Let X and Y be independent Bernoulli (p) random variables, and let $S = X+Y$. What is the conditional probability mass function of Y given $S = s$ for each of $s = 0, 1, 2$? What is the conditional expected value of Y given the random variable S?

2. Verify directly that in the previous problem, $Z = \mathrm{E}(Y \mid S)$ satisfies Equation (10.8) with X in this expression replaced by S.

3. If X and Y are independent with respective densities $f(x)$ and $g(y)$ and $\mathrm{E}(|Y|) < \infty$, what is $\mathrm{E}(Y \mid X = x)$? What about $Z = \mathrm{E}(Y \mid X)$? Verify directly that Z satisfies Equation (10.8).

4. Let U_1 and U_2 be independent observations from a uniform distribution on $[0, 1]$, and let $X = \min(U_1, U_2)$ and $Y = \max(U_1, U_2)$. What is the joint density function for (X, Y)? Using this density, find $Z = \mathrm{E}(Y \mid X)$. Verify directly that Z satisfies Equation (10.8).

5. Let Y have a discrete uniform distribution on $\{1, -1, 2, -2, \ldots, n, -n\}$. I.e., $P(Y = y) = 1/(2n)$ for $y = \pm i$, $i = 1, \ldots, n$. Define $X = |Y|$. What is $\mathrm{E}(Y \mid X = x)$? What about $Z = \mathrm{E}(Y \mid X)$? Verify directly that Z satisfies Equation (10.8).

6. Notice that Expression (10.2) assumes that (X, Y) has a density with respect to two-dimensional Lebesgue measure or counting measure. Generalize Expression (10.2) to allow (X, Y) to have a density with respect to an arbitrary product measure $\mu_X \times \mu_Y$.

7. $\uparrow$ A mixed Bernoulli distribution results from first observing the value p from a random variable P with density $f(p)$, and then observing a random variable Y from a Bernoulli (p) distribution.

(a) Determine the density function $g(p, y)$ of the pair (P, Y) with respect to the product measure $\mu_L \times \mu_C$, where μ_L and μ_C are Lebesgue measure on $[0, 1]$ and counting measure on $\{0, 1\}$, respectively.

(b) Use your result from the preceding problem to prove that $\mathrm{E}(Y \mid P) = P$ a.s.

10.2 More General Definition of Conditional Expectation

There is only one problem with defining conditional expectations with respect to densities or probability mass functions, as done in Equation (10.2): (X, Y) need not have a density function or probability mass function with respect to a two-dimensional product measure. The development at the end of the preceding section showed us a better way to define conditional expectation. Note that the collection of sets $\{X \in B, B \in \mathcal{B}\}$ is $\sigma(X)$, the sigma-field generated by X, so condition (10.8) can be rephrased as in the following definition.

Definition 10.2. Conditional expectation given a random variable *Let X and Y be random variables on $(\Omega, \mathcal{F}, P)$, with $\mathrm{E}(|Y|) < \infty$. A conditional expected value of Y given X is a random variable $Z(\omega)$ on $(\Omega, \mathcal{F}, P)$ that is measurable with respect to $\sigma(X)$ and satisfies $\mathrm{E}\{ZI(A)\} = \mathrm{E}\{YI(A)\}$ for all $A \in \sigma(X)$. Z is said to be a version of $\mathrm{E}(Y \mid X)$.*

Notice that Definition 10.2 says "**A** conditional expected value" and not "**The** conditional expected value." The definition allows more than one conditional expectation. We can change the value of $\mathrm{E}(Y \mid X)$ on a set $N \in \sigma(X)$ with $P(N) = 0$, and it will still satisfy the definition of conditional expectation.

We have proven the following result when there is a probability density function. The proof for probability mass functions is similar and left as an exercise.

Proposition 10.3. Conditional expectation when there is a density *If (X, Y) has joint density or mass function $f(x, y)$ and X has marginal density or mass function $g(x)$, then one version of $\mathrm{E}(Y \mid X)$ is given by Equation (10.6).*

One of the conditions of Definition 10.2 is that $Z = \mathrm{E}(Y \mid X)$ is $\sigma(X)$-measurable. This means that it is an extended Borel function of X, as we see in the next result.

Proposition 10.4. Conditional expectation as an extended Borel function *Let X be a random variable on $(\Omega, \mathcal{F}, P)$, and suppose that Y is $\sigma(X)$-measurable. Then $Y = \phi(X)$ for some extended Borel function $\phi : R \longmapsto \bar{R}$. Therefore, one version of $\mathrm{E}(Y \mid X)$ is $\phi(X)$ for some extended Borel function ϕ.*

Proof. Assume first that Y is a nonnegative simple random variable on $\sigma(X)$. Then $Y = \sum_{i=1}^k a_i I(F_i)$, where each $F_i \in \sigma(X)$. Each F_i is of the form $X^{-1}(B_i)$ for some Borel set $B_i \subset \mathcal{B}$. Then $Y = \sum_{i=1}^k a_i I(X \in B_i) = \phi(X)$, where $\phi(x) = \sum_{i=1}^k a_i I(x \in B_i)$ is clearly a Borel function.

Now suppose that Y is any nonnegative random variable. Then $Y = \lim_{n\to\infty} Y_n$, where each Y_n is a simple random variable on $\sigma(X)$. By what we just proved, $Y_n = \phi_n(X)$ for Borel functions ϕ_n. This means that $\phi_n\{X(\omega)\} \to Y(\omega)$ for each ω, so $\phi_n(x)$ must converge to some function $\phi(x)$ for each $x \in \chi$, the range of $X(\omega)$. The problem is that χ need not be an extended Borel set, which means that the function

$$\begin{cases} \lim_{n\to\infty} \phi_n(x)\} & x \in \chi \\ 0 & x \in \chi^C \end{cases} \tag{10.9}$$

need not be an extended Borel function. For example, $\phi(x) = \lim_{n\to\infty} \phi_n(x)$ could be 1 for all $x \in \chi$, in which case Expression (10.9) is not an extended Borel function because $\phi^{-1}(1)$ is not a Borel set. We can avoid this quandary by defining $\phi(x)$ by $\overline{\lim}\, \phi_n(x)$ for all $x \in R$. Proposition 4.9 implies that ϕ is an extended Borel function, and $Y = \phi(X)$.

Now suppose that Y is any random variable on $\sigma(X)$. Then $Y = Y^+ - Y^-$, where Y^+ and Y^- are nonnegative $\sigma(X)$-measurable random variables and $Y^+(\omega) - Y^-(\omega)$ is not of the form $\infty - \infty$. By what we have just proven, $Y^+ = \phi_1(X)$ and $Y^- = \phi_2(X)$ for some extended Borel functions ϕ_1 and ϕ_2 such that $\phi_1(x) - \phi_2(x)$ is not of the form $\infty - \infty$. Then $Y = \phi_1(X) - \phi_2(X)$, and $\phi_1 - \phi_2$ is an extended Borel function. □

Notation 10.5. *If $\phi(X)$ is a conditional expected value* $\mathrm{E}(Y \mid X)$ *of Y given X, then we write* $\mathrm{E}(Y \mid X = x)$ *for $\phi(x)$.*

Defining conditional expectation by Definition 10.2 makes clear that it depends on X only through the sigma-field generated by X. This makes sense. If X is a binary random variable, it should not and does not matter whether we condition on X or $1 - X$ because they both give the same information, and Section 4.1.1 taught us to think of information in terms of sigma-fields. The following example further illustrates this point.

Example 10.6. Conditional expectation depends only on the sigma-field generated by a random variable Let (X, Y) have probability mass function $f(x, y)$, and let $g(x)$ be the probability mass function of X, where $g(0) = 0$. Define $U = X^3$. The joint probability mass function of (U, Y) is $P(U = u, Y = y) = P(\{X = u^{1/3}\} \cap \{Y = y\}) = f(u^{1/3}, y)$, and the marginal probability mass function of U is $P(U = u) = P(X = u^{1/3}) = g(u^{1/3})$. It follows that $\sum_y f(U^{1/3}, y)/g(U^{1/3})$ is a version of $\mathrm{E}(Y \mid U)$. But of course $U^{1/3} = X$, so $\sum_y f(X, y)/g(X)$ is a version of $\mathrm{E}(Y \mid U)$. In other words, $\mathrm{E}(Y \mid X)$ is a version of $\mathrm{E}(Y \mid U)$. We are **not** saying that $\mathrm{E}(Y \mid U = u) = \mathrm{E}(Y \mid X = u)$; that is not true. However, as random variables, $\mathrm{E}(Y \mid X^3(\omega))$ and $\mathrm{E}(Y \mid X(\omega))$ are the same function of ω. We get the same information whether we condition on X or on X^3 because X and X^3 generate the same sigma-field. □

The observation that $\mathrm{E}(Y \mid X)$ depends only on the sigma-field generated by X leads us to the following generalization of Definition 10.2.

Definition 10.7. Conditional expectation with respect to a sigma-field *Let Y be a random variable on $(\Omega, \mathcal{F}, P)$ with* $\mathrm{E}(|Y|) < \infty$*, and let $\mathcal{A} \subset \mathcal{F}$ be a sigma-field. A conditional expectation of Y given $\mathcal{A}$ is an $\mathcal{A}$-measurable random variable $Z(\omega)$ satisfying* $\mathrm{E}\{ZI(A)\} = \mathrm{E}\{YI(A)\}$ *for all $A \in \mathcal{A}$. If $Y = I(B)$, then* $\mathrm{E}(Y \mid \mathcal{A})$ *is said to be a conditional probability of B given $\mathcal{A}$.*

Conditioning on a sigma-field $\mathcal{A}$ is very general because $\mathcal{A}$ could be generated by a random variable, a random vector, or even an uncountable collection of random variables. Thus, if $X_1, \ldots, X_n$ and Y are random variables, we could denote the expected value of Y given $X_1, \ldots, X_n$ by either $\mathrm{E}(Y \mid X_1, \ldots, X_n)$ or $\mathrm{E}(Y \mid \mathcal{A})$, where $\mathcal{A} = \sigma(X_1, \ldots, X_n)$ is the sigma-field generated by $X_1, \ldots, X_n$. Likewise, if Y depends on uncountably many variables X_s, $s \le t$, then we could denote the expectation of Y given X_s, $s \le t$ by either

$\mathrm{E}(Y \mid X_s,\ s \leq t)$ or $\mathrm{E}(Y \mid \mathcal{A})$, where $\mathcal{A} = \sigma(X_s,\ s \leq t)$ is the sigma-field generated by $X_s,\ s \leq t$. In many applications, we condition on a random variable or random vector. Nonetheless, it is just as easy mathematically to treat the more general case of conditioning on an arbitrary sigma-field.

We often surmise the conditional expectation $Z = \mathrm{E}(Y \mid \mathcal{A})$ from general principles, and then prove that Z is $\mathrm{E}(Y \mid \mathcal{A})$ using Definition 10.7. That is, we establish that Z is $\mathcal{A}$-measurable and satisfies $\mathrm{E}\{ZI(A)\} = \mathrm{E}\{YI(A)\}$ for all $A \in \mathcal{A}$. We illustrate this technique with the following example that is the "flip side" of Example 10.6. It shows that two random variables that are "almost the same" can generate very different sigma-fields, and therefore very different conditional expectations.

Example 10.8. Conditioning is not necessarily continuous Let X and Y be independent Bernoulli (1/2) random variables. It is intuitively clear from the independence of X and Y that one version of the conditional expectation $Z = \mathrm{E}(Y \mid X)$ is the unconditional expectation of Y, namely 1/2. We can verify this fact using Definition 10.7 as follows. The first condition is satisfied because the sigma-field generated by the constant 1/2 is $\{\emptyset, \Omega\}$. To verify the second condition, note that each $A \in \sigma(X)$ is either $X^{-1}(0)$, $X^{-1}(1)$, $\emptyset$, or Ω. Consider $A = X^{-1}(0)$. Then $\mathrm{E}\{ZI(A)\} = \mathrm{E}\{(1/2)I(A)\} = (1/2)P(A) = (1/2)P(X = 0)$. By the independence of X and Y, $\mathrm{E}\{YI(A)\} = \mathrm{E}\{YI(X = 0)\} = \mathrm{E}(Y)\mathrm{E}\{I(X = 0)\} = (1/2)P(X = 0)$. Therefore, for $A = X^{-1}(0)$, $\mathrm{E}\{(1/2)I(A)\} = \mathrm{E}\{YI(A)\}$. A similar argument can be used for the other sets in $\sigma(X)$ to show that 1/2 is a version of $\mathrm{E}(Y \mid X)$.

On the other hand, suppose we condition on something "very close to" X, namely $X + (1/1000)Y$. Intuitively, once we know $X + (1/1000)Y$, we know Y because $X + (1/1000)Y$ is an integer if and only if $Y = 0$. Once we know Y and $X + (1/1000)Y$, we also know X. Therefore, conditioning on $X + (1/1000)Y$ fixes the values of X and Y, so $\mathrm{E}\{Y \mid X + (1/1000)Y\}$ must be Y a.s. We can verify this intuition using sigma-fields. Even though $X + (1/1000)Y$ is close to X, the sigma-field generated by $X + (1/1000)Y$ is completely different from that generated by X. The sigma-field generated by X is $\{X^{-1}(0), X^{-1}(1), \emptyset, \Omega\}$. On the other hand, the four values of $X + (1/1000)Y$ corresponding to $(X = 0, Y = 0)$, $(X = 0, Y = 1)$, $(X = 1, Y = 0)$, and $(X = 1, Y = 1)$ are all distinct. Therefore, the sigma-field generated by $X + (1/1000)Y$ is the smallest sigma-field containing the sets $X^{-1}(0) \cap Y^{-1}(0)$, $X^{-1}(0) \cap Y^{-1}(1)$, $X^{-1}(1) \cap Y^{-1}(0)$, and $X^{-1}(1) \cap Y^{-1}(1)$. This sigma-field is $\{X^{-1}(0) \cap Y^{-1}(0), X^{-1}(0) \cap Y^{-1}(1), X^{-1}(1) \cap Y^{-1}(0), X^{-1}(1) \cap Y^{-1}(1), X^{-1}(0), X^{-1}(1), Y^{-1}(0), Y^{-1}(1), \emptyset, \Omega\}$. Notice that this is also the sigma-field generated by (X, Y). Therefore, conditioning on $X + (1/1000)Y$ is the same as conditioning on (X, Y). To verify that $\mathrm{E}\{Y \mid (X, Y)\} = Y$ a.s., note first that Y is clearly measurable with respect to $\sigma(X, Y)$. Also, the second condition of Definition 10.7 is satisfied trivially because $Z = Y$.

We have demonstrated that $\mathrm{E}(Y \mid X) = 1/2$ a.s., yet $\mathrm{E}\{Y \mid X + (1/1000)Y\} = \mathrm{E}\{Y \mid (X, Y)\} = Y$ a.s. In other words, even though X and $X + (1/1000)Y$ are very close to each other, conditioning on X yields a dramatically different answer than conditioning on $X + (1/1000)Y$. The same result obtains if we replace 1/1000 by $1/10^{10}$ or $1/10^{100}$, etc. □

The following result shows that two versions of $\mathrm{E}(Y \mid \mathcal{A})$ can differ only on a set of probability 0.

Proposition 10.9. Existence and almost sure uniqueness of conditional expectation *If* $\mathrm{E}(|Y|) < \infty$, *there is always at least one version of* $\mathrm{E}(Y \mid \mathcal{A})$. *Two versions,* Z_1 *and* Z_2, *of* $\mathrm{E}(Y \mid \mathcal{A})$ *are equal with probability* 1.

Proof. The existence part follows from a deep result in analysis called the Radon-Nikodym theorem. See Section 10.9 for details. To prove uniqueness, let Z_1 and Z_2 be two versions of $\mathrm{E}(Y \mid \mathcal{A})$. Because Z_1 and Z_2 are both $\mathcal{A}$-measurable, $A = I(Z_1 - Z_2 > 0) \in \mathcal{A}$. Also, Z_1 and Z_2 are integrable. Therefore,

$$\begin{aligned} \mathrm{E}\{(Z_1 - Z_2)I(A)\} &= \mathrm{E}\{Z_1 I(A)\} - \mathrm{E}\{Z_2 I(A)\} \\ &= \mathrm{E}\{YI(A)\} - \mathrm{E}\{YI(A)\} = 0. \end{aligned} \tag{10.10}$$

But $Z_1 - Z_2$ is strictly positive on A, so $\mathrm{E}\{(Z_1 - Z_2)I(A) = 0\}$ implies that $P(A) = 0$. That is, $P(Z_1 > Z_2) = 0$. The same argument with Z_1 and Z_2 reversed shows that $P(Z_2 > Z_1) = 0$. Thus, $P(Z_1 = Z_2) = 1$. □

Certain results for conditional expectation follow almost immediately from the definition. For instance, if we take $A = \Omega$, then $\mathrm{E}(Z) = \mathrm{E}\{ZI(\Omega)\} = \mathrm{E}\{YI(\Omega)\} = \mathrm{E}(Y)$. We have proven the following.

Proposition 10.10. Computing expectations by first computing conditional expectations and then "unconditioning" *If* $\mathrm{E}(|Y|) < \infty$, *then* $\mathrm{E}\{\mathrm{E}(Y \mid \mathcal{A})\} = \mathrm{E}(Y)$.

Example 10.11. Assume that whether a patient in a clinical trial experiences a drug-related adverse event is a Bernoulli random variable, but that patient i has his or her own Bernoulli parameter p_i. We can imagine the Bernoulli parameters for different patients as random draws from a distribution $F(p)$. If Y is the indicator that a randomly selected patient experiences an adverse event, then $P(Y = 1 \mid P = p)$ is Bernoulli (p), so $\mathrm{E}(Y \mid P) = P$. By Proposition 10.10, the probability that a randomly selected patient has an adverse event is $\mathrm{E}(Y) = \mathrm{E}\{E(Y \mid P)\} = \mathrm{E}(P) = \int p dF(p)$. □

Example 10.12. In imaging studies of the lungs, we may express the burden of a disease by the total volume of diseased lesions. Let N be the number of lesions for a given patient, and Y_i be the volume of lesion i. Assuming that the number of lesions is small, it may be reasonable to assume that N is bounded and independent of the Ys (this probably would not be reasonable if N is large, in which case the larger the number of lesions, the smaller their volumes must be). Let μ_Y and μ_N be the (finite) means of Y and N, respectively. The disease burden is $S_N = \sum_{i=1}^{N} Y_i$. Also, because N is bounded by some integer B, $|S_N| \leq \sum_{i=1}^{B} |Y_i|$. Therefore, $\mathrm{E}(|S_N|) < \infty$. To find $\mathrm{E}(S_N)$, first condition on $N = n$. Then $S_N = S_n$, the sum of n independent observations, each with mean μ_Y. The conditional mean $\mathrm{E}(S_N \mid N = n)$ is $n\mu_Y$, so $\mathrm{E}(S_N) = \mathrm{E}\{\mathrm{E}(S_N \mid N)\} = \mathrm{E}(N\mu_Y) = \mu_N \mu_Y$. □

Proposition 10.13. Elementary properties of conditional expectation *Let* Y *and* Y_n, $n = 1, 2, \ldots$ *be integrable random variables on* $(\Omega, \mathcal{F}, P)$, *and let* $\mathcal{A} \subset \mathcal{F}$ *be a sigma-field. Then*

1. $\mathrm{E}(c_1 Y_1 + c_2 Y_2 \mid \mathcal{A}) = c_1 \mathrm{E}(Y_1 \mid \mathcal{A}) + c_2 \mathrm{E}(Y_2 \mid \mathcal{A})$ *a.s.*

2. *If* $P(Y_1 \leq Y_2) = 1$, *then* $\mathrm{E}(Y_1 \mid \mathcal{A}) \leq \mathrm{E}(Y_2 \mid \mathcal{A})$ *a.s.*

3. *If* $Y_n \uparrow Y$ *a.s., then* $\mathrm{E}(Y_n \mid \mathcal{A}) \uparrow \mathrm{E}(Y \mid \mathcal{A})$ *a.s.*

4. *If* $Y_n \downarrow Y$ *a.s., then* $\mathrm{E}(Y_n \mid \mathcal{A}) \downarrow \mathrm{E}(Y \mid \mathcal{A})$ *a.s.*

5. **DCT for conditional expectation** *If* $Y_n \to Y$ *a.s. and* $|Y_n| \leq U$ *a.s., where* $\mathrm{E}(U) < \infty$, *then* $\mathrm{E}(Y_n \mid \mathcal{A}) \to \mathrm{E}(Y \mid \mathcal{A})$ *a.s.*

Proof. The general method of proof for conditional expectation of Y given $\mathcal{A}$ is to show that the candidate random variable Z is $\mathcal{A}$-measurable and has the same expectation as Z over sets $A \in \mathcal{A}$.

For part 1, note that $c_1\mathrm{E}(Y_1 \mid \mathcal{A}) + c_2\mathrm{E}(Y_2 \mid \mathcal{A})$ is $\mathcal{A}$-measurable because $\mathrm{E}(Y_1 \mid \mathcal{A})$ and $\mathrm{E}(Y_2 \mid \mathcal{A})$ are $\mathcal{A}$-measurable. Moreover,

$$\int_A \{c_1\mathrm{E}(Y_1 \mid \mathcal{A}) + c_2\mathrm{E}(Y_2 \mid \mathcal{A})\}dP(\omega) = c_1 \int_A \mathrm{E}(Y_1 \mid \mathcal{A})dP(\omega) + c_2 \int_A \mathrm{E}(Y_2 \mid \mathcal{A})dP(\omega)$$
$$= c_1 \int_A Y_1 dP(\omega) + c_2 \int_A Y_2 dP(\omega) = \int_A (c_1Y_1 + c_2Y_2)dP(\omega) \tag{10.11}$$

for each $A \in \mathcal{A}$, proving part 1.

For parts 3 and 4, we prove the results first for nonnegative random variables. For example, for part 3, $Z_n = \mathrm{E}(Y_n \mid \mathcal{A})$ are $\mathcal{A}$-measurable random variables and are increasing by part 2. Therefore, the limit $Z_\infty = \lim_{n\to\infty} Z_n$ exists (Proposition A.33) and is $\mathcal{A}$-measurable (Proposition 4.9). We will demonstrate that $\int Z_\infty I(A)dP(\omega) = \int Y(\omega)I(A)dP(\omega)$ for each $A \in \mathcal{A}$, which will show that Z_∞ satisfies the definition of $\mathrm{E}(Y \mid \mathcal{A})$. Note that $Z_n I(A) \uparrow Z_\infty I(A)$ a.s. and is nonnegative, so the MCT implies that $\mathrm{E}\{Z_n I(A)\} \to \mathrm{E}\{Z_\infty I(A)\}$. The MCT also implies that $\mathrm{E}\{Y_n I(A)\} \to \mathrm{E}\{Y I(A)\}$. Therefore, for each $A \in \mathcal{A}$,

$$\mathrm{E}\{Z_\infty I(A)\} = \lim_{n\to\infty} \mathrm{E}\{Z_n I(A)\} = \lim_{n\to\infty} \mathrm{E}\{Y_n I(A)\} = \mathrm{E}\{Y I(A)\}. \tag{10.12}$$

A similar argument shows that part 4 holds for nonnegative random variables. To prove parts 3 and 4 for arbitrary Y_n, write Y_n as $Y_n^+ - Y_n^-$ and use the fact that $Y_n^+ \uparrow Y^+$ and $Y_n^- \downarrow Y^-$.

Proofs of the remaining parts are left as exercises. □

When we condition on information that fixes the value of a random variable, we can essentially treat that random variable as a constant. For instance, when we condition on Y_1, then Y_1Y_2 behaves as if Y_1 were a constant: $\mathrm{E}(Y_1Y_2 \mid Y_1) = Y_1\mathrm{E}(Y_2 \mid Y_1)$ when the expectations exist. Specifically:

Proposition 10.14. Treating known random variables as constants *If* $\mathrm{E}(|Y_2|) < \infty$ *and* $\mathrm{E}(|Y_1Y_2|) < \infty$ *and* Y_1 *is* $\mathcal{A}$*-measurable, then* $\mathrm{E}(Y_1Y_2 \mid \mathcal{A}) = Y_1\mathrm{E}(Y_2 \mid \mathcal{A})$ *almost surely.*

Proof. First, $Y_1\mathrm{E}(Y_2 \mid \mathcal{A})$ is $\mathcal{A}$-measurable because Y_1 is $\mathcal{A}$-measurable by assumption, and $\mathrm{E}(Y_2 \mid \mathcal{A})$ is $\mathcal{A}$-measurable by definition of conditional expectation. It remains to prove that if $A \in \mathcal{A}$, $\int Y_1\mathrm{E}(Y_2 \mid \mathcal{A})I(A)dP(\omega) = \int Y_1Y_2 I(A)dP(\omega)$. We show first that this holds if Y_1 is a simple random variable.

If $Y_1 = \sum_{i=1}^n a_i I(A_i)$, $A_i \in \mathcal{A}$, then $Y_1\mathrm{E}(Y_2 \mid \mathcal{A})I(A) = \sum_{i=1}^n a_i I(A_i \cap A)\mathrm{E}(Y_2 \mid \mathcal{A})$, and $A_i \cap A \in \mathcal{A}$. Moreover, each $I(A_i \cap A)\mathrm{E}(Y_2 \mid \mathcal{A})$ is integrable by definition of $\mathrm{E}(Y_2 \mid \mathcal{A})$. It follows that for each $A \in \mathcal{A}$,

$$\begin{aligned}
\int Y_1\mathrm{E}(Y_2 \mid \mathcal{A})I(A)dP(\omega) &= \int \sum_{i=1}^n a_i I(A_i \cap A)\mathrm{E}(Y_2 \mid \mathcal{A})dP(\omega) \\
&= \sum_{i=1}^n a_i \int I(A_i \cap A)\mathrm{E}(Y_2 \mid \mathcal{A})dP(\omega) \\
&= \sum_{i=1}^n a_i \int I(A_i \cap A)Y_2\, dP(\omega)
\end{aligned}$$

$$
\begin{aligned}
&= \int \left\{ \sum_{i=1}^{n} a_i I(A_i) \right\} Y_2 \, I(A) dP(\omega) \\
&= \int Y_1 Y_2 \, I(A) dP(\omega), \qquad (10.13)
\end{aligned}
$$

proving the result when Y_1 is a nonnegative simple random variable.

If Y_1 is any nonnegative random variable, then $Y_1 = \lim_{n\to\infty} U_n$, where U_n are simple, $\mathcal{A}$-measurable random variables increasing to Y_1 (see Section 5.2.1). Then for each $A \in \mathcal{A}$, $\int Y_1 \mathrm{E}(Y_2 \,|\, \mathcal{A}) I(A) dP(\omega)$ is

$$
\begin{aligned}
&= \int Y_1 \mathrm{E}(Y_2^+ \,|\, \mathcal{A}) I(A) dP(\omega) - \int Y_1 \mathrm{E}(Y_2^- \,|\, \mathcal{A}) I(A) dP(\omega) \text{ (if finite)} \\
&= \int \lim_{n\to\infty} \{U_n \mathrm{E}(Y_2^+ \,|\, \mathcal{A}) I(A)\} dP(\omega) - \int \lim_{n\to\infty} \{U_n \mathrm{E}(Y_2^- \,|\, \mathcal{A}) I(A)\} dP(\omega) \\
&= \lim_{n\to\infty} \int \{U_n \mathrm{E}(Y_2^+ \,|\, \mathcal{A}) I(A)\} dP(\omega) - \lim_{n\to\infty} \int \{U_n \mathrm{E}(Y_2^- \,|\, \mathcal{A}) I(A)\} dP(\omega) \text{ (MCT)} \\
&= \lim_{n\to\infty} \int \{U_n Y_2^+ I(A)\} dP(\omega) - \lim_{n\to\infty} \int \{U_n Y_2^- I(A)\} dP(\omega) \text{ (result for simple r.v.s)} \\
&= \int \lim_{n\to\infty} \{U_n Y_2^+ I(A)\} dP(\omega) - \int \lim_{n\to\infty} \{U_n Y_2^- I(A)\} dP(\omega) \text{ (MCT)} \\
&= \int \{Y_1 Y_2^+ I(A)\} dP(\omega) - \int \{Y_1 Y_2^- I(A)\} dP(\omega) \\
&= \int \{Y_1 Y_2 I(A)\} dP(\omega). \qquad (10.14)
\end{aligned}
$$

The first line is finite if and only if the last line is finite, and the last line is finite because $\mathrm{E}(|Y_1 Y_2|)$ is assumed finite. Thus, the result holds when Y_1 is any nonnegative $\mathcal{A}$-measurable random variable.

If Y_1 is any $\mathcal{A}$-measurable random variable, then $Y_1 = Y_1^+ - Y_1^-$, and Y_1^+ and Y_1^- are nonnegative and $\mathcal{A}$-measurable. By what we have just proven, $\int_A Y_1^+ \mathrm{E}(Y_2 \,|\, \mathcal{A}) dP(\omega) = \int_A Y_1^+ Y_2 dP(\omega)$. Similarly, $\int_A Y_1^- \mathrm{E}(Y_2 \,|\, \mathcal{A}) dP(\omega) = \int_A Y_1^- Y_2 dP(\omega)$ for each $A \in \mathcal{A}$. Therefore,

$$
\begin{aligned}
\int_A Y_1 \mathrm{E}(Y_2 \,|\, \mathcal{A}) dP(\omega) &= \int_A Y_1^+ \mathrm{E}(Y_2 \,|\, \mathcal{A}) dP(\omega) - \int_A Y_1^- \mathrm{E}(Y_2 \,|\, \mathcal{A}) dP(\omega) \\
&= \int_A Y_1^+ Y_2 dP(\omega) - \int_A Y_1^- Y_2 dP(\omega) \\
&= \int_A Y_1 Y_2 dP(\omega) \qquad (10.15)
\end{aligned}
$$

for each $A \in \mathcal{A}$, completing the proof. □

Exercises

1. Let Y have a discrete uniform distribution on $\{\pm 1, \pm 2, \ldots, \pm n\}$, and let $X = Y^2$. Find $\mathrm{E}(Y \,|\, X = x)$. Does $\mathrm{E}(Y \,|\, X = x)$ match what you got for $\mathrm{E}(Y \,|\, X = x)$ for $X = |Y|$ in Problem 5 in the preceding section? Now compute $Z = \mathrm{E}(Y \,|\, X)$ and compare it with your answer for $\mathrm{E}(Y \,|\, X)$ in Problem 5 in the preceding section.

2. Let Y be as defined in the preceding problem, but let $X = Y^3$ instead of Y^2. Find $\mathrm{E}(Y \,|\, X = x)$ and $Z = \mathrm{E}(Y \,|\, X)$. Does Z match your answer in the preceding problem?

3. Tell whether the following is true or false. If it is true, prove it. If it is false, give a counterexample. If $E(Y \mid X_1) = E(Y \mid X_2)$, then $X_1 = X_2$ almost surely.

4. Let Y be a random variable defined on $(\Omega, \mathcal{F}, P)$ with $E(|Y|) < \infty$. Verify the following using Definition 10.2.

   (a) If $\mathcal{A} = \{\Omega, \emptyset\}$, then $E(Y \mid \mathcal{A}) = E(Y)$ a.s.

   (b) If $\mathcal{A} = \sigma(Y)$, then $E(Y \mid \mathcal{A}) = Y$ a.s.

5. Let $X_1, \ldots, X_n$ be iid with $E(|X_i|) < \infty$. Prove that $E(X_1 \mid \bar{X}) = \bar{X}$ a.s. Hint: it is clear that $E\{(1/n)\sum_{i=1}^n X_i \mid \bar{X}\} = \bar{X}$ a.s.

6. If Y is a random variable with $E(|Y|) < \infty$, and g is a Borel function, then $E\{Y \mid X, g(X)\} = E(Y \mid X)$ a.s.

7. Suppose that X is a random variable with mean 0 and variance $\sigma^2 < \infty$, and assume that $E(Y \mid X) = X$ a.s. Find $E(XY)$.

8. Prove Proposition 10.3 when (X, Y) has a joint probability mass function.

9. Prove part 2 of Proposition 10.13.

10.3 Regular Conditional Distribution Functions

We defined the conditional probability of an event B given the sigma-field $\mathcal{A}$ as $E\{I(B) \mid \mathcal{A}\}$. Therefore, for any random variable Y and value y, the conditional probability that $Y \le y$ given $\mathcal{A}$ is $P(Y \le y \mid \mathcal{A}) = E\{I(Y \le y) \mid \mathcal{A}\}$. The question is: will this result in a distribution function in y for fixed ω? That is, will an arbitrary version of $E\{I(Y \le y) \mid \mathcal{A}\}$ be a distribution function in y? To see that the answer is no even in a simple setting, let Y have a standard normal distribution and $X = 0$ with probability 1. One version of $E\{I(Y \le y) \mid X\}$ is $\Phi(y)$, which is, of course, a distribution function in y. On the other hand, another version of $E\{I(Y \le y \mid X\}$ is

$$\begin{cases} \Phi(y) & \text{if } X \neq 0 \\ 1 & \text{if } X = 0 \text{ and } y = 0, \\ \Phi(y) & \text{if } X = 0 \text{ and } y \neq 0, \end{cases}$$

and this is not monotone in y if $X = 0$. Therefore, using just any version of $E\{I(Y \le y) \mid \mathcal{A}\}$ does not guarantee that it will be a distribution function in y for each ω. Nonetheless, we can always find a version of $E\{I(Y \le y) \mid \mathcal{A})$ that is a distribution function in y. Such a function is called a *regular conditional distribution function of* Y *given* $\mathcal{A}$.

Theorem 10.15. Distribution function of Y given $\mathcal{A}$ *Let Y be a random variable on $(\Omega, \mathcal{F}, P)$ and $\mathcal{A} \subset \mathcal{F}$ be a sigma-field. There exists a version $F(y, \omega)$ of $E\{I(Y \le y) \mid \mathcal{A})$ that is a regular conditional distribution function of Y given $\mathcal{A}$.*

Proof. Except on a null set N_1, the following conditions hold for rational r.

$$F(r, \omega) = E\{I(Y \le r) \mid \mathcal{A}\} \uparrow \text{ in } r, \ F(r, \omega) \to 0 \text{ or } 1 \text{ as } r \to -\infty \text{ or } \infty. \tag{10.16}$$

To see the monotonicity property, note that for each pair $r_1 < r_2$ of rationals, the set $B(r_1, r_2) = \{\omega : F(r_1, \omega) > F(r_2, \omega)$ has probability 0 by part 2 of Proposition 10.13. The set of pairs (r_1, r_2) of rational numbers such that $F(r_1, \omega) > F(r_2, \omega)$ is the countable union, $\cup_{r_1, r_2} B(r_1, r_2)$, of sets of probability 0, so $P\{\cup_{r_1, r_2} B(r_1, r_2)\} \le \sum_{r_1, r_2} P\{B(r_1, r_2)\} = 0$.

This shows that except on a null set, $F(r, \omega)$ is monotone increasing in r. The limits as $r \to \pm\infty$ follow from properties 3 and 4 of Proposition 10.13 because $I(Y \le r)$ converges almost surely to 0 or 1 as $r \to -\infty$ or ∞, respectively. Thus, except on a null set N_1, conditions (10.16) are satisfied.

Except on another null set N_2,

$$F(r + 1/n, \omega) \to F(r, \omega) \text{ as } n \to \infty \tag{10.17}$$

for all rational numbers r. To see this, note that for a particular r, part 4 of Proposition 10.13 implies that condition (10.17) holds except on a null set $N_2(r)$. The set of ω for which condition (10.17) fails to hold for at least one rational r is the countable union $N_2 = \cup_r N_2(r)$ of null sets, so $P(N_2) \le \sum_r P(N_2(r)) = 0$. Therefore, outside the null set N_2, condition (10.17) holds.

We have defined the candidate distribution function $F(r, \omega)$ on the rational numbers in a way that, except for ω in the null set $N = N_1 \cup N_2$, conditions (10.16) and (10.17) hold. We now define $F(y, \omega)$ for $\omega \in N^C$ and y irrational: $F(y, \omega) = \underline{\lim}_{\, r > y} F(r, \omega)$. Then $F(y, \omega)$, being a liminf of $\mathcal{A}$-measurable random variables, is also $\mathcal{A}$-measurable. For each y, whether rational or irrational, there is a sequence of rational numbers r_n decreasing to y such that $F(r_n, \omega) \to F(r, \omega)$ as $n \to \infty$. This fact can be used to show that $F(y, \omega)$ is right-continuous on N^C for each y. To see this, let $y_n \downarrow y$. We can find a rational number $r_n \ge y_n$ such that

$$0 \le F(r_n, \omega) - F(y_n, \omega) < 1/n. \tag{10.18}$$

Therefore, $F(r_n, \omega) - 1/n < F(y_n, \omega) \le F(r_n, \omega)$ for all n. The limits of the left and right sides as $n \to \infty$ are both $F(y, \omega)$, so $F(y_n, \omega) \to F(y, \omega)$ as $n \to \infty$. We have established that $F(y, \omega)$ is right continuous. It is also easy to show that $\lim_{y \to -\infty} F(y, \omega) = 0$, $\lim_{y \to \infty} F(y, \omega) = 1$.

For $\omega \in N$, take $F(y, \omega)$ to be any fixed distribution function, such as $\mathrm{N}(0, 1)$. □

Notation 10.16. *When $\mathcal{A}$ is the sigma-field generated by a random variable X, the conditional distribution function $F(y, \omega)$ of Y given $\mathcal{A} = \sigma(X)$ is a Borel function of X. This follows from Proposition 10.4 and the fact that $F(y, \omega)$ is a version of $\mathrm{E}\{I(Y \le y \,|\, X\}$. Therefore, we sometimes denote the conditional distribution of Y given $X = x$ by $F(y \,|\, x)$.*

It seems like we have attained all we need. After all, the distribution function for a random variable Y determines the probability that $Y \in B$ for every Borel set B (see Proposition 4.22). There is only one glitch: we have shown only that the conditional distribution function $F(y, \omega)$ is $\mathcal{A}$-measurable. How can we be sure that $P(Y \in B \,|\, \mathcal{A})$ is $\mathcal{A}$-measurable for an arbitrary Borel set B? Fortunately, it is.

Theorem 10.17. Probability measure of Y given $\mathcal{A}$ *Let Y be a random variable on $(\Omega, \mathcal{F}, P)$ and $\mathcal{A} \subset \mathcal{F}$ be a sigma-field. There exists a probability measure $\mu(B, \omega)$ defined on Borel subsets B such that $\mu(B, \omega)$ is a version of $\mathrm{E}\{I(Y \in B) \,|\, \mathcal{A})$.*

Proof. We use the notation $F_\omega(y)$ for $F(y, \omega)$ to emphasize that we are fixing ω and regarding F as a function of y. Define $\mu(B, \omega)$ by $\int_B dF_\omega(y)$. It is an exercise to show that the set $\mathcal{B}'$ of Borel sets B such that $\mu(B, \omega)$ is $\mathcal{A}$-measurable is a monotone class containing the field in Proposition 3.8. By the monotone class theorem (Theorem 3.32), $\mathcal{B}'$ contains all Borel sets. □

The reason for defining a conditional probability measure for Y given $\mathcal{A}$ is to facilitate the calculation of conditional expected values of functions of Y given $\mathcal{A}$.

Proposition 10.18. Using conditional distribution functions to compute conditional expectations *If $F(y,\omega)$ is a regular conditional distribution function of Y given $\mathcal{A}$ and $g(Y)$ is a Borel function with $\mathrm{E}\{|g(Y)|\} < \infty$, then $\int g(y)dF(y,\omega)$ is a version of $\mathrm{E}\{g(Y)\,|\,\mathcal{A}\}$.*

The proof follows the familiar pattern of beginning with nonnegative simple g, then extending to all nonnegative Borel functions, then to all Borel functions (exercise).

Example 10.19. Simon and Simon, 2011: rerandomization tests protect type I error rate There are many different ways to randomize patients to treatment (T) or control (C) in a clinical trial, including: (1) simple randomization, akin to flipping a fair coin for each new patient, (2) permuted block randomization, whereby k patients in each block of size $2k$ are assigned to T, the other k to C, (3) Efron's biased coin design, whereby a fair coin is used whenever the numbers of Ts and Cs are balanced, and an unfair coin with probability, say $2/3$, favoring the under-represented treatment when the numbers of Ts and Cs are unbalanced, and (4) various covariate-adaptive schemes making it more likely that the treatment assigned to the next patient balances the covariate distributions across the arms.

A very general principle is to "analyze as you randomize." To do this, treat all data $\mathcal{D}$ in the clinical trial other than the treatment labels as fixed constants and re-generate the randomization sequence using whatever method was used to generate the original labels. For this rerandomized dataset, compute the value of the test statistic T. Repeat this process of rerandomizing the patients and computing the test statistic until all possible rerandomizations have been included. This generates the rerandomization distribution $F(t)$ of T. For a one-tailed test rejecting the null hypothesis for large values of T, determine $c_* = \inf\{c : 1 - F(c) \le \alpha\}$. By the right-continuity of distribution functions, $1 - F(c_*) \le \alpha$. Reject the null hypothesis if T_{orig} corresponding to the original randomization exceeds c_*. This test is called a rerandomization test. This is a generalization of a permutation test that accommodates any randomization scheme.

Let $\mathcal{D}$ be the sigma-field generated by all data other than the treatment labels. The rerandomization distribution is the conditional distribution function of T given $\mathcal{D}$. By construction, $1 - F(c_*) = P(T > c_*\,|\,\mathcal{D}) \le \alpha$. In other words, the conditional type I error rate given the data $\mathcal{D}$ is no greater than α. By Proposition 10.10, the unconditional type I error rate, $P(T > c_*)$, is simply $\mathrm{E}[\mathrm{E}\{I(T > c_*)\,|\,\mathcal{D})] \le \alpha$. Therefore, any rerandomization test controls the type I error rate both conditional on the observed data and unconditionally.

□

Example 10.20. Asymptotic equivalence of rerandomization test and stratified t-test under permuted block randomization Consider a clinical trial using random permuted blocks of size 4 to assign patients to treatment (T) or control (C). Let X_{i1}, X_{i2}, X_{i3}, and X_{i4} be the observations on a continuous outcome for patients in block i. The treatment less control difference in block i is $D_i = \sum_{j=1}^{4} X_{ij}Z_{ij}$, where Z_{ij} is $+1$ if patient j of block i is assigned to treatment, and -1 if assigned to control. Let $\mathcal{A} = \sigma(X_{ij},\ i = 1,2,\ldots,\ j = 1,\ldots,4)$ be the infinite set of data from which the first $4n$ patients constitute the clinical trial. Let $T_n = \sum_{i=1}^{n} D_i/(\sum_{i=1}^{n} D_i^2)^{1/2}$ be the test statistic, and suppose that we reject the null hypothesis for large values of T_n.

The rerandomization distribution is the conditional distribution of T_n given $\mathcal{A}$. We claim that under the null hypothesis, this conditional distribution converges to N(0, 1) as $n \to \infty$. To see this, first compute the conditional distribution $F_n(t \mid \mathcal{C})$, where $\mathcal{C}$ is the sigma-field $\sigma(X_{ij},\ i = 1, 2, \ldots,\ j = 1, \ldots, 4,\ |D_1|, |D_2|, \ldots)$. This conditional distribution is the distribution of $\sum_{i=1}^n \delta_i d_i / (\sum_{i=1}^n d_i^2)^{1/2}$, where the d_i are fixed constants and δ_i are iid random variables taking values ± 1 with probability 1/2. We have already seen in Example 8.20 that the conditional distribution $F_n(t \mid \mathcal{C})$ of T_n given $\mathcal{C}$ converges to N(0, 1) almost surely. Denoting the conditional distribution function of T_n given $\mathcal{A}$ by $F_n(t \mid \mathcal{A})$, we have

$$\begin{aligned} F_n(t \mid \mathcal{A}) &= \mathrm{E}\{I(T_n \le t \mid \mathcal{A})\} = \mathrm{E}[\mathrm{E}\{I(T_n \le t) \mid \mathcal{C}\} \mid \mathcal{A}] \\ &\to \mathrm{E}\{\Phi(t) \mid \mathcal{A}\} = \Phi(t) \text{ a.s.} \end{aligned} \tag{10.19}$$

The second line follows from the DCT for conditional expectation because $\mathrm{E}\{I(T_n \le t) \mid \mathcal{C}\}$ is dominated by 1. We have thus shown that the permutation test is asymptotically equivalent to rejecting the null hypothesis when $\sum_{i=1}^n D_i / (\sum_{i=1}^n D_i^2)^{1/2} > z_\alpha$, where z_α is the $(1-\alpha)$th quantile of the standard normal distribution. Under the null hypothesis, $(1/n)\sum_{i=1}^n D_i^2 \to \sigma^2 = \mathrm{var}(X_i)$. It follows that the rerandomization test is asymptotically equivalent to rejecting the null hypothesis if $\sum_{i=1}^n D_i/(n\sigma^2)^{1/2} > z_\alpha$. Also, the difference between the usual t-statistic and $\sum_{i=1}^n D_i/(n\sigma^2)^{1/2}$ converges almost surely to 0, so the rerandomization test is asymptotically equivalent to the t-test under random permuted block randomization.

Proposition 10.21. Inequalities for conditional expectation *Let X and Y be random variables defined on $(\Omega, \mathcal{F}, P)$, and let $\mathcal{A} \subset \mathcal{F}$ be a sigma-field.*

1. *Jensen's inequality: If $\phi(\cdot)$ is convex and Y and $\phi(Y)$ are integrable, then $\mathrm{E}\{\phi(Y) \mid \mathcal{A}\} \ge \phi\{\mathrm{E}(Y \mid \mathcal{A})\}$ a.s.*

2. *Markov's inequality: If C is any $\mathcal{A}$-measurable random variable, then $P(|Y| \ge C \mid \mathcal{A}) \le (1/C)\mathrm{E}(|Y| \mid \mathcal{A})$ a.s.*

3. *Chebychev's inequality: If $\mathrm{E}(Y^2) < \infty$, and C is $\mathcal{A}$-measurable, then $P\{|Y - \mathrm{E}(Y \mid \mathcal{A})| \ge C \mid \mathcal{A}\} \le (1/C^2)\mathrm{var}(Y \mid \mathcal{A})$ a.s.*

4. *Hölder's inequality: if $p > 0$, $q > 0$, $1/p + 1/q = 1$, and $\mathrm{E}(|X|^p) < \infty$, $\mathrm{E}(|Y|^q) < \infty$, then $\mathrm{E}(|XY| \mid \mathcal{A}) \le \{\mathrm{E}(|X|^p \mid \mathcal{A})\}^{1/p}\{\mathrm{E}(|Y|^q \mid \mathcal{A})\}^{1/q}$ a.s.*

5. *Schwarz's inequality: If X and Y are random variables with $\mathrm{E}(X^2) < \infty$, $\mathrm{E}(Y^2) < \infty$, then $\mathrm{E}(|XY| \mid \mathcal{A}) \le \sqrt{\mathrm{E}(X^2 \mid \mathcal{A})\mathrm{E}(Y^2 \mid \mathcal{A})}$ a.s.*

6. *Minkowski's inequality: If $p \ge 1$ and $\mathrm{E}(|X|^p) < \infty$ and $\mathrm{E}(|Y|^p) < \infty$, then $\{\mathrm{E}(|X + Y|^p \mid \mathcal{A})\}^{1/p} \le \{\mathrm{E}(|X|^p \mid \mathcal{A})\}^{1/p} + \{\mathrm{E}(|Y|^p \mid \mathcal{A})\}^{1/p}$ a.s.*

Proof. We prove only the first result. Proofs of the others are similar and left as exercises. Let $F_\omega(y)$ be a regular conditional distribution function of Y given $\mathcal{A}$. Because $F_\omega(y)$ is a distribution function in y for fixed ω, the usual Jensen's inequality implies that $\int \phi(y)dF_\omega(y) \ge \phi\{\int y dF_\omega(y)\}$ for each ω. But $\int \phi(y)dF_\omega(y)$ and $\int y dF_\omega(y)$ are versions of $\mathrm{E}\{\Phi(Y) \mid \mathcal{A}\}$ and $\mathrm{E}(Y \mid \mathcal{A})$, respectively, from which the result follows. □

Notice that with the Markov and Chebychev inequality for conditional expectation, C is allowed to be any $\mathcal{C}$-measurable random variable, whereas with the usual Markov and Chebychev inequalities, C is a constant. Remember that once we condition on $\mathcal{A}$, any $\mathcal{A}$-measurable random variable becomes constant.

Just as we defined the conditional mean of a random variable given a sigma-field $\mathcal{A}$, we can define conditional variances and covariances of random variables given $\mathcal{A}$.

Definition 10.22. Conditional variances and covariances *Let $\mathcal{A} \subset \mathcal{F}$ be a sigma-field.*

1. *If Y is a random variable with $\mathrm{E}(Y^2) < \infty$, the conditional variance of Y given $\mathcal{A}$ is $\mathrm{var}(Y \mid \mathcal{A}) = \mathrm{E}\{(Y - Z)^2 \mid \mathcal{A}\}$, where Z is a version of $\mathrm{E}(Y \mid \mathcal{A})$.*

2. *If Y_1 and Y_2 are random variables with $\mathrm{E}(Y_1^2) < \infty$ and $\mathrm{E}(Y_2^2) < \infty$, the conditional covariance of Y_1 and Y_2 given $\mathcal{A}$ is $\mathrm{cov}(Y_1, Y_2 \mid \mathcal{A}) = \mathrm{E}\{(Y_1 - Z_1)(Y_2 - Z_2) \mid \mathcal{A}\}$, where Z_i is a version of $\mathrm{E}(Y_i \mid \mathcal{A})$, $i = 1, 2$.*

The following are some elementary properties of conditional variances and covariances. Other important identities are presented in the next section.

Proposition 10.23. Elementary properties of conditional variances and covariances *Suppose that Y, Y_1, and Y_2 are random variables with finite second moments, and $\mathcal{A} \subset \mathcal{F}$ is a sigma-field.*

1. $\mathrm{var}(Y \mid \mathcal{A}) = \mathrm{E}(Y^2 \mid \mathcal{A}) - \{\mathrm{E}(Y \mid \mathcal{A})\}^2$ *a.s.*

2. $\mathrm{cov}(Y_1, Y_2 \mid \mathcal{A}) = \mathrm{E}(Y_1 Y_2 \mid \mathcal{A}) - \{\mathrm{E}(Y_1 \mid \mathcal{A})\}\{\mathrm{E}(Y_2 \mid \mathcal{A})\}$ *a.s.*

3. *If C is an $\mathcal{A}$-measurable random variable with $\mathrm{E}(C^2) < \infty$, then $\mathrm{var}(Y + C \mid \mathcal{A}) = \mathrm{var}(Y \mid \mathcal{A})$ a.s.*

4. *If C_1 and C_2 are $\mathcal{A}$-measurable random variables with $\mathrm{E}(C_1^2) < \infty$ and $\mathrm{E}(C_2^2) < \infty$, then $\mathrm{cov}(Y_1 + C_1, Y_2 + C_2 \mid \mathcal{A}) = \mathrm{cov}(Y_1, Y_2 \mid \mathcal{A})$ a.s.*

We close this section with extensions of results for conditional distribution functions of random variables to conditional distribution functions of random vectors.

Proposition 10.24. Conditional multivariate distribution function *Let $\mathbf{Y} = (Y_1, \ldots, Y_k)$ be a random vector and $\mathcal{A} \subset \mathcal{F}$ be a sigma-field. There exists a version $F(\mathbf{y}, \omega)$ of $\mathrm{E}\{I(Y_1 \le y_1, \ldots, Y_k \le y_k) \mid \mathcal{A}\}$ that is, for each ω, a distribution function in $(y_1, \ldots, y_k)$.*

Proposition 10.25. Using conditional multivariate distribution functions to compute conditional expectations *If $F(\mathbf{y}, \omega)$ is a conditional distribution function of $\mathbf{Y}$ given $\mathcal{A}$, and $g(\mathbf{Y})$ is an integrable, Borel function of $\mathbf{Y}$, then $\int g(\mathbf{y}) dF(\mathbf{y}, \omega)$ is a version of $\mathrm{E}\{g(\mathbf{Y}) \mid \mathcal{A}\}$.*

Exercises

1. Let (X, Y) take the values $(0,0)$, $(0,1)$, $(1,0)$, and $(1,1)$ with probabilities p_{00}, p_{01}, p_{10}, and p_{11}, respectively, where $p_{00} + p_{01} + p_{10} + p_{11} = 1$ and $p_{00} + p_{01} > 0$, $p_{10} + p_{11} > 0$.

   (a) What is the conditional distribution of Y given $X = 0$?

   (b) Show that $\mathrm{E}(Y \mid X)$ is linear in X, and determine the slope and intercept.

2. Roll a die and let X denote the number of dots showing. Then independently generate $Y \sim \mathrm{U}(0, 1)$, and set $Z = X + Y$.

   (a) Find a conditional distribution function of Z given $X = x$ and a conditional distribution function of Z given $Y = y$.

   (b) Find a conditional distribution function of X given $Z = z$ and a conditional distribution function of Y given $Z = z$.

3. Dunnett's one-tailed test for the comparison of k treatment means $\mu_1, \ldots, \mu_k$ to a control mean μ_0 with common known variance σ^2 and common sample size n rejects the null hypothesis if $\max_i Z_{i0} > c$, where

$$Z_{i0} = \frac{\bar{Y}_i - \bar{Y}_0}{\sqrt{2\sigma^2/n}}$$

and $\bar{Y}_i$ is the sample mean in arm i. Under the null hypothesis, $\mu_i = \mu_0$, $i = 1, \ldots, k$, and without loss of generality, assume that $\mu_i = 0$, $i = 0, 1, \ldots, k$. Therefore, assume that $\bar{Y}_i \sim \mathrm{N}(0, \sigma^2/n)$.

   (a) Find the conditional distribution of $\max_i Z_{i0}$ given $\bar{Y}_0 = y_0$.

   (b) Find the conditional distribution of $\max_i Z_{i0}$ given $\bar{Y}_0 = z_0\sigma/n^{1/2}$.

   (c) Find the unconditional distribution of $\max_i Z_{i0}$.

4. Let X, Y be iid $\mathrm{N}(0, \sigma^2)$. Find the conditional distribution of $X^2 - Y^2$ given that $X + Y = s$.

5. Fisher's least significant difference (LSD) procedure for testing whether means $\mu_1, \ldots, \mu_k$ are equal declares $\mu_1 < \mu_2$ if both the t-statistic comparing μ_1 and μ_2 and F-statistic comparing all means are both significant at level α. When the common variance σ^2 is known, this is equivalent to rejecting the null hypothesis if $Z_{12}^2 > c_{1,\alpha}$ and $R^2 > c_{k-1,\alpha}$, where

$$Z_{12} = \frac{\bar{Y}_1 - \bar{Y}_2}{\sqrt{2\sigma^2/n}}, \quad R^2 = \frac{n}{(k-1)\sigma^2} \sum_{i=1}^{k} (\bar{Y}_i - \bar{Y})^2$$

and $c_{i,\alpha}$ is the upper α point of a chi-squared distribution with i degrees of freedom. Use the result of Problem 11 of Section 8.6 to find the conditional distribution of $Z_{12}^2 + R^2$ given Z_{12}^2. Use this to find an expression for $P(Z_{12}^2 > c_1 \cap R^2 > c_{k-1,\alpha})$.

6. Let Y be a random variable with finite mean, and suppose that $\mathrm{E}\{\exp(Y)\} < \infty$. What is the probability that $\mathrm{E}\{\exp(Y) \mid \mathcal{A}\} < \exp\{\mathrm{E}(Y \mid \mathcal{A})\}$?

7. Prove Markov's inequality for conditional expectation (part 2 of Proposition 10.21).

8. Prove Chebychev's inequality for conditional expectation (part 3 of Proposition 10.21).

9. Prove Hölder's inequality for conditional expectation (part 4 of Proposition 10.21).

10. Prove Schwarz's inequality for conditional expectation (part 5 of Proposition 10.21).

11. Prove Minkowski's inequality for conditional expectation (part 6 of Proposition 10.21).

12. Prove parts 1 and 2 of Proposition 10.23.

13. Prove parts 3 and 4 of Proposition 10.23.

14. Complete the proof of Proposition 10.17 by showing that the set $\mathcal{B}'$ of Borel sets B such that $\int_B dF_\omega(y)$ is $\mathcal{A}$-measurable is a monotone class containing the field in Proposition 3.8.

15. Consider the Z-statistic comparing two means with known finite variance σ^2,

$$Z = \frac{\bar{Y} - \bar{X}}{\sigma\sqrt{1/n_X + 1/n_Y}}.$$

Suppose that n_X remains fixed, and let F be the distribution function for $\bar{X}$. Assume that $Y_1, Y_2, \ldots$ are iid with mean μ_Y. Show that the asymptotic (as $n_Y \to \infty$ and n_X remains fixed) conditional distribution function for Z given $\bar{X} = x$ is normal and determine its asymptotic mean and variance. What is the asymptotic unconditional distribution of Z as $n_Y \to \infty$ and n_X remains fixed?

16. Prove Proposition 10.18, first when g is simple, then when g is nonnegative, then when g is an arbitrary Borel function.

10.4 Conditional Expectation As a Projection

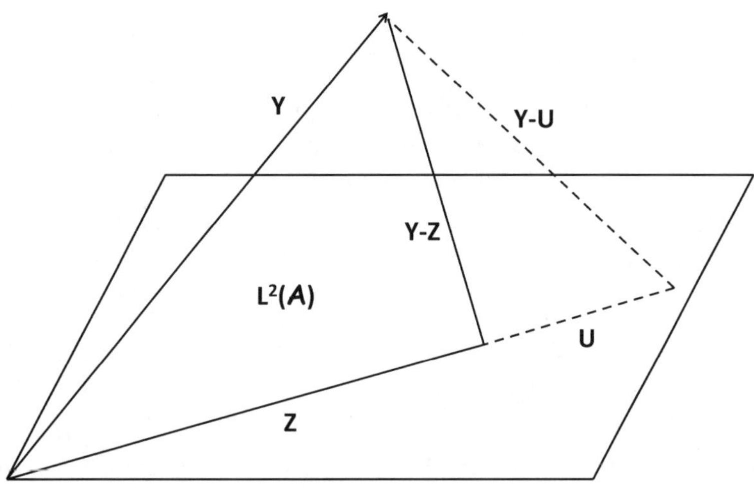

Figure 10.1: Conditional expectation as a projection. The conditional expected value $Z = \mathrm{E}(Y \mid \mathcal{A})$ is such that $Y - Z$ is orthogonal to every random variable in $L^2(\mathcal{A})$, the set of $\mathcal{A}$-measurable random variables in L^2.

We can view conditional expectation of Y given $\mathcal{A}$ in a geometric way. Assume throughout this subsection that $Y \in L^2$; i.e., $\mathrm{E}(Y^2) < \infty$. The set of L^2 random variables is a vector space with inner product $< U_1, U_2 >= \mathrm{E}(U_1 U_2)$. More generally, the set of L^2 functions on a measure space $(\Omega, \mathcal{F}, \mu)$ is a vector space with inner product $< f_1, f_2 >= \int f_1(\omega) f_2(\omega) d\mu$. This is a generalization of the dot product $\mathbf{x} \cdot \mathbf{y} = \sum_{i=1}^n x_i y_i$ of two vectors $\mathbf{x}$ and $\mathbf{y}$ because we can write this dot product as $\int_\Omega f_1(\omega) f_2(\omega) d\mu$, where μ is counting measure on $\Omega = \{\omega_1 = (x_1, y_1), \ldots, \omega_n = (x_n, y_n)\}$ and $f_1(\omega_i) = x_i$, $f_2(\omega_i) = y_i$. It suffices for our purposes to consider probability spaces. There is a close analogy between n-dimensional vectors equipped with dot products and random variables in L^2 equipped with inner products. Just as vectors $\mathbf{x}_1$ and $\mathbf{x}_2$ are orthogonal if and only if their dot product $\mathbf{x}_1 \cdot \mathbf{x}_2$ is 0, random variables U_1 and U_2 in L^2 are orthogonal if and only if their inner product $< U_1, U_2 >= \mathrm{E}(U_1 U_2)$ is 0. Just as the length of a vector $\mathbf{x}$ is $(\mathbf{x} \cdot \mathbf{x})^{1/2}$, the length of a random variable U in L^2 is $< U, U >^{1/2}= \{\mathrm{E}(U^2)\}^{1/2}$.

Now consider the conditional expectation $\mathrm{E}(Y \mid \mathcal{A})$ of a random variable $Y \in L^2$. $\mathrm{E}(Y \mid \mathcal{A})$ is an $\mathcal{A}$-measurable random variable Z such that $\mathrm{E}\{(Y - Z) I(A)\} = 0$ for each

$A \in \mathcal{A}$. That is, $Y - Z$ is orthogonal to $I(A)$ for every $A \in \mathcal{A}$. But this means that $Y - Z$ is orthogonal to any nonnegative simple, $\mathcal{A}$-measurable random variable $\sum_{i=1}^n a_i I(A_i)$.

We show next that $Y-Z$ is orthogonal to U for every nonnegative $\mathcal{A}$-measurable random variable U with $\mathrm{E}(U^2) < \infty$. Any such U is a limit of simple, increasing $\mathcal{A}$-measurable random variables U_n. Therefore, $(Y-Z)^+U_n \uparrow (Y-Z)^+U$ and $(Y-Z)^-U_n \uparrow (Y-Z)U$. By the MCT, $\mathrm{E}\{(Y-Z)^+U_n\} \to \mathrm{E}\{(Y-Z)^+U\}$ and $\mathrm{E}\{(Y-Z)^-U_n\} \to \mathrm{E}\{(Y-Z)^-U\}$. Therefore,

$$\begin{aligned} 0 = \mathrm{E}\{(Y-Z)U_n\} &= \mathrm{E}\{(Y-Z)^+U_n\} - \mathrm{E}\{(Y-Z)^-U_n\} \\ &\to \mathrm{E}\{(Y-Z)^+U\} - \mathrm{E}\{(Y-Z)^-U\} \\ &= \mathrm{E}\{(Y-Z)U\}; \end{aligned} \tag{10.20}$$

i.e., $Y - Z$ is orthogonal to any nonnegative, $\mathcal{A}$-measurable random variable $U \in L^2$.

Similarly, we can show that $Y - Z$ is orthogonal to any $\mathcal{A}$-measurable random variable $U \in L^2$. We have proven the following result.

Proposition 10.26. Conditional expectation as a projection *If* $Y \in L^2$, *then* $\mathrm{E}(Y \mid \mathcal{A})$ *is a projection of* Y *onto* $\mathcal{A}$*-measurable random variables* $U \in L^2$. *That is,* $Y - Z$ *is orthogonal to every* $\mathcal{A}$*-measurable random variable* $U \in L^2$.

Example 10.27. Now suppose that we are attempting to predict the value of a random variable Y that might be difficult to measure. For instance, Y might require invasive medical imaging. But suppose that with less invasive imaging techniques, we have a set of variables $X_1, \ldots, X_k$ with which we might be able to predict Y accurately. Let $\mathcal{A}$ be $\sigma(X_1, \ldots, X_k)$, the sigma-field generated by $X_1, \ldots, X_k$. We will estimate Y using some function of $X_1, \ldots, X_k$, hence an $\mathcal{A}$-measurable random variable U. We want to minimize the mean-squared error $\mathrm{E}\{(Y-U)^2\}$ when using U to estimate Y.

In Figure 10.1, the plane represents the vector space of $\mathcal{A}$-measurable random variables in L^2, while the vector above the plane is the random variable Y. It is clear from the figure that among all vectors U in the plane, the one minimizing the squared residual $(Y-U)^2$ is the projection of Y onto the plane. That is, $Z = \mathrm{E}(Y \mid \mathcal{A})$ is the estimator of Y that minimizes the mean-squared error $\mathrm{E}\{(Y-U)^2\}$. □

In Example 10.27, we gave a geometric argument for the fact that $\mathrm{E}(Y \mid \mathcal{A})$ mimimizes $\mathrm{E}\{(Y-U)^2\}$ among $\mathcal{A}$-measurable functions $U \in L^2$. We now give a more formal proof of this fact.

Proposition 10.28. Conditional expectation minimizes MSE *If* $Y \in L^2$, *then* $Z = \mathrm{E}(Y \mid \mathcal{A})$ *minimizes* $\mathrm{E}\{(Y-U)^2\}$ *among all* $\mathcal{A}$*-measurable functions* $U \in L^2$.

Proof.

$$\begin{aligned} \mathrm{E}\{(Y-U)^2\} &= \mathrm{E}\{(Y-Z+Z-U)^2\} \\ &= \mathrm{E}\{(Y-Z)^2\} + \mathrm{E}\{(Z-U)^2\} + 2\mathrm{E}\{(Y-Z)(Z-U)\} \\ &= \mathrm{E}\{(Y-Z)^2\} + \mathrm{E}\{(Z-U)^2\} + 2\mathrm{E}[\mathrm{E}\{(Y-Z)(Z-U) \mid \mathcal{A}\}] \\ &= \mathrm{E}\{(Y-Z)^2\} + \mathrm{E}\{(Z-U)^2\} + 2\mathrm{E}[(Z-U)\mathrm{E}\{(Y-Z) \mid \mathcal{A}\}] \text{ (Prop 10.14)} \\ &= \mathrm{E}\{(Y-Z)^2\} + \mathrm{E}\{(Z-U)^2\} \\ &\ge \mathrm{E}\{(Y-Z)^2\}. \end{aligned} \tag{10.21}$$

This shows that $Z = \mathrm{E}(Y \mid \mathcal{A})$ minimizes $\mathrm{E}(Y-U)^2$ among all $\mathcal{A}$-measurable functions $U \in L^2$. □

Another consequence of viewing $Z = \mathrm{E}(Y \mid \mathcal{A})$ as a projection is the identity: $\mathrm{E}(Y^2) = \mathrm{E}(Z^2) + \mathrm{E}(Y - Z)^2$. This is just the Pythagorean theorem in a different vector space. A less visual approach can be used to derive this identity analogously to the proof of Proposition 10.21 (exercise). Apply this identity when Y has been centered to have mean 0, so that $Z = \mathrm{E}(Y \mid \mathcal{A})$ has mean 0 as well. Then $\mathrm{E}(Z^2) = \mathrm{var}(Z) = \mathrm{var}\{\mathrm{E}(Y \mid \mathcal{A})\}$. Also,

$$\begin{aligned} \mathrm{E}(Y - Z)^2 &= \mathrm{E}[\mathrm{E}\{(Y - Z)^2 \mid \mathcal{A}\}] \\ &= \mathrm{E}\{\mathrm{var}(Y \mid \mathcal{A})\}. \end{aligned} \tag{10.22}$$

We have deduced that $\mathrm{var}(Y) = \mathrm{var}\{\mathrm{E}(Y \mid \mathcal{A})\} + \mathrm{E}\{\mathrm{var}(Y \mid \mathcal{A})\}$. Apply this identity to $Y_1 + Y_2$: $\mathrm{var}(Y_1) + \mathrm{var}(Y_2) + 2\,\mathrm{cov}(Y_1, Y_2) = \mathrm{var}(Y_1 + Y_2)$

$$\begin{aligned} &= \mathrm{var}\{\mathrm{E}(Y_1 + Y_2 \mid \mathcal{A})\} + \mathrm{E}\{\mathrm{var}(Y_1 + Y_2 \mid \mathcal{A})\} \\ &= \mathrm{var}\{\mathrm{E}(Y_1 \mid \mathcal{A})\} + \mathrm{var}\{\mathrm{E}(Y_2 \mid \mathcal{A})\} + 2\,\mathrm{cov}\{\mathrm{E}(Y_1 \mid \mathcal{A}), \mathrm{E}(Y_2 \mid \mathcal{A})\} \\ &+ \mathrm{E}\{\mathrm{var}(Y_1 \mid \mathcal{A}) + \mathrm{var}(Y_2 \mid \mathcal{A}) + 2\,\mathrm{cov}(Y_1, Y_2 \mid \mathcal{A})\} \\ &= \mathrm{var}(Y_1) + \mathrm{var}(Y_2) + 2\,\mathrm{cov}\{\mathrm{E}(Y_1 \mid \mathcal{A}), \mathrm{E}(Y_2 \mid \mathcal{A})\} + 2\,\mathrm{E}\{\mathrm{cov}(Y_1, Y_2 \mid \mathcal{A})\}. \end{aligned}$$

Therefore, $\mathrm{cov}(Y_1, Y_2) = \mathrm{cov}\{\mathrm{E}(Y_1 \mid \mathcal{A}), \mathrm{E}(Y_2 \mid \mathcal{A})\} + \mathrm{E}\{\mathrm{cov}(Y_1, Y_2 \mid \mathcal{A})\}$.

We have proven the following useful result.

Proposition 10.29. Decomposition formulas for variances and covariances *Let Y, Y_1, Y_2 be random variables on $(\Omega, \mathcal{F}, P)$ with finite variance and $\mathcal{A} \subset \mathcal{F}$ be a sigma-field. The following decomposition formulas for variances and covariances hold.*

$$\mathrm{var}(Y) = \mathrm{var}\{\mathrm{E}(Y \mid \mathcal{A})\} + \mathrm{E}\{\mathrm{var}(Y \mid \mathcal{A})\}. \tag{10.23}$$

and

$$\mathrm{cov}(Y_1, Y_2) = \mathrm{cov}\{\mathrm{E}(Y_1 \mid \mathcal{A}), \mathrm{E}(Y_2 \mid \mathcal{A})\} + \mathrm{E}\{\mathrm{cov}(Y_1, Y_2 \mid \mathcal{A})\}. \tag{10.24}$$

Helpful mnemonic devices for (10.23) and (10.24) are $V = VE + EV$ and $C = CE + EC$.

Other results are also apparent from viewing conditional expectation as a projection. For instance, if Y is already $\mathcal{A}$-measurable, then the projection of Y onto $L^2(\mathcal{A})$ is Y itself. That is $\mathrm{E}(Y \mid \mathcal{A}) = Y$ almost surely if Y is $\mathcal{A}$-measurable. Also, suppose that $\mathcal{A} \subset \mathcal{C}$. Projecting Y first onto the larger sigma-field $\mathcal{C}$, and then onto the sub-sigma-field $\mathcal{A}$, is equivalent to projecting Y directly onto $\mathcal{A}$. That is $\mathrm{E}\{\mathrm{E}(Y \mid C) \mid \mathcal{A}\} = \mathrm{E}(Y \mid \mathcal{A})$ almost surely.

Proposition 10.30. Projecting first onto a larger space and then onto a subspace is equivalent to projecting directly onto the subspace *Let Y be a random variable with $\mathrm{E}(|Y|) < \infty$. If $\mathcal{A}$ and $\mathcal{C}$ are sigma-fields with $\mathcal{A} \subset \mathcal{C}$, then $\mathrm{E}\{\mathrm{E}(Y \mid C) \mid \mathcal{A}\} = \mathrm{E}(Y \mid \mathcal{A})$ a.s.*

Exercises

1. Show that Proposition 10.10 is a special case of Proposition 10.30.

2. Let Y be a random variable with $\mathrm{E}(|Y|) < \infty$, and suppose that Z is a random variable such that $Y - Z$ is orthogonal to X (i.e., $\mathrm{E}\{(Y - Z)X\} = 0$) for each $\mathcal{A}$-measurable random variable X . Prove that $Z = \mathrm{E}(Y \mid \mathcal{A})$ a.s.

3. Suppose that $\mathrm{E}(Y^2) < \infty$, and let $Z = \mathrm{E}(Y \mid \mathcal{A})$. Prove the identity $\mathrm{E}(Y^2) = \mathrm{E}(Z^2) + \mathrm{E}(Y - Z)^2$.

10.5 Conditioning and Independence

Recall from Section 4.5.1 that we defined random variables X and Y to be independent if $P(X \in A, Y \in B) = P(X \in A)P(Y \in B)$ for all Borel sets A and B, and we saw that this was equivalent to $P(X \leq x, Y \leq y) = P(X \leq x)P(Y \leq y)$ for each x and y. The same is true if we replace random variables X and Y by random vectors $\mathbf{X}$ and $\mathbf{Y}$, A and B by k-dimensional Borel sets, and $X \leq x$ and $Y \leq y$ by $\mathbf{X} \leq \mathbf{x}$ (meaning $X_i \leq x_i$ for each i) and $\mathbf{Y} \leq \mathbf{y}$. We can also formulate independence in terms of conditional distribution functions as follows.

Proposition 10.31. X and Y are independent if and only if conditional distribution given $X = x$ des not depend on x $(\mathbf{X}, \mathbf{Y})$ *are independent if and only if there is a conditional distribution function* $\psi(\mathbf{y} \,|\, \mathbf{x})$ *of* $(\mathbf{Y} \,|\, \mathbf{X} = \mathbf{x})$ *that does not depend on* $\mathbf{x}$.

Proof. Suppose that $\mathbf{X}$ and $\mathbf{Y}$ are independent. We claim that the distribution function $G(\mathbf{y})$ of $\mathbf{Y}$ is a conditional distribution function of $\mathbf{Y}$ given $\mathbf{X} = \mathbf{x}$ that does not depend on $\mathbf{x}$. It is clearly a distribution function in $\mathbf{y}$ and does not depend on $\mathbf{x}$. We need only show that $G(\mathbf{y})$ satisfies the definition of a conditional expected value of $\{I(\mathbf{Y} \leq \mathbf{y}) \,|\, \mathbf{X}\}$. Let B be any k-dimensional Borel set, where k is the dimension of $\mathbf{X}$. We must show that

$$\mathrm{E}\{G(\mathbf{y})I(\mathbf{X} \in B)\} = \mathrm{E}\{I(\mathbf{Y} \leq \mathbf{y})I(\mathbf{X} \in B)\}. \tag{10.25}$$

The left side of Equation (10.25) is clearly $P(\mathbf{X} \in B)G(\mathbf{y})$ because $G(\mathbf{y})$ is a constant. By Proposition 5.30, the right side of Equation (10.25) is $\mathrm{E}\{I(\mathbf{Y} \leq \mathbf{y})\}\mathrm{E}\{I(\mathbf{X} \in B)\} = P(\mathbf{X} \in B)G(\mathbf{y})$. Thus, Equation (10.25) holds. This completes the proof that if $(\mathbf{X}, \mathbf{Y})$ are independent, there exists a conditional distribution of $\mathbf{Y}$ given $\mathbf{X} = \mathbf{x}$ that does not depend on $\mathbf{x}$. The proof of the reverse direction is left as an exercise. □

Another important concept is that of conditional independence given another random vector.

Definition 10.32. Conditional independence given a random vector *Random vectors* $\mathbf{X}$ *and* $\mathbf{Y}$ *are said to be conditionally independent given* $\mathbf{Z}$ *if there are conditional distribution functions* $H(\mathbf{x}, \mathbf{y} \,|\, \mathbf{z})$, $F(\mathbf{x} \,|\, \mathbf{z})$, *and* $G(\mathbf{y} \,|\, \mathbf{z})$ *of* $(\mathbf{X}, \mathbf{Y} \,|\, \mathbf{Z} = \mathbf{z})$, $(\mathbf{X} \,|\, \mathbf{Z} = \mathbf{z})$, *and* $(\mathbf{Y} \,|\, \mathbf{Z} = \mathbf{z})$ *such that* $H(\mathbf{x}, \mathbf{y} \,|\, \mathbf{z}) = F(\mathbf{x} \,|\, \mathbf{z})G(\mathbf{y} \,|\, \mathbf{z})$.

Proposition 10.33. $\mathbf{X}$ *and* $\mathbf{Y}$ *are conditionally independent given* $\mathbf{Z}$ *if and only if there is a conditional distribution function* $H(\mathbf{y} \,|\, \mathbf{x}, \mathbf{z})$ *of* $(\mathbf{Y} \,|\, \mathbf{X} = \mathbf{x}, \mathbf{Z} = \mathbf{z})$ *that does not depend on* $\mathbf{x}$.

Example 10.34. Conditionally independent but unconditionally positively associated Many statistical applications involve random variables $Y_1, \ldots, Y_n$ that are iid conditional on other information. Examples include the following.

1. The simple mixed model $Y_{ij} = \mu + b_i + \epsilon_{ij}$ for observation j on participant i, where the person-specific effects b_i are iid from some distribution and the ϵ_{ij} are iid mean 0 random errors. Once we condition on b_i, the observations are iid with mean b_i and variance σ_ϵ^2.

2. The Bayesian statistical paradigm under which, conditional on the parameter θ, the data Y_i are iid from density $f(y, \theta)$, and θ is randomly drawn from a "prior" distribution $\pi(\theta)$.

3. Comparisons of several sample means to a control mean with common sample size n: $Y_i = \bar{X}_i - \bar{X}_0$, $i = 1, \ldots, m$. Once we condition on the control sample mean $\bar{X}_0$, the Y_i are independent with mean $\mu_i - \bar{X}_0$ and variance σ_i^2/n. Under the null hypothesis that $\mu_i = \mu_0$, $i = 1, \ldots, m$, the Y_i are conditionally iid given $\bar{X}_0$.

Whenever $Y_1, Y_2, \ldots, Y_n$ are iid non-degenerate random variables conditional on other information, they are unconditionally positively associated in a certain sense. To see this, let the sigma-field $\mathcal{A} \subset \mathcal{F}$ represent the other information, and suppose that $Y_1, \ldots, Y_n$ are conditionally iid and non-degenerate given $\mathcal{A}$. Then

$$\begin{aligned} \text{cov}(Y_1, Y_2) &= \text{cov}\{\text{E}(Y_1 \mid \mathcal{A}), \text{E}(Y_2 \mid \mathcal{A})\} + \text{E}\{\text{cov}(Y_1, Y_2 \mid \mathcal{A})\}. \\ &= \text{cov}\{\text{E}(Y_1 \mid \mathcal{A}), \text{E}(Y_1 \mid \mathcal{A})\} + 0 \\ &= \text{var}\{\text{E}(Y_1 \mid \mathcal{A})\} > 0. \end{aligned} \tag{10.26}$$

Of course if the Y_i are iid given $\mathcal{A}$, so are $Z_i = I(Y_i \le y)$, $i = 1, \ldots, n$. It follows from Equation (10.26) applied to the Z_i that $\text{cov}(Z_i, Z_j) > 0$. But

$$\begin{aligned} \text{cov}(Z_i, Z_j) &= \text{E}\{I(Y_i \le y)I(Y_j \le y)\} - \text{E}\{I(Y_i \le y)\}\text{E}\{I(Y_j \le y)\} \\ &= P(Y_i \le y, Y_j \le y) - P(Y_i \le y)P(Y_j \le y). \end{aligned} \tag{10.27}$$

Therefore, $P(Y_1 \le y, Y_2 \le y) > P(Y_1 \le y)P(Y_2 \le y)$, and similarly, $P(Y_1 > y, Y_2 > y) > P(Y_1 > y)P(Y_2 > y)$. That is, Y_1 and Y_2 tend to track together more frequently than they would if they were independent. □

Example 10.35. Unconditionally independent but conditionally dependent Example 10.34 shows that random variables can be conditionally independent, but very highly dependent unconditionally. The opposite is also true: two random variables X and Y can be unconditionally independent but conditionally highly correlated. For example, let Z_1 and Z_2 be independent normal random variables with variance σ^2, and let $X = Z_1 + Z_2$ and $Y = Z_1 - Z_2$. Then (X, Y) is bivariate normal with correlation 0, so X and Y are independent. On the other hand, conditional on Z_2, X is Z_1 plus the constant Z_2, and Y is Z_1 minus the constant Z_2. Therefore, $\text{cor}(X, Y \mid Z_2) = 1$. That is, although X and Y are unconditionally independent, they are perfectly positively correlated given Z_2. Similarly, X and Y are conditionally perfectly negatively correlated given Z_1.

Independence of $X = Z_1 + Z_2$ and $Y = Z_1 - Z_2$ holds whenever Z_1 and Z_2 are bivariate normal with the same variance, even if they are not independent, and this has many important applications in statistics. For example, Z_1 and Z_2 may be measurements of the same quantity using two different devices, e.g., two different blood pressure machines, and we want to determine whether the results agree. Suppose that Z_1 is the measurement from a new, untested machine, whereas Z_2 is the measurement from the well-tested "gold standard" machine on the same patient. We regard the difference $Y = Z_1 - Z_2$ as the error of the new machine. To see if the magnitude of errors depends on the patient's blood pressure, we are tempted to plot Y against Z_2, but this is problematic. For instance, suppose that the two measurements have the same variance σ^2. Then

$$\begin{aligned} \text{cov}(Y, Z_2) &= \text{cov}(Z_1 - Z_2, Z_2) = \text{cov}(Z_1, Z_2) - \text{cov}(Z_2, Z_2) \\ &= \text{cov}(Z_1, Z_2) - \sigma^2 \\ &= \sigma^2\{\text{cor}(Z_1, Z_2) - 1\} < 0 \end{aligned} \tag{10.28}$$

unless Z_1 and Z_2 are perfectly negatively correlated. Thus, it will look like the accuracy of the new machine depends on the patient's blood pressure. We should instead plot $Y = Z_1 - Z_2$ against $X/2$, the average of Z_1 and Z_2 (Bland and Altman, 1986).

If $\text{var}(Z_1) = \text{var}(Z_2)$, X and Y are uncorrelated. The plot of Y against X should show no discernible pattern. If the plot shows a linear relationship with positive slope, that is an indication that $\text{var}(Z_1) > \text{var}(Z_2)$, whereas a negative slope indicates that $\text{var}(Z_1) < \text{var}(Z_2)$. This observation provides the basis for Pitman's test of equality of variances for paired observations: use the sample correlation coefficient between $X = Z_1 + Z_2$ and $Y = Z_1 - Z_2$ to test whether the population correlation coefficient is 0 (Pitman, 1939). This is equivalent to testing whether the slope of the regression of Y on X is 0. □

Example 10.36. Independence/dependence can sometimes depend on point of view We have seen that random variables can be conditionally, but not unconditionally, independent and vice versa. Whether we think conditionally or unconditionally depends on what we want to make inferences about. For instance, consider the simple mixed model

$$Y_{ij} = \mu + b_i + \epsilon_{ij}, \tag{10.29}$$

where Y_{ij} is the jth measurement on person i, μ is the mean over all people, and b_i is a random effect for person i. Usually, we want to make inferences about the mean μ over people. For instance, a pharmaceutical company or regulatory agency is interested in the mean effect of a drug over people. In this case, multiple measurements on each person are highly correlated. A one-patient study would never suffice to prove that the drug was effective in the general population. An individual patient has a different focus: "Does the drug help me." In this case it makes sense to condition on b_i. Under the simple mixed model (10.29), the multiple observations on person i are conditionally independent given b_i. Of course (10.29) is just a model, and may or may not be accurate. This example shows that one might view multiple measurements on the same person as dependent or (conditionally) independent, depending on whether the focus is on a population average or a single person. □

Example 10.37. Missing data There is an important distinction between conditional independence of X and Y given Z and conditional independence of X and Y given **particular** (but not all) values of z. For example, consider Rubin's (1976) seminal paper on missing data. Let $\mathbf{Y}$ be a random vector of data, and let $\mathbf{M}$ be the indicators that Y_i is nonmissing (a common, but unfortunate choice of notation, since the letter M sounds like it should indicate missing rather than nonmissing), $i = 1, \ldots, n$. Let $\mathbf{y}_{\text{obs}}$ and $\mathbf{y}_{\text{mis}}$ denote the values of the data that are observed and missing, respectively. Rubin defines the data to be *missing at random (MAR)* if

$$P(\mathbf{M} = \mathbf{m} \,|\, \mathbf{y}_{\text{obs}}, \mathbf{y}_{\text{mis}}) \text{ is the same for all possible values of } \mathbf{y}_{\text{mis}}. \tag{10.30}$$

This sounds like it means $\mathbf{M}$ and $\mathbf{Y}_{\text{mis}}$ are conditionally independent, given $\mathbf{Y}_{\text{obs}}$, but it does not because condition (10.30) might hold only for the value of $\mathbf{y}_{\text{obs}}$ actually observed. Rubin offers an example involving hospital survey data that includes blood pressure. Blood pressure is missing if and only if that blood pressure is lower than the mean blood pressure in the population, μ. Suppose all participants in the survey have blood pressure exceeding μ. Then $\mathbf{y}_{\text{obs}}$ will be the vector of all blood pressures (none are missing), and for that value of $\mathbf{y}_{\text{obs}}$, condition (10.30) is satisfied vacuously. However, if any of the participants in the survey had blood pressure readings below μ, then there would have been missing data and condition (10.30) would not have held. Therefore, the data are missing at random if and only if none are missing. □

Example 10.38. Regression to the mean Regression to the mean is a phenomenon whereby extreme measurements tend to be less extreme when they are repeated. This is most often associated with a continuous outcome like blood pressure under the assumption that the initial and later measurements Y_0 and Y_1 follow a bivariate normal distribution. If the means and variances are the same and $\mathrm{cor}(Y_0, Y_1) = \rho$, then $\mathrm{E}(Y_1 \mid Y_0 = y_0) = \rho y_0 + (1-\rho)\mu$ is a weighted average of y_0 and the mean μ. That is, Y_1 tends to "regress toward the mean" μ.

Regression to the mean can occur with discrete random variables as well, and can lead the uninitiated observer to misinterpret changes from baseline to the end of a clinical trial whose entry criteria require patients to have the disease at baseline. For example, let Y_0 and Y_1 be the indicator that a patient is diseased at baseline and the end of follow-up, respectively. Imagine that there is a "frailty" parameter P indicating the probability that the patient is diseased at a given time. Assume that Y_0 and Y_1 are conditionally independent given P, with $\mathrm{E}(Y_i \mid P) = P$, $i = 0, 1$. If the trial recruits only patients with disease at baseline, then we must condition on $Y_0 = 1$. Even if a patient receives no treatment, Y_1 will be smaller than Y_0, on average.

To determine how much smaller Y_1 tends to be than Y_0, calculate $\mathrm{E}(Y_1 \mid Y_0)$ using

$$\begin{aligned} \mathrm{E}(Y_1 \mid Y_0) &= \mathrm{E}\{\mathrm{E}(Y_1 \mid Y_0, P) \mid Y_0\} \text{ (Proposition 10.30)} \\ &= \mathrm{E}(P \mid Y_0) \text{ (conditional independence of } Y_0, Y_1 \text{ given } P). \end{aligned} \tag{10.31}$$

Assume that the frailty parameters vary from patient to patient according to a uniform distribution on $[0, 1]$. That is, the density for p is $\pi(p) = 1$ for $p \in [0, 1]$. Before conditioning on $Y_0 = 1$, the joint density of (Y_0, P) with respect to the product measure $\mu_C \times \mu_L$ of counting measure and Lebesgue measure, evaluated at $(1, p)$, is $f(1, p) = P(Y_0 = 1 \mid p)\pi(p) = p$. The marginal density of Y_0 with respect to counting measure (i.e., its marginal probability mass function) is obtained by integrating the joint density over p: $g(1) = \int_0^1 p dp = 1/2$. Therefore, the conditional expectation of P given $Y_0 = 1$ is

$$\int_0^1 p \frac{f(1,p)}{g(1)} dp = \int_0^1 p \frac{p}{1/2} dp = 2/3.$$

Therefore, on average, $\mathrm{E}(Y_1 - Y_0 \mid Y_0 = 1) = 2/3 - 1 = -1/3$. In other words, patients tend to have less disease at follow-up even in the absence of treatment. McMahon et al. (1994) showed a similar result applying a Poisson-gamma model to data from the Asymptomatic Cardiac Ischemia Pilot (ACIP) study. □

We hope the examples in this section give the readers an appreciation for the wealth of applications of conditional independence. Section 11.9 covers a useful graphical tool for understanding conditional independence relationships between variables.

Exercises

Prove that if there is a regular conditional distribution function $F(\mathbf{y} \mid \mathbf{x})$ of $\mathbf{Y}$ given $\mathbf{X} = \mathbf{x}$ that does not depend on $\mathbf{x}$, then $\mathbf{X}$ and $\mathbf{Y}$ are independent.

10.6 Sufficiency

10.6.1 Sufficient and Ancillary Statistics

Suppose we want to estimate the probability p that a patient in the treatment arm of a clinical trial has an adverse event (AE). We have a random sample of 50 patients, and let

Y_i be the indicator that patient i has the adverse event; Y_i are iid Bernoulli (p). If someone supplements our data with the weights of 100 randomly selected rocks, should we use these weights to help us estimate p? Of course not. The rock weights are *ancillary*, meaning that their distribution does not depend on p.

Now imagine that the AE data are obtained by first generating the sum $S = \sum_{i=1}^{50} Y_i$ from its distribution, namely binomial $(50, p)$, and then generating the individual Y_is from their conditional distribution given S. If $S = s$, each outcome $y_1, \ldots, y_{50}$ with sum s has the same conditional probability, namely $1/\binom{50}{s}$. Notice that this conditional probability does not depend on p. Once we observe $S = s$, no additional information can be gleaned from observations generated from a conditional distribution that does not depend on p. The situation is completely analogous to the above example of augmenting adverse event data with rock weights. This motivates the following definition.

Definition 10.39. Sufficient statistic *The vector statistic* $\mathbf{S}$ *is said to be sufficient if there is a regular conditional distribution function of of* $Y_1, \ldots, Y_n$ *given* $\mathbf{S}$ *that does not depend on* θ.

Again it is helpful to imagine generating the data $Y_1, \ldots, Y_n$ by first generating the sufficient statistic $\mathbf{S}$ from its distribution, and then drawing the individual observations from their conditional distribution given $\mathbf{S}$. As these latter draws are from a distribution that is free of θ, they cannot help us make inferences about θ. This makes clear the fact that inferences about θ should be based solely on the sufficient statistic.

10.6.2 Completeness and Minimum Variance Unbiased Estimation

An estimator $\tilde{\theta}$ of a parameter θ is said to be *unbiased* if $\mathrm{E}(\tilde{\theta}) = \theta$. Loosely speaking, an unbiased estimator is on target on average. If we restrict attention to unbiased estimators of θ, it is natural to seek one with smallest variance, called a *minimum variance unbiased estimator (MVUE)*. The heuristic discussion above suggests that we should not even consider estimators that are not functions of the sufficient statistic. The following makes this more explicit. Any unbiased estimator $\tilde{\theta}$ of θ can be improved by taking $\hat{\theta} = \mathrm{E}(\tilde{\theta} \,|\, \mathbf{S})$; then $\hat{\theta}$ is unbiased (Proposition 10.10), is an extended Borel function of the sufficient statistic (Proposition 10.4), and has variance no greater than that of $\tilde{\theta}$ (Equation 10.23 of Proposition 10.29). Without loss of generality then, we can restrict attention to functions of the sufficient statistic. If there is only one unbiased function of $\mathbf{S}$, then our search is over. We therefore seek a condition that ensures there is only one unbiased function of $\mathbf{S}$. This motivates the following definition.

Definition 10.40. Complete statistic *A statistic* $\mathbf{S}$ *with distribution function* $F(s_1, \ldots, s_k; \theta)$ *is said to be complete if* $\mathrm{E}\{f(\mathbf{S})\} \equiv 0$ *for all values of* θ *implies that* $f(\mathbf{S}) = 0$ *with probability* 1 *for each* θ.

Proposition 10.41. Only one unbiased function if S is complete *If* $\mathbf{S}$ *is complete and* $f(\mathbf{S})$ *and* $g(\mathbf{S})$ *are unbiased, Borel functions of* $\mathbf{S}$, *then* $f(\mathbf{S}) = g(\mathbf{S})$ *with probability* 1 *for each* θ.

Proof. $\mathrm{E}\{f(\mathbf{S}) - g(\mathbf{S})\} = \theta - \theta = 0$. By completeness, $f(\mathbf{S}) - g(\mathbf{S}) = 0$ with probability 1 for each θ, so $f(\mathbf{S}) = g(\mathbf{S})$ with probability 1 for each θ. □

Proposition 10.42. Unbiased functions of a complete sufficient statistic are UMVUE *If* $\mathbf{S}$ *is complete and sufficient and* $f(\mathbf{S})$ *is unbiased, then* $f(\mathbf{S})$ *is a UMVUE.*

Proof. If not, then there is an unbiased estimator $\tilde{\theta}$ with smaller variance than $f(\mathbf{S})$. But then $\mathrm{E}(\tilde{\theta}\,|\,\mathbf{S})$ is an unbiased, Borel function of $\mathbf{S}$ with smaller variance than $f(\mathbf{S})$. But this is a contradiction because, by Proposition 10.41, $\mathrm{E}(\tilde{\theta}\,|\,\mathbf{S}) = f(\mathbf{S})$ with probability 1. □

The importance of complete, sufficient statistics has led to a thorough investigation of settings admitting such statistics. Complete, sufficient statistics have been established for the very large class of exponential families.

10.6.3 Basu's Theorem and Applications

Ancillary and sufficient statistics $\mathbf{A}$ and $\mathbf{S}$ are at opposite ends of the spectrum: in a real sense, ancillary statistics tell us nothing and sufficient statistics tell us everything about θ. It makes sense, then, that $\mathbf{A}$ and $\mathbf{S}$ might be independent. A beautiful theorem by Basu (1955) asserts that this is true, provided that $\mathbf{S}$ is complete.

Theorem 10.43. Basu's theorem: Ancillary and complete sufficient statistics are independent *Let $\mathbf{S}$ be a complete, sufficient statistic and $\mathbf{A}$ be ancillary. Then $\mathbf{A}$ and $\mathbf{S}$ are independent.*

Proof. Let $F(a_1,\ldots,a_k) = P(A_1 \le a_1,\ldots,A_k \le a_k)$. Because $\mathbf{A}$ is ancillary, $F(a_1,\ldots,a_k)$ does not depend on θ. Let $G(a_1,\ldots,a_k\,|\,S)$ be a regular conditional distribution function of $\mathbf{A}$ given $\mathbf{S}$. Then $F(a_1,\ldots,a_k) = \mathrm{E}\{G(a_1,\ldots,a_k\,|\,\mathbf{S})\}$. Therefore,

$$\mathrm{E}\{G(a_1,\ldots,a_k\,|\,\mathbf{S}) - F(a_1,\ldots,a_k)\} = 0.$$

But $G(a_1,\ldots,a_k\,|\,\mathbf{S}) - F(a_1,\ldots,a_k)$ is a function of $\mathbf{S}$, and $\mathbf{S}$ is complete, so $G(a_1,\ldots,a_k\,|\,\mathbf{S}) - F(a_1,\ldots,a_k) = 0$ a,s. I.e., the conditional distribution function of $\mathbf{A}$ given $\mathbf{S}$ is the same as its unconditional distribution function. By result 10.31, $\mathbf{A}$ and $\mathbf{S}$ are independent. □

There are many applications of Basu's theorem. For example, if $Y_1,\ldots,Y_n$ are iid $\mathrm{N}(\mu,\sigma^2)$, with σ^2 known, the sample mean $\bar{Y}_n$ is sufficient and complete for μ. On the other hand, the sample variance s_n^2 is ancillary for μ. This follows from the fact that adding the same constant c to each observation does not change s_n^2, so does not change the distribution of s_n^2. Basu's theorem implies the well-known result that $\bar{Y}_n$ and s_n^2 are independent for iid normal observations with finite variance. In fact, we can say more. The set $(s_2^2, s_3^2, \ldots, s_n^2$ of consecutive sample variances with sample sizes $2, 3, \ldots, n$ is also ancillary for the same reason. Therefore, $\bar{Y}_n$ is independent of $(s_2^2, s_3^2, \ldots, s_n^2)$. This has ramifications for adaptive clinical trials in which design changes might be based on interim variances. Conditioned on those variances, the distribution of $\bar{Y}_n$ is the same as its unconditional distribution. The next example expands on the use of Basu's theorem in adaptive clinical trials.

Example 10.44. Application of Basu's theorem: adaptive sample size calculation in clinical trials Consider a clinical trial with paired data and a continuous outcome Y such as change in cholesterol from baseline to 1 year. Let D_i be the difference between treatment and control measurements on pair i, and assume that $D_i \sim N(\mu,\sigma^2)$. We are interested in testing the null hypothesis $H_0 : \mu = 0$ versus the alternative hypothesis $H_1 : \mu > 0$. Before the trial begins, we determine the approximate number of pairs required for a one-tailed t-test at $\alpha = 0.025$ and 90% power using the formula

$$n \approx \frac{(1.96 + 1.28)^2\sigma^2}{\mu^2}. \tag{10.32}$$

The two quantities we need for this calculation are the size of the treatment effect, μ, and the variance σ^2. We can usually determine the treatment effect more easily than the variance because we argue that if the effect is not at least a certain magnitude, the treatment is not worth developing. The variance, on the other hand, must be estimated. Sometimes there is good previous data on which to base σ^2, other times not. It would be appealing if we could begin with a pre-trial estimate σ_0^2 of the variance, but then modify that estimate and the sample size after seeing data from the trial itself.

Consider the following two-stage procedure. The first stage consists of half ($n_1 = n_0/2$) of the originally planned number of observations. From this first-stage data, we use $\hat{\sigma}^2 = (1/n)\sum D_i^2$ to estimate the variance σ^2. This is slightly different from the usual sample variance, which subtracts $\bar{D}$ from each observation and uses $n-1$ instead of n in the denominator. Nonetheless, $\hat{\sigma}^2$ is actually very accurate for typical clinical trials in which the treatment effect μ is not very large. We then substitute $\hat{\sigma}^2$ for σ^2 in (10.32) and compute the sample size $n = n(\hat{\sigma}^2)$. If $n \le n_0/2$, we collect no additional observations. Otherwise, the second stage consists of the number of additional observations required, $n_2 = n(\hat{\sigma}^2) - n_0/2$. It is tempting to pretend that the sample size had been fixed in advance, compute the usual t-statistic, and refer it to a t-distribution with $n-1$ degrees of freedom. This is actually a very good approximation, but the resulting test statistic is not exactly t_{n-1}.

We can construct an exact test as follows. Let T_1 be the t-statistic computed on the first-stage data. Under the null hypothesis that $\mu = 0$, the first-stage data are iid $N(0, \sigma^2)$, and $\hat{\sigma}^2$ is a complete, sufficient statistic for σ^2. On the other hand, T_1 is ancillary for σ^2. This follows from the fact that dividing each observation by σ does not change T_1. By Basu's theorem, T_1 and $\hat{\sigma}^2$ are independent. This implies that, conditional on $\hat{\sigma}^2$, T_1 has a t-distribution with $n_1 - 1 = n_0/2 - 1$ degrees of freedom. If there is a second stage, then conditional on $\hat{\sigma}^2$, the t-statistic T_2 using only data from stage 2 has a t-distribution with $n_2 - 1 = n - n_0/2 - 1$ degrees of freedom and is independent of T_1. Let P_1 and P_2 denote the p-values corresponding to T_1 and T_2. Conditional on $\hat{\sigma}^2$, P_1 and P_2 are independent uniforms, so the inverse probability transformations $Z_1 = \Phi^{-1}(1-P_1)$ and $Z_2 = \Phi^{-1}(1-P_2)$ are independent standard normals. If there is no second stage, we can generate a superfluous standard normal deviate Z_2. Then, conditional on $\hat{\sigma}^2$,

$$Z = \frac{\sqrt{n_1}Z_1 + \sqrt{n_2}Z_2}{\sqrt{n_1 + n_2}}$$

has a standard normal distribution (when there is no second stage, $n_2 = 0$, and the superfluous random variable Z_2 receives zero weight). We reject the null hypothesis when $Z > z_\alpha$, the $(1-\alpha)$th quantile of a standard normal distribution. The conditional type I error rate given $\hat{\sigma}^2$ is exactly α. Therefore, the unconditional type I error rate is $E\{P(Z > z_\alpha \mid \hat{\sigma}^2)\} = E(\alpha) = \alpha$. That is, this two-stage procedure provides an exact level α test. □

10.6.4 Conditioning on Ancillary Statistics

It is good statistical practice to condition on the values of ancillary statistics. For instance, in a one-sample t-test, suppose that we flipped a coin to decide whether to use a sample size of 50 or 100. Thus, the estimated mean is $\bar{Y}_N$, where N is the random sample size. Suppose that as a result of the coin flip, $N = 50$. It would be silly to treat the sample size as random and compute the variance of $\bar{Y}_N$ as

$$
\begin{aligned}
\mathrm{var}(\bar{Y}_N) &= \mathrm{E}\{\mathrm{var}(\bar{Y}_N \mid N)\} + \mathrm{var}\{\mathrm{E}(\bar{Y}_N \mid N)\} \\
&= (1/2)\mathrm{var}(\bar{Y}_{50}) + (1/2)\mathrm{var}(\bar{Y}_{100}) + \mathrm{var}(\mu) \\
&= (1/2)(\sigma^2/50) + (1/2)(\sigma^2/100). \qquad (10.33)
\end{aligned}
$$

This unconditional approach gives a misleading estimate of the variance for the sample size actually used, 50. We would instead use $\mathrm{var}(\bar{Y}_{50}) = \sigma^2/50$. That is, we would condition on $N = 50$ because N is an ancillary statistic.

The above example may seem artificial because we usually do not flip a coin to decide whether to double the sample size. But the same issue arises in a more realistic setting. In a clinical trial with n patients randomly assigned to treatment or control, analyses condition on the numbers actually assigned to treatment and control. For instance, suppose that the sample sizes in the treatment and control arms are 22 and 18, respectively. When we use a permutation test, we consider all $\binom{40}{22}$ different ways to assign 22 of the 40 patients to treatment; we do not consider all 2^{40} possibilities that would result if we did not fix the sample sizes. It makes sense to condition on the sample sizes actually used because they are ancillary. They give us no information about the treatment effect. However, the next example is a setting in which the sample sizes give a great deal of information about the treatment effect.

Example 10.45. Sample size is not always ancillary: ECMO The Extracorporeal Membrane Oxygenation (ECMO) trial (Bartlett et al., 1985) was in infants with primary pulmonary hypertension, a disease so serious that the mortality rate using the standard treatment, placing the baby on a ventilator, was expected to be 80%. The new treatment was extracorporeal membrane oxygenation (ECMO), an outside the body heart and lung machine used to allow the baby's lungs to rest and heal. Because of the very high mortality expected on the standard treatment, the trial used a nonstandard urn randomization technique that can be envisioned as follows. Place one standard (S) therapy ball and one ECMO (E) ball in an urn. For the first baby's assignment, randomly draw one of the two balls. If the ball is ECMO and the baby survives, or standard therapy and the baby dies, then "stack the deck" in favor of ECMO by replacing the ball and then adding another ECMO ball. On the other hand, if the first baby is assigned to ECMO and dies, or to the standard treatment and survives, replace the ball and add a standard therapy ball. That way, the second baby has probability 2/3 of being assigned to the therapy doing better so far. Likewise, after each new assignment, replace that ball and add a ball of the same or opposite treatment depending on whether that baby survives or dies. This is called a *response-adaptive* randomization scheme.

Table 10.1: Data from the ECMO trial; 0 and 1 denote survival and death, and E and S denote ECMO and standard treatment.

Outcome	0	1	0	0	0	0	0	0	0	0	0	0
Assignment	E	S	E	E	E	E	E	E	E	E	E	E
Probability	$\frac{1}{2}$	$\frac{1}{3}$	$\frac{3}{4}$	$\frac{4}{5}$	$\frac{5}{6}$	$\frac{6}{7}$	$\frac{7}{8}$	$\frac{8}{9}$	$\frac{9}{10}$	$\frac{10}{11}$	$\frac{11}{12}$	$\frac{12}{13}$

The actual data in order are shown in Table 10.1, where 0 and 1 denote alive and dead, and E and S denote ECMO and standard therapy. The first baby was assigned to ECMO and survived. The next baby was assigned to the standard therapy and died. Then the next 10 babies were all assigned to ECMO and survived. At that point, randomization was

Table 10.2: Summary 2×2 table for the ECMO trial.

	Dead	Alive	
ECMO	0	11	11
Standard	1	0	1
	1	11	

discontinued. Table 10.2 summarizes the outcome data. If we use Fisher's exact test, which is equivalent to a permutation test on binary data, the one-tailed p-value is

$$\frac{\binom{11}{0}\binom{1}{1}}{\binom{12}{1}} = 1/12 = 0.083.$$

But the above calculation assumes that the 12 randomizations leading to 11 patients assigned to ECMO and 1 to standard therapy are equally likely. This is not true if we condition on the data and order of entry of patients shown in Table 10.1. To see this, consider the probability of the actual treatment assignments shown in Table 10.1. The probability that the first baby is assigned to E is 1/2. Because the first baby survived on E, there are 2 Es and 1 S in the urn when the second baby is randomized. Therefore, the conditional probability that the second baby is assigned to S is 1/3. Because that baby died on S, there are 3 Es and 1 S when the third baby is assigned. The conditional probability that the third baby is assigned to E is 3/4, etc. The probability of the observed assignment, given the outcome vector, is $(1/2)(2/3)(3/4)\ldots(12/13) = 1/26$. On the other hand, the randomization assignment (S,E,E,E,E,E,E,E,E,E,E,E) has probability 1/1716 given the observed outcome vector. The probability of each of the other 10 randomization sequences leading to the marginal totals of Table 10.2 is 1/429 (exercise). Therefore, the sum of probabilities of treatment sequences leading to these marginals is $1/26+1/1716+10(1/429) = 107/1716$. To obtain the p-value conditional on marginal totals of Table 10.2, we must sum the conditional probabilities of assignment vectors leading to tables at least as extreme as the observed one, and divide by 107/1716. But the actual assignment vector produces the most extreme table consistent with the given marginals, so the p-value conditional on marginal totals is $p = (1/26)/(107/1716) = 66/107 = 0.62$. This hardly seems to reflect the level of evidence in favor of ECMO treatment!

The problem with conditioning on the sample sizes is that they are not ancillary. They are quite informative about the treatment effect. Eleven of 12 babies were assigned to ECMO precisely because ECMO was working. Therefore, it makes no sense to condition on information that is informative about the treatment effect. It would be like arguing that a z-score of 3.1 is not at all unusual, conditioned on the fect that the z-score exceeded 3; conditioning on $Z > 3$ makes no sense because we are conditioning away the evidence of a treatment effect. That is why Wei (1988) did not condition on the marginals. He summed the probabilities of the actual assignments and the probability of (E,E,E,E,E,E,E,E,E,E,E,E), which yielded a p-value of 0.051. In the end, the trial generated substantial controversy (see Begg, 1990 and commentary, or Section 2 of Proschan and Nason, 2009) and did not convince the medical community. A subsequent larger trial showed that ECMO was superior to the standard treatment. There are many valuable lessons from the original ECMO trial, but the one we stress here is that the sample sizes in clinical trials are not always ancillary. When sample sizes are informative about the treatment effect, the analysis should not condition on them. □

Exercises

1. A common test statistic for the presence of an outlier among iid data from $N(\mu, \sigma^2)$ is the maximum normed residual

$$U = \max_{1 \le i \le n} \frac{|X_i - \bar{X}|}{\sqrt{\sum_{i=1}^{n}(X_i - \bar{X})^2}}.$$

   Using the fact that the sample mean and variance $(\bar{X}, s^2)$ is a complete, sufficient statistic, prove that U is independent of s^2.

2. Let $Y \sim \mathrm{N}(\mu, 1)$, and suppose that A is a set such that $P(Y \in A)$ is the same for all μ. Use Basu's theorem to prove that $P(Y \in A) = 0$ for all μ or $P(Y \in A) = 1$ for all μ.

3. In the ECMO example, consider the set of possible treatment assignment vectors that are consistent with the marginals of Table 10.2. Show that the probability of each of the 10 assignment vectors other than (E,S,E,E,E,E,E,E,E,E,E,E) and (S,E,E,E,E,E,E,E,E, E,E,E) is 1/429.

4. Let X be exponential (λ). It is known that X is a complete and sufficient statistic for λ. Use this fact to deduce the following result on the uniqueness of the *Laplace transform* $\psi(t) = \int_0^\infty f(x)\exp(tx)dx$ of a function $f(x)$ with domain $(0, \infty)$. If $f(x)$ and $g(x)$ are two functions whose Laplace transforms agree for all $t < 0$, then $f(x) = g(x)$ except on a set of Lebesgue measure 0.

10.7 Expect the Unexpected from Conditional Expectation

This section covers some common pitfalls and errors in reasoning with conditional expectation. In more elementary courses that assume there is an underlying density function, such reasoning works and is often encouraged. But our new, more general definition of conditional expectation applies whether or not there is a density function. With this added generality comes the opportunity for errors, as we shall see. In other cases, errors result from a simple failure to compute the correct conditional distribution, as in the two envelope paradox of Example 1.4.

Example 10.46. Return to the two envelopes: a simulation Recall that in the two envelope paradox of Example 1.4, one envelope has twice the amount of money as the other. The amounts in your and my envelopes are X and Y, respectively. Simulate this experiment as follows. Generate a random variable T_1 from a continuous distribution on $(0, \infty)$. For simplicity, let T_1 be exponential with parameter 1. Generate a Bernoulli (1/2) random variable Z_1 independent of T_1; if $Z_1 = 0$, set $T_2 = (1/2)T_1$, while if $Z_1 = 1$, set $T_2 = 2T_1$. Now generate another independent Bernoulli Z_2; if $Z_2 = 0$, set $(X, Y) = (T_1, T_2)$, while if $Z_2 = 1$, set $(X, Y) = (T_2, T_1)$. Repeat this experiment a thousand times, recording the (X, Y) pairs for each. See which of the following steps is the first not to hold.

1. All pairs have $Y = X/2$ or $Y = 2X$.

2. Approximately half the pairs have $Y = X/2$ and half have $Y = 2X$ (that is, $Y = X/2$ with probability 1/2 and $Y = 2X$ with probability 1/2).

3. Regardless of the value x of X, approximately half the pairs (x, Y) have $Y = x/2$, half have $Y = 2x$ (that is, given $X = x$, $Y = x/2$ or $2x$, with probability $1/2$ each).

4. $\mathrm{E}(Y \mid X = x) = (x/2)(1/2) + (2x)(1/2) = (5/4)x$.

Figure 10.2 shows (X, Y) pairs from a thousand simulations. The points all lie on one of two lines, $Y = (1/2)X$ or $Y = 2X$. Thus, statement 1 is true. In this simulation, 491 of the 1000 pairs have $X < Y$, so statement 2 is true. However, among the 27 pairs with $X > 5$, none produced $X < Y$. It is clear that the conditional probability that $Y = x/2$ given that $X = x$ is not $1/2$ for large values of x. The problem in the two envelopes paradox is that we incorrectly conditioned on $X = x$. The statement that Y is equally likely to be $X/2$ or $2X$ is true unconditionally, but not conditional on X. □

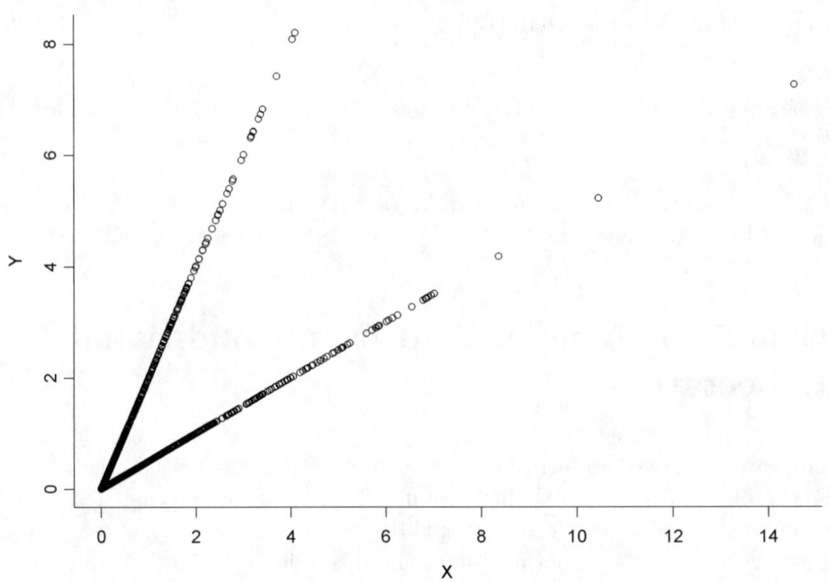

Figure 10.2: Plot of 1000 simulations of (X, Y) from the two envelopes paradox.

10.7.1 Conditioning on Sets of Probability 0

Most of the problems with conditional expectation stem from conditioning on sets of probability 0. The following is a case in point.

Example 10.47. Distribution of random draw from $[0, 1]$ given it is rational Let $(\Omega, \mathcal{F}, P)$ be $([0, 1], \mathcal{B}, \mu_L)$. Then ω corresponds to drawing a number randomly from the unit interval. What is the conditional distribution of ω given that ω is rational? It seems that ω must be uniformly distributed on the set of rationals. If not, then which rationals should be more likely than others? But Section 3.6 shows that there is no uniform distribution on a countable set. On the one hand, the distribution seems like it must be uniform, and on the other hand, it cannot be uniform. □

The problem with Example 10.47 is that conditional expectation conditions on a random variable or a sigma-field, not on a single set of probability 0. At this point the reader may be wondering (1) whether this kind of problem could arise in practice and (2) whether one could solve the problem by formulating it as a conditional expectation given a random variable. The next example answers both of these questions.

Example 10.48. Conditioning on the equality of two continuous random variables Borel paradox In a clinical trial of an HIV vaccine, investigators discovered participants with genetically extremely similar viruses, suggesting that some trial participants either had sex with each other or with a common partner. We will refer to this as "sexual sharing." This raises a question about whether the usual methods of analyzing independent data can still be applied. Proschan and Follmann (2008) argue that under certain reasonable assumptions, a permutation test is still valid. In response to a reviewer's query, the authors pointed out that even in clinical trials with independent data, one can view the observations as conditionally dependent given certain information. For instance, a certain gene might protect someone against acquiring HIV even if that person engages in risky behavior. Conditioned on the presence/absence of that gene, the HIV indicators Y_i of different patients are correlated: the Y_i of two patients who both have the gene or both do not have the gene (concordant patients) are positively correlated, whereas the Y_i of two patients who have different gene statuses (discordant patients) are negatively correlated.

The gene status random variable is fairly simple because it takes only two values, but we might try to extend the same reasoning to a setting with continuous random variables. This is exactly the setting of Example 1.3. In that example, we postulate a linear model $Y = \beta_0 + \beta_1 X + \epsilon$ relating the continuous outcome Y to the continuous covariate X. For simplicity, we assume that $X \sim N(0,1)$. We then attempt to obtain the relationship between Y values of two people with the same X value. In formulation 2 of the problem, we imagine continually observing pairs (X, Y) until we find two pairs with the same X value. The first step toward determining the relationship between the two Y values is determining the distribution of the common X, i.e., the distribution of X_1 given that $X_1 = X_1$. Formulating the event $X_1 = X_2$ in different ways, $X_2 - X_1 = 0$ or $X_2/X_1 = 1$, gives different answers.

The problem is that in the plane, sets of the form $x_2/x_1 \leq a$ look quite different from sets of the form $x_2 - x_1 \leq b$, and this leads to different conditional distributions given X_2/X_1 versus given $X_2 - X_1$. The conditional distribution of X_1 given $X_1 = X_2$ is not well-defined. Again we cannot think in terms of conditioning on **sets** of probability 0, but only on random variables or sigma-fields. Because we can envision sets of probability 0 in more than one way as realizations from random variables, in general, there is not a unique way to define conditioning on an arbitrary set of probability 0. □

10.7.2 Substitution in Conditioning Expressions

Who among us has never used the following reasoning? To compute the distribution of $X + Y$, where X and Y are independent with respective distributions F and G, we argue that

$$\begin{aligned} P(X+Y \leq z) &= \int P(X+Y \leq z \,|\, X = x) dF(x) \\ &= \int P(x+Y \leq z \,|\, X = x) dF(x) \\ &= \int P(Y \leq z - x \,|\, X = x) dF(x) \end{aligned}$$

$$= \int G(z-x)dF(x). \tag{10.34}$$

The last step follows from the independence of X and Y. The second step replaces the random variable X with its value, x. Such arguments abound in statistics, and they can help us deduce results. Nonetheless care is needed in carrying out substitution in conditional expectations. This was not the case in elementary statistics courses because the definition of conditional expectation was much more specific—Expression (10.2). But the more general Definition 10.2 admits different versions of conditional expectation, and this can lead to confusion.

Example 10.49. Confusion from substitution of $X = x$ Compute $P(X \le x)$ by first conditioning on $X = x$: $P(X \le x \,|\, X = x)$. Given that $X = x$, $X \le x$ is guaranteed, so $P(X \le x \,|\, X = x) = 1$. By Proposition 10.10, $P(X \le x) = \mathrm{E}(1) = 1$. Something is clearly amiss if we have shown that for an arbitrary random variable X and value x, $P(X \le x) = 1$. The confusion stems from the two different xs, the argument of the distribution function and the value of the random variable. Had we used, say x_0 and x to denote the argument of the distribution function and value of X, respectively, we would have concluded that

$$P(X \le x_0 \,|\, X = x) = I(x \le x_0), \quad P(X \le x_0 \,|\, X) = I(X \le x_0).$$

When we take the expected value of this expression over the distribution of X, we reach the correct conclusion that $P(X \le x_0) = \mathrm{E}\{I(X \le x_0)\}$. □

There is potential for confusion whenever there is a random variable X and its value x in the same expression, as in $\mathrm{E}\{f(x,Y) \,|\, X = x\}$. The following example makes this more clear.

Example 10.50. More confusion in substitution Let $f(x,y) = x + y$, and let X and Y be iid standard normals. Consider the calculation of $\mathrm{E}\{f(X,Y) \,|\, X = 0\}$ using the "rule" $\mathrm{E}\{f(0,Y) \,|\, X = 0\} = \mathrm{E}(0 + Y \,|\, X = 0) = 0 + \mathrm{E}(Y \,|\, X = 0)$. One version of $\mathrm{E}(Y \,|\, X)$ is $\mathrm{E}(Y) = 0$ because X and Y are independent. Using this version, we get $\mathrm{E}\{f(x,Y) \,|\, X = 0\} = 0 + 0 = 0$. But we can change the definition of $\mathrm{E}(Y \,|\, X)$ at the single value $X = 0$, and it will remain a version of $\mathrm{E}(Y \,|\, X)$. For instance,

$$\mathrm{E}(Y \,|\, X) = \begin{cases} 0 & \text{if } X \ne 0 \\ 1 & \text{if } X = 0. \end{cases} \tag{10.35}$$

Using this version, we get $\mathrm{E}\{f(0,Y) \,|\, X = 0\} = 0 + 1 = 1$. Of course, we could replace the value 1 in Equation (10.35) with any other value, so $\mathrm{E}\{f(0,Y) \,|\, X = 0)$ could be any value whatsoever.

The same problem holds if we replace the conditioning value 0 by any value x. In this example: $\mathrm{E}\{f(x,Y) \,|\, X = x\} = \mathrm{E}(x + Y \,|\, X = x) = x + \mathrm{E}(Y \,|\, X = x)$. Let $g(x)$ be an arbitrary Borel function. One version of $\mathrm{E}(Y \,|\, X)$ is

$$\mathrm{E}(Y \,|\, X) = \begin{cases} 0 & \text{if } X \ne x \\ g(x) - x & \text{if } X = x. \end{cases} \tag{10.36}$$

With this version, $\mathrm{E}\{f(x,Y) \,|\, X = x\} = x + g(x) - x = g(x)$. But $g(x)$ was arbitrary, so $\mathrm{E}\{f(x,Y) \,|\, X = x\}$ could literally be any Borel function of x. □

In light of the above examples, which of the following two statements is incorrect?

1. If $G(y \mid X = x)$ is a regular conditional distribution function of Y given $X = x$, then $\mathrm{E}\{f(X,Y) \mid X = x\} = \int f(x,y)dG(y \mid x)$.

2. $\mathrm{E}\{f(X,Y) \mid X = x\} = \mathrm{E}\{f(x,Y) \mid X = x\}$.

The statements seem equivalent, but they are not. The first involves a **specific** version of $\mathrm{E}\{f(x,Y) \mid X = x\}$, whereas $\mathrm{E}\{f(x,Y) \mid X = x\}$ in the second statement is **any** version of $\mathrm{E}\{f(x,Y) \mid X = x\}$. Thus, the second statement asserts that $\mathrm{E}\{f(X,Y) \mid X = x\} = \mathrm{E}\{f(x,Y) \mid X = x\}$ for **any** version of $\mathrm{E}\{f(x,Y) \mid X = x\}$. Example 10.50 is a counterexample.

Fortunately, the first statement is correct. More generally:

Proposition 10.51. Substitution is permitted once we select a specific version of the conditional distribution function *Let* $\mathbf{X}$ *and* $\mathbf{Y}$ *be random vectors and* $\lambda(X,Y)$ *be a function such that* $|\lambda(\mathbf{X},\mathbf{Y})|$ *is integrable. If* $G(\mathbf{y} \mid \mathbf{x})$ *is a regular conditional distribution function of* $\mathbf{Y}$ *given* $\mathbf{X} = \mathbf{x}$*, then one version of* $\mathrm{E}\{\lambda(\mathbf{X},\mathbf{Y}) \mid \mathbf{X} = \mathbf{x}\}$ *is* $\int \lambda(\mathbf{x},\mathbf{y})dG(\mathbf{y} \mid \mathbf{x})$.

An immediate consequence of Proposition 10.51 is the following.

Corollary 10.52. Substitution and independent vectors *Suppose that* $\mathbf{X}$ *and* $\mathbf{Y}$ *are independent with* $\mathbf{Y} \sim G(\mathbf{y})$*. If* $|\lambda(\mathbf{X},\mathbf{Y})|$ *is integrable, then one version of* $\mathrm{E}\{\lambda(\mathbf{X},\mathbf{Y}) \mid \mathbf{X} = \mathbf{x}\}$ *is* $\int \lambda(\mathbf{x},\mathbf{y})dG(\mathbf{y})$.

Example 10.53. Permutation tests condition on outcome data In a clinical trial comparing means in the treatment and control arms, the treatment effect estimate is

$$\hat{\delta}(\mathbf{Z},\mathbf{Y}) = (1/n_T)\sum_{i=1}^{n} Z_i Y_i - (1/n_C)\sum_{i=1}^{n}(1 - Z_i)Y_i \qquad (10.37)$$

where Z_i is 1 if patient i is assigned to treatment and 0 if control, and n_T and n_C are the numbers of patients assigned to treatment and control, respectively. That is, $\sum_{i=1}^{n} Z_i = n_T$. Under the null hypothesis, treatment has no effect on outcome, and the treatment vector $\mathbf{Z}$ is assumed independent of the outcome vector $\mathbf{Y}$. A regular conditional distribution function $F(\mathbf{z} \mid \mathbf{Y} = \mathbf{y})$ of $\mathbf{Z}$ given $\mathbf{Y} = \mathbf{y}$ is its unconditional distribution with mass function $p(\mathbf{z}) = \binom{n}{n_T}^{-1}$ for each string $\mathbf{z}$ of zeroes and ones with exactly n_T ones. Therefore, a regular conditional distribution function of $\hat{\delta}$ given $\mathbf{Y} = \mathbf{y}$ is

$$\begin{aligned} P\{\hat{\delta}(\mathbf{Z},\mathbf{Y}) \le d \mid \mathbf{Y} = \mathbf{y}\} &= \int I\{\hat{\delta}(\mathbf{z},\mathbf{y}) \le d\}dF(\mathbf{z} \mid \mathbf{y}) \\ &= \sum_{\mathbf{z}} I\{\hat{\delta}(\mathbf{z},\mathbf{y}) \le d\}p(\mathbf{z}) \text{ (Corollary 10.52)} \\ &= \frac{1}{\binom{n}{n_T}} \sum_{\mathbf{z}} I\{\hat{\delta}(\mathbf{z},\mathbf{y}) \le d\}. \end{aligned} \qquad (10.38)$$

This is the permutation distribution of $\hat{\delta}$. □

It is sometimes possible to bypass arguments using conditional probability. This is considered better form, just as a direct proof is considered better form than a proof by contradiction. The following example illustrates how to bypass conditioning.

Example 10.54. An illustration of avoiding conditioning: Stein's method In a one-sample t-test setting with iid normal observations, the standard 95% confidence interval for the mean μ is $\bar{Y}_n \pm t_{n-1,\alpha/2} s_n/n^{1/2}$, where n is the sample size, s_n is the sample standard deviation, and $t_{n-1,\alpha/2}$ is the upper $\alpha/2$ point of a t-distribution with $n-1$ degrees of freedom. Notice that the width of the confidence interval is random because it depends on s_n. Even though the width tends to 0 almost surely as $n \to \infty$, even for large sample size n, there is some small probability that s_n is large enough to make the interval wide. Stein (1945) showed how to construct a confidence interval with a fixed and arbitrarily small width.

The first step of Stein's method is to take a subsample of size m, say $m = 30$. Let s_m be the sample standard deviation for this subsample. Once we observe the subsample, s_m is a fixed number. We choose the final sample size $N = N(s_m)$ as a certain Borel function of s_m to be revealed shortly. We then observe $N - m$ additional observations. Consider the distribution of

$$T = \frac{\bar{Y}_N - \mu}{s_m/N^{1/2}}.$$

We claim that T has a t-distribution with $m-1$ degrees of freedom, so that

$$P\left(\bar{Y}_N - t_{m-1,\alpha/2}\frac{s_m}{\sqrt{N}} \leq \mu \leq \bar{Y}_N + t_{m-1,\alpha/2}\frac{s_m}{\sqrt{N}}\right) = 1-\alpha.$$

That is,

$$\bar{Y}_N \pm t_{m-1,\alpha/2}\frac{s_m}{\sqrt{N}}$$

is a $100(1-\alpha)\%$ confidence interval for μ with width $2t_{m-1,\alpha/2}(s_m/n^{1/2})$. We can construct a confidence interval of arbitrarily small width w or less as follows. For the observed value s_m, simply choose N to be the smallest integer n such that $2t_{m-1,\alpha/2}(s_m/n^{1/2}) < w$. The confidence interval will then have width w or less. This device can also be used to construct a test whose power does not depend on the unknown value of σ, hence the title of Stein's (1945) paper.

To deduce that T follows a t-distribution with $m-1$ degrees of freedom, take $\mu = 0$ without loss of generality. Consider the conditional distribution of $\bar{Y}_N$ given s_m. Given s_m, $N = n$ is a fixed number. Thus, the conditional distribution of Y_N given s_m such that $N = n$ is the same as the conditional distribution of Y_n given s_m such that $N = n$. But we saw from Basu's theorem that $\bar{Y}_n$ is independent of $s_2^2, s_3^2, \ldots s_n^2$, so the conditional distribution of $Z_N = \bar{Y}_N/(\sigma/N^{1/2})$ given s_m such that $N = n$ is the same as the unconditional distribution of Z_n, namely standard normal. Therefore, unconditionally, Z_N is standard normal and independent of s_m. It follows that

$$T = \frac{\bar{Y}_N}{s_m/N^{1/2}} = \frac{\frac{\bar{Y}_N}{\sigma/N^{1/2}}}{s/\sigma} = \frac{Z_N}{\sqrt{\frac{(m-1)s_m^2/\sigma^2}{m-1}}} \tag{10.39}$$

is the ratio of a standard normal and the square root of an independent chi-squared $(m-1)$ random variable divided by its number of degrees of freedom. By definition, T has a t-distribution with $m-1$ degrees of freedom.

The above development is very helpful to **deduce** the distribution of T, but is somewhat awkward as a proof. Once we know the right answer, we can circumvent conditioning and provide a more appealing proof:

$$\begin{aligned}
P(\{Z_N \le z\} \cap \{s_m^2 \le u\}) &= \sum_{n=m}^{\infty} P\left(\left\{\frac{\bar{Y}_N}{\sigma/N^{1/2}} \le z\right\} \cap \{s_m^2 \le u\} \cap \{N = n\}\right) \\
&= \sum_{n=m}^{\infty} P\left(\left\{\frac{\bar{Y}_n}{\sigma/n^{1/2}} \le z\right\} \cap \{s_m^2 \le u\} \cap \{N = n\}\right) \\
&= \sum_{n=m}^{\infty} P\left(\frac{\bar{Y}_n}{\sigma/n^{1/2}} \le z\right) P\left(\{s_m^2 \le u\} \cap \{N = n\}\right) \\
&= \sum_{n=m}^{\infty} \Phi(z) P\left(\{s_m^2 \le u\} \cap \{N = n\}\right) \\
&= \Phi(z) \sum_{n=m}^{\infty} P\left(\{s_m^2 \le u\} \cap \{N = n\}\right) \\
&= \Phi(z) P(s_m^2 \le u). \qquad (10.40)
\end{aligned}$$

The third line follows from the independence of $\bar{Y}_n$ and s_m and the fact that $N = N(s_m)$ is a function of s_m. Equation (10.40) shows that the joint distribution function of (Z_N, s_m^2) is that of two independent random variables, the first of which is standard normal. Also, we know that $(m-1)s_m^2/\sigma^2$ is chi-squared with $m-1$ degrees of freedom. It follows that (10.39) is the ratio of a standard normal deviate to the square root of a chi-squared $(m-1)$ divided by its degrees of freedom. Therefore, T has a t-distribution with $m-1$ degrees of freedom.

Exercises

1. In Example 10.47, let $Y(\omega) = \omega$ and consider two different sigma-fields. The first, $\mathcal{A}_1$, is the sigma-field generated by $I(\omega$ is rational). The second, $\mathcal{A}_2$, is the sigma-field generated by Y. What are $\mathcal{A}_1$ and $\mathcal{A}_2$? Give regular conditional distribution functions for Y given $\mathcal{A}_1$ and Y given $\mathcal{A}_2$.

2. Show that the (X, Y) pairs in Example 10.48 exhibit quirky behavior even when X is a binary gene status random variable. More specifically, consider the following two ways of simulating pairs (X, Y). Method 1: generate a gene status (present or absent) for person 1, then assign that same value to person 2. Then generate the two Ys using the regression equation, with ϵ having a $\mathrm{N}(0, \sigma^2)$ distribution. Method 2: continue generating (X, Y) pairs until you find two pairs with the same X value. Show that the distribution of the common value of X is different using Method 1 versus Method 2.

10.7.3 Weak Convergence of Conditional Distributions

In this section we investigate whether weak convergence of joint distributions translates into weak convergence of conditional distributions and vice versa.

Let (X_n, Y_n) be a sequence of random variables with joint distribution function $H_n(x, y)$ and marginal distribution functions $F_n(x)$ and $G_n(y)$. Write $H_n(x, y)$ as

$$H_n(x, y) = F_n(x) P(Y_n \le y \,|\, X_n \le x). \qquad (10.41)$$

We can define $P(Y_n \le y \,|\, X_n \le x)$ to be $\Phi(Y)$ when $F_n(x) = 0$, so factorization (10.41) holds even when $F_n(x) = 0$ because the left and right sides are both 0. Furthermore, if H_n

converges weakly to some joint distribution function $H(x, y)$, then the marginal distribution $F_n(x)$ also converges weakly to the corresponding marginal distribution function $F(x)$ by the Mann-Wald theorem (Theorem 6.59) because $\lambda(x, y) = x$ is a continuous function of (x, y) (alternatively, one could invoke the Cramer-Wold device, Corollary 8.43). By this fact and the factorization (10.41), if $H_n(x, y) \stackrel{D}{\to} H(x, y)$ and (x, y) is a continuity point of $H(x, y)$ such that $H(x, y) > 0$ (which implies that x is a continuity point of $F(x)$ and $F(x) > 0$), then the left side of Equation (10.41) converges to $H(x, y)$ if and only if the right side converges to $F(x)H(x, y)$. That is, weak convergence of H_n is equivalent to convergence of the marginal distribution $F_n(x)$ plus convergence of $P(Y_n \le y \,|\, X_n \le x)$ at each continuity point (x, y) of $H(x, y)$ such that $H(x, y) > 0$.

The next question is whether weak convergence of the joint distribution function is equivalent to weak convergence of the marginal distribution function plus weak convergence of the conditional distribution function $K_n(y \,|\, X_n = x)$. More specifically, our question is:

$$\text{Is } H_n(x, y) \stackrel{D}{\to} H(x, y) \text{ equivalent to } F_n(x) \stackrel{D}{\to} F(x) \text{ plus } K_n(y \,|\, x) \stackrel{D}{\to} K(y \,|\, x)?$$

This differs from the development of the preceding paragraph because we are now considering the conditional distribution of Y_n given $X_n = x$ rather than the conditional distribution of $Y_n \,|\, X_n \le x$. Before proceeding, we note that we have already seen one setting in which this holds, namely when (X_n, Y_n) are independent.

Let $H_n(x, y)$ and $K_n(y \,|\, x)$ denote the joint and conditional distribution of (X_n, Y_n) and Y_n given $X_n = x$ respectively. Suppose that F_n converges weakly to a distribution function F and $K_n(y \,|\, x)$ converges weakly to a conditional distribution function $K(y \,|\, x)$. Does it follow that the joint distribution function $H_n(x, y)$ converges weakly to $F(x)K(y \,|\, x)$? Likewise, suppose that $H_n(x, y)$ converges weakly to the joint distribution function $H(x, y)$. Does it follow that the conditional distribution function $K_n(y \,|\, x)$ converges weakly to the corresponding conditional distribution function $K(y \,|\, x) = H(x, y)/\int_{x=-\infty}^{\infty} dH(x, y)$? Unfortunately, the answer to both of these questions is no. The following example shows that even in a relative simple setting in which X_n and Y_n take on only two possible values and the limiting joint and conditional distributions are point masses, those point masses may disagree.

Example 10.55. Possible disconnect between convergence of joint and conditional distributions Let

$$X_n = \begin{cases} 0 & \text{with probability } 1/n \\ 1 & \text{with probability } 1 - 1/n \end{cases}$$

and let $Y_n = I(X_n > 0)$. The conditional distribution of Y_n given $X_n = 0$ is a point mass at 0. Therefore, the conditional distribution of Y_n given $X_n = 0$ converges weakly to a point mass at 0. On the other hand, (X_n, Y_n) converges in probability to $(1, 1)$. Therefore, the joint distribution of (X_n, Y_n) converges weakly to a point mass at $(1, 1)$. □

Definition 10.56. Strong convergence *A sequence of distribution functions F_n is said to converge strongly to distribution F if $F_n(x) \to F(x)$ for every $x \in R$.*

Theorem 10.57. *(Sethuraman)* **Weak convergence of marginal distribution and strong convergence of conditional distribution implies weak convergence of joint distribution** *If F_n converges weakly to F and $K_n(y \,|\, x)$ converges strongly to $K(y \,|\, x)$, then $H_n(x, y)$ converges weakly to $H(x, y)$.*

10.8 Conditional Distribution Functions As Derivatives

There is another intuitive way to try to define a conditional distribution function $H(y\,|\,x)$ of Y given $X = x$. Condition on $x - \Delta < X < x + \Delta$ and let $\Delta \to 0$:

$$\begin{aligned} H(y\,|\,x) &= \lim_{\Delta\to 0} P(Y \le y\,|\,x - \Delta < X < x + \Delta) \\ &= \lim_{\Delta\to 0} \frac{P(\{Y \le y\} \cap \{x - \Delta < X < x + \Delta\})}{P(x - \Delta < X < x + \Delta)}. \end{aligned} \tag{10.42}$$

The first question is whether this limit always exists and is finite. Notice that both the numerator and denominator are increasing functions of Δ, so each decreases to some limit as $\Delta \to 0$. If $P(X = x) = p > 0$, then by the continuity property of probability, the denominator tends to p, while the numerator tends to $P(X = x, Y \le y)$. Thus, the limit in (10.42) is $P(X = x, Y \le y)/p$. On the other hand, if $P(X = x) = 0$, then the numerator and denominator of Expression (10.42) both tend to 0 as $\Delta \to 0$. Nonetheless, the ratio cannot "blow up" because the numerator can never exceed the denominator. That is, the ratio in Expression (10.42) cannot exceed 1, so cannot have an infinite limit. Still, there could be two or more distinct limit points as $\Delta \to 0$, in which case the limit would not exist.

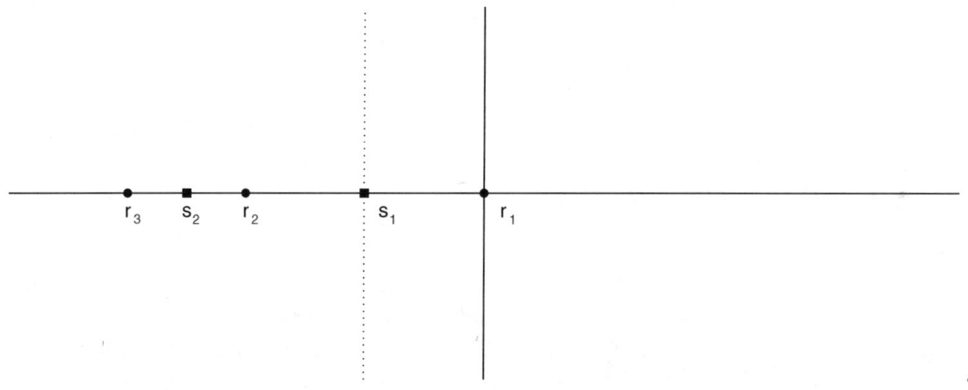

Figure 10.3: Points r_n (circles) and s_n (squares) in Example 10.58. Points strictly to the left of vertical lines represent $X < \Delta_n$ when Δ_n is r_n (solid line) or s_n (dashed line).

Example 10.58. Let $r_n = 1/2^n$, $n = 1, 2, \ldots$, and let $s_n = (r_n + r_{n+1})/2$, so that $r_{n+1} < s_n < r_n$, $n = 1, 2, \ldots$ (see Figure 10.3). Let X take value r_n with probability $1/2^{n+1}$ and s_n with probability $1/2^{n+1}$, $n = 1, 2, \ldots$ Define Y to be -1 if $X = r_n$ for some n, and 1 if $X = s_n$ for some n. Consider Expression (10.42) for $x = 0$, $y = 0$. Suppose that $\Delta_n = r_n$. Conditioned on $0 - \Delta_n < X < 0 + \Delta_n$, X can take any of the values $r_{n+1}, r_{n+2}, \ldots$ or $s_n, s_{n+1}, \ldots$ The conditional probability that X is one of $r_{n+1}, r_{n+2}, \ldots$ is

$$\begin{aligned} P(X \in \{r_{n+1}, r_{n+2}, \ldots\}\,|\,X < r_n) &= \frac{\sum_{i=n+1}^{\infty} P(X = r_i)}{\sum_{i=n+1}^{\infty} P(X = r_i) + \sum_{i=n}^{\infty} P(X = s_i)} \\ &= \frac{\sum_{i=n+1}^{\infty} 1/2^{i+1}}{\sum_{i=n+1}^{\infty} 1/2^{i+1} + \sum_{i=n}^{\infty} 1/2^{i+1}} \end{aligned}$$

$$= \frac{1/2^{n+1}}{1/2^{n+1} + 1/2^n} = 1/3.$$

Thus, $P(Y \le 0 \,|\, -\Delta_n < X < \Delta_n) = 1/3$ for $\Delta_n = r_n$. If we now let $n \to \infty$, $P(Y \le 0 \,|\, -\Delta_n < X < \Delta_n) \to 1/3$.

On the other hand, if $\Delta_n = s_n$, then X can take any of the values $r_{n+1}, r_{n+2}, \ldots$ or $s_{n+1}, s_{n+2}, \ldots$ Moreover, X is equally likely to be one of the r_i or one of the s_i. Therefore, if $\Delta_n = s_n$, then $P(Y \le 0 \,|\, -\Delta_n < X < \Delta_n) = 1/2$. We have shown that $P(Y \le 0 \,|\, -\Delta < X < 0)$ tends to 1/3 if $\Delta \to 0$ along the path r_n, and tends to 1/2 if $\Delta \to 0$ along the path s_n. Thus, the limit in Expression (10.42) does not exist at $x = 0$, $y = 0$. The fact that the limit may not exist highlights one problem with using Expression (10.42) as the definition of the conditional distribution of Y given $X = x$. □

Even if the limit in Expression (10.42) exists for a given x, it need not be a distribution function in y. The following example illustrates this.

Example 10.59. Let X be uniformly distributed on $(0, 1)$, and let $r \in (0, 1)$. Define Y to be $1/(X - r)$ if $X \neq r$, and 0 if $X = r$. Consider the conditional distribution of Y given $r - \Delta < X < r + \Delta$. Given $r - \Delta < X < r + \Delta$, X is as likely to be in $(r - \Delta, r)$ as it is to be in $(r, r + \Delta)$. Either way, $|Y| = 1/|X - r|$ is very large if Δ is very small; its sign is negative if $X \in (r - \Delta, r)$ and positive if $X \in (r, r + \Delta)$. Thus, Y is virtually guaranteed to be $\le y$ if $X \in (r - \Delta, r)$ and virtually guaranteed to be $> y$ if $X \in (r, r + \Delta)$ for tiny Δ. Therefore, $P(Y \le y \,|\, r - \Delta < X < r + \Delta) \to 1/2$ as $\Delta \to 0$ for each y. Clearly, $H(y \,|\, x)$ as defined by Expression (10.42) is not a distribution function in y because it does not satisfy $\lim_{y \to -\infty} H(y \,|\, x) = 0$ and $\lim_{y \to \infty} H(y \,|\, x) = 1$.

Because the limit in Expression (10.42) may either not exist or not be a distribution function in y for some x, it is problematic to define the conditional distribution function by Expression (10.42). Nonetheless, in Example 10.58, the set of x points such that the limit in Expression (10.42) does not exist has probability 0. In Example 10.59, the set of x points at which Expression (10.42) fails to converge to a distribution function has probability 0. These are not accidents.

Proposition 10.60. *(Pfanzagl, 1979)* **Existence of $H(y \,|\, x)$ of Expression (10.42)** *For any random variables (X, Y), the set N of x points such that (10.42) either fails to exist or fails to be a distribution function in y has $P(X \in N) = 0$.*

Thus, one actually could take Expression (10.42) as a definition of the conditional distribution function, and define the conditional distribution function to be some arbitrary F if the limit either does not exist or is not a distribution function in y for a given x. Although intuitive, this definition is avoided because it makes proofs more difficult.

Exercises

1. Suppose that (X, Y) has joint distribution function $F(x, y)$, and X has marginal distribution function $G(x)$. Suppose that, for a given (x, y), $\partial F/\partial x$ exists, and $G'(x)$ exists and is nonzero. What is Expression (10.42)?

10.9 Appendix: Radon-Nikodym Theorem

Theorem 10.61. Radon-Nikodym theorem *Let $(\Omega, \mathcal{F}, \mu)$ be a measure space with $\mu(\Omega) < \infty$. If ν is a measure that is absolutely continuous with respect to μ (i.e., $A \in \mathcal{F}$ and $\mu(A) = 0$ implies that $\nu(A) = 0$ for each set $A \in \mathcal{F}$), then there exists a nonnegative measurable function $f(\omega)$ such that $\nu(A) = \int_A f(\omega) d\mu(\omega)$ for each $A \in \mathcal{F}$. The function f is unique in the sense that if g also has this property, then $g(\omega) = f(\omega)$ except on a set of μ measure 0.*

The function f described in the theorem is called the *Radon-Nikodym derivative* of ν with respect to μ.

To apply this theorem in the setting of conditional expectation, note that if $\mathrm{E}(|Y|) < \infty$, $\nu(A) = \mathrm{E}\{|Y| I(A)\}$ for $A \in \mathcal{F}$ defines a finite measure on $\mathcal{F}$. Therefore, there exists a nonnegative, measurable function $Z(\omega)$ such that $\nu(A) = \int_A Z(\omega) d\mu(\omega)$. That is, $\mathrm{E}\{|Y| I(A)\} = \mathrm{E}\{Z I(A)\}$. The same argument can be used for $\mathrm{E}\{Y^+ I(A)\}$ and $\mathrm{E}\{Y^- I(A)\}$.

10.10 Summary

Let Y be a random variable with $\mathrm{E}(|Y|) < \infty$, $\mathcal{A} \subset F$ be a sigma-field, and $Z = \mathrm{E}(Y \mid \mathcal{A})$.

1. **Definition** Z is unique up to equivalence with probability 1, and is defined by:
   (a) Z is $\mathcal{A}$-measurable.
   (b) $\mathrm{E}\{Z I(A)\} = \mathrm{E}\{Y I(A)\}$ for all $A \in \mathcal{A}$.
2. **Function of X** If $\mathcal{A} = \sigma(\mathbf{X})$, the sigma-field generated by the random vector $\mathbf{X}$, then there is an extended Borel function f such that $Z = f(\mathbf{X})$ a.s.
3. **Geometry and prediction** Suppose that $Y \in L^2$, and let $L^2(\mathcal{A})$ be the space of $\mathcal{A}$-measurable random variables with finite second moment.
   (a) Z is the projection of Y onto $L^2(\mathcal{A})$. I.e., $Y - Z \perp X$ for each $X \in L^2(\mathcal{A})$.
   (b) Z minimizes $\mathrm{E}(Y - \hat{Y})^2$ over all $\hat{Y} \in L^2(\mathcal{A})$.
4. **Conditional distributions**
   (a) There is a probability measure $\mu(B, \omega)$ such that:
      i. For fixed $B \in \mathcal{B}$, $\mu(B, \omega)$ is a version of $P(Y \in B \mid \mathcal{A})$.
      ii. For fixed ω, $\mu(B, \omega)$ is a probability measure on the Borel sets $B \in \mathcal{B}$.
   (b) One version of $\mathrm{E}\{f(Y) \mid \mathcal{A}\}$ is $\int f(y) d\mu(y, \omega)$.
   (c) Inequalities for expectation hold for conditional expectation as well, but we must add "almost surely."
5. **Important conditional mean, variance, and covariance identities**
   If $\mathrm{E}(|Y|) < \infty$,

$$
\begin{aligned}
\mathrm{E}\{\mathrm{E}(Y \mid \mathcal{A})\} &= \mathrm{E}(Y) \\
\mathrm{E}(Y \mid \mathcal{A}) &= \mathrm{E}\{\mathrm{E}(Y \mid \mathcal{C}) \mid \mathcal{A}\} \quad \text{for } \mathcal{A} \subset \mathcal{C}.
\end{aligned}
$$

If $\mathrm{E}(Y^2) < \infty$, $\mathrm{E}(Y_1^2) < \infty$, $\mathrm{E}(Y_2^2) < \infty$,

$$\begin{aligned}\mathrm{var}(Y) &= \mathrm{var}\{\mathrm{E}(Y \mid \mathcal{A})\} + \mathrm{E}\{\mathrm{var}(Y \mid \mathcal{A})\} \\ \mathrm{cov}(Y_1, Y_2) &= \mathrm{cov}\{\mathrm{E}(Y_1 \mid \mathcal{A}), \mathrm{E}(Y_2 \mid \mathcal{A})\} + \mathrm{E}\{\mathrm{cov}(Y_1, Y_2 \mid \mathcal{A})\}.\end{aligned}$$

6. **Paradoxes**

   (a) Conditioning on a null set N makes sense only in a wider context of random variables/sigma-fields. N might be expressible in different ways (e.g., $X - Y = 0$ or $X/Y = 1$) that give different answers. Therefore, we condition on random variables or sigma-fields, not on a specific set of probability 0.

   (b) Substituting the value of a random variable in a conditioning statement is valid using a pre-specified regular conditional distribution function, but need not be valid for every version of the conditional expectation.

Chapter 11

Applications

This chapter takes an in-depth look at the use of probability in practical applications we have encountered over the course of our career. They range from questions about the validity of permutation tests in different settings to conditional independence and path diagrams to asymptotic arguments.

11.1 $F(X) \sim U[0, 1]$ and Asymptotics

In Section 4.6, we took great pains to show that the only probability space we ever really need is $(\Omega, \mathcal{F}, P) = ((0, 1), \mathcal{B}_{(0,1)}, \mu_L)$, the unit interval equipped with Lebesgue measure. We showed how to generate a random variable $Y(\omega)$ with distribution function F via $Y = F^{-1}(\omega)$. More generally, regardless of the probability space on which a random variable U is defined, if U is uniformly distributed on $(0, 1)$, $F^{-1}(U)$ has distribution F.

One application of $F(X) \sim$ uniform $(0, 1)$ is in flow cytometry. Researchers can use a technique called intracellular cytokine staining using a flow cytometry machine to find out whether individual cells respond to pieces of a virus or other antigens by producing regulatory proteins called cytokines. Researchers add a fluorescent dye that attaches to the cytokine of interest, so by shining a fluorescent light and recording the light intensity, they can measure the level of the cytokine. Each cell is classified as responding or not on the basis of its brightness when the light is shone through it. The problem is that there is no concensus on how bright the cell must be to be declared responsive. Researchers point out that there can be day-to-day shifts in the fluoresence of cells. Also, cells can sometimes respond under background conditions without being exposed to the antigen. It is therefore essential to study a control group of cells that are not exposed to the antigen. The goal is to compare the immune response to the antigen, as measured by the number of responding cells, to the immune response under background conditions. See Nason (2006) for more details and statistical issues.

One method for determining the positivity threshold for cell response is to "eyeball" the responses of the control cells and pick a value that seems to separate two different distributions of light intensities. This cutpoint determined from the control sample is then applied to the cells exposed to the antigen, and a comparison is made to determine whether the probability of response is different between the exposed and control cells. How do we test whether the probability of response to the antigen is greater than the probability of response to background conditions? Fisher's exact test does not really apply because Fisher's exact test conditions on the total number of responding cells in both samples. That would be appropriate if the threshold had been determined using all cells instead of just the exposed cells. Immunologists would be extremely reluctant to set a threshold based on the combined control and stimulated cells; this would not make sense to them from a scientific standpoint. Given that the threshold was set using the control sample, is there a valid way to test whether the probability of response is greater among stimulated cells?

While there is no method that is guaranteed to be valid when the threshold is determined in a subjective way, some methods are more valid than others. For instance, we can imagine that the threshold corresponds to selecting a certain order statistic from the control sample. We can then determine whether the number of exposed cells exceeding that order statistic is very large. We next determine the distribution of the number of exposed cells exceeding a given order statistic of the control sample.

Let $X_1,\ldots,X_m$ and $Y_1,\ldots,Y_n$ be the fluorescence levels of control and exposed cells, respectively. Under the null hypothesis, these are iid from some continuous distribution. Imagine that the threshold in the control sample corresponds to r cells being at least as large as it, which corresponds to selecting the $m-r+1$st order statistic $X_{(m-r+1)}$. What is the distribution of the number of Ys that exceed $X_{(m-r+1)}$? Without loss of generality, we can assume that the underlying distribution of the data is uniform $[0,1]$. To see this, let F be the common distribution function of the Xs and Ys. Then $F(X_i)$, $i=1,\ldots,m$ and $F(Y_1),\ldots,F(Y_n)$ are iid uniform $[0,1]$, and the number of Y_i exceeding $X_{(m-r+1)}$ is the same as the number of $F(Y_i)$ exceeding $F(X_{(m-r+1)})$. Therefore, we can assume, without loss of generality, that $X_1,\ldots,X_m$ and $Y_1,\ldots,Y_n$ are iid uniform $[0,1]$. Conditioned on $X_{(m-r+1)}=x$, $I(Y_1>x),\ldots,I(Y_n>x)$ are iid Bernoulli random variables with probability $p=(1-x)$, Therefore, $P\{\#(Y_i>X_{(m-r+1)})=s\,|\,X_{(m-r+1)}=x\}=\binom{n}{s}(1-x)^s x^{n-s}$. Now integrate over the density of $X_{(m-r+1)}$ to deduce that

$$\begin{aligned}
P\{\#(Y_i>X_{(m-r+1)})=s\} &= \int_0^1 \binom{n}{s}(1-x)^s x^{n-s}\left\{\frac{m!x^{m-r}(1-x)^{r-1}}{(m-r)!(r-1)!}\right\}dx \\
&= \frac{\binom{n}{s}m!}{(m-r)!(r-1)!}\int_0^1 x^{m+n-r-s+1-1}(1-x)^{r+s-1}dx \\
&= \frac{\binom{n}{s}m!}{(m-r)!(r-1)!}\frac{\Gamma(m+n-r-s+1)\Gamma(r+s)}{\Gamma(m+n-r-s+1+r+s)}dx \\
&= \left(\frac{r}{m+n-r-s+1}\right)\frac{\binom{n}{s}\binom{m}{r}}{\binom{m+n}{r+s-1}}. \qquad (11.1)
\end{aligned}$$

We can use this distribution to compute a p-value. In flow cytometry applications, m and n are usually very large, whereas a relatively small number of points exceed the threshold. To approximate the above distribution, assume that as $n\to\infty$, $m=m_n\to\infty$ such that $m_n/(m_n+n)\to p$. Use the fact that, as $n\to\infty$, $\binom{n}{s}\sim n^s/s!$, $\binom{m_n}{r}\sim m_n^r/r!\sim \{np/(1-p)\}^r/r!$, and $\binom{m_n+n}{r+s-1}\sim (m_n+n)^{r+s-1}/(r+s-1)!\sim n^{r+s-1}/\{(r+s-1)!(1-p)^{r+s-1}\}$. We conclude that Expression (11.1) tends to

$$\binom{r+s-1}{r-1}p^r(1-p)^s \qquad (11.2)$$

as $n \to \infty$. This is the negative binomial probability mass function (Johnson, Kotz, and Kemp, 1992). We can view Expression (11.2) as the probability of exactly s failures by the time of the rth success in a sequence of iid Bernoulli trials with success probability p. Equivalently, it is the probability that the total number of trials required to achieve r successes is $r+s$. In summary, Expression (11.1) is a density with respect to counting measure on the nonnegative integers, and converges to Expression (11.2), another density with respect to counting measure on the nonnegative integers. By Scheffé's theorem (Theorem 9.6), the probability of any set A of nonnegative integers converges uniformly to $\sum_{s \in A} \binom{r+s-1}{r-1} p^r (1-p)^s$. This can be used to very closely approximate the p-value.

Exercises

1. Suppose there are n red and n blue balls in a container, and we draw k balls without replacement. Let Y_n be the number of balls in the sample that are red. For each step of the following argument, tell whether it is true or false. If it is true, prove it.

   (a) As k remains fixed and $n \to \infty$, $Y_n \xrightarrow{D} Y$, where $Y \sim \text{bin}(k, 1/2)$.

   (b) The probability mass function for Y_n converges to the probability mass function of Y.

   (c) By Scheffe's theorem, $\sup_B |P(Y_n \in B) - P(Y \in B)| \to 0$ as $n \to \infty$.

   (d) If $k = n/2$, $|P(Y_n \le y) - P(Y \le y)| \to 0$ as $n \to \infty$, where $Y \sim \text{bin}(n/2, 1/2)$.

11.2 Asymptotic Power and Local Alternatives

In commonly used statistical tests like the t-test or test of proportions, power under a fixed alternative tends to 1 as the sample size tends to ∞. This is because these test statistics are of the form

$$Z_n = \frac{\hat{\delta}_n}{\sqrt{\hat{V}_n}},$$

where $\hat{\delta}_n$ is the difference of two independent sample means of iid data (either $\bar{Y}_T - \bar{Y}_C$ or $\hat{p}_T - \hat{p}_C$) and its estimated variance $\hat{V}_n$ tends to 0 in probability. For a one-tailed test at level α, we reject the null hypothesis if $Z_n > z_\alpha$, where z_α is the $(1-\alpha)$th quantile of a standard normal distribution. By the WLLN, each sample mean tends to its true mean in probability, so the treatment effect estimator $\hat{\delta}_n$ tends to the true treatment effect δ in probability. Assume that we have parameterized things such that $\delta > 0$. Because the numerator of Z_n tends to a fixed positive number in probability and the denominator tends to 0 in probability, Z_n tends to ∞ in probability. Power, namely $P(Z_n > z_\alpha)$, tends to 1 (exercise). To make the power tend to a value other than 1, we must consider values of δ under the alternative hypothesis that decrease to 0 as $n \to \infty$ at such a rate that Z_n converges in distribution to a non-degenerate random variable Z. These alternative hypothesis values δ_n close to the null value of 0 are called *local alternatives.* Our goal in this section is to use the CLT to derive power formulas that are asymptotically valid under a local alternative with very few assumptions. This requires a bit of care because both the random variables and parameters are changing with n.

11.2.1 T-Test

Consider a two-sample t-test of the null hypothesis that $\mu_T - \mu_C = 0$ versus the alternative that $\mu_T - \mu_C > 0$. Assume first that data in each treatment arm are iid normal with common, known variance σ^2. The z-statistic,

$$Z_n = \frac{\bar{Y}_T - \bar{Y}_C}{\sqrt{2\sigma^2/n}},$$

is normally distributed with mean $\delta/(2\sigma^2/n)^{1/2}$ and variance 1. Power is

$$\begin{aligned} P(Z_n > z_\alpha) &= P\left(\frac{\bar{Y}_T - \bar{Y}_C}{\sqrt{2\sigma^2/n}} > z_\alpha\right) \\ &= P\left(\frac{\bar{Y}_T - \bar{Y}_C}{\sqrt{2\sigma^2/n}} - \frac{\delta}{\sqrt{2\sigma^2/n}} > z_\alpha - \frac{\delta}{\sqrt{2\sigma^2/n}}\right) \\ &= 1 - \Phi\left(z_\alpha - \frac{\delta}{\sqrt{2\sigma^2/n}}\right) = \Phi\left(\frac{\delta}{\sqrt{2\sigma^2/n}} - z_\alpha\right), \end{aligned} \tag{11.3}$$

If $n \to \infty$ and everything else remains fixed, power tends to 1. For power to tend to something other than 1 as $n \to \infty$, the size of the treatment effect must tend to 0 as $n \to \infty$. If $\delta = \delta_n$ is such that $\delta_n/(2\sigma^2/n)^{1/2} \to z_\alpha + z_\beta$, then Expression (11.3) tends to $\Phi(z_\alpha + z_\beta - z_\alpha) = \Phi(z_\beta) = 1 - \beta$. For example, if we want 90% power, set $\beta = 0.1$ and $z_\beta = 1.28$; power tends to $\Phi(1.28) = 0.90$. The local alternative in this two-sample t-test setting is $\delta = (z_\alpha + z_\beta)/(2\sigma^2/n)^{1/2}$.

Readers may feel that local alternatives are very strange. After all, clinical trials are often powered for the smallest effect thought to be clinically relevant, or an effect similar to what was observed in another study. The idea that the alternative should become closer to the null value as the sample size gets larger is perplexing. Actually, local alternatives are realistic; we would not power a study at a value extremely close to 1. Rather, we might fix power at, say, 90%, and determine the per-arm sample size n. If n is fairly large, then the approximation obtained by assuming that $\delta/(2\sigma^2/n)^{1/2}$ is close to a constant is likely to be quite accurate. This is reminiscent of approximating binomial probabilities with n large and p_n small by Poisson probabilities with parameter $\lambda = np_n$ (see Proposition 6.24).

We originally assumed that data were iid normal with common known variance, but we now show that the power formula (11.3) holds asymptotically regardless of the underlying (continuous) distribution of data, provided that the variance is finite. We also relax the assumption that σ^2 is known. Because the local alternative changes with n, it may appear that we must invoke the Lindeberg Feller CLT rather than the ordinary CLT. Nonetheless, we will show that only the ordinary CLT is required.

Let $Y_1, \ldots, Y_{2n}$ denote the observations, with the first n being control data. Assume that $(Y_i - \mu_C)/\sigma \sim F$ and $(Y_i - \mu_T)/\sigma \sim F$ in the control and treatment arms, respectively, where F is a continuous distribution function with finite variance. This is a so-called "shift" alternative. Consider first the null hypothesis that $\mu_T = \mu_C$. Generate $2n$ iid observations from $F\{(y - \mu_C)/\sigma\}$, and let Z_n be the t-statistic

$$Z_n = \frac{\bar{Y}_T - \bar{Y}_C}{\sqrt{2s^2/n}}, \tag{11.4}$$

where $\bar{Y}_T$ and $\bar{Y}_C$ are the treatment and control means and s^2 is the pooled variance. By the CLT coupled with Slutsky's theorem, Z_n tends to $N(0, 1)$.

To see what happens under a local alternative hypothesis, add $(z_{\alpha+z_\beta})(2\sigma^2/n)^{1/2}$ to each treatment observation. The sample variance in each arm does not change, so neither does the pooled variance. The resulting value of the t-statistic is $Z_n + (z_{\alpha+z_\beta})(\sigma/s)$. But $s^2 \xrightarrow{p} \sigma^2$, In fact, $s^2 \xrightarrow{a.s.} \sigma^2$ (see Example 7.6). By Slutsky's theorem, the distribution of the t-statistic tends to $Z + z_\alpha + z_\beta$, where $Z \sim N(0,1)$. That is, the t-statistic tends in distribution to $N(z_\alpha + z_\beta, 1)$. We have shown that under a local alternative in which

$$\frac{\delta}{\sqrt{\frac{2\sigma^2}{n}}} = (z_\alpha + z_\beta), \tag{11.5}$$

power, namely $P(Z_n > z_\alpha)$, tends to $P(\mathrm{N}(z_\alpha + z_\beta, 1) > z_\alpha) = 1 - \Phi(z_\alpha - z_\alpha - z_\beta) = 1 - \beta$ as $n \to \infty$. Therefore, power formula (11.3) holds approximately for a shift alternative whenever the underlying (continuous) distribution has finite variance. The per-arm sample size required for power $1 - \beta$ is obtained by solving (11.5) for n:

$$n = \frac{2(z_\alpha + z_\beta)^2\sigma^2}{\delta^2}.$$

Incidentally, the reason for specifying that the distribution underlying the Y_i be continuous is that we added $(z_{\alpha+z_\beta})(2\sigma^2/n)^{1/2}$ to $Y_{n+1}, \ldots, Y_{2n}$. If the Y_i take only values 0 or 1, this would not make sense.

Exercises

1. Prove that if X_n converges in probability to the positive number a and $Y_n \xrightarrow{p} 0$, then $X_n/|Y_n| \xrightarrow{p} \infty$, meaning that $P(X_n/|Y_n| > B) \to 1$ as $n \to \infty$ for each positive number B. What are the implications of this result for a test rejecting the null hypothesis when $\hat{\theta}_n/\{\hat{\text{var}}(\hat{\theta}_n)\}^{1/2} > z_\alpha$, where $\hat{\theta}_n$ is a consistent estimator of a parameter $\theta > 0$ and $\hat{\text{var}}(\hat{\theta}_n)$ is an estimator of its variance such that $\hat{\text{var}}(\hat{\theta}_n) \xrightarrow{p} 0$?

2. State and prove a result about the asymptotic distribution of the one-sample t-statistic $n^{1/2}\bar{Y}_n/s_n$ under a local alternative.

11.2.2 Test of Proportions

Imagine a clinical trial with a binary outcome and again, for simplicity, assume the same sample size n in the treatment and control arms. In the continuous outcome setting above, the local alternative was proportional to $n^{1/2}$. Therefore, in the binary setting, we consider a local alternative value for $p_C - p_T$ of $\lambda/n^{1/2}$. First generate $2n$ Bernoulli (p_C) observations $Y_1, \ldots, Y_{2n}$. The first n are the control observations. If we want to generate data under the null hypothesis, we leave $Y_{n+1}, \ldots, Y_{2n}$ alone and compute the test statistic

$$U_n = \frac{\sum_{i=1}^n Y_i - \sum_{i=n+1}^{2n} Y_i}{\sqrt{2n}}. \tag{11.6}$$

Note that U_n is $\{\hat{p}(1-\hat{p})\}^{1/2} Z_n$, where Z_n is the usual z-statistic comparing proportions and $\hat{p} = (2n)^{-1}\sum_{i=1}^{2n} Y_i$. Also, $U_n \xrightarrow{D} N\{0, p_C(1-p_C)\}$ (recall that $Y_1, \ldots, Y_{2n}$ are iid Bernoulli (p_C) because we have yet to consider the alternative hypothesis).

We now generate data under a local alternative. We cannot use exactly the same technique as for the continuous outcome case because addition of $(z_{\alpha+z_\beta})(2\sigma^2/n)^{1/2}$ to each observation leads to values outside the support of the binary variables Y_i. Instead we adopt the following approach. Examine the observations $Y_{n+1}, \ldots, Y_{2n}$ and leave any $Y_i = 0$ alone. If $Y_i = 1$, then switch it to 0 with probability π_n to be specified shortly. This will create new iid observations $Y'_{n+1}, \ldots, Y'_{2n}$ in the treatment group with $P(Y'_i = 1) = P(Y_i = 1)(1-\pi_n) = p_C(1-\pi_n)$. Now choose π_n to make this probability $p_C - \lambda/n^{1/2}$. That is, $\pi_n = \lambda/(n^{1/2}p_C)$. Now compute the statistic U'_n, namely U_n with Y_i replaced by Y'_i for $i = n+1, \ldots, 2n$. Then

$$U'_n - U_n = \frac{\sum_{i=n+1}^{2n} I(Y_i = 1, Y'_i = 0)}{\sqrt{2n}}$$

$$\mathrm{E}(U'_n - U_n) = \frac{\sum_{i=n+1}^{2n} p_C \pi_n}{\sqrt{2n}} = \frac{np_C\left(\frac{\lambda}{p_C\sqrt{n}}\right)}{\sqrt{2n}} = \frac{\lambda}{\sqrt{2}}.$$

$$\mathrm{var}(U'_n - U_n) = \frac{n(p_C\pi_n)(1-p_C\pi_n)}{2n} = (p_C/2)\pi_n(1-p_C\pi_n) \to 0 \text{ as } n \to 0.$$

It follows that $U'_n - U_n$ converges to $\lambda/2^{1/2}$ in L^2, and therefore in probability (Proposition 6.32). By Slutsky's theorem, $U'_n \xrightarrow{D} N(\lambda/2^{1/2}, p_C(1-p_C))$. Also, $\hat{p}$ converges in probability to p_C, so another application of Slutsky's theorem implies that the z-statistic $U'_n/\{\hat{p}(1-\hat{p}\}^{1/2}$ converges in distribution to $\mathrm{N}[\lambda/\{2p_C(1-p_C)\}^{1/2}, 1]$. Suppose that $\lambda/\{2p_C(1-p_C)\}^{1/2} = z_\alpha + z_\beta$, so that

$$\frac{\delta}{\sqrt{\frac{2p_C(1-p_C)}{n}}} = (z_\alpha + z_\beta). \tag{11.7}$$

Then power, $P(Z_n > z_\alpha)$, tends to $P(\mathrm{N}(z_\alpha + z_\beta, 1) > z_\alpha) = 1 - \Phi(z_\alpha - z_\alpha - z_\beta) = 1 - \beta$.

We have shown rigorously that an asymptotically valid sample size formula to achieve power $1-\beta$ in a one-tailed test of proportions at level α can be obtained by solving (11.7) for n. This yields

$$n = \frac{2(z_\alpha + z_\beta)^2 p_C(1-p_C)}{\delta^2}$$

per arm.

11.2.3 Summary

We have shown rigorously that in both the continuous and binary settings, asymptotically valid sample size formulas can be obtained by equating the expected z-score to $z_\alpha + z_\beta$ (Equations (11.5) and (11.7)) and solving for n. Conversely, for a given per-arm sample size n, approximate power can be obtained by solving Equations (11.5) and (11.7) for $1-\beta$ by first subtracting z_α from both sides and then applying the normal distribution function Φ to both sides.

11.3 Insufficient Rate of Convergence in Distribution

Safety signals have been observed for some vaccines. For example, there was an increased incidence of Guillain-Barré syndrome (GBS), a very rare neurological disorder, following administration of the 1976 H1N1 vaccine (Schonberger et al., 1979). An increase was also observed during the 2009–2010 season (Salmon et al., 2013). Different types of studies have been used to assess whether a vaccine increases the risk of a disease. These studies include comparisons of vaccinees with non-vaccinees or comparisons of recipients of the vaccine of interest with recipients of another vaccine. But there is no guarantee that the control group of non-vaccinees or recipients of a different vaccine are comparable.

An alternative self-controlled design considers people who received the vaccine and developed the disease within a certain period of time, say 84 days. The idea is to see whether the disease occurred close to the time of vaccination. If the vaccine does not increase the risk of disease, then on any given day, there is a tiny and equal probability of disease onset. Therefore, we are interested in testing the null hypothesis that time of disease onset of participants in the self-controlled design is uniformly distributed on $[0, 84]$. One alternative hypothesis is that time from vaccination to disease onset in the general population follows a Weibull distribution function $F(t) = 1 - \exp(-\lambda t)^\beta$ with $\beta < 1$. Under this distribution, there is higher risk of disease soon after vaccination. Under the null hypothesis, $\beta = 1$ and λ is tiny because the condition is very rare; λ is also tiny under realistic alternative values of β close to, but smaller than, 1. Because we are studying only people who developed the disease within 84 days of vaccination, we must condition on $T \le 84$. The conditional distribution function for T given that $T \le 84$ is

$$G(t) = \frac{F(t)}{F(84)} = \frac{1 - \exp\{-(\lambda t)^\beta\}}{1 - \exp\{-(84\lambda)^\beta\}}, \; t \le 84. \tag{11.8}$$

We can approximate the distribution for tiny λ by taking the limit of (11.8) as $\lambda \to 0$. It is an exercise to prove that this limit is $(t/84)^\beta$. With the transformed variable $U = T/84$, we would like to construct a powerful test of the null hypothesis that U is uniform on $[0, 1]$ versus the alternative that the distribution function for U is $G(u) = u^\beta$, $\beta < 1$.

A very powerful way to test a null hypothesis against an alternative hypothesis is based on the likelihood ratio. The likelihood is the probability mass or density function of the data. We take the ratio of the likelihood assuming the alternative hypothesis to the likelihood under the null hypothesis. A large likelihood ratio means that the data are more consisten with having arisen from the alternative, rather than the null, hypothesis. It is usually more convenient to compute the logarithm of the likelihood ratio, rejecting the null hypothesis for large values. The log likelihood for a sample of n observations from $G(u) = u^\beta$ is $n \ln \beta + (\beta - 1) \sum_{i=1}^n \ln(U_i)$. The likelihood ratio test of $\beta = 1$ versus $\beta < 1$ rejects the null hypothesis for large values of

$$\sum_{i=1}^n - \ln(U_i). \tag{11.9}$$

The null distribution of Expression (11.9) is gamma with parameters $(n, 1)$. Thus, we reject the null hypothesis if Expression (11.9) exceeds the upper α point of a gamma $(n, 1)$ distribution.

A practical concern is that we do not really have the precise time of onset of disease. We have only the day of onset. Thus, we actually observe a discrete version of U. It seems that there should not be a problem because we have seen that a discrete uniform converges to a continuous uniform as the number of equally-spaced support points tends to ∞ (see Example 6.22 for the setting of dyadic rationals). Using the continuous uniform approximation works well in some circumstances. For example, try simulating 10 observations from a discrete uniform distribution on $\{1, 2, \ldots, 1000\}$. Do this thousands of times, computing the test statistic (11.9) each time. You will find that the proportion of times the test statistic exceeds the 0.05 point of a gamma $(10, 1)$ distribution is close to 0.05 (exercise). But now modify the simulation by sampling 500 observations on $\{1, 2, \ldots, 100\}$. You will find that the proportion of simulated experiments such that the test statistic (11.9) exceeds the 0.05 point of a gamma $(500, 1)$ distribution is much less than 0.05 (exercise).

What went wrong? The test statistic (11.9) "blows up" at $u = 0$. That is, the test statistic when U has a continuous uniform distribution can be very large at times. On the other hand, the discrete uniform never produces an arbitrarily large value. Approximating the discrete uniform with a continuous uniform works well if the number of participants is much smaller than the number of days, but not the other way around. The reader is invited to confirm this using simulation.

Exercises

1. Prove that the limit of (11.8) as $\lambda \to 0$ is $(t/84)^{\beta}$.

2. Simulate 10 discrete uniforms on $\{1, 2 \ldots, 1000\}$ and compute the test statistic (11.9). Repeat this thousands of times and calculate the proportion of times the statistic exceeds the upper 0.05 quantile of a gamma $(10, 1)$ distribution. Is it close to 0.05?

3. Repeat the preceding exercise, but with 500 discrete uniforms on $\{1, 2 \ldots, 100\}$. What proportion of simulated experiments resulted in test statistic (11.9) exceeding the upper 0.05 quantile of a gamma $(500, 1)$ distribution?

11.4 Failure to Condition on All Information

In tuberculosis, it is important to have many different treatment regimens that are typically several months long because some people have developed multi-drug resistance (MDR) or extensively drug resistant (XDR) tuberculosis. Also, it is desirable to try to shorten regimens. If separate trials were run comparing each new combination and duration of treatment to a control, the time to develop new effective treatments would be prohibitive. Multi-arm multi-stage (MAMS) trials seek to shorten development time by comparing several regimens to the same active control. These involve dropping poorly performing arms at an interim analysis. At the end of the trial, only the treatments that meet a minimum level of efficacy versus the control are compared. Arguments have been offered to explain why the penalty one must pay for comparing treatments meeting a minimum threshold is not as great as the penalty required if we had picked the best treatment.

Suppose you begin a clinical trial comparing a control to 5 new regimens with respect to a continuous outcome. For simplicity, assume that the common variance σ^2 in different arms is known. Let $Z_i = (\bar{Y}_i - \bar{Y}_0)/(2\sigma^2/n)^{1/2}$ be the z-score comparing arm i to the control, and assume that a large z-score indicates that arm i is effective compared to the control. It is common practice to use two-tailed tests in clinical trials, so the ith arm is declared significantly different from the control if $|Z_i| > c$, where c is a critical value. If we use the Bonferroni method to adjust for multiple comparisons, we would divide 0.05 by 5 and use a two-tailed test at level 0.01. The required critical value is $c = 2.576$, the upper 0.005 point of a standard normal distribution.

See if you agree with the following "improvement" over the Bonferroni method. We first determine how many of the 5 z-scores exceed 0. After all, we would not be interested in treatments that were not even better than control. Suppose that 2 of the 5 z-scores are positive. Now divide 0.05 by 2 instead of 5, and use a two-tailed test at level 0.025. That is, we reject the null hypothesis if $|Z_i| > 2.241$. The argument is as follows. The distribution of each positive z-score is that of $|Z_i|$ given that $Z_i > 0$. Under the null hypothesis, before conditioning on $Z_i > 0$, Z_i is standard normal and $|Z_i|$ has the distribution of the absolute value of a standard normal deviate. Conditioning on $Z_i > 0$ does not change the distribution of $|Z_i|$ because the distribution of Z_i is symmetric about 0. Using critical value 2.241 for the positive z-scores means that the two-tailed type I error rate for each comparison is 0.025. By the Bonferroni inequality, the probability of falsely rejecting at least one of the two null hypotheses is at most $2(0.25/2) = 0.05$. It does not matter whether the z-statistics are independent or dependent because the Bonferroni inequality does not require independence.

Interestingly, simulation results seem to confirm that the type 1 error rate is nearly controlled at level 0.05. If the total number of arms exceeds 3, the type I error rate is controlled at level 0.05 or less. With three arms, the type 1 error rate is slightly elevated (see Proschan and Dodd, 2014). However, if the above argument were really valid, then the **conditional** type 1 error rate given the number of positive z-scores should also be controlled. But the conditional type 1 error rate can be greatly inflated. Therefore, the argument given above must be incorrect.

Did you detect the flaw in the above reasoning? We conditioned on $Z_i > 0$ one at a time, as if the only relevant information about $|Z_i|$ is that $Z_i > 0$. But other Z_js also are informative about $|Z_i|$. For example, if several z-scores are positive and the global null hypothesis is true, the control sample mean was probably unusually small. Therefore, information from other Z_js is informative about the control sample mean, which is informative about all of the $|Z_j|$s. To control the type 1 error rate, we must account for the screening out of $Z_i < 0$, which is informative about all comparisons.

11.5 Failure to Account for the Design

11.5.1 Introduction and Simple Analysis

People with hypertension are advised to reduce their salt intake to reduce their blood pressure. Some claim that there are people whose blood pressure consistently **increases** when they decrease their salt intake. Data from the Dietary Approaches to Stop Hypertension (DASH)-Sodium trial provided an opportunity to test the hypothesis that there is substantial variability in different people's responses to salt decreases (see Obarzanek et al., 2003; Proschan and Nason, 2009). The trial compared two different dietary patterns, but within each pattern, used a crossover design of low, medium, and high salt levels. That is, each participant received all three salt levels. In addition, there was a "run-in" period at the beginning of the trial during which everyone received the high salt diet to make sure it could be tolerated. Therefore, in a subset of DASH-Sodium participants, we were able to see the effect of reducing the salt level on two separate occasions.

One simple way to test whether there is large between-person variability in response to reducing salt is as follows. Under the hypothesis that the between-person blood pressure variability in response to sodium reduction is large, someone with a big response on the first occasion would be expected to have a big response on the second because that person is likely to be a hyper-responder. Under the null hypothesis, someone's response on the second occasion would be independent of his or her response on the first occasion. One could define a threshold such as 5 mmHg or 10 mmHg; people whose blood pressures decrease by more than that amount would be classified as hyper-responders. We could then see if the number of people declared hyper-responders is greater than what would be expected if responses on the two occasions are independent. But the threshold for being a hyper-responder is arbitrary. It seems preferable to use the median responses as cutpoints. Thus, we compute the median response M_1 over all participants on occasion 1, and the median response M_2 on occasion 2. The probability of exceeding M_i on occasion i is 1/2. Under the null hypothesis, the indicators I_i of exceeding M_i on occasion i, $i = 1, 2$, are independent. The probability that $I_1 = 1$ and $I_2 = 1$ is $(1/2)^2 = 1/4$. It seems that one could test the independence assumption by referring the number of people with $I_1 = 1, I_2 = 1$ to a binomial distribution with parameters n and 1/4, where n is the number of participants.

11.5.2 Problem with the Proposed Method

The problem with the above analysis is that by using the medians as thresholds, we are fixing the marginal totals of the 2×2 table. For instance, suppose that the total number of people is 1000, and let X and Y denote the blood pressure responses on occasions 1 and 2. Each marginal total has to be approximately 500 (Table 11.1). Under the null hypothesis of independence of responses on the two occasions, the conditional distribution of X given marginal totals of 500 is hypergeometric instead of binomial. This distribution describes what happens if we draw a sample of 500 balls without replacement from an urn containing 500 red balls and 500 blue balls. The number of red balls in the sample has the same hypergeometric distribution as the number of people declared hyper-responders in the salt experiment. It can be shown that this distribution is approximately normal with mean 250 and variance 62.5. On the other hand, the binomial (1000, 1/4) distribution is approximately normal with mean $1000(1/4) = 250$ and variance $1000(1/4)(3/4) = 187.5$. The hypergeometric distribution is much less variable, so erroneously using the binomial (1000, 1/4) distribution would make it more difficult to declare salt response variability.

Table 11.1: Numbers of people with blood pressure response categorized by exceeding the median or not exceeding the median on two separate occasions in a hypothetical group of 1000 participants.

	$>$ Median$_2$	$\leq$ Median$_2$	
$>$Median$_1$	X		500
$\leq$ Median$_1$	Y		500
	500	500	

11.5.3 Connection to Fisher's Exact Test

What is interesting about this example is that one might erroneously surmise that it should not matter whether we use the hypergeometric or binomial distribution. After all, when the marginal totals are large, Fisher's exact test and the usual z-test of proportions give nearly the same answer. Fisher's exact test conditions on the column totals, whereas the z-test of proportions does not. Although this seems analogous to the salt responder story, it is not. Here is a heuristic explanation. We will make this argument rigorous shortly. The z-test of proportions is a standardized version of $X_n - Y_n$, whereas the column total is $X_n + Y_n$. Note that we have included the subscript n because we intend to study properties as $n \to \infty$. The joint distribution of $(X_n - Y_n, X_n + Y_n)$ is asymptotically bivariate normal by the multivariate CLT. Also, $\text{cov}(X_n - Y_n, X_n + Y_n) = \text{var}(X_n) - \text{var}(Y_n) = 0$ under the null hypothesis. Therefore, $X_n - Y_n$ and $X_n + Y_n$ are asymptotically independent. This is a loose justification of the fact that we get approximately the same answer whether or not we condition on $X_n + Y_n$.

We now make the above argument rigorous. The first step is to show that properly standardized versions of $X_n - Y_n$ and $X_n + Y_n$ converge in distribution to independent standard normals. Let $p \in (0,1)$ be the common Bernoulli parameter under the null hypothesis. Let $U_{n1} = (X_n - np)/\{np(1-p)\}^{1/2}$ and $U_{n2} = (Y_n - np)/\{np(1-p)\}^{1/2}$. Then $(U_{n1}, U_{n2}) \stackrel{D}{\to} (U_1, U_2)$, where (U_1, U_2) are iid standard normals. Let $Z_{n1} = (U_{n1} - U_{n2})/2^{1/2}$ and $Z_{n2} = (U_{n1} + U_{n2})/2^{1/2}$. By the Mann-Wald theorem, $(Z_{n1}, Z_{n2}) \stackrel{D}{\to} (Z_1, Z_2)$, where $Z_1 = (U_1 - U_2)/2^{1/2}$ and $Z_2 = (U_1 + U_2)/2^{1/2}$. But (Z_1, Z_2) are iid standard normals because they are bivariate normal with correlation 0. Note also that Z_{n1} is the standardized difference, $(X_n - Y_n)/\{2np(1-p)\}^{1/2}$, between X_n and Y_n, and similarly for Z_{n2}. Therefore, $[(X_n - Y_n)/\{2np(1-p)\}^{1/2}, (X_n + Y_n - 2np)/\{2np(1-p)\}^{1/2}] \stackrel{D}{\to} (Z_1, Z_2)$.

We would like to conclude that because (Z_{n1}, Z_{n2}) converges to (Z_1, Z_2) and Z_1 and Z_2 are independent, the conditional distribution of Z_{n_1} given Z_{n2} converges to the conditional distribution of Z_1 given Z_2, which is standard normal. But recall from Section 10.7.3 that weak convergence of conditional distributions does not follow automatically from weak convergence of joint distributions, so we need to prove that $P(Z_{n1} \le z_1 \,|\, Z_{n2} = z_2) \to \Phi(z_1)$. We first present a slightly flawed argument that the reader should attempt to debunk. We will then repair the flaw.

11.5.4 A Flawed Argument that $P(Z_{n1} \le z_1 \,|\, Z_{n2} = z_2) \to \Phi(z_1)$

It is not difficult to show that the conditional distribution $F_n(z_1 \,|\, z_2) = P(Z_{n1} \le z_1 \,|\, Z_{n2} = z_2)$ is a decreasing function of z_2 over the support points z_2 of Z_2. Therefore, for $\epsilon > 0$,

$$\begin{aligned} F_n(z_1 \,|\, z_2) &\ge P\{Z_{n1} \le z_1 \,|\, Z_{n2} \in [z_2, z_2 + \epsilon]\} \\ &= \frac{P\{(Z_{n1} \le z_1) \cap (Z_{n2} \in [z_2, z_2 + \epsilon])\}}{P\{Z_{n2} \in [z_2, z_2 + \epsilon]\}}. \end{aligned} \quad (11.10)$$

Now take the liminf of both sides as $n \to \infty$, and note that by the independence of Z_1 and Z_2, the liminf of the right side is $P(Z_1 \le z_1)P(Z_2 \in [z_2, z_2 + \epsilon]/P(Z_2 \in [z_2, z_2 + \epsilon]) = \Phi(z_1)$. Therefore, $\underline{\lim}\, F_n(z_1 \,|\, z_2) \ge \Phi(z_1)$. A similar argument shows that $\overline{\lim}\, F_n(z_1 \,|\, z_2) \le \Phi(z_1)$. It follows that $F_n(z_1 \,|\, z_2) \stackrel{D}{\to} \Phi(z_1)$.

Did you find the flaw? We cannot fix z_2 and condition on $Z_{n_2} = z_2$ because the support of the distribution of Z_{n2} changes with n.

11.5.5 Fixing the Flaw: Polya's Theorem to the Rescue

To fix the above flaw, we must condition not on $Z_{n2} = z_2$, but on $Z_{n2} = z_{n2}$, where z_{n2} is a support point of the distribution of Z_{n2} such that $z_{n2} \to z_2$. But replacing z_2 by z_{n2} in Expression (11.10) is potentially problematic: how do we know that $P(Z_{n2} \in [z_{n2}, z_{n2} + \epsilon]) \to P(Z_2 \in [z_2, z_2 + \epsilon])$? Both Z_{n2} and z_{n2} depend on n, so ordinary convergence in distribution is not sufficient. The key is to use Polya's theorem in R^2 (Theorem 9.5 applied to $k = 2$). Because the limiting distribution of (Z_{n1}, Z_{n2}) is $\Phi(z_1)\Phi(z_2)$, a continuous function of (z_1, z_2), Theorem 9.5 implies that $P(Z_{n1} \le z_1, Z_{n2} \in [z_{n2}, z_{n2} + \epsilon]) = \Phi(z_1)\{\Phi(z_{n2} + \epsilon) - \Phi(z_{n2})\} + a_n$, where $a_n \to 0$ as $n \to \infty$. Similarly, Polya's theorem in R^1 implies that $P(Z_{n2} \in [z_{n2}, z_{n2} + \epsilon]) = \Phi(z_{n2} + \epsilon) - \Phi(z_{n2}) + b_n$, where $b_n \to 0$ as $n \to \infty$. Therefore, Expression (11.10) becomes:

$$
\begin{aligned}
F_n(z_1 \,|\, z_{n2}) &\ge P\{Z_{n1} \le z_1 \,|\, Z_{n2} \in [z_{n2}, z_{n2} + \epsilon]\} \\
&= \frac{P\{(Z_{n1} \le z_1) \cap (Z_{n2} \in [z_{n2}, z_{n2} + \epsilon])\}}{P\{Z_{n2} \in [z_{n2}, z_{n2} + \epsilon]\}} \\
&= \frac{\Phi(z_1)\{\Phi(z_{n2} + \epsilon) - \Phi(z_{n2})\} + a_n}{\Phi(z_{n2} + \epsilon) - \Phi(z_{n2}) + b_n} \\
\underline{\lim}_{\,n\to\infty} F_n(z_1 \,|\, z_{n2}) &\ge \frac{\Phi(z_1)\{\Phi(z_2 + \epsilon) - \Phi(z_2)\}}{\Phi(z_2 + \epsilon) - \Phi(z_2)} = \Phi(z_1). \qquad (11.11)
\end{aligned}
$$

A similar argument shows that $\overline{\lim}_{\,n\to\infty} F_n(z_1 \,|\, z_{n2}) \le \Phi(z_1)$ (exercise). This mends the hole in the above argument and completes the proof that $F_n(z_1 \,|\, Z_{n2} = z_{n2}) \to \Phi(z_1)$ whenever z_{n2} is in the support of the distribution of Z_{n2} and $z_{n2} \to z_2$.

11.5.6 Conclusion: Asymptotics of the Hypergeometric Distribution

We can recast what we have just proven as follows. Note that $Z_{n1} = (X_n - Y_n)/\{2np(1-p)\}^{1/2} = (X_n - S_n/2)/\{np(1-p)/2\}^{1/2}$, where $S_n = X_n + Y_n$. The conditional distribution of X_n given that $Z_{n_2} = z_{n2}$ is the conditional distribution of Z_{n1} given that $X_n + Y_n = s_n$, where $(s_n - 2np)/\{2np(1-p)\}^{1/2} = z_{n2}$. The latter conditional distribution is hypergeometric with parameters (n, n, s_n). Therefore, we have proven that if X_n is hypergeometric (n, n, s_n), where $(s_n - 2np)/(2np)^{1/2} \to z_2$, then $(X_n - s_n/2)/\{np(1-p)/2\}^{1/2} \xrightarrow{D} \mathrm{N}(0,1)$. It turns out that this latter condition can be relaxed to $s_{2n}/2n \to p$ as $n \to \infty$. This shows that for the data of Table 11.1, the null distribution of the number of hyper-responders is approximately normal with mean $s_{1000}/2 = 250$ and variance $500(1/2)(1/2)/2 = 62.5$, as mentioned earlier.

Incidentally, the asymptotic normality of the hypergeometric distribution is a special case of the following (see page 194 of Feller, 1968).

Proposition 11.1. *Let X_N have a hypergeometric distribution with parameters (m_N, n_N, s_N):*

$$H_N(x) = P(X_N \le x) = \frac{\sum_{i=0}^{x} \binom{m_N}{i}\binom{n_N}{s_N - i}}{\binom{m_N + n_N}{s_N}}.$$

Suppose that, as $N \to \infty$, m_N, n_N, and s_N tend to ∞ in such a way that $s_N/(m_N + n_N) \to p$ and $m_N/(m_N + n_N) \to \lambda$. Then

$$\frac{X_N - \lambda s_N}{\sqrt{(m_N + n_N)p(1-p)\lambda(1-\lambda)}} \xrightarrow{D} \mathrm{N}(0,1).$$

The result we deduced corresponds to $m_N = n_N$, and our proof is under the additional condition that $(s_n - 2np)/(2np)^{1/2} \to z_2$. Nonetheless, the essential argument of our proof, that $X_n - Y_n$ is asymptotically independent of $X_n + Y_n$, is quite simple. It is an exercise to modify our reasoning above to deduce Proposition 11.1 under the additional condition that $\{s_N - (m_N + n_N)p\}/\{(m_N + n_N)p(1-p)\} \to z_2$.

Exercises

1. We asserted that $P(X \le x \mid X + Y = s)$ is a decreasing function of s. Prove this using the representation that a hypergeometric random variable U is the number of red balls in a sample of size s drawn without replacement from a population of m red balls and n blue balls.

2. Modify the argument leading to Expression (11.11) to prove that $\overline{\lim}\, F_n(z_{n1} \mid z_{n2}) \le \Phi(z_1)$.

3. Review the argument leading to Proposition 11.1 in the special case $m_N = n_N = n$. Modify it to allow $m_N \neq n_N$ using the fact that, under the null hypothesis, $x_N/m_N - Y_N/n_N$ is uncorrelated with, and asymptotically independent of, $X_N + Y_N$.

4. Suppose that X_N is hypergeometric (m_N, n_N, s), where s is a fixed integer. Assume that, as $N \to \infty$, m_N and n_N tend to ∞ in such a way that $m_N/(m_N + n_N) \to \lambda$. Is X_N still asymptotically normal?

11.6 Validity of Permutation Tests: I

Permutation tests offer a virtually assumption-free way to test hypotheses about parameters. Let $\mathbf{Y}$ be the outcome vector and $\mathbf{Z}$ be the treatment indicator vector in a clinical trial. A permutation test constructs the null distribution of a statistic $T(\mathbf{Y}, Z)$ by fixing $\mathbf{Y} = \mathbf{y}$, and assuming that the conditional distribution of $\mathbf{Z}$ given $\mathbf{Y} = \mathbf{y}$ is the same as its unconditional distribution induced by randomization alone. In most cases we condition also on the per-arm sample sizes. For instance, with simple randomization, all treatment assignments consistent with the given sample sizes are treated as equally likely. For permuted block randomization, all treatment assignments consistent with the block constraints are treated as equally likely, etc. The conditional distribution of $\mathbf{Z}$ given $\mathbf{Y} = \mathbf{y}$ is the same as its unconditional distribution for all possible realizations $\mathbf{y}$ if and only if $\mathbf{Z}$ and $\mathbf{Y}$ are independent (Proposition 10.31). Thus, the permutation approach tests the null hypothesis that $\mathbf{Z}$ and $\mathbf{Y}$ are independent.

Because permutation tests condition on the outcome data, they can be very attractive in clinical trials with unplanned changes. An important principle in clinical trials is that the primary outcome variable and analyses methods should be pre-specified in the protocol (Friedman, Furberg, and DeMets, 2010). Changes are frowned upon because of concern that they might be driven by trends seen in the data, in which case the type 1 error rate could be inflated. But things sometimes go wrong. For instance, investigators in a lung trial discovered that their original primary outcome, defined in terms of X-ray findings, could not be measured. At the time they realized that they had to change the outcome, they had not yet broken the treatment blind. That is, they did not know the treatment assignments of patients. Would it be permissible under these circumstances to change the primary outcome and perform an ordinary permutation test as if that outcome had been pre-specified as primary?

Here is an argument justifying a permutation test in the above setting. Suppose we look at data on k potential outcomes, and let $\mathbf{Y}_i$ be the data for the n patients on outcome

i, $i = 1, \ldots, k$. After looking at the data, we allow a change of the primary outcome. A permutation test is still a valid test of the strong null hypothesis that $\mathbf{Z}$ is independent of $\mathbf{Y}_1, \ldots, \mathbf{Y}_k$. This hypothesis says that treatment has no effect on any of the potential outcomes, either alone or in concert. Under this strong null hypothesis, the conditional distribution of $\mathbf{Z}$ given $\mathbf{Y}_i = \mathbf{y}_i$, $i = 1, \ldots, k$, is its unconditional distribution dictated by the randomization method. Therefore, even though we changed the primary outcome after looking at data, the type 1 error rate cannot be inflated if we use a permutation test.

If the above argument is correct, then what goes wrong in the following apparent counterexample, given in Posch and Proschan (2012), to the proposition that the type I error rate cannot be inflated? Suppose that three potential outcomes are being considered: (1) coronary heart disease, (2) cardiovascular disease, and (3) level of study drug in the blood. Never mind that no one would use the third outcome; it still illustrates an important point. Even though we look at $(\mathbf{Y}_1, \mathbf{Y}_2, \mathbf{Y}_3)$ and not $\mathbf{Z}$, the third outcome variable clearly gives us information about $\mathbf{Z}$. Only treated patients will have nonzero levels of the study drug. Once we observe $\mathbf{Y}_3$, we know the full set of treatment assignments. Once we know the treatment assignments, we can compute the values of the standardized test statistics (z-scores) for each of the candidate outcome variables and pick the one with the largest z-score. This will clearly inflate the type 1 error rate if we use an ordinary permutation test without accounting for the fact that we are selecting the most statistically significant of the outcomes. The problem is compounded with a greater number of potential outcomes. Westfall and Young (1993) describe a valid way to account for picking the smallest p-value, but our question concerns the validity of an ordinary permutation test treating the selected outcome as if it had been pre-specified.

What went wrong with the argument that the type 1 error rate could not be inflated with a permutation test on the modified primary outcome? Surprisingly, nothing went wrong! Remember that the permutation test is testing the **strong** null hypothesis that $\mathbf{Z}$ is independent of $(\mathbf{Y}_1, \mathbf{Y}_2, \mathbf{Y}_3)$, i.e., that treatment has no effect on **any** of the outcomes. When we reject this null hypothesis, we are not making a type 1 error because the strong null hypothesis is false: treatment does have an effect on outcome 3, level of study drug in the blood. Therefore, the type I error rate is not inflated.

Do not dismiss the above example because of its extremeness. It illustrates the potential danger when changing the primary outcome after examining data but before breaking the treatment blind. We may unwittingly glean information about the treatment labels from the data examined. For instance, it is tempting, when choosing among several potential outcome variables, to select the one with less missing data. But how can we be sure that the amount of missing data does not give us information about the treatment assignments? It could be that patients in the treatment group are more likely than patients in the control group to be missing data on a given outcome. If so, we could become at least partially unblinded. Just as in the above example, this unblinding could inflate the type I error rate for the selected outcome. It still provides a valid test of the strong null hypothesis that treatment has no effect on any of the information examined, including the amount of missing data. But rejecting that null hypothesis is not meaningful if the only difference between the treatment arms is the amount of missing data. Therefore, a permutation test may not test the scientifically relevant question.

Exercises

1. Many experiments involving non-human primates are very small because the animals are quite expensive. Consider an experiment with 3 animals per arm, and suppose that one of two binary outcomes is being considered. You look at outcome data blinded

to treatment assignment, and you find that 5 out of 6 animals experienced outcome 1, whereas 3 animals experienced outcome 2. You argue as follows. With outcome 1, it is impossible to obtain a statistically significant result at 1-tailed $\alpha = 0.05$ using Fisher's exact test. With outcome 2, if all 3 events are in the control arm, the 1-tailed p-value using Fisher's exact test will be 0.05. Therefore, you select outcome 2 and use Fisher's exact test (which is a permutation test on binary data).

(a) Is this adaptive test valid? Explain.

(b) Suppose your permutation test did not condition on the numbers of animals per arm. For example, your permutation test treats as equally likely all $2^6 = 64$ treatment assignments corresponding to flipping a fair coin for each person. Your test statistic is the difference in proportions, which is not defined if all animals are assigned to the same treatment. Therefore, your permutation distribution excludes these two extreme assignments. Is the resulting test a valid permutation test?

11.7 Validity of Permutation Tests: II

11.7.1 A Vaccine Trial Raising Validity Questions

Recall that in the HIV vaccine trial discussed in Example 10.48, some trial participants had genetically very similar viruses, suggesting that they had common sexual partners. This raises serious concerns because different patients' HIV indicators are not independent; someone with HIV who has sex with another participant increases the probability that the other participant will have HIV. The usual z-test of proportions assumes independent observations. Is there a valid way to analyze the data? Is a permutation test valid? Proschan and Follmann (2008) addressed these questions.

11.7.2 Assumptions Ensuring Validity

As noted in Section 11.6, a permutation test for a statistic $T(\mathbf{Y}, \mathbf{Z})$ fixes the outcome data $\mathbf{Y} = \mathbf{y}$ and tests whether the conditional distribution of the vaccine indicators $\mathbf{Z}$ given $\mathbf{Y} = \mathbf{y}$ depends on $\mathbf{y}$. That is, it tests whether $\mathbf{Z}$ and $\mathbf{Y}$ are independent. Let A be the matrix summarizing information on "sexual sharing;" i.e., A_{ij} is the indicator that participants i and j had sex with each other or with a common partner. It is reasonable to assume that, under the null hypothesis, the distribution of $\mathbf{Y}$ might depend on A, but not additionally on $\mathbf{Z}$. Formally, consider the following modified null hypothesis:

$$\mathbf{Y} \text{ and } \mathbf{Z} \text{ are conditionally independent given } A. \tag{11.12}$$

Another way to express this condition is that the conditional distribution of $\mathbf{Y}$ given A and $\mathbf{Z}$ depends only on A. Thus, in a network of people who have sex with each other or who have a common partner, the HIV indicators may be positively correlated, but their joint distribution is not further affected by knowledge of $\mathbf{Z}$.

This relationship is depicted in the top of the *path diagram*, Figure 11.1. There is an arrow leading from A to $\mathbf{Y}$, indicating that sexual sharing could affect the distribution of the number of people infected. In other words, the conditional distribution $F(\mathbf{y} \mid A)$ of $\mathbf{Y}$ given A could depend on A. Also, there is an arrow leading from $\mathbf{Z}$ to A, meaning that treatment might affect the amount of sexual sharing. For instance, participants assigned to vaccine might be more likely to have sex than participants assigned to placebo. Later, we assume this does not happen. Notice that the only path connecting **Z** to **Y** goes through A, meaning that the only effect of **Z** on **Y** is through the effect of **Z** on A. Once we condition on A, **Z** and **Y** are independent. This is reflected by hypothesis (11.12). Pictorially, we can envision conditioning on A as blocking the path from **Z** to **Y**; there is no longer a way to get from **Z** to **Y** or vice versa. We will expand on path diagrams and their role in deducing conditional independence relationships in Section 11.9.

The conditional independence of **Z** and **Y** given A means that we can construct a valid *stratified* permutation test. For example, suppose participants 1,5,14,21, and 100 had sex with each other or with a common partner, whereas the other participants did not have sex with other trial participants or with a common partner. If 3 of participants (1,5,14,21,100) received the vaccine and simple randomization was used, we would treat each of the $\binom{5}{3}$ sets of 3 participants as equally likely to have received vaccine. For the complementary set of $n-5$ participants who did not have sex with each other or with a common partner, we would likewise condition on the number m receiving vaccine, and treat each set of $\binom{n-5}{m}$ as equally likely. A major problem with this plan is that there is only very limited information on sexual sharing among trial participants. Only when participants contracted HIV with genetically similar viruses was sexual sharing discovered. Without more information, a stratified permutation test is not possible.

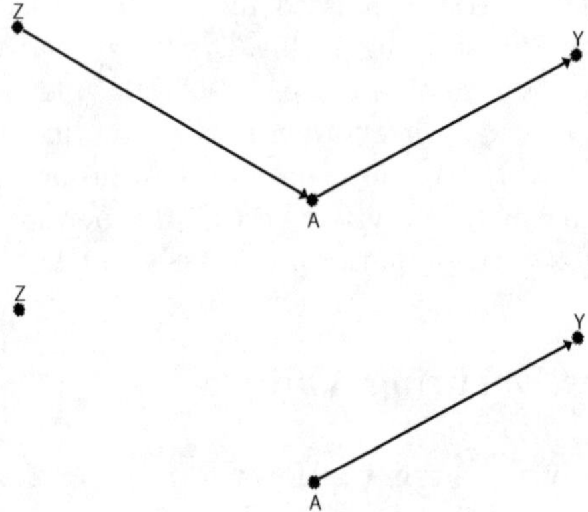

Figure 11.1: Path diagrams illustrating two different sets of assumptions about **Z** (treatment indicators), A (sexual sharing indicators) and **Y** (outcome). The top diagram is under the sole assumption that **Z** and **Y** are conditionally independent given A. The bottom diagram is under the additional assumption that **Z** is independent of A.

Given that a stratified permutation test is not feasible, we seek additional conditions under which an ordinary (unstratified) permutation test is valid. Consider the following assumption.

$$\mathbf{Z} \text{ and } A \text{ are independent.} \tag{11.13}$$

This also seems reasonable unless the vaccine somehow makes participants more sexually attractive or empowered. This is an unlikely scenario in such a blinded trial. If the trial were not blinded, people in the vaccine arm might feel like they are protected, which could cause them to engage in riskier behavior. This is called a dis-inhibition effect. But in a blinded trial, participants do not know whether they are receiving the vaccine or placebo. This should equalize the dis-inhibition in the two arms and render the assumption of independence of **Z** and A reasonable. Under assumption (11.13), there is no arrow connecting **Z** to either A or **Y** (bottom of Figure 11.1). Because of this, **Z** and $(A, \mathbf{Y})$ are independent. This can be proven as follows:

$$\begin{aligned} P(\mathbf{Z} \in B \,|\, A, \mathbf{Y}) &= P(\mathbf{Z} \in B \,|\, A) \text{ (Assumption 11.12)} \\ &= P(\mathbf{Z} \in B). \quad \text{(Assumption 11.13)} \end{aligned} \tag{11.14}$$

We have shown that the conditional distribution of **Z** given $(A, \mathbf{Y})$ is the unconditional distribution of **Z**. That is, **Z** and $(A, \mathbf{Y})$ are independent. This, of course, implies that **Z** and **Y** are independent. It follows that a permutation test is valid under assumptions (11.12) and (11.13).

11.7.3 Is a Z-Test of Proportions Asymptotically Valid?

The fact that a permutation test is still valid under certain assumptions begs the question: is a z-test of proportions also valid? We have seen that permutation tests and t-tests are asymptotically equivalent in certain settings (see Examples 8.20 and 10.20 for 1- and 2-sample settings, and van der Vaart, 1998 for a more general result). That is, with probability 1, the permutation distribution of the test statistic tends to N(0, 1). The key question, then, is whether, for almost all infinite sequences $y_1, y_2, \ldots$ of 0s and 1s,

$$\frac{(1/n)\sum_{i=1}^{2n} y_i Z_i - (1/n)\sum_{i=1}^{2n} y_i(1 - Z_i)}{\sqrt{2\hat{p}(1-\hat{p})/n}} \xrightarrow{D} \mathrm{N}(0, 1), \tag{11.15}$$

where $\hat{p} = (2n)^{-1}\sum_{i=1}^{2n} y_i$ and the treatment indicators Z_i satisfy $\sum_{i=1}^{2n} Z_i = n$. If so, then the permutation test is asymptotically equivalent to an ordinary z-test of proportions. We have already shown that a permutation test is valid, so this would imply that a z-test of proportions is also asymptotically valid. It is immaterial whether the y_i arose as realizations from independent or non-independent random variables. Once we condition on them, they are fixed constants. However, certain sets of infinite strings are exceedingly unlikely under independence, but not under dependence. Therefore, it might be the case that certain infinite realizations can be ignored under independence because they have probability 0, but do not have probability 0 and cannot be ignored if the observations are not independent. That is, it might be the case that the permutation test is asymptotically equivalent to the z-test for almost all infinite sequences of data under independence, but not under dependence.

Given $y_1, \ldots, y_{2n}$, the distribution of $\sum_{i=1}^{2n} y_i Z_i$ is hypergeometric. It is as if we randomly draw n balls without replacement from an urn containing $\sum_{i=1}^{2n} y_i$ red balls and $\sum_{i=1}^{2n}(1 - y_i)$ blue balls. The number of red balls in the sample of n is $\sum_{i=1}^{2n} y_i Z_i$. We can use this perspective to concoct scenarios under which conclusion (11.15) fails to hold. For instance, suppose that $Y_1 = Y_2 = Y_3 \ldots$ with probability 1. Then the denominator of condition (11.15) is 0. Similarly, if there is nonzero probability that $\sum_{i=1}^{\infty} Y_i$ is finite, then the numerator of condition (11.15) will not be asymptotically normal. For instance, suppose that $\sum_{i=1}^{\infty} y_i = k$. For $2n$ sufficiently large, all k red balls will appear somewhere among the $2n$; each red ball has probability approaching 1/2 of being among the first n of the $2n$ balls. Therefore, the number of red balls among the first n converges in distribution to the sum of k Bernoulli random variables with parameter 1/2, namely a binomial $(k, 1/2)$.

The following argument tells us when conclusion (11.15) holds. Imagine that the probability space Ω is $\mathcal{Y} \times \mathcal{Z}$, where $\mathcal{Y}$ is the space from which we draw the infinite string $y_1, y_2, \ldots$ and $\mathcal{Z}$ is the space from which we draw the treatment indicators. For each n, we draw

$Z_{1,2n}, \ldots, Z_{2n,2n}$ with $\sum_{i=1}^{2n} Z_{i,2n} = n$. By construction, $(Z_{1,2n}, \ldots, Z_{2n,2n})$, $n = 1, 2, \ldots$ is independent of $Y_1, Y_2, \ldots$. For a given set $y_1, \ldots, y_{2n}$, the same hypergeometric distribution obtains for $\sum_{i=1}^{2n} y_i Z_{i,2n}$ whether the Y_i were originally independent or dependent. Also, that hypergeometric distribution depends on $y_1, \ldots, y_{2n}$ only through the sample proportion of ones, $(2n)^{-1} \sum_{i=1}^{2n} y_i$. We saw in Section 11.5 that if this quantity converges to a number, then the hypergeometric random variable $\sum_{i=1}^{2n} y_i Z_{i,2n}$ is asymptotically normal with parameters proscribed by Proposition 11.1 and conclusion (11.15) holds.

Proposition 11.2. *Let $Y_1, Y_2, \ldots$ be (possibly correlated) Bernoulli random variables such that $(2n)^{-1} \sum_{i=1}^{2n} Y_i(\omega) \stackrel{a.s.}{\to} T(\omega)$ for some random variable $T(\omega)$, $0 < T(\omega) < 1$. Then (11.15) holds and the permutation test applied to the usual z-test of proportions is asymptotically equivalent to referring the z-score to a standard normal distribution.*

Exercises

1. Make rigorous the above informal argument that if $\sum_{i=1}^{\infty} y_i = k$, then $\sum_{i=1}^{2n} Z_i y_i$ is asymptotically binomial with parameters $(k, 1/2)$.

2. Suppose the clustering from the sexual sharing results in n pairs of patients. Within each pair, the two Ys are correlated, but if we select a single member from each pair, they are iid. Prove that the condition in Proposition 11.2 holds. Extend the result to triplets, quadruplets, etc.

3. Now suppose the $2n$ patients represent one giant cluster and were generated as follows. We flip a coin once. We generate $Y_1, Y_2, \ldots$ as iid Bernoulli(p), where $p = 0.2$ if the coin flip is heads and 0.8 if the coin flip is tails. Does the condition in Proposition 11.2 hold?

11.8 Validity of Permutation Tests III

11.8.1 Is Adaptive Selection of Covariates Valid?

When analysis of covariance (ANCOVA) is used to analyze data from a clinical trial with a continuous outcome Y, the covariates are pre-specified in the protocol. They are usually selected because they are highly correlated with the outcome. Adjusting for between-arm differences in these covariates at baseline gains efficiency over using a simple t-test on Y. But inclusion of too many covariates can be counterproductive because each one uses up a degree of freedom, and because the model may be more complicated than it needs to be. Therefore, we consider here inclusion of only a single covariate. But what if we pre-specify a covariate that turns out not to be correlated with the outcome, and a covariate that was not pre-specified turns out to be highly correlated with the outcome? It would be nice to be able to use ANCOVA with the covariate more highly correlated with Y. Better yet, suppose we had instead specified the following adaptive procedure for determining the best covariate. Without breaking the treatment blind, evaluate the sample correlation coefficient between Y and each candidate covariate X_i. There is no restriction on the number of potential covariates from which we are selecting one. Let X_{best} be the covariate most correlated (either negatively or positively) with Y, and let $\hat{\beta}$ be the slope coefficient in the univariate regression of Y on X_{best}. Now use $Y - \hat{\beta} X_{\text{best}}$ as the outcome, as if X_{best} had been pre-specified and $\hat{\beta}$ were a fixed constant instead of a random variable. After all, if X_{best} had been pre-specified, then analysis of covariance would be nearly identical to performing a t-test on $Y - \hat{\beta}X_{\text{best}}$, which gains power over a t-test on Y alone. We will perform an ordinary permutation test on $Y' = Y - \hat{\beta}X_{\text{best}}$ as if X_{best} had been pre-specified and $\hat{\beta}$ were a fixed constant. This is equivalent to a permutation test on the treatment effect estimator

$$\bar{Y}_T - \bar{Y}_C - \hat{\beta}(\bar{X}_{\text{best},T} - \bar{X}_{\text{best},C}). \tag{11.16}$$

Is this procedure valid?

Our assumption under the null hypothesis is that the conditional distribution of $\mathbf{Y}$ given $\mathbf{X}_1 = \mathbf{x}_1, \ldots, \mathbf{X}_k = \mathbf{x}_k, \mathbf{Z} = \mathbf{z}$ depends on $\mathbf{x}_1, \ldots, \mathbf{x}_k$, but not additionally on the treatment indicator vector $\mathbf{z}$. That is, $\mathbf{Y}$ and $\mathbf{Z}$ are conditionally independent given $\mathbf{X}_1, \ldots, \mathbf{X}_k$. Moreover, $\mathbf{X}_1, \ldots, \mathbf{X}_k$ are measured at baseline, so treatment can have no effect on them. Therefore, $\mathbf{Z}$ is independent of $(\mathbf{X}_1, \ldots, \mathbf{X}_k)$. The situation is probabilistically identical to the example in Section 11.7, where A represented information on sexual sharing and Y_i was the indicator that patient i had HIV. In the current example, take A to be the matrix whose columns are $\mathbf{X}_1, \ldots, \mathbf{X}_k$. In both examples, $\mathbf{Y}$ and $\mathbf{Z}$ are conditionally independent given A, and $\mathbf{Z}$ and A are independent. We showed in Section 11.6 that these two facts imply that $\mathbf{Z}$ is independent of $(A, \mathbf{Y})$. Moreover, $\mathbf{Y} - \hat{\beta}\mathbf{X}_{\text{best}}$ is a Borel function of $(A, \mathbf{Y})$. Therefore, $\mathbf{Z}$ is independent of $\mathbf{Y} - \hat{\beta}\mathbf{X}_{\text{best}}$. It follows that a permutation test is valid for this adaptive regression. The key was that the covariate selection was blinded to treatment assignment. Had we instead broken the blind and picked the covariate that minimizes the p-value for treatment, an ordinary permutation test that did not account for picking the smallest p-value would not have controlled the type 1 error rate.

11.8.2 Sham Covariates Reduce Variance: What's the Rub?

Closely related to the above permutation test is a t-test on the adjusted outcome $Y - \hat{\beta}X_{\text{best}}$ (again treating $\hat{\beta}$ as a fixed number). If one or more of the original covariates $X_1, \ldots, X_k$ is highly correlated with Y, the adaptive procedure is likely to select it or another variable that is quite correlated with Y. The trouble is, even if $X_1, \ldots, X_k$ are completely independent of Y, our selection process is likely to identify a covariate that **appears** to be correlated with Y. Therefore, it seems that we gain efficiency. If this is true, then we could generate a huge number of covariates having nothing to do with Y, pick the one most correlated with Y, and reduce residual variability and increase power through apparent sleight of hand.

We now investigate whether we really can gain an advantage by generating sham covariates. Assume for simplicity that all potential covariates take values ± 1, with an equal number of $+1$ and -1. In fact, suppose we generate every possible such $\mathbf{X}$ vector completely independently from $\mathbf{Y}$, so any observed correlation between $\mathbf{Y}$ and any of the $\mathbf{X}$s is purely by chance. Assume also that Y_i are iid standard normals. Without loss of generality, assume that we have ordered the Ys so that $Y_1 < Y_2 < \ldots < Y_n$. Assume also that n is divisible by 4. It is easy to see that the covariate most correlated with $\mathbf{Y}$ takes the value $X_{\text{best}\,i} = -1$ for $i = 1, \ldots, n/2$ (i.e., for the smallest $n/2$ Ys), and $X_{\text{best}\,i} = 1$ for $i = n/2+1, \ldots, 2n$ (i.e., for the largest $n/2$ Ys). The coefficient $\hat{\beta}$ for the regression of $\mathbf{Y}$ on $\mathbf{X}_{\text{best}}$ is

$$\begin{aligned}
\hat{\beta} &= \frac{\sum_{i=1}^{n} X_{\text{best}\,i} Y_i}{\sum_{i=1}^{n} X_{\text{best}\,i}^2} = \frac{\sum_{i=1}^{n/2}(-1)Y_i + \sum_{i=n/2+1}^{n}(+1)Y_i}{\sum_{i=1}^{n/2}(-1)^2 + \sum_{i=n/2+1}^{n}(+1)^2} \\
&= \frac{\sum_{i=n/2+1}^{n} Y_i - \left(n\bar{Y} - \sum_{i=n/2+1}^{n} Y_i\right)}{n}
\end{aligned}$$

$$= (2/n) \sum_{i=n/2+1}^{n} Y_i - \bar{Y}$$

$$= (2/n) \sum_{i=1}^{n} Y_i I(Y_i \geq \hat{\theta}) - \bar{Y}, \tag{11.17}$$

where $\hat{\theta}$ is the sample median of $Y_1, \ldots, Y_n$. It is an exercise to prove that $\hat{\beta}$ converges almost surely to $2\mathrm{E}\{YI(Y > 0)\} = (2/\pi)^{1/2}$. Keep in mind that the covariates were generated independently of Y, so the true β_i for the univariate regression of Y on X_i is 0 for $i = 1, \ldots, k$. However, we just showed that when we select the covariate with largest sample correlation with Y, the estimated β for the regression of Y on that covariate will be close to $(2/\pi)^{1/2}$ instead of 0. The resulting residual variance can be shown to be approximately $\mathrm{var}(Y \mid Y > 0) = 2\int_0^\infty y^2\phi(y)dy - \{2\int_0^\infty y\phi(y)dy\}^2 = 1 - 2/\pi$.

We now recap. Had we used a simple t-test on Y, the residual variance would have been close to 1 because the Y_i are iid standard normals. Through generating sham covariates and adjusting for the one most correlated with Y, we reduced the residual variance to $1 - 2/\pi$. This suggests we are getting some benefit from generating sham covariates! Fortunately, this impression of benefit is an illusion; it can be shown that under the alternative hypothesis, the adjusted treatment effect estimate is attenuated when we select the most correlated of artificially generated covariates. This attenuation creates bias that more than offsets any gain in precision.

11.8.3 A Disturbing Twist

If we are really intent on gaining an unfair advantage through sham covariates, we can try the following sneaky alternative procedure. Restrict attention to the set of **balanced** binary covariates with half -1 and half $+1$. That is, among participants with covariate value $X = -1$, exactly half are assigned to treatment, and similarly for participants with covariate value $X = 1$. Again generate all possible such covariates, and select the one most correlated with $\mathbf{Y}$. Is a t-test on the adjusted outcome $\mathbf{Y} - \hat{\beta}\mathbf{X}_{\text{best}}$ valid now?

The first thing to notice is that, because the covariates are all perfectly balanced, $\bar{X}_{\text{best},T} = \bar{X}_{\text{best},C}$ and Expression (11.16) is $\bar{Y}_T - \bar{Y}_C$. In other words, the adjusted treatment effect estimator is exactly the same as the unadjusted estimator. This circumvents the problem at the end of the preceding subsection, namely an attenuated treatment effect. The real variance of the treatment effect estimator is $2\sigma^2/(n/2)$. On the other hand, it can be shown using arguments similar to the one above that the denominator of the covariate-adjusted t-statistic is close to $\{2(1 - 2/\pi)/(n/2)\}^{1/2}$ if n is large. Therefore, a t-test on the adjusted outcome $\mathbf{Y} - \hat{\beta}\mathbf{X}_{\text{best}}$ has an inflated type 1 error rate; its actual type 1 error rate is approximately 0.24 if the intended alpha is 0.025 and the sample size is large.

The preceding paragraph shows that an ordinary t-test does not work. Given the close connection between permutation and t-tests, this suggests, but does not prove, that a permutation test is invalid when we pick the balanced binary covariate most correlated with Y. Let us return to the argument in Section 11.8.1 justifying the validity of a permutation test when we did not restrict ourselves to balanced covariates. We argued $\mathbf{Y}$ is conditionally independent of $\mathbf{Z}$ given A (recall that A is the matrix whose columns comprise the set of all covariates under consideration). This is still correct under the null hypothesis. What is not correct is the assumption that $\mathbf{Z}$ and A are independent. To see this, let $\mathbf{Y} = (Y_1, Y_2, Y_3, Y_4)'$ Suppose we observe the following two balanced covariates.

$$\begin{pmatrix} -1 & +1 \\ -1 & -1 \\ +1 & +1 \\ +1 & -1 \end{pmatrix}. \tag{11.18}$$

It is an exercise to show that if these two columns represent balanced covariates, then the only possible randomizations are (C,T,T,C) or (T,C,C,T). Therefore, $\mathbf{Z}$ and A cannot be independent. Thus, one of the key assumptions that we used to validate a permutation test in Sections 11.8 and 11.8.1, namely Assumption 11.13, is not satisfied.

Exercises

1. From Expression (11.17) prove that $\hat{\beta} \stackrel{a.s.}{\to} (2/\pi)^{1/2}$.

2. Prove that if two columns of A are as shown in (11.18), then the only possible treatment assignments are (C,T,T,C) or (T,C,C,T). Why does this prove that $\mathbf{Z}$ cannot be independent of A?

11.9 A Brief Introduction to Path Diagrams

In Section 11.7, we used a path diagram to help understand the relationships between random variables. Here we briefly expand on the use of these diagrams. A more detailed presentation may be found in Pearl (2000), for example. We begin with a heuristic presentation to illustrate ideas in causal inference applied to medical studies. Thinking in terms of causal inference helps explain definitions and rules of path diagrams that we later make rigorous. We then apply these rules to deduce conditional independence of random variables. Remarkably, this simple tool can be formally justified as a way to deduce complex relationships between variables that are induced by conditioning.

Medical advances often begin by noticing the relationship between a risk factor and a clinical outcome. For example, we notice that people with high blood pressure have a higher risk of stroke. We then develop and test a medicine to see if it reduces blood pressure. After showing that it does, we conduct a much larger randomized trial to see if the stroke probability is lower in the treatment group than in the placebo group. When we see that it is, we theorize that the treatment is reducing strokes through its effect on blood pressure. When the effect of treatment on the clinical outcome of interest is through its effect on an intermediate outcome like blood pressure, we call the intermediate outcome a *surrogate outcome* (Prentice, 1989).

We can depict this relationship graphically through the *path diagram* of Figure 11.2. The *nodes* of the diagram are the dots representing the variables T (treatment indicator), X (blood pressure), and Y (stroke indicator). The graph is also called a *directed acyclic graph (DAG)*. It is directed because the arrows lead from one node to another. Acyclic means that there is no directed path from one node back to itself. There is an arrow leading from T to X, indicating that the treatment has an effect on blood pressure. The arrow leading from X to Y means that blood pressure affects stroke risk. Therefore, treatment has an effect on stroke through its effect on blood pressure. There is no arrow leading from T directly to Y.

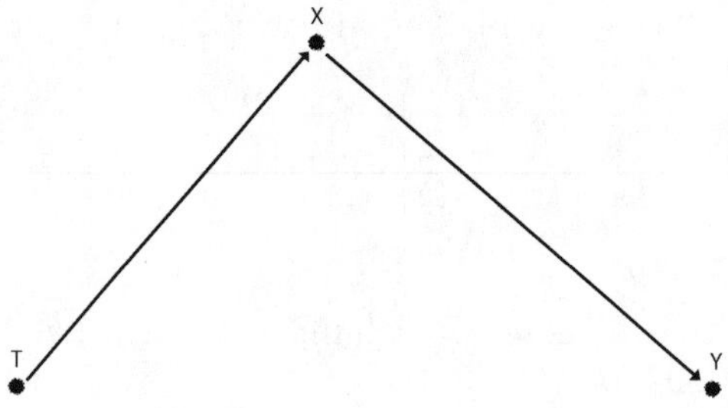

Figure 11.2: Path diagram indicating that the full effect of treatment assignment T on outcome Y is through its effect on the surrogate X.

This is because we believe that the entire effect of the treatment on reducing stroke risk is through its effect on blood pressure. If this is correct, then two things should happen. First, there should be a relationship between T and Y. That is, the probability of stroke should be lower given that $T = 1$ than given that $T = 0$. Second, the probability of stroke given T and X should depend only on X. That is, T and Y should be conditionally independent given X (Prentice, 1989). In other words, X is a surrogate outcome for Y.

For the remainder of this section, assume that the random variables have a density with respect to Lebesgue measure or counting measure. We continue to think informally, deferring a rigorous presentation until later. We move away from causal inference and think of the path diagram as simply depicting a way to factor the joint density function $f(t, x, y)$ of random variables (T, X, Y). We can always factor a density function $f(t, x, y)$ into

$$f(t, x, y) = f_1(t) f_2(x \,|\, t) f_3(y \,|\, t, x).$$

But we get a bit of simplification in the above example involving stroke and blood pressure because the conditional density function $f_3(y \,|\, t, x)$ does not depend on t. Therefore, the joint density factors as follows in this example.

$$f(t, x, y) = f_1(t) f_2(x \,|\, t) f_3(y \,|\, x).$$

We can traverse the arrows from T to X and from X to Y. The fact that there is a path connecting T and Y means that the conditional distribution of Y given T may depend on T. That is, T and Y may be correlated. If they are correlated, then of course the conditional distribution of T given Y also depends on Y. This suggests that we can travel against the arrows as well; starting at Y, we may go against the arrow to X, and then against the arrow to T. Even though T and Y are connected, they become independent once we condition on X. Think of conditioning on X as applying a roadblock at the node at X. This roadblock prevents us from traveling from T to Y or vice versa. The nodes T and Y are now disconnected, meaning that they are conditionally independent given X. This is what we noticed earlier in the stroke and blood pressure example, that $f_3(y \,|\, t, x) = f_3(y \,|\, x)$ for any t, x, y.

Although we pointed out that we can travel either in the direction of the arrow or against the arrow, there is an exception. Suppose that X and Y both affect an outcome Z.

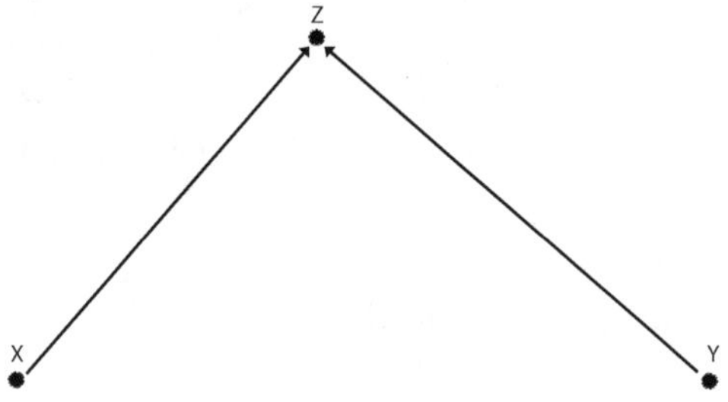

Figure 11.3: Vertex Z is an "inverted fork," also known as a collider.

For instance, suppose that $Z = X + Y$. Suppose further that X and Y are independent. We depict this relationship with arrows from X to Z and from Y to Z (Figure 11.3). Vertex Z is called a *collider* or *inverted fork* (had the direction of arrows been reversed, there would have been a *fork*: Z could lead to X or to Y). If we allowed travel from X to Z and then to Y, it would imply that X and Y might be dependent. But we stated that X and Y are independent. Thus, we disallow travel through the inverted fork **unless we condition on Z.** Here, $Z = X + Y$, so conditioning on $Z = z$ means $X + Y = z$. Equivalently, $Y = z - X$. This induces a correlation between X and Y. That is, even though X and Y are independent, they are conditionally dependent given Z. Therefore, the rule for inverted forks is reversed: the path is blocked (indicating independence) unless we condition on the node at the inverted fork, in which case the path becomes open (indicating possible dependence).

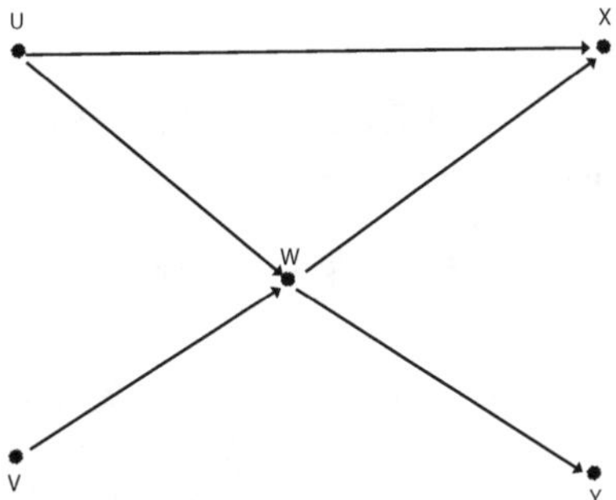

Figure 11.4: Path diagram for random variables (U, V, W, X, Y).

Each DAG for a set of random variables corresponds to a factoring of the joint density function of the variables. For each node (a random variable or random vector), we incorporate a term for the conditional density function of that variable given the variables pointing to it. For Figure 11.4, this corresponds to

$$f_1(u)f_2(v)f_3(w \mid u,v)f_4(x \mid w,u)f_5(y \mid w).$$

There are no arrows pointing to U and V, so they are called *root nodes.* We can factor the density function of the subset (U,V) as the product $f_1(u)f_2(v)$ of marginal densities. Therefore, U and V are independent. Any two root nodes in a DAG are independent.

The only arrows pointing to X are from U and W. This means that the conditional distribution of X given U,V and W depends only on U and W. That is, V and X are conditionally independent given (U,W). Because there is still another path from U to X that does not go through W, we cannot conclude that X is independent of U, given W. The only arrow pointing to Y is from W. If we block that path by conditioning on W, then nodes (U,V) are separated from Y. Therefore, the set (U,V) are conditionally independent of Y, given W.

In the figure, nodes W and X are called *descendents* of node U because there are directed paths from U to W and X. Likewise, we call U an *ancestor* of W and X. When we condition on a descendent, that gives information about its ancestor. Therefore, suppose we condition on X in the figure. That gives information about W, so we must treat W as also given. But W is an inverted fork, so that opens up the path from U to W to V. Therefore, we can no longer be assured that U and V are independent once we condition on X. They are unconditionally independent because they are root nodes, but they are not necessarily conditionally independent given X.

We are now in a position to formalize the rules and key result about DAGs.

Definition 11.3. Blocked path *A path p is said to be blocked by a set of nodes* $\mathbf{Z}$ *if*

1. *p contains a chain $i \to j \to k$ or a fork $i \leftarrow j \to k$ such that the middle node j is in* $\mathbf{Z}$ *or*

2. *p contains an inverted fork $i \to j \leftarrow k$ such that neither j nor any of its descendants is in* $\mathbf{Z}$.

Recall that when we condition on descendants of j, it could reveal information about j, That is why Definition 11.3 includes the clause about any descendant of j being in $\mathbf{Z}$.

Definition 11.4. d-separation *A set* $\mathbf{Z}$ *is said to d-separate sets* $\mathbf{X}$ *and* $\mathbf{Y}$ *if* $\mathbf{Z}$ *blocks every path from an X_i in* $\mathbf{X}$ *to a Y_j in* $\mathbf{Y}$.

Theorem 11.5. (Verma and Pearl, 1988) *In a DAG, if* $\mathbf{X}$ *and* $\mathbf{Y}$ *are d-separated by* $\mathbf{Z}$, *then* $\mathbf{X}$ *and* $\mathbf{Y}$ *are conditionally independent given Z for every distribution compatible with the graph. On the other hand, if there is at least one path from a node in* $\mathbf{X}$ *to a node in* $\mathbf{Y}$ *that is not blocked by* $\mathbf{Z}$, *then there is some joint distribution compatible with the graph such that* $\mathbf{X}$ *and* $\mathbf{Y}$ *are not conditionally independent given* $\mathbf{Z}$.

This result can be used to construct graphical counterexamples to erroneous identities on conditional independence. We begin with some notation.

Notation 11.6. *We use the symbol $\amalg$ to denote independence or conditional independence. For instance, $X \amalg Y$ means that X and Y are independent, while $(X \amalg Y \mid Z)$ means that X and Y are conditionally independent given Z.*

Suppose you are trying to decide whether the following proposition is true:

$$\{X \amalg Y \mid (U,V)\} \Rightarrow (X \amalg Y \mid U) \text{ or } (X \amalg Y \mid V\}. \tag{11.19}$$

Figure 11.5 shows a path diagram illustrating a situation in which both paths from X to Y are blocked by U and V, so $(X \amalg Y \mid (U,V))$. However, it is not true that $(X \amalg Y \mid U)$ because conditioning on U blocks only one path from X to Y; the path X to V to Y remains open. The same reasoning shows that it is not necessarily true that $(X \amalg Y \mid V)$. The fact that we are able to find a graphical counterexample shows, in conjunction with Theorem 11.5, that there exist random variables (U,V,X,Y) such that (11.19) is false. In fact, the following is a counterexample. Let U and V be iid standard normals, and let $X = U + V$, $Y = U - V$. Conditioned on (U,V), X and Y are constants, hence independent. But conditioned on U alone, X and Y have correlation -1, and conditioned on V alone, X and Y have correlation $+1$.

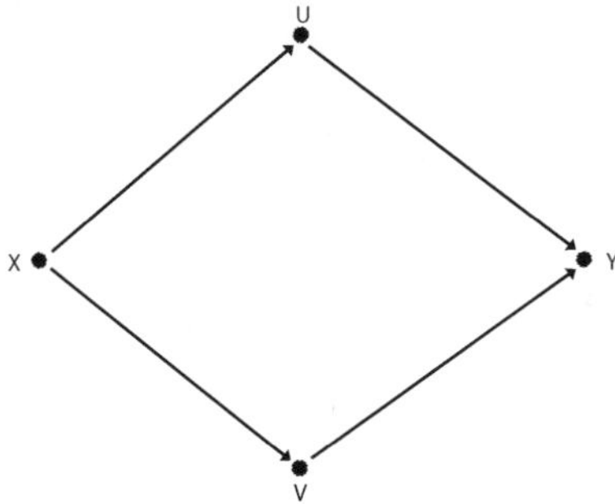

Figure 11.5: Path diagram.

Another graphical counterexample involves the following proposition

$$\{X \amalg Y \mid (U,V)\} \text{ and } (X \amalg Y \mid U) \Rightarrow (X \amalg Y \mid V) \tag{11.20}$$

Figure 11.6 shows a path diagram in which X and Y are conditionally independent given (U,V) and conditionally independent given U because conditioning on U blocks both paths from X to Y. However, X and Y are not conditionally independent given V because the path X to U to Y remains open. A concrete example takes $X = U$ and $Y = U + V$. Given (U,V), X and Y are constants, and hence conditionally independent. Also, given U, X is constant and therefore conditionally independent of Y. On the other hand, given V, X and Y have correlation 1.

Exercises

1. Explain why Figure 11.7 is a graphical counterexample to the following proposition:

   $$(X \amalg Y \mid U) \text{ and } (X \amalg Y \mid W) \Rightarrow \{X \amalg Y \mid (U,W)\}.$$

   Then show that the following is a specific counterexample. Let U, W, X be iid Bernoulli (1/2). If $(U,W) = (0,0)$ or $(1,1)$, take $Y = X$. If $(U,W) = (0,1)$ or $(1,0)$, take $Y = 1 - X$.

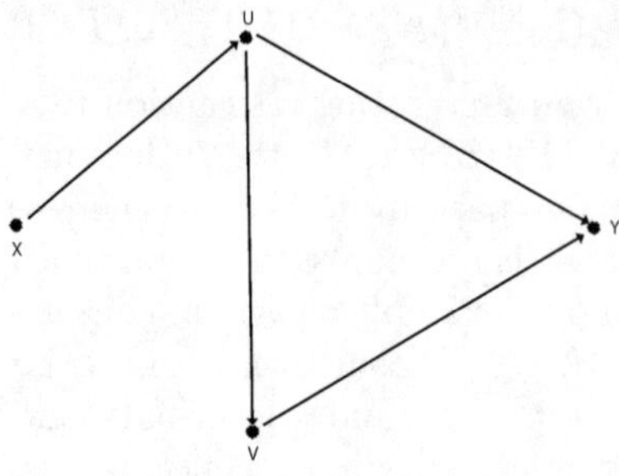

Figure 11.6: Path diagram.

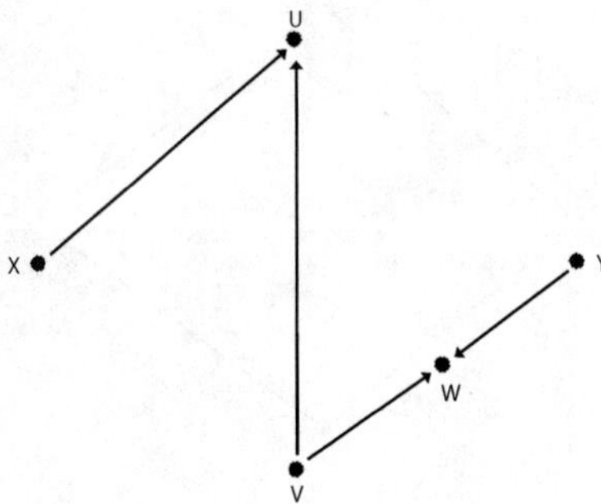

Figure 11.7: Path diagram.

2. Decide which of the following are true and prove the true ones.

   (a) $\{X \amalg (Y, W) \,|\, Z\} \Rightarrow (X \amalg Y \,|\, Z)$.

   (b) $\{X \amalg (Y, W) \,|\, Z\} \Rightarrow \{X \amalg Y \,|\, (Z, W)\}$.

   (c) $(X \amalg Y \,|\, Z)$, and $Y \amalg W \Rightarrow (X \amalg W \,|\, Z)$.

   (d) $(X \amalg Y \,|\, Z)$ and $\{X \amalg W \,|\, (Z, Y)\} \Rightarrow \{X \amalg (Y, W) \,|\, Z\}$.

3. In the path diagram of Figure 11.8, determine which of the following are true and factor the joint density function for (X, U, V, W, Y, Z).

   (a) U and V are independent.

   (b) X and Y are independent.

   (c) X and Y are conditionally independent given Z.

   (d) X and Y are conditionally independent given (U, V).

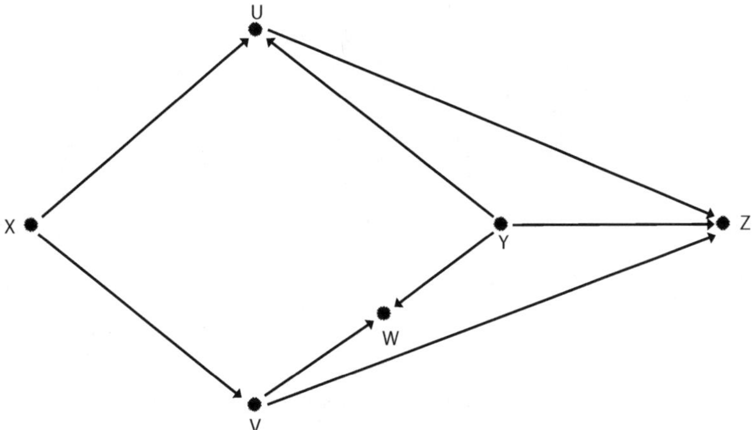

Figure 11.8: Path diagram.

11.10 Estimating the Effect Size

11.10.1 Background and Introduction

In some clinical trials, there is debate about what the primary outcome should be. For instance, in a trial evaluating the effect of an intervention on blood pressure, should diastolic or systolic blood pressure be the primary outcome? Investigators designing the Dietary Approaches to Stop Hypertension (DASH) feeding trial had to make that decision. The trial compared two different dietary patterns to a control dietary pattern (Appel et al., 1997). There are many considerations for choice of primary outcome, the most important being which outcome is more predictive of clinically important outcomes like stroke and coronary heart disease. At the time of the DASH trial, both systolic and diastolic blood pressure appeared to be strongly linked to important clinical outcomes of interest. Another important consideration is which variable the intervention is expected to affect more. This has important consequences for sample size. But sample size for a continuous outcome depends not only on the difference between treatment and control means, $\delta = \mu_T - \mu_C$, but on the standard deviation σ as well. In fact, the crucial parameter for the per-arm sample size, $n = \{2\sigma^2(1.96 + 1.28)^2/\delta^2\}$, for a two-tailed t-test with 80% power and $\alpha = 0.05$ is the effect size δ/σ. Consequently, to evaluate which outcome results in a smaller sample size, we must estimate the effect size from other studies.

The same issue arises in paired settings using the t-statistic $T = \bar{D}/(s^2/n)^{1/2}$ on paired differences D_i. Sample size depends on the effect size $E = \mu/\sigma$, where μ and σ are the mean and standard deviation of the D_i. Therefore, interest centers on estimation of E. A closely related issue arises when comparing two methods of measuring the same quantity. For example, one might compare blood pressure measured using a tried and true "gold standard" machine to that measured by a newer, untested device. A standard technique after measuring each person using both methods is to plot the error $D_i = Y_1 - Y_2$ against the average $(Y_1 + Y_2)/2$ (Bland-Altman, 1986). The resulting graph might show a funnel pattern with the larger part of the funnel occurring for larger mean values (top panel of Figure 11.9). This indicates greater error variability for larger values. This can sometimes be corrected by plotting the relative error against the mean (bottom panel of Figure 11.9). The absolute value of each relative error is $|Y_1 - Y_2|/\{(Y_1 + Y_2)/2\} = 2^{1/2}s/\bar{Y}$, where s is

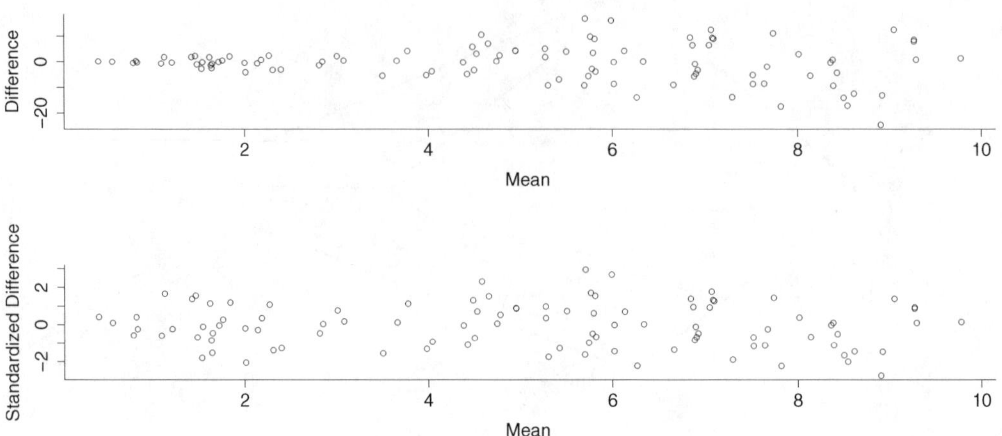

Figure 11.9: The Bland-Altman plot of errors against means in the top panel shows a funnel indicating larger variability of errors for larger values. In such cases, the relative error may be constant (bottom panel).

the standard deviation of (Y_1, Y_2) and $\bar{Y}$ is $(Y_1 + Y_2)/2$. That is, the absolute value of each relative error is proportional to $1/E$, where E is the effect size.

11.10.2 An Erroneous Application of Slutsky's Theorem

Consider the estimation of effect size $E = \mu/\sigma$ in a one-sample setting using $\hat{E} = \bar{Y}/s$, where $\bar{Y}$ and s are the sample mean and standard deviation. Assume that the observations are iid from a normal distribution with finite variance and the sample size n is large. Here is an erroneous line of reasoning leading to the wrong asymptotic distribution. Because $\bar{Y}$ is asymptotically normal with mean μ and variance σ^2/n, and s converges to σ in probability (even almost surely), $\hat{E}$ should have the same asymptotic distribution as $\bar{Y}/\sigma$, namely $N(\mu/\sigma, 1/n)$. Therefore, it seems that an approximate 95% confidence interval for E should be

$$\left(\hat{E} - \frac{1.96}{\sqrt{n}},\ \hat{E} + \frac{1.96}{\sqrt{n}}\right). \tag{11.21}$$

To see if (11.21) has 95% coverage, simulate 500 observations from a normal distribution with mean $\mu = 1$ and $\sigma = 1$, and calculate the confidence interval (11.21). Repeat this process 100,000 times to see that approximately 89% of the intervals contain the true effect size, $E = 1/1 = 1$. Thus, the interval does not have the correct coverage probability. Increasing the mean makes the coverage even worse; with $\mu = 3$ and $\sigma = 1$, only about 60% of the intervals cover the true effect size 3. On the other hand, the coverage probability is 95% if $\mu = 0$, and close to 95% if μ is very close to 0.

11.10.3 A Correct Application of Slutsky's Theorem

We take a closer look at what went wrong and why interval (11.21) is close to being correct if μ is close to 0. Interval (11.21) resulted from a misunderstanding of Slutsky's theorem, which requires one of the terms to converge in distribution. The limiting distribution cannot involve n. Here, $\bar{Y}$ was asymptotically normal with mean μ and variance σ^2/n, which does depend on n. The actual asymptotic distribution of $\hat{E}$ can be obtained as follows.

$$\begin{aligned}
\sqrt{n}(\hat{E}-E) &= \sqrt{n}\left(\frac{\bar{Y}-\mu}{s}+\frac{\mu}{s}-\frac{\mu}{\sigma}\right) \\
&= \sqrt{n}\left(\frac{\bar{Y}-\mu}{s}\right)+\sqrt{n}\mu\left(\frac{1}{s}-\frac{1}{\sigma}\right) \\
&= T_{n-1}+\frac{\sqrt{n}\mu(\sigma-s)}{s\sigma} \\
&= T_{n-1}-\frac{\sqrt{n}\mu(s^2-\sigma^2)}{s\sigma(\sigma+s)} \\
&= T_{n-1}-\frac{\mu\sigma\sqrt{\frac{2n}{n-1}}}{s(\sigma+s)}\left\{\frac{(n-1)s^2/\sigma^2-(n-1)}{\sqrt{2(n-1)}}\right\},
\end{aligned} \tag{11.22}$$

where T_{n-1} is the usual one-sample t-statistic for testing whether the mean is μ. Asymptotically, this t-statistic is N$(0,1)$. Also, $(n-1)s^s/\sigma^2$ has a chi-squared distribution with $n-1$ degrees of freedom, so it may be represented as $\sum_{i=1}^{n-1} Z_i^2$, where the Z_i are iid $N(0,1)$. The variance of a χ^2_{n-1} random variable is $2(n-1)$. Therefore, by the CLT,

$$\frac{\frac{(n-1)s^2}{\sigma^2}-(n-1)}{\sqrt{2(n-1)}}=\frac{\sum_{i=1}^{n-1}\{Z_i^2-\mathrm{E}(Z_i^2)\}}{\sqrt{\mathrm{var}\{\sum_{i=1}^{n-1}(Z_i^2)\}}}\xrightarrow{D} N(0,1).$$

Also, $\mu\sigma\{2n/(n-1)\}^{1/2}/\{s(\sigma+s)\}\xrightarrow{p} 2^{-1/2}\mu/\sigma$. Therefore, by Slutsky's theorem, the term to the right of T_{n-1} in (11.22) converges in distribution to a normal with mean 0 and variance $\mu^2/(2\sigma^2)=E^2/2$. But T_{n-1} is independent of the term to its right in Expression (11.22) because s^2 and $\bar{Y}$ are independent. Therefore, $n^{1/2}(\hat{E}-E)$ converges in distribution to a N$(0,1)$ plus an independent N$(0,1+E^2/2)$; the sum is asymptotically N$(0,1+E^2/2)$. That is,

$$\hat{E}\sim N\left(E,\frac{1+E^2/2}{n}\right). \tag{11.23}$$

Using approximation (11.23), we can verify the simulation result that the coverage probability of (11.21) is only 89% and 60% when $\mu=1$ and $\mu=3$, respectively. Note that we could have reached the correct asymptotic distribution (11.23) using the delta method (exercise).

11.10.4 Transforming to Stabilize The Variance: The Delta Method

One problem with Expression (11.23) is that the asymptotic variance of $\hat{E}$ depends on the parameter we are trying to estimate, $E=\mu/\sigma$. In such situations it is helpful to consider a transformation $f(\hat{E})$. By the univariate delta method, the asymptotic distribution of $f(\hat{E})$ is normal with mean $f(E)$ and variance $\{f'(E)\}^2(1+E^2/2)/n$. If we want this variance not to depend on E, we set $\{f'(E)\}^2(1+E^2/2)=1$. This results in $f'(E)=(1+E^2/2)^{-1/2}$, so $f(E)=\int(1+E^2/2)^{-1/2}dE=2^{1/2}\ln\{E+(E^2+2)^{1/2}\}$. We can dispense with the $2^{1/2}$ and conclude that

$$\ln\left\{\hat{E}+\sqrt{\hat{E}^2+2}\right\}\approx N\left[\ln\left\{E+\sqrt{E^2+2}\right\},1/(2n)\right].$$

The probability is approximately 95% that $\ln\{E+(E^2+2)^{1/2}\}$ is between $\ln\{\hat{E}+(\hat{E}^2+2)^{1/2}\}-1.96/(2n)^{1/2}$ and $\ln\{\hat{E}+(\hat{E}^2+2)^{1/2}\}+1.96/(2n)^{1/2}$. Exponentiating, we find that

$$\left(\hat{E}+\sqrt{\hat{E}^2+2}\right)\exp\left(-\frac{1.96}{\sqrt{2n}}\right)\leq E+\sqrt{E^2+2}\leq\left(\hat{E}+\sqrt{\hat{E}^2+2}\right)\exp\left(\frac{1.96}{\sqrt{2n}}\right).$$

Denoting the left and right sides by L and U and solving for E, we find that

$$\frac{L^2-2}{2L}\leq E\leq\frac{U^2-2}{2U},$$

where

$$L=\left(\hat{E}+\sqrt{\hat{E}^2+2}\right)\exp\left(-\frac{1.96}{\sqrt{2n}}\right),\quad U=\left(\hat{E}+\sqrt{\hat{E}^2+2}\right)\exp\left(\frac{1.96}{\sqrt{2n}}\right).$$

Exercises

1. Use the delta method to verify Expression (11.23).

2. In the effect size setting, the variance of the asymptotic distribution depended on the parameter we were trying to estimate, so we made a transformation. Find the appropriate transformations in the following settings to eliminate the dependence of the asymptotic distribution on the parameter we are trying to estimate.

   (a) Estimation of the probability of an adverse event. We count only the first adverse event, so the observations are iid Bernoulli random variables with parameter p, and we use the sample proportion with events, $\hat{p}=X/n$, where X is binomial (n,p).

   (b) Estimation of the mean number of adverse events per person. We count multiple events per person. Assume that the total number of events across all people follows a Poisson distribution with parameter λ, where λ is very large.

3. Suppose that the asymptotic distribution of an estimator $\hat{\delta}$ is $\mathrm{N}(\delta, f(\delta)/n)$ for some function f. How can we transform the estimator such that the variance of the asymptotic distribution does not depend on the parameter we are trying to estimate?

11.11 Asymptotics of An Outlier Test

11.11.1 Background and Test

Outliers are extreme observations that can cause serious problems in statistical inference. This is illustrated by the story of a statistician who is given a set of data presumed to be iid $\mathrm{N}(\mu,\sigma^2)$, and asked to confirm with a t-test that $\mu>0$. The statistician verifies this, at which time the investigator notices that one of the observations is erroneous; the actual value should be 100.0 instead of 1.0. "That should make the results even more statistically significant," thinks the investigator. But the recalculated t-statistic becomes **non-significant.** The problem is that the outlying value increases not only the sample mean, but the sample standard deviation as well. In fact, it is easy to verify that if one fixes all but one observation and sends the remaining observation to infinity, the t-statistic tends to 1, which is not statistically significant at typical alpha levels.

Statistical methods have been developed to test for the presence of outliers. The simplest setting is that of n iid $N(\mu, \sigma^2)$ observations, one of which is suspected to be an outlier caused by some sort of error. To support this, we must account for the fact that we are flagging the most discrepant observation. If we knew μ and σ^2, things would be easy. For instance, suppose we knew that $\mu = 0$ and $\sigma = 1$. In that case, we could compute $Y_{\max} = \max(Y_i)$; its distribution function is $P(Y_{\max} \le y) = P\{\cap_{i=1}^n (Y_i \le y)\} = \{\Phi(y)\}^n$. Therefore, we could declare $Y_{\max}$ an outlier if $Y_{\max} > y_n$, where $1 - \{\Phi(y_n)\}^n = \alpha$. That is, $y_n = \Phi^{-1}\{(1-\alpha)^{1/n}\}$. For instance, if $n = 500$ and $\alpha = 0.025$, then $y_n = \Phi^{-1}\{(.975)^{1/500}\} = 3.89$. We would declare the largest observation an outlier if it exceeds 3.89.

We could also have used the asymptotic distribution of $Y_{\max}$ derived in Example 6.25 concerning the Bonferroni approximation because $P(Y_{\max} \ge y_n) = P(\cup_{i=1}^n \{Y_i \ge y_n\})$. The indicators $I(Y_i \ge y_n)$ are iid Bernoulli with probability $p_n = P(Y_1 \ge y_n)$. If we choose y_n such that $n\{1 - \Phi(y_n)\} \to \beta$, then by the law of small numbers (Proposition 6.24), the number of Y_i at least as large as y_n converges in distribution to a Poisson with parameter β. Therefore,

$$
\begin{aligned}
P\left[n\left\{1 - \Phi(Y_{\max})\right\} \le \alpha\right] &= P\{Y_{\max} \ge \Phi^{-1}(1 - \alpha/n))\} \\
&= P\left[\cup_{i=1}^n \{Y_i \ge \Phi^{-1}(1 - \alpha/n)\}\right] \\
&\to 1 - \frac{\exp(-\alpha)\alpha^0}{0!} \\
&= 1 - \exp(-\alpha). \qquad (11.24)
\end{aligned}
$$

That is, $n\{1 - \Phi(Y_{\max})\} \xrightarrow{D}$ exponential (1). This approximate method declares $Y_{\max}$ an outlier if $n\{1 - \Phi(Y_{\max})\}$ exceeds the $(1 - \alpha)$th quantile of an exponential (1) distribution, namely $-\ln(1 - \alpha)$. Equivalently, $Y_{\max} > \Phi^{-1}\{1 + \ln(1 - \alpha)/n\}$. In the numerical example above with $n = 500$ and $\alpha = 0.025$, we get $Y_{\max} > \Phi^{-1}\{1 + \ln(.975)/500\} = 3.89$. That is, we get the same answer as we did using the exact method above.

But now suppose we do not know μ or σ^2. Substituting the sample analogs, $\bar{Y}$ and s, for μ and σ leads to

$$
U = \frac{\max(Y_1, \ldots, Y_n) - \bar{Y}}{s} = \frac{Y_{\max} - \bar{Y}}{s}, \qquad (11.25)
$$

To use U, we must determine its null distribution. Assume that the sample size is large. The distribution of U does not depend on μ or σ because because U has the same value if we replace Y_i by $(Y_i - \mu)/\sigma$. Therefore, without loss of generality, we may assume that Y_i are iid $N(0, 1)$.

We already showed that $n\{1 - \Phi(Y_{\max})\} \xrightarrow{D}$ exponential (1). We must now show that this continues to hold if $Y_{\max}$ is replaced by $(Y_{\max} - \bar{Y})/s$. The first step toward that end is to prove a very useful result in its own right.

11.11.2 Inequality and Asymptotics for $1 - \Phi(x)$

Theorem 11.7. *The standard normal distribution function $\Phi(x)$ and density $\phi(x)$ satisfy:*

$$
\frac{\phi(x)/x}{1 + 1/x^2} \le 1 - \Phi(x) \le \phi(x)/x \quad \text{for } x > 0. \qquad (11.26)
$$

Accordingly, $1-\Phi(x)\sim\phi(x)/x$ *as* $x\to\infty$.

Proof. Note that

$$\begin{aligned}
1-\Phi(x) &= \int_x^\infty \frac{\exp(-t^2/2)}{\sqrt{2\pi}}dt = \int_x^\infty \frac{t\exp(-t^2/2)}{t\sqrt{2\pi}}dt \\
&\le \int_x^\infty \frac{t\exp(-t^2/2)}{x\sqrt{2\pi}}dt = \left(\frac{1}{x\sqrt{2\pi}}\right)\left\{-\exp(-t^2/2)\Big|_x^\infty\right\} \\
&= \phi(x)/x, \qquad (11.27)
\end{aligned}$$

which is the right inequality in Equation (11.26).

For the left side, integrate by parts as follows:

$$\begin{aligned}
1-\Phi(x) &= \left(\frac{1}{\sqrt{2\pi}}\right)\int_x^\infty (1/t)\{t\exp(-t^2/2)\}dt \\
&= \left(\frac{1}{\sqrt{2\pi}}\right)\left[(1/t)\{-\exp(-t^2/2\}\Big|_x^\infty - \int_x^\infty \frac{\exp(-t^2/2)}{t^2}dt\right] \\
&= \phi(x)/x - \int_x^\infty \frac{\exp(-t^2/2)}{t^2\sqrt{2\pi}}dt \\
&\ge \phi(x)/x - \int_x^\infty \frac{\exp(-t^2/2)}{x^2\sqrt{2\pi}}dt \\
&= \phi(x)/x-(1/x^2)\{1-\Phi(x)\}, \qquad (11.28)
\end{aligned}$$

from which the left side of the inequality in Equation (11.26) follows.

That $1-\Phi(x)\sim\phi(x)/x$ as $x\to\infty$ follows by dividing all sides of Equation (11.26) by $\phi(x)/x$ and noting that the left and right sides have limit 1. □

11.11.3 Relevance to Asymptotics for the Outlier Test

We next show that Equation (11.24) holds when the left side is replaced by $n\{1-\Phi(Y_{\max}-\bar{Y})\}$. By Slutsky's theorem, it suffices to show that

$$\frac{n\{1-\Phi(Y_{\max})\}}{n\{1-\Phi(Y_{\max}-\bar{Y})\}} \xrightarrow{p} 1 \text{ as } n\to\infty. \qquad (11.29)$$

By Theorem 11.7,

$$\begin{aligned}
\frac{n\{1-\Phi(Y_{\max})\}}{n\{1-\Phi(Y_{\max}-\bar{Y})\}} &\sim \frac{\frac{n\phi(Y_{\max})}{Y_{\max}}}{\frac{n\phi(Y_{\max}-\bar{Y})}{Y_{\max}-\bar{Y}}} \\
&= \left(\frac{Y_{\max}-\bar{Y}}{Y_{\max}}\right)\exp\left[-\left\{Y_{\max}^2-(Y_{\max}-\bar{Y})^2\right\}/2\right]
\end{aligned}$$

$$\begin{aligned} &= \left(\frac{Y_{\max} - \bar{Y}}{Y_{\max}}\right) \exp\left[-\left\{Y_{\max}\bar{Y} - \bar{Y}^2/2\right\}\right] \\ &= \left(1 - \frac{\bar{Y}}{Y_{\max}}\right) \exp\left[-\left\{Y_{\max}\bar{Y} - \bar{Y}^2/2\right\}\right] \qquad (11.30) \end{aligned}$$

Also, $(1 - \bar{Y}/Y_{\max}) \xrightarrow{p} 1$ because $\bar{Y} \xrightarrow{a.s.} 0$. To prove that $\exp\left[-\left\{Y_{\max}\bar{Y} - \bar{Y}^2/2\right\}\right] \xrightarrow{p} 1$, we need only prove that $Y_{\max}\bar{Y} \xrightarrow{p} 0$ because $\bar{Y}^2/2 \xrightarrow{a.s.} 0$. Write $Y_{\max}\bar{Y}$ as $(Y_{\max}/n^{1/2})(n^{1/2}\bar{Y})$ and note that $n^{1/2}\bar{Y}$ converges in distribution by the CLT. Therefore, it suffices to prove that $Y_{\max}/n^{1/2} \xrightarrow{p} 0$. To that end, note that

$$\begin{aligned} P\left(\frac{Y_{\max}}{n^{1/2}} > \epsilon\right) &= P\left(\bigcup_{i=1}^{n} \{Y_i > n^{1/2}\epsilon\}\right) \\ &\leq nP(Y_1 > n^{1/2}\epsilon) = nP(Y_1^2 > n\epsilon^2) \to 0 \qquad (11.31) \end{aligned}$$

because $\mathrm{E}(Y_1^2) < \infty$.

We have shown that (11.29) holds. A similar argument shows that

$$\frac{n\{1 - \Phi(Y_{\max})\}}{n\left\{1 - \Phi\left(\frac{Y_{\max} - \bar{Y}}{s}\right)\right\}} \xrightarrow{p} 1 \text{ as } n \to \infty.$$

By Slutsky's theorem,

$$P\left[n\left\{1 - \Phi\left(\frac{Y_{\max} - \bar{Y}}{s}\right)\right\} \leq y\right] \to 1 - \exp(-y).$$

Take $y = -\ln(1 - \alpha)$ to conclude that

$$P\left\{\frac{Y_{\max} - \bar{Y}}{s} \geq \Phi^{-1}\left(1 + \frac{\ln(1 - \alpha)}{n}\right)\right\} \to \alpha.$$

We have established that when the sample size n is large, an approximately valid test declares the largest observation an outlier if $(Y_{\max} - \bar{Y})/s$ exceeds $\Phi^{-1}\{1 + \ln(1 - \alpha)/n\}$. Equivalently, if the largest sample standardized residual exceeds $\Phi^{-1}\{1 + \ln(1 - \alpha)/n\}$, we declare the observation producing that residual an outlier.

Exercises

1. Let $x_1, \ldots, x_n$ be a sample of data. Fix n and $x_2, \ldots, x_n$, and send x_1 to ∞. Show that the one-sample t-statistic $n^{1/2}\bar{x}/s$ converges to 1 as $x_1 \to \infty$. What does this tell you about the performance of a one-sample t-test in the presence of an outlier?

2. Let $Y_1, \ldots, Y_n$ be iid $\mathrm{N}(\mu, \sigma^2)$, and $Y_{\max}$, $\bar{Y}$ and s^2 be their sample maximum, mean and variance, respectively. Suppose that a_n is a sequence of numbers converging to ∞ such that $a_n(Y_{\max} - \mu)/\sigma \xrightarrow{D} U$ for some non-degenerate random variable U. Prove that $a_n(Y_{\max} - \bar{Y})/s \xrightarrow{D} U$ as well.

3. Show that if $Y_1, \ldots, Y_n$ are iid standard normals and $Y_{\min}$ is the smallest order statistic, then $n\Phi(Y_{\min})$ converges in distribution to an exponential with parameter 1.

4. Prove that if $Y_1, \ldots, Y_n$ are iid standard normals and $Y_{\min}$ and $Y_{\max}$ are the smallest and largest order statistics,

$$P\left[n\Phi(Y_{\min}) > \alpha_1 \bigcap n\{1-\Phi(Y_{\max})\} > \alpha_2\right] = (1-\alpha_1/n - \alpha_2/n)^n$$

for $(\alpha_1+\alpha_2)/n < 1$. What does this tell you about the asymptotic joint distribution of $[n\Phi(Y_{\min}), n\{1-\Phi(Y_{\max})\}]$?

5. ↑ Let $Y_1, \ldots, Y_n$ be iid $\mathrm{N}(\mu, \sigma^2)$, with μ and σ^2 known. Declare the smallest order statistic to be an outlier if $n\Phi\{(Y_{(1)}-\mu)/\sigma\} \le a$, and the largest order statistic to be an outlier if $n[1-\Phi\{(Y_{(n)}-\mu)/\sigma\}] \le a$. Determine a such that the probability of erroneously declaring an outlier is approximately 0.05 when n is large.

11.12 An Estimator Associated with the Logrank Statistic

11.12.1 Background and Goal

Table 11.2: Two-by-two table at the ith death time.

	Dead		
Treatment	X_i		n_{Ti}
Control			n_{Ci}
	1		n_i

The logrank statistic is commonly used to compare two survival distributions (namely one minus the distribution functions) when the treatment-to-control hazard ratio, $\theta = [f_T(t)/\{1-F_T(t)\}]/[f_C(t)/\{1-F_C(t)\}]$, is assumed constant. Here, $f(t)$ and $F(t)$ are the density and distribution functions of time to death, and T and C denote treatment and control. The hazard function $f(t)/\{1-F(t)\}$ is the instantaneous mortality rate at time t, conditional on surviving to time t. Consider a clinical trial in which all patients start at the same time and no one is lost to follow-up. In actuality, patients arrive in staggered fashion, and some of them drop out. Our purpose is to show that even in the admittedly over-simplified case we consider, use of elementary arguments to determine the approximate distribution of the logrank statistic is problematic.

Table 11.2 shows that just prior to the ith death time, there are n_{Ti} and n_{Ci} patients alive in the treatment and control arms. We say that these patients are "at risk" of dying. The random variable X_i is the indicator that the ith death came from the treatment arm. Under the null hypothesis, the expected value of X_i, given n_{Ti} and n_{Ci}, is the proportion of the $n_{Ti}+n_{Ci}$ at-risk patients who are in the treatment arm, namely, $E_i = n_{Ti}/(n_{Ti}+n_{Ci})$. Likewise, the null conditional variance of the Bernoulli random variable X_i, given n_{Ti} and n_{Ci}, is $V_i = E_i(1-E_i)$. The logrank z-statistic and a closely related estimator are:

$$Z_n = \frac{\sum_{i=1}^D (X_i - E_i)}{\sqrt{\sum_{i=1}^D V_i}}, \quad \hat{\tau}_n = \frac{\sum_{i=1}^D (X_i - E_i)}{\sum_{i=1}^D V_i},$$

where D is the total number of deaths. Readers familiar with survival methods may recognize that $\hat{\tau}_n$ is an estimator of the logarithm of the hazard ratio.

The logrank z-statistic and its associated estimator are analogous to the one-sample t-statistic and its associated estimator. In the one-sample t-statistic setting with iid $N(\mu, \sigma^2)$ observations Y_i, we are interested in testing whether $\mu = 0$. The t-statistic is $\bar{Y}_n/(s/n^{1/2})$, where $\bar{Y}_n$ and s are the sample mean and standard deviation, respectively. It is easy to show using a similar approach to that in Section 11.2.1 that under a local alternative $\mu_n = a/n^{1/2}$, the one-sample t-statistic $n^{1/2}(\bar{Y}_n - \mu_n)$ converges in distribution to $\mathrm{N}(a, \sigma^2)$. This implies that the estimator $\bar{Y}_n$ has the following property. When $\mu_n = a/n^p$ with $0 < p < 1/2$, the relative error $(\bar{Y}_n - \mu_n)/\mu_n$ converges in probability to 0 as $n \to \infty$. Equivalently, $\bar{Y}_n/\mu_n \xrightarrow{p} 1$.

Our goal is to show a similar result about the estimator $\hat{\tau}_n$ in the logrank setting. Specifically, we will show that under the local alternative $\theta_n = 1 - a/n^p$, where $0 < p < 1/2$,

$$\frac{\hat{\tau}_n - \ln(\theta_n)}{\ln(\theta_n)} \xrightarrow{p} 0. \text{ Equivalently, } \frac{\hat{\tau}_n}{\ln(\theta_n)} \xrightarrow{p} 1.$$

We will accomplish this in several steps, the first of which is to derive the distributions of $(X_i \,|\, n_{Ti}, n_{Ci})$ and $X_i - \mathrm{E}(X_i \,|\, n_{Ti}, n_{Ci})$.

11.12.2 Distributions of $(X_i \,|\, n_{Ti}, n_{Ci})$ and $X_i - \mathrm{E}(X_i \,|\, n_{Ti}, n_{Ci})$

Consider the expected value $E_i^* = E(X_i \,|\, n_{Ti}, n_{Ci})$ under an alternative hypothesis. Note that this is already a bit strange because we are conditioning on the numbers at risk "just prior" to the ith death. To accomplish this, imagine conditioning on the numbers at risk Δ units of time prior to the i death, where Δ is infinitesimally small. Each treatment patient has survived $t_i - \Delta$. The probability of dying in the time $(t_i - \Delta, t_i]$ is asymptotic to $f_T(t_i)\Delta/\{1 - F_T(t_i)\} = \lambda_T(t_i)\Delta$ as $\Delta \to 0$, where $\lambda_T(t_i) = f_T(t_i)/\{1 - F_T(t_i)\}$ is the hazard function in the treatment group at time t_i. Likewise, the probability of a control patient dying in the interval $(t_i - \Delta, t_i]$ is asymptotic to $\lambda_C(t_i)\Delta$ as $\Delta \to 0$, where $\lambda_C(t_i)$ is the hazard rate in the control group at time t_i. The conditional probability that the death came from the treatment arm is

$$\begin{aligned} E_i^* &\sim \frac{n_{Ti}\lambda_T(t_i)\Delta}{n_{Ti}\lambda_T(t_i)\Delta + n_{Ci}\lambda_C(t_i)\Delta} \\ &= \frac{n_{Ti}\{\lambda_T(t_i)/\lambda_C(t_i)\}}{n_{Ti}\{\lambda_T(t_i)/\lambda_{Ci}(t_i)\} + n_{Ci}} \\ &= \frac{\theta n_{Ti}}{\theta n_{Ti} + n_{Ci}}, \end{aligned} \tag{11.32}$$

where θ is the treatment to control hazard ratio, which is the same for all t. Let $\Delta \to 0$ to conclude that $E_i^* = \theta n_{Ti}/(\theta n_{Ti} + n_{Ci})$. Under the null hypothesis, $E_i^* = E_i = n_{Ti}/(n_{Ti} + n_{Ci})$.

Some authors say that the X_i may be treated as if they were independent Bernoulli random variables with probability parameters E_i^* (e.g., Schoenfeld, 1980). It is true that, conditioned on $D = d$, the probability of any given string $x_1, \ldots, x_d$ of zeros and ones is $\prod_{i=1}^{d}\{(E_i^*)^{x_i}(1-E_i^*)^{1-x_i}\}$, which matches that of independent Bernoullis. However, for different strings $x_1,\ldots,x_d$, the Bernoulli parameters E_i^* differ. Thus, the X_i are not independent. Neither does it make sense to say that they are conditionally independent given the set $(n_{Ci},n_{Ti})_{i=1}^{d}$. Once we condition on the full set of (n_{Ti},n_{Ci}), there is no more randomness! For instance, if $(n_{T1},n_{C1})=(100,96)$, then the first death came from the control arm if $(n_{T2},n_{C2})=(100,95)$, and from the treatment arm if $(n_{T2},n_{C2})=(99,96)$.

Even though the X_i are not independent, the following argument shows that $Y_i = X_i - E_i^*$, $i=1,\ldots,d$ are uncorrelated. First note that the Y_i have conditional mean 0 given n_{Ci},n_{Ti}, and therefore unconditional mean 0. Thus,

$$\begin{aligned}
\text{cov}(Y_i,Y_j) &= \text{E}(Y_iY_j)=\text{E}\{\text{E}(Y_iY_j\,|\,\text{death},n_{Tj},n_{Cj},Y_i)\} \text{ (Proposition 10.10)}\\
&= Y_i\,\text{E}(Y_j\,|\,\text{death},n_{Tj},n_{Cj},Y_i) \text{ (Proposition 10.14)}\\
&= Y_i\,\text{E}(Y_j\,|\,\text{death},n_{Tj},n_{Cj})\\
&= Y_i[\text{E}\{(X_j-E_j^*)\,|\,\text{death},n_{Tj},n_{Cj}\}]\\
&= Y_i(E_j^*-E_j^*)=0. \qquad (11.33)
\end{aligned}$$

This completes the argument that the Y_i are uncorrelated.

11.12.3 The Relative Error of $\hat{\tau}_n$

We next consider the relative error of the estimator $\hat{\tau}=\hat{\tau}_n$. First consider:

$$\begin{aligned}
\frac{\hat{\tau}_n}{\theta_n-1} &= \frac{\sum_{i=1}^{D}(X_i-E_i)}{(\theta_n-1)\sum_{i=1}^{D}V_i}\\
&= \frac{\sum_{i=1}^{D}(X_i-E_i^*)}{(\theta_n-1)\sum_{i=1}^{D}V_i}+\frac{\sum_{i=1}^{D}(E_i^*-E_i)}{(\theta_n-1)\sum_{i=1}^{D}V_i},\\
&= \frac{\sum_{i=1}^{D}Y_i}{(\theta_n-1)\sum_{i=1}^{D}V_i}+\frac{\sum_{i=1}^{D}(E_i^*-E_i)}{(\theta_n-1)\sum_{i=1}^{D}V_i}, \qquad (11.34)
\end{aligned}$$

where θ_n is the local alternative

$$\theta_n = 1-\frac{a}{n^p},\quad 0<p<1/2,\ \ a>0. \qquad (11.35)$$

We will show that the first term of Expression (11.34) tends to 0 in probability, whereas the second term tends to 1 in probability.

To show that the first term of Expression (11.34) is small, we show first that its denominator is large. Consider

$$\sum_{i=1}^{d}V_i=\sum_{i=1}^{d}E_i(1-E_i).$$

Note that, for $0\le p\le 1$, the parabola $f(p)=p(1-p)$ is maximized at $p=1/2$ and decreases as $|p-1/2|$ increases from 0 to 1/2. Therefore, $E_i(1-E_i)$ decreases as $|E_i-1/2|$ increases. Suppose that the proportion of people who die tends to a limit that is less than

1/2. Then $E_i(1-E_i)$ is at least as large as $\{(n/2-d+1)/n\}[1-\{(n/2-d+1)/n\}] = n\{1/2-(d-1)/n\}\{1/2+(d-1)/n\}$, so

$$\sum_{i=1}^{d} E_i(1-E_i) \geq n\left(\frac{1}{2}-\frac{d-1}{n}\right)\left(\frac{1}{2}+\frac{d-1}{n}\right).$$

It follows that

$$\left|\frac{\sum_{i=1}^{d} Y_i}{(\theta_n-1)\sum_{i=1}^{d} E_i(1-E_i)}\right| \leq \left|\frac{\sum_{i=1}^{d} Y_i}{(\theta_n-1)n\left(\frac{1}{2}-\frac{d-1}{n}\right)\left(\frac{1}{2}+\frac{d-1}{n}\right)}\right|. \tag{11.36}$$

Conditioned on $D=d$, the expression within the absolute value sign on the right of Expression (11.36) has mean 0 and variance

$$\begin{aligned}
\frac{\text{var}\left(\sum_{i=1}^{d} Y_i\right)}{(\theta_n-1)^2 n^2\left(\frac{1}{2}-\frac{d-1}{n}\right)^2\left(\frac{1}{2}+\frac{d-1}{n}\right)^2} &= \frac{\sum_{i=1}^{d} \text{var}(Y_i)}{(\theta_n-1)^2 n^2\left(\frac{1}{2}-\frac{d-1}{n}\right)^2\left(\frac{1}{2}+\frac{d-1}{n}\right)^2} \\
&= \frac{\sum_{i=1}^{d} \text{E}(Y_i^2)}{(\theta_n-1)^2 n^2\left(\frac{1}{2}-\frac{d-1}{n}\right)^2\left(\frac{1}{2}+\frac{d-1}{n}\right)^2} \\
&\leq \frac{\sum_{i=1}^{d} 1}{(\theta_n-1)^2 n^2\left(\frac{1}{2}-\frac{d-1}{n}\right)^2\left(\frac{1}{2}+\frac{d-1}{n}\right)^2} \\
&= \frac{d}{(\theta_n-1)^2 n^2\left(\frac{1}{2}-\frac{d-1}{n}\right)^2\left(\frac{1}{2}+\frac{d-1}{n}\right)^2} \\
&= \frac{d/n}{\left(\frac{a^2}{n^{2p}}\right) n\left(\frac{1}{2}-\frac{d-1}{n}\right)^2\left(\frac{1}{2}+\frac{d-1}{n}\right)^2}
\end{aligned}$$

In the last step we substituted the local alternative value (11.35) for θ_n.

Because d/n converges and $(a^2/n^{2p})n = a^2 n^{1-2p} \to \infty$ as $n \to \infty$ for $p < 1/2$, we conclude that the variance of the term within the absolute value sign on the right side of Expression (11.36) tends to 0. From this we conclude that

$$P\left\{\left|\frac{\sum_{i=1}^{D} Y_i}{(\theta_n-1)\sum_{i=1}^{D} E_i(1-E_i)}\right| > \epsilon \,\middle|\, D\right\} \to 0.$$

That is, conditioned on the number of deaths, the first term of Expression (11.34) tends to 0 in probability. The next step is to "uncondition" on the number of deaths. By Proposition 10.10 and the bounded convergence theorem,

$$\begin{aligned}
P\left[\left|\frac{\sum_{i=1}^{D}(X_i-E_i^*)}{(\theta_n-1)\sum_{i=1}^{D} E_i(1-E_i)}\right| > \epsilon\right] &= \text{E}\left[P\left\{\left|\frac{\sum_{i=1}^{D}(X_i-E_i^*)}{(\theta_n-1)\sum_{i=1}^{D} E_i(1-E_i)}\right| > \epsilon \,\middle|\, D\right\}\right] \\
&\to 0.
\end{aligned} \tag{11.37}$$

Therefore, the first term of (11.34) tends to 0 in probability.

Now consider the second term of Expression (11.34).

$$\frac{\sum_{i=1}^{d}(E_i^*-E_i)}{(\theta_n-1)\sum_{i=1}^{d} E_i(1-E_i)} = \frac{\sum_{i=1}^{d}\left(\frac{\theta_n n_{Ti}}{\theta_n n_{Ti}+n_{Ci}} - \frac{n_{Ti}}{n_{Ti}+n_{Ci}}\right)}{(\theta_n-1)\sum_{i=1}^{d} \frac{n_{Ti}n_{Ci}}{(n_{Ti}+n_{Ci})^2}}$$

$$= \frac{\sum_{i=1}^{d} \frac{n_{Ti}n_{Ci}}{(\theta_n n_{Ti}+n_{Ci})(n_{Ti}+n_{Ci})}}{\sum_{i=1}^{d} \frac{n_{Ti}n_{Ci}}{(n_{Ti}+n_{Ci})^2}}. \quad (11.38)$$

Note that $\theta_n \uparrow 1$ as n$\rightarrow \infty$, so

$$1 \leq \frac{\sum_{i=1}^{d} \frac{n_{Ti}n_{Ci}}{(\theta_n n_{Ti}+n_{Ci})(n_{Ti}+n_{Ci})}}{\sum_{i=1}^{d} \frac{n_{Ti}n_{Ci}}{(n_{Ti}+n_{Ci})^2}} \leq \frac{\sum_{i=1}^{d} \frac{n_{Ti}n_{Ci}}{(\theta_n n_{Ti}+\theta_n n_{Ci})(n_{Ti}+n_{Ci})}}{\sum_{i=1}^{d} \frac{n_{Ti}n_{Ci}}{(n_{Ti}+n_{Ci})^2}} = \frac{1}{\theta_n}.$$

The left and right sides both tend to 1 as $n \rightarrow \infty$. Therefore, the second term of Expression (11.34) tends to 1 in probability.

This completes the proof that

$$\frac{\hat{\tau}_n}{\theta_n - 1} = \frac{\sum_{i=1}^{D}(X_i - E_i)}{(\theta_n - 1)\sum_{i=1}^{D} V_i} \xrightarrow{p} 1.$$

Also, recall that $(\theta_n - 1)/\ln(\theta_n) \rightarrow 1$ as $n \rightarrow \infty$. Therefore, $\hat{\tau}_n/\tau \xrightarrow{p} 1$, where $\tau = \ln(\theta)$.

It was a very involved process to show that $\hat{\tau}_n/\ln(\theta_n) \xrightarrow{p} 1$ as $n \rightarrow \infty$ even under an unrealistically simplified scenario that all patients are available at the beginning of the trial and there is no censoring. Monitoring presents further complications. For example, suppose that at a given time, patient A has the shortest time to death, so his follow-up time is X_1 in the above notation. If we monitor the trial 6 months from now, a new patient may have a shorter time to death. Therefore, X_1 no longer corresponds to the original patient A. Likewise, the original X_2 also changes, etc. Trying to use simple arguments like the one above becomes impossible. One must rely instead on more advanced techniques using stochastic processes. Each patient has an array of data, including time of arrival into the trial, time to event, and time to censoring. The time to event is observed if and only if it is less than the time to censoring. Therefore, data from different patients are independent processes, and the numerator of the logrank statistic can be approximated by a sum of iid processes. Powerful central limit theorems for stochastic processes can then be used.

The simplification from looking at independent processes instead of conditioning on arrival times results from taking an unconditional, rather than conditional, view. The situation is analogous to what happens in a one-sample t-test of paired differences $D_1, D_2, \ldots$ If we take an unconditional view, the D_i are iid, so we can use the ordinary CLT to deduce that $\sum_{i=1}^{n} D_i$ is asymptotically normal under the assumption that var$(D) < \infty$. If we had instead conditioned on $|D_1| = d_1^+, |D_2| = d_2^+, \ldots$, then the D_i would no longer be iid. The D_i would still be independent, but the distribution of D_i would be $-d_i^+$ with probability 1/2 and $+d_i^+$ with probability 1/2. The ordinary CLT would no longer apply, although the Lindeberg-Feller CLT would.

Appendix A

Whirlwind Tour of Prerequisites

We present a brief review of results from advanced calculus and real analysis that are most useful in the study of probability theory. Most results are stated without proof, though we give some proofs if they illustrate techniques that will be useful in subsequent material.

A.1 A Key Inequality

One tool that you will use repeatedly is the triangle inequality. If $\mathbf{x} \in R^k$, let $||\mathbf{x}||$ denote its Euclidean length, $||\mathbf{x}|| = \sqrt{\sum_{i=1}^k x_i^2}$.

Proposition A.1. Triangle and reverse triangle inequality *If* $\mathbf{x}$ *and* $\mathbf{y}$ *are* k-*dimensional vectors, then* $|\ ||\mathbf{x}|| - ||\mathbf{y}||\ | \le ||\mathbf{x} \pm \mathbf{y}|| \le ||\mathbf{x}|| + ||\mathbf{y}||$.

The rightmost inequality is the most commonly used of the two, and is referred to as the triangle inequality. It gets its name from the fact that the vectors $\mathbf{x}$ and $\mathbf{y}$, together with their resultant, $\mathbf{x}+\mathbf{y}$, form the three sides of a triangle. Geometrically, $||\mathbf{x}+\mathbf{y}|| \le ||\mathbf{x}||+||\mathbf{y}||$ says that the shortest distance between two points is a line (see Figure A.1). In R^1, the triangle inequality says that $|x+y| \le |x| + |y|$ for all $x, y \in R$.

The leftmost inequality is called the reverse triangle inequality. It also has a geometric interpretation in terms of a triangle: $|\ ||\mathbf{x}|| - ||\mathbf{y}||\ | \le ||\mathbf{x} - \mathbf{y}||$ says that the length of any triangle side is at least as great as the absolute value of the difference in lengths of the other two sides. In R^1, the reverse triangle inequality says that $|\ |x| - |y|\ | \le |x \pm y|$ for all $x, y \in R$.

A.2 The Joy of Sets

A.2.1 Basic Definitions and Results

A *set* A is a collection of objects called *elements*. These elements can be numbers, but they need not be. For instance, $A = \{\text{couch, chair, bed}\}$ is a set. Likewise, the elements of a set may themselves be sets. For instance, the elements of $A = \{(t, \infty),\ t \in R\}$ are the sets

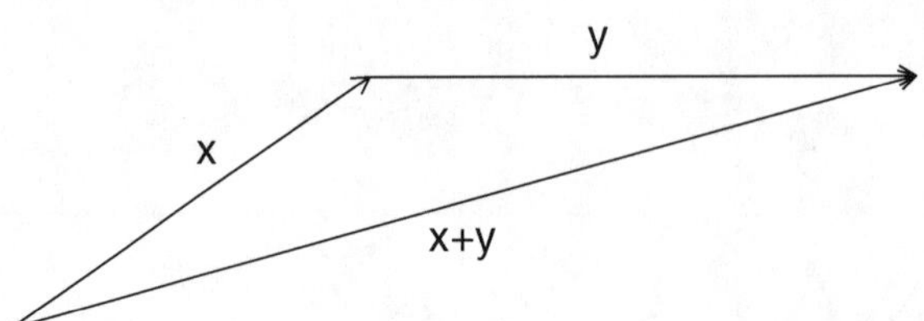

Figure A.1: The triangle inequality illustrated in two dimensions. The length of the resultant vector $\mathbf{x}+\mathbf{y}$ cannot be greater than the sum of the lengths of $\mathbf{x}$ and $\mathbf{y}$.

(t, ∞). The set containing no elements is called the empty set and denoted $\emptyset$ A *subset* of A is a set whose elements all belong to A. In the first example, $B = \{\text{couch, chair}\}$ is a subset of $A = \{\text{couch, chair, bed}\}$, written $B \subset A$, because each element in B also belongs to A. In the second example, $B = \{(q, \infty),\ q \text{ rational}\} \subset A = \{(t, \infty),\ t \in R\}$ because each element, in this case, set, in B is also in A. This second example involves infinitely many sets $A_t = (t, \infty)$, where t ranged over all real numbers or all rationals. More generally, consider a collection A_t of sets, where t ranges over some index set that could be countably or uncountably infinite (see Section 2.1).

Definition A.2. Intersection and union *If A_t, $t \in I$, is any collection of sets:*

1. *indexset!intersection The intersection $\cap A_t$ is the set of elements that belong to all of the A_t.*

2. *The union $\cup A_t$ is the set of elements that belong to at least one of the A_t.*

Thus, if A_1 is the set of rational numbers and $A_2 = [0, 1]$, then $A_1 \cap A_2$ is the set of rational numbers in the interval $[0, 1]$, whereas $A_1 \cup A_2$ is the set of numbers that are either rational or between 0 and 1. That is, $A_1 \cup A_2$ consists of all real numbers in $[0, 1]$, plus the rational numbers outside $[0, 1]$.

If $A_n = (0, 1 + 1/n)$ for $n = 1, 2, 3, \ldots$, then $\cap_{n=1}^{\infty} A_n$ consists of all numbers x that are in $(0, 1 + 1/n)$ for every $n = 1, 2, \ldots$ That is, x must be in $(0, 2)$ and $(0, 3/2)$ and $(0, 4/3)$, etc. The only numbers in all of these intervals are in $(0, 1]$. Therefore, $\cap_{n=1}^{\infty} A_n = (0, 1]$. On the other hand, $\cup_{n=1}^{\infty} A_n = (0, 2)$.

Remark A.3. *Throughout this book we will consider subsets of a set Ω called the sample space.*

If $A \subset \Omega$, the set of elements in Ω that are not in A is called the *complement* of A, denoted by A^C or by $\Omega \setminus A$. For instance, if Ω is the real line and A is the set of rational numbers, then A^C is the set of irrational numbers.

Proposition A.4. *If $A_t \subset \Omega$ for $t \in I$ and $B \subset \Omega$:*

1. **DeMorgan's law** $(\cap A_t)^C = \cup A_t^C$.

2. **DeMorgan's law** $(\cup A_t)^C = \cap A_t^C$.
3. $B \cap (\cup_s A_s) = \cup_s (B \cap A_s)$.
4. $B \cup (\cap_s A_s) = \cap_s (B \cup A_s)$.

The first two items are known as DeMorgan's laws. To prove item 1, we will show that every element of $(\cap A_t)^C$ is an element of $\cup A_t^C$, and vice versa. Let $x \in (\cap A_t)^C$. Then x is not in all of the A_t. In other words, there is at least one t such that x is outside of A_t, and hence $x \in A_t^C$ for at least one t. By definition, $x \in \cup A_t^C$. Therefore, $x \in (\cap A_t)^C \Rightarrow x \in \cup A_t^C$.

Now suppose that $x \in \cup A_t^C$. Then x is in A_t^C for at least one t. That is, x lies outside A_t for at least one t, so x cannot belong to $\cap A_t$. In other words, $x \in (\cap A_t)^C$. Therefore, $x \in \cup A_t^C \Rightarrow x \in (\cap A_t)^C$.

Having shown that every element of $(\cap A_t)^C$ is an element of $\cup A_t^C$ and vice versa, we conclude that the two sets are the same. This completes the proof of item 1. Items 2–4 can be proven in a similar way.

□

A.2.2 Sets of Real Numbers: Inf and Sup

Up to now we have been considering arbitrary sets, but here we restrict attention to sets of numbers. We would like to generalize the notions of minimum and maximum to infinite sets.

Definition A.5. Lower and upper bounds *A number b is said to be a lower bound of a set A of real numbers if $b \le x$ for each $x \in A$. A number B is said to be an upper bound of A if $x \le B$ for all $x \in A$.*

Thus, for example, -1 is a lower bound and 2 is an upper bound of $A = [0, 1)$ because each point in A is at least as large as -1 and no greater than 2.

Definition A.6. Infimum and supremum *If A is a set of real numbers:*

1. *b is said to be the greatest lower bound (glb) of A, also called infimum and denoted* $\inf(A)$, *if b is a lower bound and no number larger than b is a lower bound of A.*
2. *B is said to be the least upper bound (lup) of A, also called supremum and denoted* $\sup(A)$, *if B is an upper bound and no number smaller than B is an upper bound of A.*

It is easy to see from the definition that there can be no more than one infimum or supremum of a set. See Figure A.2 for an illustration of lower and upper bounds and the infimum and supremum.

Remark A.7. Infimum and supremum of the empty set *The infimum of the empty set is $+\infty$ because every number is a lower bound of the empty set. That is, if x is any number, it is vacuously true that $y \ge x$ for all $y \in \emptyset$. Similarly, the supremum of the empty set is $-\infty$ because every number is an upper bound of the empty set.*

Axiom A.8. *Every set with a lower bound has a greatest lower bound, and every set with an upper bound has a least upper bound.*

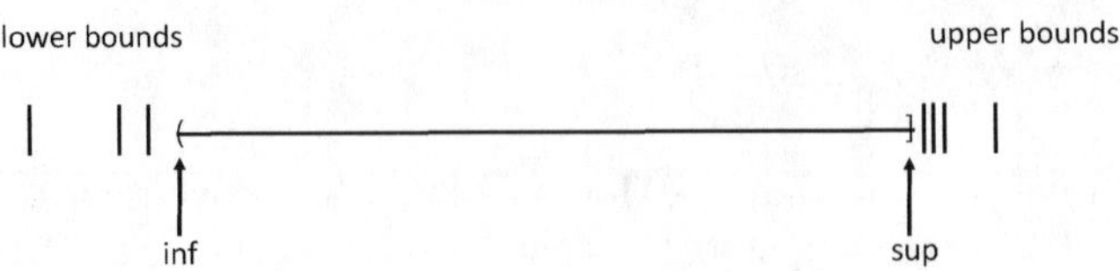

Figure A.2: A bounded set A has infinitely many lower and upper bounds (hash marks), but only one inf and one sup.

Notice that infs or sups may or may not be in the set. In the above example with $A = [0, 1)$, $\inf(A) = 0 \in A$, whereas $\sup(A) = 1 \notin A$. It is easy to see that the infimum and supremum of any finite set are the minimum and maximum, respectively. If a set A has no lower bound, then its infimum is $-\infty$, and if a set has no upper bound, its supremum is $+\infty$. Thus, if A is the set of integers, then $\inf(A) = -\infty$ and $\sup(A) = +\infty$.

A.3 A Touch of Topology of R^k

A.3.1 Open and Closed Sets

Some concepts from topology are very helpful in the study of probability. We are all familiar with an open interval (a, b). Its center is $c = (a + b)/2$, and (a, b) is the set of points x such that $|x - c| < r$, where $r = (b - a)/2$. The k-dimensional generalization of this region is an open ball (Figure A.3).

Definition A.9. Open ball *The open ball $B(\mathbf{c}, r)$ centered at $\mathbf{c} \in R^k$ with radius r is the set of points $\mathbf{x} \in R^k$ such that $||\mathbf{x} - \mathbf{c}|| < r$. The closed ball $\bar{B}(\mathbf{c}, r)$ replaces $<$ by $\leq$ in this definition.*

A generalization of an open ball is an open set defined as follows.

Definition A.10. Open set *A set O in R^k is said to be open if for each point $\mathbf{x} \in O$, there is a sufficiently small $\epsilon > 0$ such that $B(\mathbf{x}, \epsilon) \subset O$ (see Figure A.4) .*

An open ball $B(\mathbf{c}, r)$ is an open set. To see this, let $\mathbf{x} \in B(\mathbf{c}, r)$. We will show that there is a sufficiently small number ϵ such that $B(\mathbf{x}, \epsilon) \subset B(\mathbf{c}, r)$. Because $||\mathbf{x} - \mathbf{c}|| < r$, $||\mathbf{x} - \mathbf{c}|| = r - d$ for some $d > 0$. Set $\epsilon = d$. We will prove that $B(\mathbf{x}, d) \subset B(\mathbf{c}, r)$ by proving that if $\mathbf{z} \in B(\mathbf{x}, d)$, then $||\mathbf{z} - \mathbf{c}|| < r$ (Figure A.5).

$$\begin{aligned} ||\mathbf{z} - \mathbf{c}|| &= ||\mathbf{z} - \mathbf{x} + \mathbf{x} - \mathbf{c}|| \\ &\leq ||\mathbf{z} - \mathbf{x}|| + ||\mathbf{x} - \mathbf{c}|| \text{ (triangle inequality)} \\ &< d + (r - d) = r, \end{aligned} \quad (A.1)$$

completing the proof.

□

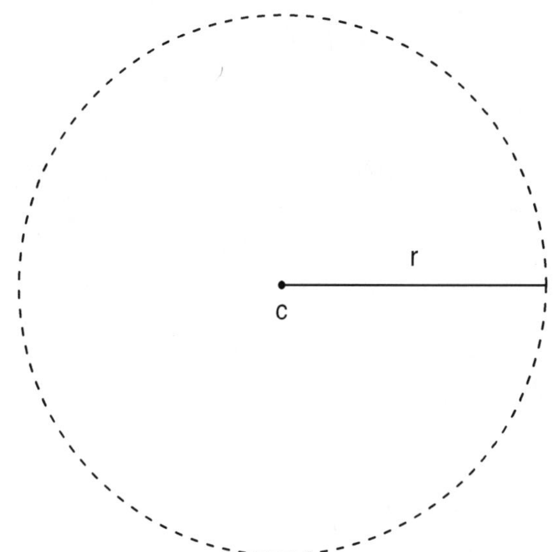

Figure A.3: The two-dimensional open ball $B(\mathbf{c}, r)$, whose circular boundary is not included.

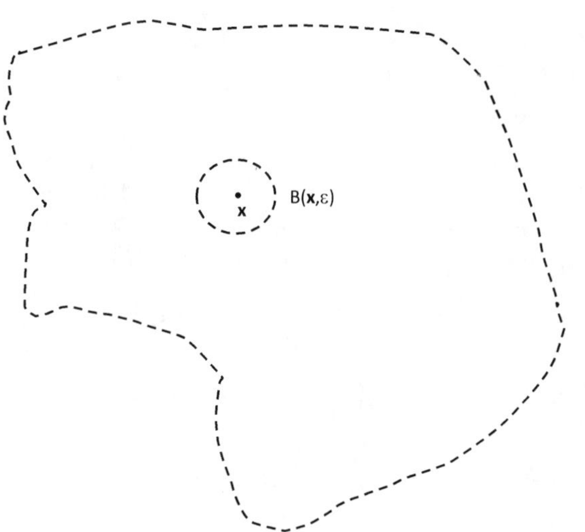

Figure A.4: An open set O. Each point $x \in O$ is in an open ball $B(\mathbf{x}, \epsilon)$ contained in O.

There are many more open sets than just open balls. For example, the union of open balls, even infinitely many, is also an open set. Also, the empty set $\emptyset$ is open because it satisfies the definition vacuously.

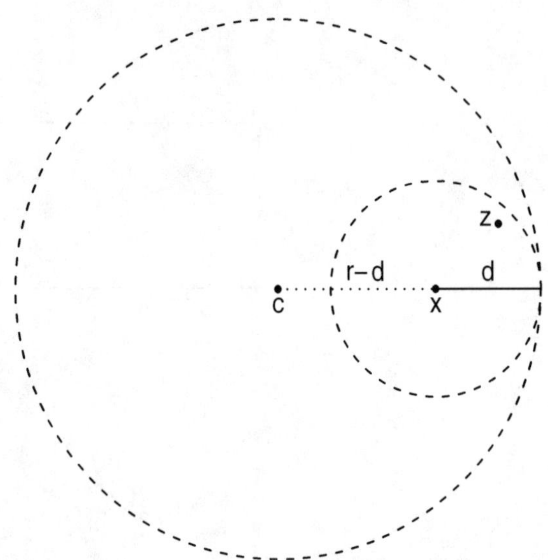

Figure A.5: Finding an ϵ such that $B(\mathbf{x},\epsilon) \subset B(\mathbf{c},r)$. If $||\mathbf{x}-\mathbf{c}|| = r-\delta$, set $\epsilon = \delta$.

An example of a set in R^1 that is not open is the half-open interval $[a,b)$. It is not open because there is no open ball $B(a,\epsilon)$, no matter how small ϵ is, that is entirely contained in $[a,b)$; there will always be points in $B(a,\epsilon)$ that are smaller than a. Likewise, $(a,b]$ and $[a,b]$ are not open. Another non-open set is a finite set of points such as $A = \{1,2,3\}$. Any open ball centered at one of these points will necessarily contain points not in A, so A is not open.

Is the intersection of an infinite number of open sets necessarily open? The intersection $\cap_{n=1}^{\infty} A_n$ of the open sets $A_n = (-1/n, 1+1/n)$ is the closed interval $[0,1]$, which is not open. Thus, the intersection of infinitely many open sets need not be open, although the intersection of a *finite* number of open sets *is* open; see Proposition A.14.

Let B be a subset of R^k. Any point in R^k is either in the interior, exterior, or boundary of B (Figure A.6), where these terms are defined as follows.

Definition A.11. Interior, exterior, and boundary points of a set

1. *Interior of* B*:* $\{\mathbf{x}$*: such that there is an open ball centered at* $\mathbf{x}$ *that is entirely contained in* $B\}$.
2. *Exterior of* B*:* $\{\mathbf{x}$*: such that there is an open ball centered at* $\mathbf{x}$ *that is entirely outside of* $B\}$.
3. *Boundary of* B*:* $\{\mathbf{x}$*: such that every open ball centered at* $\mathbf{x}$ *contains at least one point in* B *and one point outside* $B\}$.

For example, if $B = (a,b)$, then each point in B is in the interior of B, a and b are boundary points of B, and each point in $(-\infty,a) \cup (b,\infty)$ is in the exterior of B. In R^2, let B be the open ball $B(\mathbf{c},r)$. Then each $(x,y) \in B$ is an interior point of B; each (x,y) with

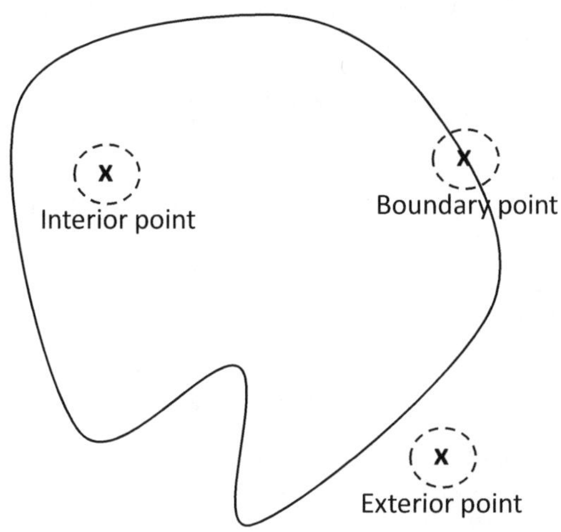

Figure A.6: Interior, exterior, and boundary points of a set.

$(x-c_1)^2+(y-c_2)^2=r^2$ is a boundary point of B; each (x,y) with $(x-c_1)^2+(y-c_2)^2>r^2$ is in the exterior of B.

Just as we generalized the idea of an open interval to an open set, we can generalize the idea of a closed interval $[a,b]$. Similarly to what we did above, we may regard this interval as the set of points x such that $|x-c|\leq r$, where $c=(a+b)/2$ and $r=(b-a)/2$. The analogy in R^k is the closed ball $\bar{B}(c,r)=\{\mathbf{x}:||\mathbf{x}-\mathbf{c}||\leq r\}$. Notice that the complement of this closed ball is the set of $\mathbf{x}$ such that $||\mathbf{x}-\mathbf{c}||>r$, which is an open set. This tells us how to generalize the notion of closed intervals and closed balls.

Definition A.12. Closed set *A set C of points in R^k is said to be closed if its complement is open.*

Any closed ball is closed. Moreover, any set A of R^k that contains all of its boundary points is closed. This is because each point in A^C is in the exterior of A, and therefore can be encased in an open ball in A^C.

Proposition A.13. Characterization of closed sets *A set C is closed if and only if it contains all of its boundary points.*

A set can be neither open nor closed. For example, we have seen that the half open interval $[a,b)$ is not open, but neither is it closed because the complement of $[a,b)$ is $(-\infty,a)\cup[b,\infty)$, which is not open. Note that $[a,b)$ contains only one of its two boundary points. More generally, a set that contains some, but not all, of its boundary points is neither open nor closed. Another example in R is the set Q of rational numbers. All real numbers are boundary points of Q, so Q contains some, but not all, of its boundary points. It is neither open nor closed. An example in R^2 is $A=\{(x,y):x^2+y^2\leq 1,\ x\in[0,1]\}\cup\{(x,y):x^2+y^2<1,\ x\in[-1,0)\}$. Note that A is the two-dimensional region bounded by the circle

$x^2+y^2=1$ with center $(0,0)$ and radius 1, and containing its right-side, but not its left-side, boundary. The set A is neither open nor closed.

There are precisely two sets in R^k that are both open and closed, namely the empty set $\emptyset$ and R^k itself.

Proposition A.14. Properties of open sets *Open sets have the following properties.*

1. *The union of any collection of open sets is open.*
2. *The intersection of a finite collection of open sets is open.*

Closed sets have the following properties.

1. *The union of any finite collection of closed sets is closed.*
2. *The intersection of any collection of closed sets is closed.*

Notice that the union of an infinite collection of closed sets need not be closed. For instance, $\cup[1/n,1]=(0,1]$.

The following result characterizing the open sets in R^1 is sometimes useful.

Proposition A.15. Characterization of open sets in R^1 *A set $A \in R^1$ is open if and only if it is the union of a countable (see Section 2.1 for definition of countable) collection of disjoint open intervals.*

Definition A.16. Cluster point *A point $\mathbf{x} \in R^k$ is said to be a cluster point, also called a point of accumulation, of a set A if every open ball containing $\mathbf{x}$ contains at least one point of A other than $\mathbf{x}$.*

Theorem A.17. Bolzano-Weierstrass theorem *Every bounded infinite set A in R^k contains a cluster point.*

The basic idea behind the proof of this result is illustrated in Figure A.7 for R^2. We encase A in a square and pick a point $\mathbf{x}_1 \in A$. Now divide the square into four equal cells. At least one of the cells must contain infinitely many points of A. From a cell with infinitely many points, pick another point $\mathbf{x}_2 \in A$. Now divide that cell into four equal cells. One of these must contain infinitely many points of A, so pick a point $\mathbf{x}_3 \in A$, etc. Continuing this process indefinitely. The intersection of all of the cells is nonempty and contains a cluster point $\mathbf{x}$.

A.3.2 Compact Sets

We have seen that closed sets are a generalization of closed intervals, but a closed interval $[a,b]$ has the additional property of being bounded, meaning that every $x \in [a,b]$ has $|x| \le r$ for some r. In particular, every $x \in [a,b]$ has $|x| \le \max(|a|,|b|)$. On the other hand, a closed set need not be bounded. For instance, the set of integers is closed (its complement is easily seen to be open) but not bounded.

To obtain a better generalization of a closed interval, note first that every subset of R^k has an *open covering*, meaning that it is contained within a union of open sets. For instance,

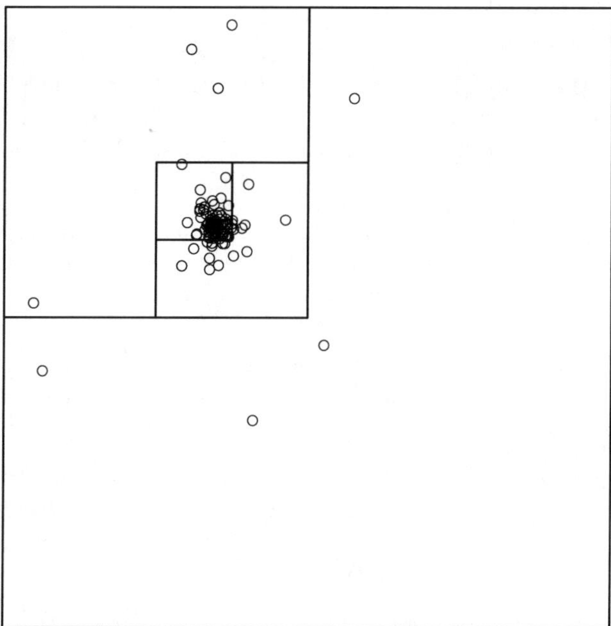

Figure A.7: Idea behind the proof of the Bolzano-Weierstrass theorem. We continue subdividing, each time choosing a cell that contains infinitely many points of A, and then picking a point from that cell. The intersection of all of the cells contains a cluster point.

if A is a subset of R^k, $\cup_{\mathbf{x}\in A}B(\mathbf{x},1)$ covers A; i.e., $A\subset\cup_{\mathbf{x}\in A}B(\mathbf{x},1)$. But for every such covering, can we find a finite subcovering? That is, can A be covered by a union of a finite number of the Bs? To see that the answer is no for some A, note that $A=(a,b)$ is covered by the sets $B_n=(a+1/n,b-1/n)$ because $\cup_{n=1}^{\infty}B_n=(a,b)$. On the other hand, no finite subcollection of the B_n covers (a,b). But it can be shown that the closed interval $[a,b]$ does have the property that every open covering has a finite subcovering. This provides us with the generalization we need of a closed interval.

Definition A.18. Compact set *A set $C\in R^k$ is called compact if every open covering of C has a finite subcovering.*

Compact sets are important because some key results about functions on closed intervals also hold more generally for functions on compact sets.

We close this subsection with an extremely useful characterization of compact sets in R^k.

Theorem A.19. Heine-Borel theorem *A set $C\in R^k$ is compact if and only if C is closed and bounded.*

A.4 Sequences in R^k

Limits are crucial to the study of probability theory, so it is essential to have a firm grasp of this topic. Loosely speaking, a sequence $(\mathbf{x}_n)$ of k-dimensional vectors has limit $\mathbf{x} \in R^k$ if the $\mathbf{x}_n$ are sufficiently close to $\mathbf{x}$ for all sufficiently large n. The following definition makes this more precise.

Definition A.20. Limit *A sequence* $(\mathbf{x}_n) \in R^k$ *is said to have limit* $\mathbf{x}$*, written* $\lim_{n\to\infty}(\mathbf{x}_n) = \mathbf{x}$ *or* $\mathbf{x}_n \to \mathbf{x}$*, if for each number* $\epsilon > 0$*, no matter how small, there is a natural number* $N = N_\epsilon$ *such that* $||\mathbf{x}_n - \mathbf{x}|| < \epsilon$ *whenever* $n \geq N$*.*

Thus, to show that $\mathbf{x}_n$ has limit $\mathbf{x}$, we must find, for any specified $\epsilon > 0$, an N such that from that N onward, all elements of the sequence are within a distance ϵ of $\mathbf{x}$.

For instance, suppose we want to prove that $(1/n)\sin(n\pi/2) \to 0$ using the definition of a limit. Given $\epsilon > 0$, we must specify an N such that $|(1/n)\sin(n\pi/2) - 0| < \epsilon$ whenever $n \geq N$. But $|(1/n)\sin(n\pi/2) - 0| = |(1/n)\sin(n\pi/2)| = |1/n|\,|\sin(n\pi/2)| \leq 1/n$ because $|\sin(x)| \leq 1$ for any x. For the given ϵ, choose N large enough that $1/N < \epsilon$. Then whenever $n \geq N$, $|(1/n)\sin(n\pi/2) - 0| \leq 1/n \leq 1/N < \epsilon$. By Definition A.20, $(1/n)\sin(n\pi/2) \to 0$.

Proposition A.21. Uniqueness of limits *Limits are unique; i.e., if* $\mathbf{x}_n \to \mathbf{x}$ *and* $\mathbf{x}_n \to \mathbf{x}'$*, then* $\mathbf{x} = \mathbf{x}'$*.*

Proof. Note that for any $\epsilon > 0$, there exist N_1 and N_2 such that $||\mathbf{x}_n - \mathbf{x}|| < \epsilon/2$ for $n \geq N_1$ and $||\mathbf{x}_n - \mathbf{x}'|| < \epsilon/2$ for $n \geq N_2$. Thus, if $N = \max(N_1, N_2)$

$$\begin{aligned} ||\mathbf{x} - \mathbf{x}'|| &= ||\mathbf{x} - \mathbf{x}_N + \mathbf{x}_N - \mathbf{x}'|| \\ &\leq ||\mathbf{x} - \mathbf{x}_N|| + ||\mathbf{x}_N - \mathbf{x}'|| \quad \text{(triangle inequality)} \\ &< \epsilon/2 + \epsilon/2 = \epsilon. \end{aligned} \tag{A.2}$$

Because ϵ is arbitrary, $\mathbf{x} = \mathbf{x}'$. □

An equivalent way to express the fact that $\mathbf{x}_n \to \mathbf{x}$ is the following.

Proposition A.22. Equivalent definition of convergence of $\mathbf{x}_n$ to $\mathbf{x}$ $\mathbf{x}_n \to \mathbf{x}$ *if and only if for each number* $\epsilon > 0$*, there are only finitely many* n *such that* $||\mathbf{x}_n - \mathbf{x}|| \geq \epsilon$*.*

To see that this is an equivalent definition, suppose first that $\mathbf{x} \to \mathbf{x}$ by Definition A.20. Then for given $\epsilon > 0$, there is an N such that $||\mathbf{x}_n - \mathbf{x}|| < \epsilon$ for $n \geq N$. That means that the only possible values n such that $||\mathbf{x}_n - \mathbf{x}|| \geq \epsilon$ are $n = 1, \ldots, N-1$. That is, there are only finitely many n such that $||\mathbf{x}_n - \mathbf{x}|| \geq \epsilon$. To prove the reverse implication, suppose for each ϵ there are only finitely many n such that $||\mathbf{x}_n - \mathbf{x}|| \geq \epsilon$. Then $M = \max\{n : ||\mathbf{x}_n - \mathbf{x}|| \geq \epsilon\}$ is finite. Set $N = M + 1$. Then for the given ϵ and $n \geq N$, $||\mathbf{x}_n - \mathbf{x}|| < \epsilon$. Therefore $\mathbf{x}_n \to \mathbf{x}$ by Definition A.20.

□

Proposition A.23. *In either Definition A.20 or Proposition A.22, we can replace* ϵ *by* $1/k$ *and require the stated condition to hold for each natural number* k*.*

Proposition A.23 holds because for fixed $\epsilon > 0$, there is a natural number k such that $1/k < \epsilon$, and for fixed k, there is an ϵ with $\epsilon < 1/k$.

It is also helpful to think about the negation of $\mathbf{x}_n \to \mathbf{x}$, which by Proposition A.22 is the negation of the statement that for each number ϵ, there are only finitely many n such that $||\mathbf{x}_n - \mathbf{x}|| \geq \epsilon$. The negation of "for each $\epsilon > 0$, condition C holds" is that there exists an $\epsilon > 0$ such that C does not hold. Therefore, $\mathbf{x}_n$ does not converge to $\mathbf{x}$ if and only if there exists an $\epsilon > 0$ such that $||\mathbf{x}_n - \mathbf{x}|| \geq \epsilon$ for infinitely many n. Let n_1 be the smallest n such that $||\mathbf{x}_n - \mathbf{x}|| \geq \epsilon$, n_2 be the second smallest such n, etc. Then for the subsequence $x_{n_1}, x_{n_2}, \ldots$, $||\mathbf{x}_{n_i} - \mathbf{x}|| \geq \epsilon$ for all $i = 1, 2, \ldots$

Proposition A.24. The negation of $\mathbf{x}_n \to \mathbf{x}$ *The negation of $\mathbf{x}_n \to \mathbf{x}$ is that there exists an $\epsilon > 0$ and a subsequence $x_{n_1}, x_{n_2}, \ldots$ such that $||\mathbf{x}_{n_i} - \mathbf{x}|| \geq \epsilon$ for all $i = 1, 2, \ldots$*

Proposition A.25. Convergence of vectors is equivalent to component-wise converge *If $\mathbf{x}_n \in R^k$, then $\mathbf{x}_n \to \mathbf{x} \in R^k$ if and only if each component of $\mathbf{x}_n$ converges to the corresponding component of $\mathbf{x}$.*

We encourage the reader to supply the details of the proof.

Proposition A.26. Convergence sequences are bounded *If $\mathbf{x}_n \to \mathbf{x} \in R^k$, then $\mathbf{x}_n$ is bounded; i.e., there exists a number B such that $||\mathbf{x}_n|| \leq B$ for all n.*

To prove Proposition A.26, note that because $\mathbf{x}_n \to \mathbf{x}$ as $n \to \infty$, there is a natural number M such that $||\mathbf{x}_n - \mathbf{x}|| < 1$ for $n \geq M$. This follows from the definition of a limit with $\epsilon = 1$. This fact and the reverse triangle inequality imply that $||\mathbf{x}_n|| - ||\mathbf{x}|| \leq ||\mathbf{x}_n - \mathbf{x}|| < 1$ for $n \geq M$. That is, $||\mathbf{x}_n|| \leq 1 + ||\mathbf{x}||$ for $n \geq M$. If we take $B = \max\{||\mathbf{x}_1||, \ldots, |\mathbf{x}_{M-1}||, ||\mathbf{x}|| + 1\}$, then $||\mathbf{x}_n|| \leq B$ for all n.

□

Proposition A.27. Elementary properties of limits *Suppose $\mathbf{x}_n \to x \in R^k$ and $\mathbf{y}_n \to y \in R^k$, then:*

1. $\mathbf{x}_n \pm \mathbf{y}_n \to \mathbf{x} \pm \mathbf{y}$.
2. $\mathbf{x}_n \cdot \mathbf{y}_n \to \mathbf{x} \cdot \mathbf{y}$, *where $\cdot$ denotes dot product.*
3. *If $k = 1$, $x_n/y_n \to x/y$ if $y \neq 0$.*

We illustrate the use of the triangle inequality by proving the second item. By item 1 and Proposition A.25, it suffices to prove the result for $k = 1$. Start with $|x_n y_n - xy| = |x_n y_n - xy_n + xy_n - xy|$. Application of the triangle inequality yields

$$\begin{aligned} |x_n y_n - xy| &\leq |x_n y_n - xy_n| + |xy_n - xy| \\ &= |(x_n - x)y_n| + |x(y_n - y)| \\ &= |x_n - x||y_n| + |x||y_n - y|. \end{aligned} \tag{A.3}$$

By Proposition A.26, there is a bound B such that $|x_n y_n - xy| \leq B|x_n - x| + |x||y_n - y|$ for all n. For $\epsilon > 0$, we must find an N such that $B|x_n - x| + |x||y_n - y| < \epsilon$ for $n \geq N$. Suppose first that $x \neq 0$. Because $x_n \to x$ and $y_n \to y$, there exists an N_1 such that $|x_n - x| < \epsilon/(2B)$ for $n \geq N_1$, and an N_2 such that $|y_n - y| < \epsilon/(2|x|)$ for $n \geq N_2$. Take $N = \max(N_1, N_2)$. Then for $n \geq N$, $B|x_n - x| + |x||y_n - y| < B\epsilon/(2B) + |x|\epsilon/(2|x|) = \epsilon/2 + \epsilon/2 = \epsilon$. If $x = 0$, we can take $N = N_1$. This shows that $|x_n y_n - xy| < \epsilon$ for $n \geq N$.

□

Readers unfamiliar with proving things like Proposition A.27 are encouraged to prove the remaining parts.

Definition A.28. Infinite limits *If (x_n) is a sequence in R^1, we say $x_n \to \infty$ as $n \to \infty$ if for every $A > 0$, there is a natural number N such that $x_n > A$ for all $n \geq N$. We say $x_n \to -\infty$ if for every $A < 0$, there is a natural number N such that $x_n < A$ for all $n \geq N$.*

Many sequences have no limits. For example, in R^1:

1. Let $x_n = (-1)^n$. Then x_n oscillates indefinitely between -1 and 1, so it cannot have a limit.

2. Let $x_n = \sin(n\pi/2)$. Then x_n oscillates indefinitely between 1, 0, and -1, again with no limit.

In the examples above, although the sequences oscillate and do not converge, we can find subsequences that converge. For instance, along the subsequence $n_1 = 1, n_2 = 5, n_3 = 9, \ldots$, $x_n = \sin(n\pi/2)$ is $(1, 1, \ldots, 1, \ldots)$ Therefore, $\lim_{i\to\infty} x_{n_i} = 1$. Similarly, along the subsequence $n_1 = 3, n_2 = 7, n_3 = 11, \ldots$, $x_{n_i} \to -1$ as $i \to \infty$, while for $n_1 = 2, n_2 = 4, n_3 = 6, \ldots$, $x_{n_i} \to 0$ as $i \to \infty$.

Definition A.29. Limit point $\mathbf{x}$ *is a limit point of* $(\mathbf{x}_n)$ *if there exists a subsequence* $n_1, n_2, \ldots$ *such that* $\lim_{i\to\infty}(\mathbf{x}_{n_i}) = \mathbf{x}$.

The smallest and largest limit points of a sequence are defined as follows.

Definition A.30. Liminf and limsup *The limit inferior of a sequence (x_n), denoted $\underline{\lim}\,(x_n)$, is $\inf\{x^* : x^*$ is a limit point of $(x_n)\}$. The limit superior of a sequence (x_n), denoted $\overline{\lim}\,(x_n)$, is $\sup\{x^* : x^*$ is a limit point of $(x_n)\}$ (See Figure A.8).*

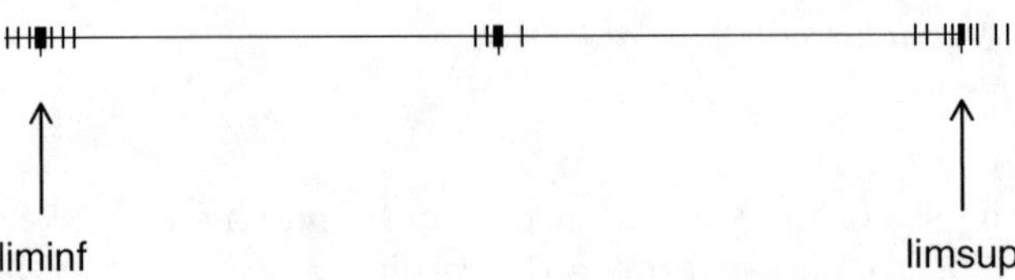

Figure A.8: The sequence depicted has 3 limit points. The smallest and largest of these are the liminf and limsup, respectively.

Some properties of liminfs and limsups that readers unfamiliar with this material should verify are as follows.

Proposition A.31. Elementary facts about liminfs and limsups

1. *Liminfs and limsups always exist, though they might be infinite.*

2. $\underline{\lim}(-x_n) = -\overline{\lim}(x_n)$.

3. *For $c \geq 0$, $\underline{\lim}(cx_n) = c\,\underline{\lim}(x_n)$ and $\overline{\lim}(cx_n) = c\,\overline{\lim}(x_n)$.*

4. $\overline{\lim}(x_n + y_n) \leq \overline{\lim}(x_n) + \overline{\lim}(y_n)$.

5. $\underline{\lim}(x_n + y_n) \geq \underline{\lim}(x_n) + \underline{\lim}(y_n)$.

6. *If* $x_n \leq y_n$ *for all* $n \geq N$ *for some* N, *then* $\underline{\lim}(x_n) \leq \underline{\lim}(y_n)$ *and* $\overline{\lim}(x_n) \leq \overline{\lim}(y_n)$.

7. $\underline{\lim}(x_n) = x$ *if and only if for each* $\epsilon > 0$, *there are infinitely many* n *such that* $x_n < x + \epsilon$, *but only finitely many* n *such that* $x_n < x - \epsilon$.

8. $\overline{\lim}(x_n) = x$ *if and only if for each* $\epsilon > 0$, *there are infinitely many* n *such that* $x_n > x - \epsilon$, *but only finitely many* n *such that* $x_n > x + \epsilon$.

Working with liminfs and limsups means that we no longer must qualify statements like "let $x = \lim(x_n)$" by "if the limit exits." Instead, we can work directly with liminfs and limsups, which always exist, and then use the following result.

Proposition A.32. Limit exists if and only if the liminf and limsup agree $\lim(x_n)$ *exists if and only if* $\underline{lim}(x_n) = \overline{lim}(x_n) = x$ *(where* x *could be* $\pm\infty$*) in which case* $\lim_{n\to\infty}(x_n) = x$.

Monotone sequences (x_n) (i.e., $x_n \leq x_{n+1}$ for all n or $x_n \geq x_{n+1}$ for all n) cannot exhibit the oscillating behavior we observed with $(-1)^n$ or $\sin(n\pi/2)$.

Proposition A.33. Monotone sequences have a (finite or infinite) limit *Every monotone sequence* x_n *has a finite or infinite limit.*

This follows from the Bolzano-Weierstrass theorem because if the sequence is bounded and contains infinitely many values, there must be a cluster point, which is the limit (there cannot be more than one cluster point for a monotone sequence). On the other hand, if the sequence is unbounded, then there is an infinite limit.

We conclude with a useful necessary and sufficient condition for a sequence to converge to a finite limit.

Definition A.34. Cauchy sequence *A sequence* $\mathbf{x}_n$ *is said to be a Cauchy sequence if for each* $\epsilon > 0$ *there exists a natural number* N *such that* $||\mathbf{x}_n - \mathbf{x}_m|| < \epsilon$ *whenever* $m, n \geq N$.

In other words, the terms of the sequence are all arbitrarily close to (within ϵ of) each other from some point onward.

Proposition A.35. A sequence is Cauchy if and only if it is convergent *A sequence* $\mathbf{x}_n \in R^k$ *is Cauchy if and only if there exists an* $\mathbf{x} \in R^k$ *such that* $\mathbf{x}_n \to \mathbf{x}$.

A.5 Series

An important part of probability theory involves infinite sums of random variables or probabilities. Therefore, we need to have a working knowledge of tools that will help us determine whether these infinite sums, called infinite series, exist and are finite.

Definition A.36. Infinite series, their convergence and divergence *If* (x_i) *is a sequence of real numbers, we define* $\sum_{i=1}^{\infty} x_i$ *to be* $\lim_{n\to\infty} \sum_{i=1}^{n} x_i$, *provided this limit exists. If the limit is finite, we say that* $\sum_{i=1}^{\infty} x_i$ *is convergent. If there is no limit or the limit is infinite, we say that* $\sum_{i=1}^{\infty} x_i$ *is divergent.*

Proposition A.37. Convergence of series implies that its terms tend to 0 *If $\sum_{i=1}^{\infty} x_i$ is convergent, then $x_i \to 0$ as $i \to \infty$. Therefore, if x_i does not converge to 0, then $\sum_{i=1}^{\infty} x_i$ is divergent.*

This follows from the fact that if $\sum x_i$ converges to a number s, then $x_n = \sum_{i=1}^{n} x_i - \sum_{i=1}^{n-1} x_i \to s - s = 0$ as $n \to \infty$.

Two classes of infinite series are particularly useful, the geometric series because we can compute the sum explicitly, and the Riemann-Zeta series because it includes, as a special case, one of the best known divergent series, the harmonic series.

Proposition A.38. Geometric series *$\sum_{i=1}^{\infty} r^i$ converges if and only if $|r| < 1$, in which case $\sum_{i=1}^{\infty} r^i = r/(1-r)$.*

Proposition A.39. Riemann-Zeta series *$\sum_{i=1}^{\infty}(1/i^r)$ is convergent if and only if $r > 1$. With $r = 2$, $\sum_{i=1}^{\infty}(1/i^2) = \pi^2/6$.*

The case $r = 1$ is known as the harmonic series. To see that it is divergent, write the sum as $1+(1/2)+(1/3+1/4)+(1/5+1/6+1/7+1/8)+\ldots$ and note that all of the terms in parentheses are at least 1/2 because $1/3+1/4 \geq 1/4+1/4 = 1/2$, $(1/5+1/6+1/7+1/8) \geq 1/8+1/8+1/8+1/8 = 1/2$, etc. Thus, $\sum_{i=1}^{\infty}(1/i) \geq 1 + 1/2 + 1/2 + \ldots + 1/2 + \ldots = \infty$.

Even though the harmonic series diverges, the alternating series $\sum_{i=1}^{\infty}(-1)^i(1/i)$ converges. This follows from Proposition A.40:

Proposition A.40. Convergence of alternating series *If (x_i) is a decreasing sequence of positive numbers with $x_i \to 0$. Then the alternating series $\sum_{i=1}^{\infty}(-1)^i x_i$ is convergent.*

Proposition A.41. The integral test *If $x_i = f(i)$, where f is nonnegative, decreasing, and continuous, then $\sum_{i=1}^{\infty} x_i$ is convergent if and only if $\int_1^{\infty} f(x)dx < \infty$. In fact, $\sum_{i=2}^{\infty} x_i \leq \int_1^{\infty} f(x)dx \leq \sum_{i=1}^{\infty} x_i$ (see Figure A.9).*

This provides another way to see that the harmonic series diverges because $\int_1^{\infty}(1/x)dx = \ln(x)|_1^{\infty} = \infty$.

Definition A.42. Absolute convergence *A series $\sum_{i=1}^{\infty} x_i$ is said to be absolutely convergent if $\sum_{i=1}^{\infty} |x_i|$ is convergent.*

Absolute convergence is important for several reasons, one of which is the following.

Proposition A.43. Absolute convergence implies convergence *If $\sum_{i=1}^{\infty} x_i$ is absolutely convergent, then it is convergent.*

Another important consequence of absolute convergence is that the sum remains the same even if we rearrange the terms. Let $\pi_1, \pi_2, \ldots$ be a rearrangement of $(1, 2, 3, \ldots)$, meaning that each natural number $1, 2, \ldots$ appears exactly once among $\pi_1, \pi_2, \ldots$ If a series $\sum_{i=1}^{\infty} x_i$ converges but not absolutely, then for any number c, we can rearrange the terms to make $\sum_{i=1}^{\infty} x_{\pi_i} = c$. A heuristic argument for this is the following. If the series converges, but not absolutely, then the sum of the positive terms must be ∞ and the sum of the negative terms must be $-\infty$ (if only one of these sums were infinite, then the series would not be convergent, whereas if neither were infinite, then the series would be absolutely convergent). Take positive terms until the sum exceeds c, then add negative terms until the sum is less than c, then positive terms until the sum exceeds c, etc. Because $x_i \to 0$, this process makes $\sum_{i=1}^{\infty} x_{\pi_i} = c$. Thus, we can get literally any sum by rearranging terms.

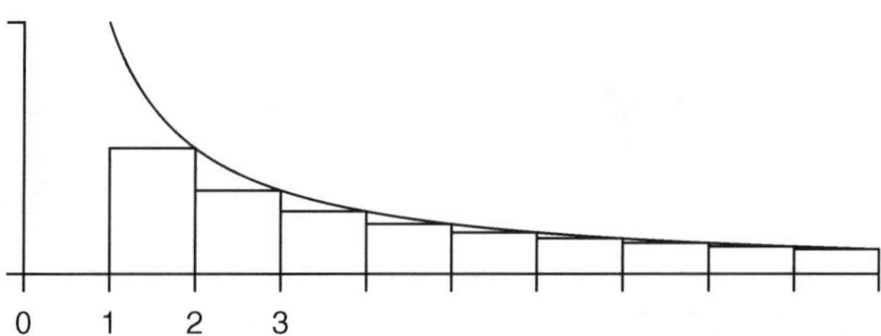

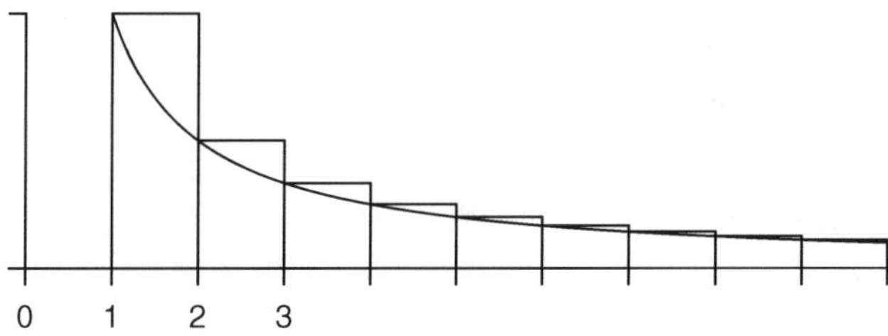

Figure A.9: The integral $\int_1^\infty f(x)dx$ is trapped between $\sum_{i=2}^\infty f(i)$ (area of bars in top graph) and $\sum_{i=1}^\infty f(i)$ (area of bars in bottom graph). It follows that $\sum_{i=1}^\infty f(i) < \infty$ if and only if $\int_1^\infty f(x)dx < \infty$.

Some rearrangements result in non-convergent sums as well. These anomalies do not occur for absolutely convergent sums, as we see in Proposition A.44.

Proposition A.44. Absolute convergence implies invariance to rearrangement of terms *If $\sum_{i=1}^\infty x_i$ is absolutely convergent, then $\sum_{i=1}^\infty x_{\pi_i}$ is convergent and has the same value for any rearrangement $\pi_1, \ldots, \pi_n, \ldots$ of the terms.*

Next we present useful criteria for determining whether a series converges absolutely. Take the absolute value of the terms and then apply the following result.

Proposition A.45. Some tests for convergence of infinite series *If x_i and y_i are nonnegative, then:*

1. *If $\lim(x_i/y_i)$ exists and is nonzero, then $\sum_{i=1}^\infty x_i$ converges if and only if $\sum_{i=1}^\infty y_i$ converges.*

2. **Comparison test** *If $x_i \le y_i$ for all $i \ge N$ and some N, then: $\sum_{i=1}^\infty x_i$ converges if $\sum_{i=1}^\infty y_i$ converges. This also implies that $\sum_{i=1}^\infty y_i$ diverges if $\sum_{i=1}^\infty x_i$ diverges.*

3. **Ratio test**

   (a) *Suppose that $x_{i+1}/x_i \to r$ or $x_{i+1}/x_i \le r$ for all $i \ge N$, where $r < 1$. Then $\sum_{i=1}^\infty x_i$ is convergent.*

(b) Suppose that $x_{i+1}/x_i \to r$ or $x_{i+1}/x_i \geq r$ for all $i \geq N$, where $r > 1$. Then $\sum_{i=1}^{\infty} x_i$ is divergent.

An example of the use of Proposition A.45 is the following. We know that $\sum_{i=1}^{\infty}(1/i^2)$ converges and $\{i(i-1)\}^{-1}/i^{-2} \to 1$. Therefore, item 1 of Proposition A.45 implies that $\sum_{i=1}^{\infty}\{i(i-1)\}^{-1} < \infty$.

Next we discuss some fundamental results about power series, which arise in a number of ways in probability and statistics. For example, we often use a first order Taylor approximation to obtain the asymptotic distribution of an estimator. This technique is called the delta method. You may have also encountered power series in elementary probability through generating functions.

Definition A.46. Power series *$P(x)$ is said to be a power series if $P(x) = \sum_{i=0}^{\infty} a_i(x-c)^i$ for constants c and a_i, $i = 0, 1, \ldots$*

Theorem A.47. Cauchy-Hadamard theorem *There is a finite or infinite value r, called the radius of convergence of $P(x)$, such that $P(x)$ converges absolutely if $|x-c| < r$ and diverges if $|x-c| > r$. In fact, $r = 1/\overline{\lim}\,\{|a_n|^{1/n}\}$.*

Proposition A.48. Uniqueness and term-by-term differentiation and integration *A power series $P(x) = \sum_{i=0}^{\infty} a_i(x-c)^i$:*

1. *Is differentiable for x in the radius of convergence. The derivative $P'(x)$ is $\sum_{i=0}^{\infty} i a_i (x-c)^{i-1}$, and it has the same radius of convergence as $P(x)$.*

2. *Can by integrated over any closed interval $[a, b]$ contained in the radius of convergence, and $\int_a^b P(x)dx = \sum_{i=0}^{\infty} a_i\{(b-c)^{i+1} - (a-c)^{i+1}\}/(i+1)$.*

3. *Is unique in the sense that if $\sum_{i=0}^{\infty} a_i(x-c)^i = \sum_{i=0}^{\infty} b_i(x-c)^i$ for all x in an interval $(c-\epsilon, c+\epsilon)$ for some $\epsilon > 0$, then $a_i = b_i$ for all $i = 0, 1, \ldots$*

If $f(x)$ is a function whose nth derivative at $x = c$ (denoted $f^{(n)}(c)$) exists, the nth order Taylor polynomial for f expanded about c is defined to be $T(x, n, c) = \sum_{i=0}^{n} f^{(i)}(c)(x-c)^i/i!$, where $f^{(0)}(c)$ is defined as $f(c)$.

Theorem A.49. Taylor's theorem *Suppose that $f^{(n)}(c)$ exists. Then $f(x) = T(x, n, c) + R_n$, where the remainder term R_n satisfies $R_n/|x-c|^n \to 0$ as $x \to c$. If additionally, $f^{(n+1)}(x)$ exists for $x \in I = [a, b]$ and $c \in [a, b]$, then $R_n = f^{(n+1)}(\eta)(x-c)^{(n+1)}/(n+1)!$ for some η lying on the line segment joining x and c.*

A.6 Functions

A.6.1 Mappings

If $f(x)$ is a function, we use the notation $f : X \longmapsto Y$ to denote that f maps the set X *into* Y, meaning that $f(x) \in Y$ for each $x \in X$. This does not necessarily mean that each point of Y is the image of some point in X. For instance, $f(x) = \sin(x)$ maps R into R, even though $\sin(x)$ is always between -1 and 1. Therefore, we could also write $f : R \longmapsto [-1, 1]$. Restricting Y to the range of the function—the set of image points $\{f(x) : x \in X\}$—means that every point $y \in Y$ is the image of some point in X.

Definition A.50. Onto and $1-1$ functions

1. *If every point $y \in Y$ is the image of at least one point $x \in X$, we say that f maps X onto Y.*

2. *If no $y \in Y$ is the image of more than one $x \in X$, f is said to be $1-1$.*

Theorem A.51. Inverse theorem *If f is a $1-1$ function from X onto Y, then there is a unique inverse function f^{-1} from Y onto X, namely $f^{-1}(y)$ is the unique x value whose image is y.*

Definition A.52. Direct and indirect images of sets *Let $f(x)$ be a function from a set X into another set Y. If $A \subset X$, then $f(A) = \{f(x) : x \in A\}$, the set of images of points in A. If $B \subset Y$, then $f^{-1}(B) = \{x \in X : f(x) \in B\}$, the set of points that get mapped into B.*

Proposition A.53. f^{-1} preserves unions, intersections, and complements *If $f : X \longmapsto Y$ and A_s is any collection of subsets of Y:*

1. $f^{-1}(\cup_s A_s) = \cup_s\{f^{-1}(A_s)\}$.

2. $f^{-1}(\cap_s A_s) = \cap_s\{f^{-1}(A_s)\}$.

3. $f^{-1}(A_s^C) = \{f^{-1}(A_s)\}^C = X \setminus f^{-1}(A_s)$.

A.6.2 Limits and Continuity of Functions

If $f(\mathbf{x}) : R^k \longmapsto R^p$ is a function, one way to define $\lim_{\mathbf{x}\to\mathbf{x}_0} f(\mathbf{x})$ is to consider each sequence $\mathbf{x}_n \to \mathbf{x}_0$ with $\mathbf{x}_n \neq \mathbf{x}_0$ and use the definition of limits of sequences.

Definition A.54. Limit of a function as $\mathbf{x} \to \mathbf{x}_0$ *The function $f(\mathbf{x}) : R^k \longmapsto R^p$ is said to have limit $\mathbf{L}$ as $\mathbf{x} \to \mathbf{x}_0$, written $\lim_{\mathbf{x}\to\mathbf{x}_0} f(\mathbf{x}) = \mathbf{L}$ or $f(\mathbf{x}) \to \mathbf{L}$ as $\mathbf{x} \to \mathbf{x}_0$, if $\mathbf{y}_n = f(\mathbf{x}_n) \to \mathbf{L}$ for each sequence $(\mathbf{x}_n)$ converging to $\mathbf{x}_0$ such that $\mathbf{x}_n \neq \mathbf{x}_0$ for any n,.*

Definition A.55. Left-handed and right-handed limits *If $f(x) : R^1 \longmapsto R^p$, the left-handed limit $\lim_{x\to x_0^-} f(x) = \mathbf{L}$ means that $\mathbf{y}_n = f(x_n) \to \mathbf{L}$ for each sequence (x_n) such that $x_n < x_0$ and $x_n \to x_0$. Similarly, the right-handed limit $\lim_{x\to x_0^+} f(x) = \mathbf{L}$ means that $\mathbf{y}_n = f(x_n) \to \mathbf{L}$ for each sequence (x_n) such that $x_n > x_0$ and $x_n \to x_0$.*

Remark A.56. *In verifying left-handed or right-handed limits, we can assume that the sequence x_n converging to $\mathbf{x}$ is monotone. For instance, to prove that $f(x_n) \to \mathbf{L}$ for every sequence $x_n \to x^-$, we need only show that $f(x_n) \to L$ for every increasing sequence $x_n \uparrow x$. The reader is encouraged to supply a proof of this fact.*

Proposition A.57. Limit exists if and only if left- and right-handed limit exists and are equal *If $f : R^1 \longmapsto R^p$, then $\lim_{x\to x_0} f(x) = \mathbf{L}$ if and only if $\lim_{x\uparrow x_0} f(x) = \mathbf{L}$ and $\lim_{x\downarrow x_0} f(x)$
$= \mathbf{L}$.*

It is sometimes convenient to use the following equivalent definition of $\lim_{\mathbf{x}\to\mathbf{x}_0} f(\mathbf{x}) = \mathbf{L}$.

Proposition A.58. Equivalent condition for convergence of a function as $\mathbf{x} \to \mathbf{x}_0$ *$\lim_{\mathbf{x}\to\mathbf{x}_0} f(\mathbf{x}) = \mathbf{L}$ if and only if for each $\epsilon > 0$, there exists a $\delta = \delta_{\mathbf{x}_0,\epsilon}$ such that $||f(\mathbf{x}) - \mathbf{L}|| < \epsilon$ whenever $0 < ||\mathbf{x} - \mathbf{x}_0|| < \delta$.*

That is, $f(\mathbf{x})$ can be made arbitrarily close to (within ϵ of) $\mathbf{L}$ if $\mathbf{x}$ is sufficiently close to (within δ of), but not equal to, $\mathbf{x}_0$. Note that the required δ may depend on $\mathbf{x}_0$ and ϵ.

Definition A.59. Continuous, left-continuous, right-continuous *A function $f(x) : R^k \longmapsto R^p$ is said to be continuous at $\mathbf{x} = \mathbf{x}_0$ if $\lim_{\mathbf{x}\to\mathbf{x}_0} f(\mathbf{x}) = f(\mathbf{x}_0)$. If $k = 1$, $f(x)$ is left-continuous at $x = x_0$ if $\lim_{x\uparrow x_0} f(x) = f(x_0)$, and right-continuous at $x = x_0$ if $\lim_{x\downarrow x_0} f(x) = f(x_0)$.*

Proposition A.60. Continuous at a point if and only if left and right continuous *A function $f(x) : R^1 \longmapsto R^p$ is continuous at x_0 if and only if $f(x)$ is left-continuous and right-continuous at x_0.*

Again the δ required to make $f(\mathbf{x})$ close to $f(\mathbf{x}_0)$ may depend on $\mathbf{x}_0$. If we can find, for each $\epsilon > 0$, a single δ that works for each $\mathbf{x}_0$, we call f uniformly continuous.

Definition A.61. Uniform continuity *A function $f(\mathbf{x}) : R^k \longmapsto R^p$ is said to be uniformly continuous on a set C if for each $\epsilon > 0$, there exists a $\delta > 0$ such that $||f(\mathbf{x}) - f(\mathbf{x}_0)|| < \epsilon$ for all $\mathbf{x} \in C$ and $\mathbf{x}_0 \in C$ such that $||\mathbf{x} - \mathbf{x}_0|| < \delta$.*

Figure A.10 illustrates the difference between pointwise and uniform continuity for $f(x) = 1/x$. Continuity at x' means that for any given $\epsilon > 0$, we can find a horizontal interval I_h of length $2\delta'$ centered at x' such that whenever $x \in I_h$, $f(x) \in I_v$, a vertical interval of length 2ϵ centered at $f(x')$. For $f(x) = 1/x$, the width of I_h ensuring $f(x) \in I_v$ is smaller at x' than at $x'' > x'$. Over a restricted domain $A \le x \le 1$, $A > 0$, we can find a single δ that works for all x because $f(x)$ does not become arbitrarily steep as $x \to A$. But over the domain $0 < x \le 1$, we can find no single δ that works for all x because $f(x)$ is arbitrarily steep as $x \to 0$. Thus, f is uniformly continuous on $[A, 1]$, but not on $(0, 1]$.

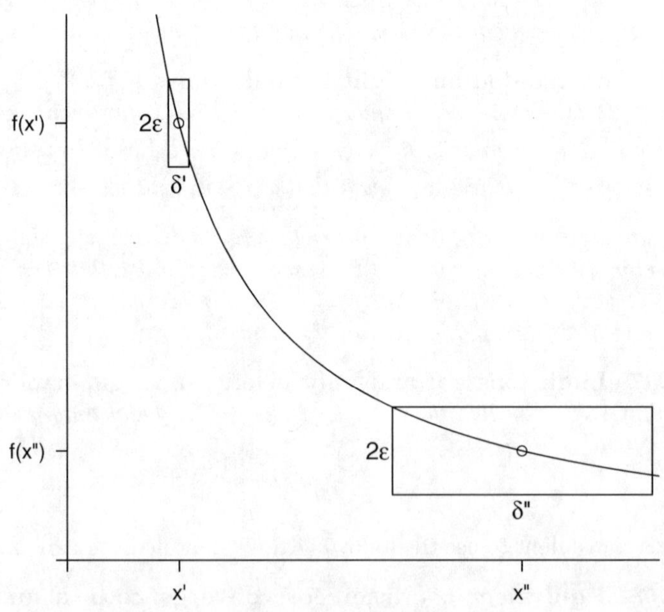

Figure A.10: Continuity of $f(x) = 1/x$.

Figure A.10 shows that a function f can be continuous on an open interval, yet still be arbitrarily steep as we approach an endpoint of the interval. This makes it impossible to find a single δ that works for all $\mathbf{x}$. This kind of anomalous behavior cannot happen on a closed interval. More generally, it cannot happen on a compact set, as we see in Proposition A.62 below.

Proposition A.62. Continuity on a compact set implies uniform continuity on that set *If f is continuous on a compact set C, then f is uniformly continuous on C.*

Definition A.63. Pointwise and uniform convergence *A sequence $f_n(\mathbf{x}) : R^k \longmapsto R^p$ converges to $f(\mathbf{x})$:*

1. *Pointwise on a set C if the sequence $f_n(\mathbf{x})$ converges to $f(\mathbf{x})$ as $n \to \infty$ for each $\mathbf{x} \in C$.*

2. *Uniformly on a set C if for each $\epsilon > 0$, there exists an N such that $||f_n(\mathbf{x}) - f(\mathbf{x})|| < \epsilon$ for all $n \geq N$ and all $\mathbf{x} \in C$.*

Example A.64. Pointwise, but not uniform convergence An example of a function that converges pointwise, but not uniformly, on the interval $[0, 1]$ is $f_n(x) = x^n$, which converges to $f(x) = 0$ if $0 \leq x < 1$ and 1 if $x = 1$. If the convergence were uniform, then for any given $\epsilon > 0$, $|f_n(x) - f(x)| = |x^n - 0| = x^n$ would have to be smaller than ϵ for all $n \geq N$ and all $x \in [0, 1]$. But for any given $n \geq N$, $|f_n(x) - f(x)| = x^n$ approaches 1 as $x \to 1$. Therefore, it cannot be less than ϵ for all $x \in [0, 1]$. Thus, f_n converges pointwise, but not uniformly, to $f(x)$ on $[0, 1]$.

Note that in this example, even though $f_n(\mathbf{x})$ is continuous for each n and $f_n(\mathbf{x}) \to f(\mathbf{x})$, the limit function $f(x)$ is discontinuous. This could not have happened if the convergence of f_n to f had been uniform, as the following result shows. □

Proposition A.65. Continuous functions converging uniformly implies the limit function is continuous *If each $f_n(\mathbf{x}) : R^k \longmapsto R^p$ is continuous on C and $f_n(\mathbf{x}) \to f(\mathbf{x})$ uniformly on C, then $f(\mathbf{x})$ is continuous on C.*

Proof. Let $x_0 \in C$ and $\epsilon > 0$ be given. We must determine a δ such that $|f(x) - f(x_0)| < \epsilon$ for all $x \in C$ such that $|x - x_0| < \delta$. For any n,

$$\begin{aligned} |f(x) - f(x_0)| &= |f(x) - f_n(x) + f_n(x) - f_n(x_0) + f_n(x_0) - f(x_0)| \\ &\leq |f(x) - f_n(x)| + |f_n(x) - f_n(x_0)| + |f_n(x_0) - f(x_0)| \end{aligned} \tag{A.4}$$

by the triangle inequality. By uniform convergence of $f_n(x)$ to $f(x)$, there exists an N such that $n \geq N \Rightarrow |f_n(x) - f(x)| < \epsilon/3$ for all $x \in C$ (including, of course, x_0). Using this fact and substituting N for n in (A.4), we see that $|f(x) - f(x_0)| < 2\epsilon/3 + |f_N(x) - f_N(x_0)|$. By the continuity of f_N, there exists a δ_N such that $|x - x_0| < \delta_N \Rightarrow |f_N(x) - f_N(x_0)| < \epsilon/3$. Therefore, for $\delta = \delta_N$, $|f(x) - f(x_0)| < \epsilon/3 + \epsilon/3 + \epsilon/3 = \epsilon$ for all $|x - x_0| < \delta$, proving that f is continuous at x_0. □

Another reason uniform convergence is so important can be seen from the following example. Suppose that $f_n(x)$ is a real valued function converging pointwise to $f(x)$, and $\int_A^B f_n(x)dx$ and $\int_A^B f(x)dx$ exist; can we conclude that $\int_A^B f_n(x)dx \to \int_A^B f(x)dx$? To see that the answer is no, let $f_n(x) = n$ if $0 < x \leq 1/n$ and 0 if $x > 1/n$. Then $f_n(x) \to 0$ for $x > 0$, yet $\int_0^1 f_n(x)dx = 1$ for all n. Thus, $f_n(x) \to f(x) \equiv 0$ for all $x > 0$, but $\int_0^1 f_n(x)dx$

does not converge to $\int_0^1 f(x)dx = 0$. On the other hand, if $f_n(x)$ converges uniformly to f, then there is an N such that $-\epsilon < f_n(x) - f(x) < \epsilon$ for all $n \geq N$ and all x, Therefore,

$$\int_A^B (-\epsilon)dx \leq \int_A^B \{f_n(x) - f(x)\}dx \leq \int_A^B (\epsilon)dx, \text{ so}$$

$$-(B-A)\epsilon \leq \int_A^B f_n(x)dx - \int_A^B f(x)dx < (B-A)\epsilon,$$

$$\left| \int_A^B f_n(x)dx - \int_A^B f(x)dx \right| \leq (B-A)\epsilon \text{ for all } n \geq N. \tag{A.5}$$

Therefore, $\int_A^B f_n(x)dx \to \int_A^B f(x)dx$. This proves the following result.

Proposition A.66. Uniform convergence and interchange of limit and integral *If $f_n \to f$ uniformly on $[A, B]$, and $\int_A^B f_n(x)dx$ and $\int_A^B f(x)dx$ exist, then $\int_A^B f_n(x)dx \to \int_A^B f(x)dx$.*

We close this section with some results about how continuous functions map sets into other sets.

Proposition A.67. Equivalent conditions for continuity *The following are equivalent for a function $f : R^k \longmapsto R^p$:*

1. *$f(x)$ is continuous at each point.*
2. *$f^{-1}(O)$ is an open set in R^k for each open set $O \in R^p$.*
3. *$f^{-1}(C)$ is a closed set in R^k for each closed set C in R^p.*

Proposition A.68. Preservation of compactness *If $f : R^k \longmapsto R^p$ is continuous at each point in a compact set $C \in R^k$, then $f(C)$ is a compact subset of R^p.*

A.6.3 The Derivative of a Function of k Variables

The reader is assumed to be familiar with derivatives of functions of a single variable. Such derivatives are invaluable because they allow us to approximate a function $f(x)$ near $x = x_0$ using a line. The generalization of derivative to a real-valued function of k variables is straightforward; we still approximate $f(\mathbf{x})$ near $\mathbf{x}_0$ with a linear function, but now it is a linear function of k variables.

Definition A.69. Derivative *A function $f : R^k \longmapsto R$ is said to be differentiable at $\mathbf{x} = \mathbf{y}$ if there exists a linear function L such that*

$$\lim_{\mathbf{x} \to \mathbf{y},\ \mathbf{x} \neq \mathbf{y}} \frac{f(\mathbf{x}) - f(\mathbf{y}) - L(\mathbf{x} - \mathbf{y})}{||\mathbf{x} - \mathbf{y}||} = 0 \tag{A.6}$$

The linear function L, represented by its k coefficients, is called the derivative or the total differential.

Implied by this definition is that the limit in (A.6) must be 0 irrespective of the direction of approach of $\mathbf{x}$ to $\mathbf{y}$. If all but the first component of $\mathbf{x}$ equal the corresponding components of $\mathbf{y}$, then

$$\frac{f(\mathbf{x}) - f(\mathbf{y}) - L(\mathbf{x} - \mathbf{y})}{||\mathbf{x} - \mathbf{y}||} = \frac{f(x_1, y_2, \ldots, y_k) - f(y_1, y_2, \ldots, y_k) - a_1(x_1 - y_1)}{|x_1 - y_1|},$$

where a_1 is the first coefficient of the derivative. But this implies that

$$\lim_{x_1 \to y_1} \frac{f(x_1, y_2, \ldots, y_k) - f(y_1, y_2, \ldots, y_k)}{x_1 - y_1}$$

exists. That is, the partial derivative $f_{x_1}(\mathbf{y})$ exists and equals the first coefficient a_1 of the derivative of $f(\mathbf{x})$ at the point $\mathbf{y}$. A similar argument shows that if $f(\mathbf{x})$ is differentiable at $\mathbf{x} = \mathbf{y}$, then all partial derivatives of f exist at $\mathbf{y}$, and the coefficients of L are $f_{x_1}(\mathbf{y}), \ldots, f_{x_k}(\mathbf{y})$. That is, the linear approximation to $f(\mathbf{x})$ near $\mathbf{x} = \mathbf{y}$ is

$$f(\mathbf{y}) + f_{x_1}(x_1 - y_1) + \ldots + f_{x_k}(x_k - y_k).$$

Appendix B

Common Probability Distributions

B.1 Discrete Distributions

1. Several probability distributions arise from independent Bernoulli random variables defined in part (a) below.

   (a) Y is a **Bernoulli random variable** (Bernoulli (p)) if it has probability mass function

   $$\begin{aligned} f(y;p) &= p^y(1-p)^{1-y},\ y=0,1,\ 0 \le p \le 1. && \text{(B.1)} \\ \mathrm{E}(Y) &= p,\ \mathrm{var}(Y) = p(1-p). \\ \text{Ch.f.} &= 1-p+pe^{it}. \end{aligned}$$

   We call $Y=1$ a "success" and $Y=0$ a "failure."

   (b) If $Y=\sum_{i=1}^n X_i$, the number of successes in n iid Bernoulli (p) trials $X_1,\ldots,X_n$, then Y has the **binomial probability mass function**

   $$\begin{aligned} f(y;n,p) &= \binom{n}{y} p^y(1-p)^{n-y},\ y=0,1,\ldots,n, && \text{(B.2)} \\ & \quad n \text{ a nonnegative integer},\ 0 \le p \le 1. \end{aligned}$$

   $$\begin{aligned} \mathrm{E}(Y) &= np,\ \mathrm{var}(Y) = np(1-p). \\ \text{Ch.f.} &= (1-p+pe^{it})^n. \end{aligned}$$

   (c) If Y is the number of failures before the sth success, then Y has the **negative binomial probability mass function**

   $$\begin{aligned} f(y;s,p) &= \binom{s+y-1}{y} p^s(1-p)^y,\ y=0,1,2\ldots, && \text{(B.3)} \\ & \quad s \text{ a positive integer},\ 0 \le p \le 1. \end{aligned}$$

$$\begin{aligned} \mathrm{E}(Y) &= \frac{s(1-p)}{p}, \ \mathrm{var}(Y) = \frac{s(1-p)}{p^2}. \\ \mathrm{Ch.f.} &= \left\{\frac{p}{1-(1-p)e^{it}}\right\}^s. \end{aligned}$$

i. The case $s = 1$ of the negative binomial is the **geometric probability mass function**

$$f(y; p) = p(1-p)^y, \ y = 0, 1, 2, \ldots, \ 0 \le p \le 1. \tag{B.4}$$

$$\begin{aligned} \mathrm{E}(Y) &= \frac{1-p}{p}, \ \mathrm{var}(Y) = \frac{1-p}{p^2}. \\ \mathrm{Ch.f.} &= \frac{p}{1-(1-p)e^{it}}. \end{aligned}$$

(d) The generalization of the binomial probability mass function is the arising as follows. Randomly and independently throw each of n balls into one of k different bins such that the probability of landing in bin i is p_i, $i = 1, \ldots, k$. Let $Y_1, \ldots, Y_{k-1}$ be the numbers of balls landing in bins $1, \ldots, k-1$. The probability mass function of $(Y_1, \ldots, Y_{k-1})$ is multinomial $(n, p_1, \ldots, p_{k-1})$:

$f(y_1, \ldots, y_{k-1}; p_1, \ldots, p_{k-1})$

$$= \frac{n! p_1^{y_1} \ldots p_{k-1}^{y_{k-1}} (1 - p_1 - \ldots - p_{k-1})^{n-y_1-\ldots-y_{k-1}}}{y_1! \ldots y_{k-1}! (n - y_1 \ldots - y_{k-1})!}, \tag{B.5}$$

$$y_i \text{ nonnegative integers}, \ \sum_{i=1}^{k-1} y_i \le n, \ p_i \ge 0, \ \sum_{i=1}^{k-1} p_i \le 1.$$

$$\begin{aligned} \mathrm{E}(Y_i) &= np_i, \ \mathrm{var}(Y_i) = np_i(1-p_i), \ \mathrm{cov}(Y_i Y_j) = -np_i p_j. \\ \mathrm{Ch.f.} &= (1 - p_1 - \ldots - p_{k-1} + p_1 e^{it_1} + \ldots + p_{k-1} e^{it_{k-1}})^n. \end{aligned}$$

2. The **(central) hypergeometric probability mass function** is

$$f(y; M, N, K) = \frac{\binom{M}{y}\binom{N}{K-y}}{\binom{M+N}{K}}, \ \max(0, K-N) \le y \le \min(K, M), \tag{B.6}$$

$$K, M, N \text{ nonnegative integers}, \ K \le M + N.$$

$$\begin{aligned} \mathrm{E}(Y) &= \frac{MK}{M+N}, \ \mathrm{var}(Y) = \frac{MNK(M+N-K)}{(M+N)^2(M+N-1)}. \\ \mathrm{Ch.f.} &= \text{not useful.} \end{aligned}$$

3. The **Poisson probability mass function** is

$$f(y; \lambda) = \frac{\exp(-\lambda)\lambda^y}{y!}, \ y = 0, 1, \ldots, \ \lambda > 0. \tag{B.7}$$

$$\begin{aligned} \mathrm{E}(Y) &= \lambda, \ \mathrm{var}(Y) = \lambda. \\ \mathrm{Ch.f.} &= \exp\{\lambda(e^{it} - 1)\}. \end{aligned}$$

B.2 Continuous Distributions

1. The **beta density function** is

$$f(y;\alpha,\beta) = \frac{\Gamma(\alpha+\beta)y^{\alpha-1}(1-y)^{\beta-1}}{\Gamma(\alpha)\Gamma(\beta)}, \quad 0 \le y \le 1,\ \alpha > 0,\ \beta > 0. \tag{B.8}$$

$$\mathrm{E}(Y) = \frac{\alpha}{\alpha+\beta},\ \mathrm{var}(Y) = \frac{\alpha\beta}{(\alpha+\beta)^2(\alpha+\beta+1)}.$$

$$\text{Ch.f} = \text{not useful.}$$

2. The **Cauchy density function** is

$$f(y;\theta,\lambda) = \frac{1}{\pi\lambda[1+\{(x-\theta)/\lambda\}^2]}, \quad -\infty < y < \infty,\ -\infty < \theta < \infty,\ \lambda > 0. \tag{B.9}$$

$\mathrm{E}(Y)$ and $\mathrm{var}(Y)$ do not exist, median $= \theta$.

$$\text{Ch.f.} = \exp(it\theta - |t|\lambda).$$

3. The **chi-squared density function** with k degrees of freedom is

$$f(y,k) = \frac{y^{k/2-1}\exp(-y/2)}{2^{k/2}\Gamma(k/2)}, \quad y \ge 0,\ k = 1,2,\ldots \tag{B.10}$$

$$\mathrm{E}(Y) = k,\ \mathrm{var}(Y) = 2k.$$

$$\text{Ch.f.} = \frac{1}{(1-2it)^{k/2}}.$$

4. The **exponential density function** is

$$f(y;\theta) = \theta\exp(-\theta y),\ y \ge 0,\ \theta > 0. \tag{B.11}$$

$$\mathrm{E}(Y) = \frac{1}{\theta},\ \mathrm{var}(Y) = \frac{1}{\theta^2}.$$

$$\text{Ch.f.} = \frac{1}{1-it/\theta}.$$

5. The **F-density function** is

$$f(y;m,n) = \frac{\Gamma\{(m+n)/2\}(m/n)^{m/2}y^{(m-2)/2}}{\Gamma(m/2)\Gamma(n/2)\{1+(m/n)y\}^{(m+n)/2}}, \quad y \ge 0,\ m,n = 1,2,\ldots \tag{B.12}$$

$$\mathrm{E}(Y) = \frac{n}{n-2} \text{ for } n > 2,\ \mathrm{var}(Y) = \frac{2n^2(m+n-2)}{m(n-2)^2(n-4)} \text{ for } n > 4.$$

$$\text{Ch.f.} = \text{not useful.}$$

6. The **gamma density function** is

$$f(y;\theta,r) = \frac{\theta^r y^{r-1}\exp(-\theta y)}{\Gamma(r)}, \quad y \ge 0,\ \theta > 0,\ r > 0. \tag{B.13}$$

$$\mathrm{E}(Y) = \frac{r}{\theta},\ \mathrm{var}(Y) = \frac{r}{\theta^2}.$$

$$\text{Ch.f.} = \frac{1}{(1-it/\theta)^r}.$$

7. The **lognormal density function** is

$$\begin{aligned} f(y;\mu,\sigma^2) &= \frac{\exp[\{-\ln(y)-\mu\}^2/(2\sigma^2)]}{y\sqrt{2\pi\sigma^2}}, \; y\geq 0, \; -\infty<\mu<\infty, \; \sigma>0. \quad \text{(B.14)} \\ \mathrm{E}(Y) &= \exp(\mu+\sigma^2/2), \; \mathrm{var}(Y)=\exp(2\mu+2\sigma^2)-\exp(2\mu+\sigma^2). \\ \text{Ch.f.} &= \text{not useful.} \end{aligned}$$

8. The **(univariate) normal density function** is

$$\begin{aligned} f(y;\mu,\sigma^2) &= \frac{\exp\left\{-\frac{(y-\mu)^2}{2\sigma^2}\right\}}{\sqrt{2\pi\sigma^2}}, \quad -\infty<y<\infty, \; -\infty<\mu<\infty, \; \sigma>0. \quad \text{(B.15)} \\ \mathrm{E}(Y) &= \mu, \; \mathrm{var}(Y)=\sigma^2. \\ \text{Ch.f.} &= \exp(i\mu t-\sigma^2t^2/2). \end{aligned}$$

9. The **(bivariate) normal density function** is

$$\begin{aligned} & f(x,y;\mu_X,\mu_Y,\sigma_X^2,\sigma_Y^2,\rho) \\ &= \frac{\exp\left[-\frac{1}{2(1-\rho^2)}\left\{\frac{(x-\mu_X)^2}{\sigma_X^2}-\frac{2\rho(x-\mu_X)(y-\mu_Y)}{\sigma_X\sigma_Y}+\frac{(y-\mu_Y)^2}{\sigma_Y^2}\right\}\right]}{2\pi\sigma_X\sigma_Y\sqrt{1-\rho^2}}, \quad \text{(B.16)} \\ & \quad x,y,\mu_X,\mu_Y\in R, \; \sigma_X,\sigma_Y>0, \; -1<\rho<+1. \\ \mathrm{E}(X) &= \mu_X, \; \mathrm{E}(Y)=\mu_Y, \; \mathrm{var}(X)=\sigma_X^2, \; \mathrm{var(Y)}=\sigma_Y^2, \; \mathrm{cor}(X,Y)=\rho. \\ \text{Ch.f.} &= \exp\{i(t_1\mu_X+t_2\mu_Y)-(1/2)(t_1^2\sigma_X^2+2t_1t_2\rho\sigma_X\sigma_Y+t_2^2\sigma_Y^2)\}. \end{aligned}$$

10. The **t-density function** with k degrees of freedom is

$$\begin{aligned} f(y;k) &= \frac{\Gamma\{(k+1)/2\}}{\Gamma(k/2)\sqrt{k\pi}(1+y^2/k)^{(k+1)/2}}, \quad -\infty<y<\infty, \; k>0. \quad \text{(B.17)} \\ \mathrm{E}(Y) &= 0, \text{ for } k>1, \; \mathrm{var}(Y)=\frac{k}{k-2}, \text{ for } k>2. \\ \text{Ch.f.} &= \text{not useful.} \end{aligned}$$

11. The **uniform density function on** $[a,b]$ is

$$\begin{aligned} f(y;a,b) &= \frac{1}{b-a}, \; a\leq y\leq b. \quad \text{(B.18)} \\ \mathrm{E}(Y) &= \frac{a+b}{2}, \; \mathrm{var}(Y)=\frac{(a-b)^2}{12}. \\ \text{Ch.f.} &= \frac{e^{ibt}-e^{iat}}{(b-a)it}. \end{aligned}$$

12. The **Weibull density function** is

$$\begin{aligned} f(y;a,b) &= aby^{b-1}\exp(-ay^b), \; y\geq 0, \; a>0, \; b>0. \quad \text{(B.19)} \\ \mathrm{E}(Y) &= a^{-\frac{1}{b}}\Gamma\left(1+\frac{1}{b}\right), \; \mathrm{var}(Y)=a^{-\frac{2}{b}}\left\{\Gamma\left(1+\frac{2}{b}\right)-\Gamma^2\left(1+\frac{1}{b}\right)\right\}. \\ \text{Ch.f.} &= \text{not useful.} \end{aligned}$$

B.3 Relationships between Distributions

1. Results from substituting specific parameter values:
   (a) A beta $(1,1)$ distribution is uniform on $[0,1]$.
   (b) A binomial $(1,p)$ distribution is Bernoulli (p).
   (c) A chi-squared distribution with 2 degrees of freedom is exponential $(\theta = 1/2)$.
   (d) A gamma distribution with $r = 1$ is exponential (θ).
   (e) A gamma distribution with $\theta = 1/2$ and $r = k/2$ is chi-squared (k).
   (f) A multinomial distribution with $k = 2$ is binomial (n,p).
   (g) A negative binomial distribution with $s = 1$ is geometric (p).
   (h) A t-distribution with 1 degree of freedom is Cauchy with $\theta = 0$ and $\beta = 1$.
   (i) A Weibull distribution with $b = 1$ is exponential (a).

2. Connections between binomial and hypergeometric distributions:
   (a) If Y_1 and Y_2 are independent binomials with respective parameters (M,p) and (N,p), the conditional distribution of Y_1 given $Y_1 + Y_2 = K$ is hypergeometric with parameters (M,N,K).
   (b) Sampling K balls from an urn with M red and N blue with or without replacement: the number of red balls in the sample has a binomial $(M, p = M/(M+N))$ distribution if the sampling is with replacement, and a hypergeometric (M,N,K) distribution if the sampling is without replacement.

3. Results involving multinomial random variables:
   (a) If $(Y_1,\ldots,Y_{k-1})$ is multinomial $(n,p_1,\ldots,p_{k-1})$, each subset is multinomial. E.g., $(Y_1,\ldots,Y_{j-1})$ is multinomial $(n,p_1,\ldots,p_{j-1})$.
   (b) If $(Y_1,\ldots,Y_{k-1})$ is multinomial $(n,p_1,\ldots,p_{k-1})$, the conditional distribution of $(Y_1,\ldots,Y_{j-1})$ given $Y_j = y_j,\ldots,Y_{k-1} = y_{k-1}$ is multinomial $\{n - y_j - \ldots - y_{k-1}, p_1/(p_1 + \ldots + p_{j-1} + p_k),\ldots,p_{j-1}/(p_1 + \ldots + p_{j-1} + p_k)\}$, where $p_k = 1 - p_1 - \ldots - p_{k-1}$.
   (c) If $\mathbf{Y}_1,\ldots,\mathbf{Y}_r$ are independent multinomial $(n_1,\mathbf{p}),\ldots,(n_r,\mathbf{p})$, then $\sum_{i=1}^r \mathbf{Y}_i$ is multinomial $(\sum_{i=1}^r n_i, \mathbf{p})$. In particular, the sum of independent binomials with respective parameters $(n_1,p),\ldots,(n_r,p)$ is binomial $(\sum_{i=1}^r n_i, p)$.

4. Results involving Poisson random variables:
   (a) If $Y_1,\ldots,Y_k$ are independent Poisson with parameters $\lambda_1,\ldots,\lambda_k$, the conditional distribution of $(Y_1,\ldots,Y_{k-1})$ given $\sum_{i=1}^k Y_i$ is multinomial $(k,p_1,\ldots,p_{k-1})$, where $p_i = \lambda_i/\sum_{j=1}^k \lambda_j$.
   (b) If $Y_1,\ldots,Y_k$ are independent Poisson with parameters $\lambda_1,\ldots,\lambda_k$, then $\sum_{i=1}^k Y_i$ is Poisson $\left(\sum_{i=1}^k \lambda_i\right)$.
   (c) **Law of small numbers** If $n \to \infty$ and $p_n \to 0$ such that $np_n \to \lambda$, the binomial (n,p_n) mass function converges to the Poisson (λ) mass function.

5. Results involving chi-squared random variables:

(a) If $Z_1, \ldots, Z_k$ are iid standard normals, then $\sum_{i=1}^{k} Z_i^2$ is chi-squared with k degrees of freedom.

(b) If Z and V are independent, with $Z \sim N(0,1)$ and V chi-squared with k degrees of freedom, then

$$\frac{Z}{\sqrt{V/k}}$$

has a t-distribution with k degrees of freedom.

(c) If Y_1 and Y_2 are independent chi-square random variables with respective degrees of freedom m_1 and m_2, then

$$\frac{Y_1/m_1}{Y_2/m_2}$$

has an F-distribution with m_1 and m_2 degrees of freedom.

6. Results involving the Cauchy distribution:

   (a) If Y_1 and Y_2 are iid standard normals, Y_1/Y_2 has a Cauchy distribution with $\theta = 0$ and $\beta = 1$.

   (b) If $Y_1, \ldots, Y_n$ are iid Cauchy (θ, β), the sample mean $\bar{Y}$ is also Cauchy (θ, β).

7. If X is $N(\mu, \sigma^2)$, then $Y = \exp(X)$ is lognormal (μ, σ^2).

8. Results involving Weibull random variables:

   (a) If X is Weibull (a, b), then $Y = X^b$ is exponential (a).

   (b) If $X_1, \ldots, X_n$ are iid Weibull (a, b), then $Y = \min(X_i)$ is Weibull (an, b).

9. Results involving exponential random variables:

   (a) If $X_1, \ldots, X_n$ are iid exponential (θ), then $\min(X_1, \ldots, X_n)$ is exponential $(n\theta)$. This is a special case of item 8b with $b = 1$.

   (b) If $Y_1, \ldots, Y_k$ are independent exponentials with parameter θ, then $\sum_{i=1}^{k} Y_i$ is gamma with parameters θ and $r = k$.

   (c) **Lack of memory property of exponential** If Y is exponential (θ), then the conditional distribution of $Y - t$, given that $Y \geq t$, is exponential (θ) for each $t \geq 0$.

10. **Lack of memory property of geometric** If Y is geometric (p), then the conditional distribution of $Y - k$, given that $Y \geq k$, is geometric (p) for each $k = 0, 1, \ldots$

11. **Conjugate priors** The following relationships are useful in Bayesian inference, whereby a "prior" density $\pi(\theta)$ for a parameter θ is specified and then updated to a "posterior" density $f(\theta \,|\, \mathbf{D})$ after observing data $\mathbf{D}$. The prior density is called a *conjugate prior* if the posterior density is in the same family.

    (a) If the conditional distribution of Y given p is binomial with parameters n and p, and the ("prior") density $\pi(p)$ for p is beta (α, β), then the conditional density of p given $Y = y$ (the "posterior" density) is beta $(\alpha + y, \beta + n - y)$.

    (b) If the conditional distribution of Y given λ is exponential (λ), and the ("prior") density $\pi(\lambda)$ for λ is gamma (θ, r), then the conditional density of λ given $Y = y$ (the "posterior" density) is gamma $(\theta + 1, y + r)$.

(c) If the conditional distribution of Y given μ is $\mathrm{N}(\mu, \sigma^2)$, and the ("prior") density $\pi(\mu)$ for μ is $\mathrm{N}(\mu_0, \sigma_0^2)$, then the conditional density of μ given $Y = y$ (the "posterior" density) is

$$\mathrm{N}\left\{\frac{y/\sigma^2 + \mu_0/\sigma_0^2}{1/\sigma^2 + 1/\sigma_0^2}, \left(1/\sigma^2 + 1/\sigma_0^2\right)^{-1}\right\}.$$

Appendix C

References

1. Appel, L.J., Moore, T.J., Obarzanek, E. et al. (1997). A clinical trial of the effects of dietary patterns on blood pressure. *The New England Journal of Medicine* **336**, 1117-1124.

2. Bartlett, R.H., Roloff, D.W., Cornell, R.G. et al. (1985). Extracorporeal circulation in neonatal respiratory failure: a prospective randomized study. *Pediatrics* **76**, 479-487.

3. Begg, C.B. (1990). On inferences from Wei's biased coin design for clinical trials, *Biometrika* **77**, 467-484.

4. Billingsley, P. (2012). *Probability and Measure Anniversary Ed.* John Wiley & Sons, New York.

5. Bland, J.M. and Altman, D.G. (1986). Statistical methods for assessing agreement between two methods of clinical measurement. *The Lancet* **327**, 307-310.

6. Boland, P. and Proschan, M. (1990). The use of statistical evidence in allegations of exam cheating. *Chance* **3**, 10-14.

7. Chatterjee, S. and Hadi, A.S. (2006). *Regression Analysis by Example,* 4th ed. John Wiley & Sons, New York.

8. Chung, K.L. (1974). *A Course in Probability Theory,* 2nd ed. Academic Press, Stanford.

9. Cohen, J. (2003). AIDS vaccine trial causes disappointment and confusion. *Science* **299**, 1290-1291.

10. Drosnin, M. (1998). *The Bible Code.* Touchstone (Simon & Schuster, Inc.), New York.

11. Feller, W. (1968). *An Introduction to Probability Theory and Its Applications, vol I,* 3rd ed. John Wiley & Sons, New York.

12. Follmann, D.A. (1997). Adaptively changing subgroup proportions in clinical trials. *Statistica Sinica* **7**, 1085-1102.

13. Friedman, L.M., Furberg, C.D., and DeMets, D.L. (2010). *Fundamentals of Clinical Trials,* 4th ed. Springer, New York.

14. Gelman, A., Carlin, J.B., Stern, H.S., and Rubin, D.B. (2004). *Bayesian Data Analysis.* Chapman & Hall/CRC, Boca Raton, FL.

15. Heath, D. and Sudderth, W. (1976). De Finetti's theorem on exchangeable variables. *The American Statistician* **30**, 188-189.

16. Hollander, M. and Wolfe, D.A. (1973). *Nonparametric Statistical Methods.* John Wiley & Sons, New York.

17. Johnson, N.L., Kotz, S., and Kemp, A.W. (1992). *Univariate Discrete Distributions,* 2nd ed. John Wiley and Sons, New York.

18. Kadane, J.B. and O'Hagan, A. (1995). Using finitely additive probability: Uniform distributions on the natural numbers. *Journal of the American Statistical Association* **90**, 626-631.

19. Mann, H.B. and Wald, A. (1943). On stochastic limit and order relationships. *Annals of Mathematical Statistics* **14**, 217-226..

20. McMahon, R.P., Proschan, M., Geller, N.L., Stone, P.H., and Sopko, G. (1994). Sample size calculation for clinical trials in which entry criteria and outcomes are counts of events. *Statistics in Medicine* **13**, 859-870.

21. Muller, J.E., Stone, P.H., Turi, Z.G., et al (1985). Circadian variation in the frequency of onset of acute myocardial infarction. *The New England Journal of Medicine* **313**, 1315-1322.

22. Nason, M. (2006). Patterns of immune response to a vaccine or virus as measured by intracellular cytokine staining in flow cytometry: Hypothesis generation and comparison of groups. *Journal of Biopharmaceutical Statistics* **16**, 483-498.

23. Nelsen, R.B. (1999). *An Introduction to Copulas.* Springer, New York.

24. Obarzanek, E., Proschan, M., Vollmer, W., Moore, T., et al. (2003). Individual blood pressure responses to changes in salt intake: Results from the DASH-Sodium trial. *Hypertension* **42**, 459-467.

25. Pearl, J. (2000). *Causality: Models, Reasoning, and Inference.* Cambridge University Press, Cambridge, England.

26. Pfanzagl, J. (1979). Conditional distributions as derivatives. *Annals of Probability* **7**, 1046-1050.

27. Pitman, E.J. (1939). A note on normal correlation. *Biometrika* **31**, 9-12.

28. Posch, M. and Proschan, M.A. (2012). Unplanned adaptations before breaking the blind. *Statistics in Medicine* **31**, 4146-4153.

29. Prentice, R.L. (1989). Surrogate endpoints in clinical trials: Definition and operational criteria. *Statisics in Medicine* **8**, 431-440.

30. Proschan, M.A. (1994). Influence of selection bias on type I error rate under random permuted block designs. *Statistica Sinica* **4**, 219-231.

31. Proschan, M.A. and Dodd, L.E. (2014). A modest proposal for dropping poor arms in clinical trials. *Statistics in Medicine* **33**, 3241-3252.

32. Proschan, M. and Follmann, D. (2008). Cluster without fluster: The effect of correlated outcomes on inference in randomized clinical trials. *Statistics in Medicine* **27**, 795-809.

33. Proschan, M.A. and Nason, M. (2009). Conditioning in 2×2 tables. *Biometrics* **65**, 316-322.

34. Proschan, M.A. and Rosenthal, J.S. (2010). Beyond the quintessential quincunx. *The American Statistician* **64**, 78-82.

35. Proschan, M.A. and Shaw, P.A. (2011). Asymptotics of Bonferroni for dependent normal test statistics. *Statistics and Probability Letters* **81**, 739-748.

36. RGP120 HIV Vaccine Study Group (2005). Placebo-controlled phase III trial of a recombinant glycoprotein 120 vaccine to prevent HIV infection. *Journal of Infectious Diseases* **191**, 654-665.

37. Royden, H.L. (1968). *Real Analysis,* 2nd ed. Macmillan, New York.

38. Rubin, D.B. (1976). Inference and missing data. *Biometrika* **63**, 581-592.

39. Salmon, D.A., Proschan, M.A., Forshee, R. et al. (2013). Association between Guillain Barré syndrome and influenza A (H1N1) 2009 monovalent inactivated vaccines in the USA: a meta-analysis. *Lancet* **381**, 1461-1468.

40. Schoenfeld, D. (1980). The asymptotic properties of nonparametric tests for comparing survival distributions. *Biometrika* **68**, 316-319.

41. Schonberger, L.B., Bregman, D.J., Sullivan-Bolyai, et al. (1979). Guillain-Barré syndrome following vaccination in the National Influenza Immunization Program, United States, 1976-1977. *American Journal of Epidemiology* **110**, 105-123.

42. Serfling, R.J. (1980). *Approximation Theorems of Mathematical Statistics.* John Wiley & Sons, New York.

43. Sethuraman, J. (1961). Some limit theorems for joint distributions. *Sankya A* **23**, 379-386.

44. Simon, R. and Simon, N.A. (2011). Using randomization tests to preserve type 1 error with response adaptive and covariate adaptive randomization. *Statistics and Probability Letters* **81**, 767-772.

45. Sklar, A. (1959). Fonctions de répartition à n dimensions et leurs marges. *Publications de l'Institut de Statistique de l'Université de Paris* **8**, 229-231.

46. Stein, C. (1945). A two-sample test for a linear hypothesis whose power is independent of the variance. *Annals of Mathematical Statistics* **16**, 243-258.

47. Stewart, S.F., Nast, E.P., Arabia, F.A., Talbot, T.L., et al. (1991). Errors in pressure gradient measurement by continuous wave Doppler ultrasound: Type, size and age effects in bioprosthetic aortic valves. *Journal of the American College of Cardiology* **18**, 769-779.

48. The History Channel (2003). *The Bible Code: Predicting Armageddon.* A&E Television Networks.

49. van der Vaart, A.W. (1998). Asymptotic Statistics. Cambridge University Press, Cambridge, England.

50. Verma, T. and Pearl, J. (1988). Causal networks: Semantics and expressiveness, Proceedings of the 4th Workshop on Uncertainty in Artificial Intelligence (Mountain View, CA), 352-359.

51. Wei, L.J. (1988). Exact two-sample permutation tests based on the randomized play-the-winner rule. *Biometrika* **75**, 603-606.

52. Westfall, P.H. and Young, S.S. (1993). *Resampling-Based Multiple Testing: Examples and Methods for p-value Adjustment.* John Wiley & Sons, New York.

53. Witztum, D., Rips, E., and Rosenberg, Y. (1994). Equidistant letter sequences in the book of Genesis. *Statistical Science* **9**, 429-438.

Appendix D

Mathematical Symbols and Abbreviations

Book Conventions

*** signifies an important technique

$\star$ next to an exercise highlighted an important problem (demonstrating key idea)

$\uparrow$ next to an exercise signifies that this exercise builds on the immediately preceding exercise.

Sets

$\cap$ = intersection

$\cup$ = union

$A \subset B$ = A is a subset of B

$\in$ = an element of

$\notin$ = not an element of

$:$ = such that

$A^C = \{x : x \notin A\}$

$A \setminus B = \{x \in A : x \notin B\}$

$A \times B = A$ direct product with B

$A_n \uparrow A$ means $\cup_{n=1}^{\infty} A_n = A$ and $A_n \subset A_{n+1}$ for all n

$A_n \downarrow A$ means $\cap_{n=1}^{\infty} A_n = A$ and $A_n \supset A_{n+1}$ for all n

$\text{card}(\cdot)$ = cardinality of

$I_A(x) = 1$ if $x \in A$ and $I_A(x) = 0$ otherwise.

$I(\cdot) = 1$ if true, 0 if false

$\sigma(\mathcal{A})$ = sigma-field generated by $\mathcal{A}$

$\emptyset$ = empty set

$\mathcal{B}$ = Borel sets

$\mathcal{B}^k$ =k-dimensional Borel sets

L = Lebesgue sets

$L^p = \{X : \int X^p(\omega) d\mu(\omega) < \infty\}$

R = set of real numbers

R^k = k-dimensional space of real numbers

Limits

$\inf(x_n)$ or $\inf_x f(x)$ =greatest lower bound

$\sup(x_n)$ = or $\sup_x f(x)$, least upper bound

$\underline{\lim}\,(x_n)$ = liminf = $\inf\{x^* : x^*$ is a limit point of $(x_n)\}$.

$\overline{\lim}\,(x_n)$ = limsup = $\sup\{x^* : x^*$ is a limit point of $(x_n)\}$.

x^- = left-hand limit

x^+ = right-hand limit

$x_n \to x$ means $\lim_{n\to\infty} x_n = x$

$x_n \uparrow x$ means $\lim_{n\to\infty} x_n = x$ and $x_n \leq x_{n+1}$ for all n

$x_n \downarrow x$ means $\lim_{n\to\infty} x_n = x$ and $x_n \geq x_{n+1}$ for all n

(M) = subsequence defined by $i \in M$

$\stackrel{a.s.}{\to}$ = almost sure convergence

$\stackrel{a.e.}{\to}$ = convergence almost everywhere

$\stackrel{p}{\to}$ = convergence in probability

$\stackrel{D}{\to}$ = convergence in distribution

Probability

$(\Omega, \mathcal{F}, \mathcal{P})$ = (sample space, sigma algebra, probability measure)

$\Pr(\omega) = P(\omega)$ = probability of event ω

$E(\cdot) = \int X(\omega)dP(\omega)$ = expected value

P = probability measure

μ_C = Counting measure

μ_L = Lebesgue measure

$\mu^\star$ = outer measure

$X^{-1}(A) = \{\omega : X(\omega) \in A\}$

$F_X^{-1}(y) = \inf\{x : F_X(x) \geq y\}$

$\sim$ = distributed as

II = independent of

$\hat{\theta}$ = estimate of θ

$N(\mu, \sigma^2)$ = normal distribution with mean μ and variance σ^2

$\Phi(t)$ = distribution function for N(0,1)

$\phi(t)$ = density function for N(0,1)

$\bar{x} = \Sigma_{i=1}^n x_i/n$

$S_n = \Sigma_{i=1}^n x_n$

$S(t) = P(X > t)$, the survival function

Elementary

$e = \lim_{n\to\infty}(1 + 1/n)^n \approx 2.718282\ldots$

$i = \sqrt{-1}$

π = ratio of circumference to diameter of a circle ≈ 3.141592

∞ = infinity

$1-1$ = one to one

$\longmapsto$ = maps to

$\Rightarrow$ = implies

$\Leftrightarrow$ = if and only if

$\approx$ = approximately equal to

$\equiv$ = identical to

$\neq$ not equal to

$\pm$

Σ = sum

$\prod$ = product

A^t = transpose of matrix A

$x \cdot y = \sum_{i=1}^n x_i y_i$, dot product

$|x| = x$ if $x \geq 0$ and $-x$ if $x < 0$

$||x|| = \sqrt{\sum_{i=1}^n x_i^2}$, Euclidean distance

$< f_1, f_2 >= \int f_1(\omega) f_2(\omega) d\mu$, inner product

n! $= 1 \cdot 2 \cdot 3 \cdot \ldots n$

$\binom{n}{x} = \frac{n!}{x!(n-x)!}$

$\Gamma(\alpha) = \int_0^\infty x^{\alpha-1}e^{-x}dx$, for $\alpha > 0$

$\Gamma(n) = (n-1)!$, for positive integer n

$f^+(\omega) = f(\omega)I(f(\omega) \geq 0)$

$f^-(\omega) = -f(\omega)I(f(\omega) < 0)$

$f^{(n)}$ = n-th derivative of f

$\int_A f du = \int f(u)I\{u \in A\}du$

$\frac{\partial f}{\partial x} = f_x$ = partial derivative with respect to x

cos = cosine function

sin = sine function

arctan = inverse tangent function

$\exp(\cdot) = e^{(\cdot)}$

$\ln = \log_e$

$\square$ = end of proof or example

Abbreviations, in alphabetical order

AN = asymptotically normal

ANCOVA = analysis of covariance

a.s. = almost surely

BCT = bounded convergence theorem

CLT = central limit theorem

ch.f. = characteristic function

cov = covariance

DCT = dominated convergence theorem

d.f. = distribution function

iid = independent and identically distributed

inf = infimum = greatest lower bound

i.o. = infinitely often

FWE = familywise error rate

max = maximum

MCT = monotone convergence theorem

m.g.f = moment generating function

min =minimum

MLE = maximum likelihood estimate

MSE = mean squared error

SLLN = strong law of large numbers

sup = supremum = least upper bound

UI = uniformly integrable

var = variance

WLLN = weak law of large numbers

Index

Printed in the United States
by Baker & Taylor Publisher Services